Abstract Convexity and Global Optimization

Nonconvex Optimization and Its Applications

Volume 44

The titles published in this series are listed at the end of this volume.

Abstract Convexity and Global Optimization

by

Alexander Rubinov
School of Information Technology and Mathematical Sciences, University of Ballarat, Victoria, Australia

SPRINGER-SCIENCE+BUSINESS MEDIA, B.V.

A C.I.P. Catalogue record for this book is available from the Library of Congress.

ISBN 978-1-4419-4831-1 ISBN 978-1-4757-3200-9 (eBook)
DOI 10.1007/978-1-4757-3200-9

Printed on acid-free paper

Contents

Preface

Special tools are required for examining and solving optimization problems. The main tools in the study of local optimization are classical calculus and its modern generalizions which form *nonsmooth analysis.* The gradient and various kinds of generalized derivatives allow us to accomplish a local approximation of a given function in a neighbourhood of a given point. This kind of approximation is very useful in the study of local extrema. However, local approximation alone cannot help to solve many problems of *global optimization*, so there is a clear need to develop special *global* tools for solving these problems.

The simplest and most well-known area of global and simultaneously local optimization is convex programming. The fundamental tool in the study of convex optimization problems is the subgradient, which actually plays both a local and global role. First, a subgradient of a convex function f at a point x carries out a local approximation of f in a neighbourhood of x. Second, the subgradient permits the construction of an affine function, which does not exceed f over the *entire* space and coincides with f at x. This affine function h is called a *support* function. Since $f(y) \geq h(y)$ for *all* y, the second role is global. In contrast to a *local approximation*, the function h will be called a *global affine support.* Generalizations of the definition of the convex subdifferential, based on the notion of local approximation, have led to the development of *nonsmooth analysis.* Generalizations of the definition of the convex subdifferential, based on the notion of global affine support, can be studied in the framework of *abstract convexity.*

The existence of global affine supports follows from the well-known separation theorem for convex sets: each point which does not belong to a convex closed set can be separated from this set by a linear function. This theorem also leads to the following fundamental result of convex analysis: each lower semicontinuous convex function f is the upper envelope (the point-wise supremum) of the set of all affine functions which are minored by f. This result is closely related to the existence of global affine supports. Since an affine function is a linear function plus a constant, the convexity can be presented as a combination of linearity and the envelope presentation.

In some instances the use of the envelope representation is very convenient, even when we consider the upper envelope of non-affine functions. In other words, if an affine global support does not exist, we can try to use some kinds of non-affine global supports. In particular, global supports form a very useful tool for examining some problems of global optimization.

Thus the idea arises to examine the main notions of convexity in non-convex environments. First, to examine *abstract convex* functions; that is, functions which can be presented as upper envelopes of subsets of a given set of not necessarily linear *elementary functions*. Second, to study *abstract convex sets*; that is, sets which enjoy the following not necessarily linear separation property: each point, which does not belong to this set, can be separated from it by an elementary function. This idea underlies the rich theory of *convexity without linearity*, known also as *abstract convexity* (see [118, 175]). As it turns out, many results from convex analysis related to subdifferentials can be transformed in abstract convex environments. The conjugation theory, which is the heart of convex analysis, also can be examined in the very general framework of abstract convexity. Excellent books [118] by D. Pallaschke and S. Rolewicz and [175] by I. Singer, which were published in 1997, contain detailed presentations of many results of *abstract convex analysis*, which are concentrated around notions of subdifferentials, conjugations and dualities. These books (especially [175]) contain detailed historical surveys of the main ideas and results from abstract convexity and its applications. It is interesting to note that the intersection of these large books is almost empty.

Abstract convexity has found many applications in the study of problems of mathematical analysis and optimization. The first book on abstract convexity and some its applications to functional analysis in ordered spaces, Choquet theory and approximation theory (see [100]) was published by S. S. Kutateladze and A. M. Rubinov in 1976 in Russian. (Actually this book was written in 1971 and its short version [99], without proofs and some details, was published in 1972 in Russian and English.) Abstract convexity also found an interesting application to the theory of inequalities. However, the development of abstract convex analysis, like the development of its forerunner, convex analysis, was mainly driven by applications to optimization.

For applications of abstract convexity, we need to describe the main objects of this theory in some concrete situations. Just as the set of linear function leads to a theory of convex functions and convex sets, each class of elementary functions should lead to the theory of abstract convex functions and abstract convex sets with respect to this class. The reverse problems are also very interesting. One of them is to find for a given broad class of functions a small set of simple elementary functions, which generated this class by means of the upper envelope. This problem leads to the notion of a *supremal generator* (see [99]).

In the present work we focus mainly on two aspects of abstract convexity. First we describe abstract convex functions and sets, and subdifferentials for some concrete sets of elementary functions. Second, we examine applications of the results obtained to the study of global optimization problems, and to some different problems. We did not use more advanced results from abstract convexity which can be found in books [118] and [175]. However, we do demonstrate that even a description of abstract convex objects in concrete situations is very useful in the study of some problems.

A short Chapter 1 contains the main definitions related to abstract convexity, some simple corollaries from these definitions and some simple examples.

Then we study abstract convexity based on classes of elementary functions, consisting of the so-called min-type functions. Instead of linear functions $x \mapsto \sum_{i=1}^{n} l_i x_i$ defined on $\mathbb{R}^n$, we consider functions of the form $x \mapsto \min_{i:l_i>0} l_i x_i$ defined either on the cone $\mathbb{R}^n_{++}$ of vectors with positive coordinates or on the cone $\mathbb{R}^n_+$ of vectors with nonnegative coordinates. Abstract convexity with respect to this class of elementary functions leads to the theory of *monotone functions* defined on cones $\mathbb{R}^n_{++}$ and $\mathbb{R}^n_+$, and to the theory of *normal* subsets of these cones. The totality of developed approaches and results obtained in this direction can be considered as elements of *monotonic analysis.* It is interesting to note that monotonic analysis is even simpler than convex analysis in some aspects. Elements of monotonic analysis are examined in Chapter 2 and Chapter 3.

Min-type functions of a general form: $l(x) = \min_{i=1,\dots,k}[l_i, x]$, where k is equal to either n or $n+1$, leads to the theory of *star-shaped* subsets of R^n and *positively homogeneous functions of degree one* defined on R^n. (Here $[l, x]$ stands for the inner product of vectors l and x.) The totality of developed approaches and results obtained in this direction can be considered as elements of *star-shaped analysis*. Elements of star-shaped analysis are examined in Chapter 5.

We present here only the foundations of the theory of monotonic analysis and of the theory of star-shaped analysis. However, we believe that the results presented demonstrate that such theories can be developed on the basis of abstract convexity.

Applications of monotonic analysis to optimization are presented in Chapter 4. Special kinds of nonlinear analogues of Lagrange and penalty functions are studied in this chapter.

Supremal generators and their applications to various problems from mathematical analysis are examined in Chapter 6. We are mainly concerned with supremal generators of the space of all continuous functions and positively homogeneous functions.

Some theoretical results from abstract convexity are presented in Chapter 7. We discuss questions related to the Fenchel-Moreau conjugation and the Minkowski duality. Special attention is paid to spaces of positively homogeneous functions. Results from this chapter are applied in the examination of some dual concepts of global optimization in Chapter 8. We consider two approaches to duality in this chapter based on solvability theorems and the Toland-Singer formula, respectively.

We focus on numerical methods of global optimization based on abstract convexity in final Chapter 9. We describe some numerical methods, in particular various versions of the cutting angle method, a version of branch-and-bound method and a method for monotonic optimization. We report the results of numerical experiments, which show that these methods are fairly efficient for solving some classes of problems.

The main part of the book does not require a knowledge of advanced branches of mathematics. It is assumed that the reader is familiar only with the foundations of the theory of finite-dimensional linear spaces, convex analysis and mathematical programming. However, more advanced results from functional analysis are exploited in some sections, where some definitions and results from the theory of metric and Banach spaces, the theory of lattices and from nonsmooth analysis are used. As a rule we formulate these definitions and results as required.

The end of each proof is denoted by Δ.

The author hopes that this book will help to promote some methods of convex analysis for examination of many nonconvex problems, especially problems of multi-extremal global optimization, and will provide researchers with powerful tools for investigation of these problems.

It is a very difficult task to get a global minimizer (even if the global minimum itself is known). The author tried hard to minimize the number of misprints and inaccuracies, but he is not quite sure that he was successful and that the global minimum (which is equal to zero in this case) has been attained. For this reason the author would be very grateful to readers for indicating any remaining errata in the book.

ALEX RUBINOV

ACKNOWLEDGMENTS

Many principal ideas, which appear in this book have been discussed with Barney Glover and many results from this book have been obtained jointly with him. Barney and I had a very fruitful and useful collaboration, which favoured the preparation of the book. I am very much obliged to Barney for his invaluable advice and help.

Some sections of this book are based on papers written by myself and T. M. Abasov, M.Yu. Andramonov, A. M. Bagirov, S.S. Kutateladze, V. Jeyakumar, H. Mays, C.E.M. Pearce, A. P. Shveidel, I. Singer, H. Tuy, A.A. Vladimirov and X.Q. Yang. I am very thankful to all these colleagues. My special thanks are due to Mikhail Andramonov and Adil Bagirov, who took part in the preparation of some sections of Chapter 9.

In the process of preparation of this book, some of its parts were discussed with A. M. Bagirov, J.-P. Penot, A. P. Shveidel, H. Tuy, A. Zaffaronni and V. G. Zhadan. I would like to express my sincere thanks to them for their constructive and valuable comments. I am also very grateful to V.F. Demyanov, who read the entire manuscript and proposed many useful corrections and improvements.

I am thankful to the Australian Research Council for its financial support of this project (under Large Research Grants A69701407 and A49906152). I am also grateful to School of Information Technology and Mathematical Sciences, the University of Ballarat, for the support of this project and for providing the facilities for preparation of a camera-ready copy of the book.

The idea to publish this book was supported and promoted by Dr. John Martindale, Kluwer Academic Publishers, and by Prof. P. Pardalos, University of Florida, to whom I am greatly indebted.

English is not my native language. I thanks Allan Adair and Beata Wysocka for their valuable suggestions concerning aspects of the English presentation of the work.

The preparation of such a book is very time-consuming. I am deeply indebted to my wife Zari and sons Eldar and Michael for their understanding and patience.

Abstract Classical methods of nonconvex mathematical programming, being based on a local approximation, cannot be used to examine and solve many problems of global optimization, and so there is a clear need to develop special global tools for solving these problems. Some of these tools are based on abstract convexity, that is, on the representation of a function of a rather complicated nature as the upper envelope of a set of fairly simple functions.

The present book consists of two parts. First, the main notions of abstract convexity and their applications in the study of some classes of functions and sets are presented. Second, both theoretical and numerical aspects of global optimization based on abstract convexity are examined. Most of the book does not require a knowledge of advanced mathematics.

The book would be of interest to specialists in global optimization, mathematical programming, convex analysis, as well as engineers using mathematical tools and optimization techniques and specialists in mathematical modelling.

Chapter 1

AN INTRODUCTION TO ABSTRACT CONVEXITY

1.1. OVERVIEW

One of the main results of convex analysis states that an arbitrary lower semicontinuous convex function f (perhaps admitting the value $+\infty$) is the *upper envelope* of the set of all its affine minorants:

$$f(x) = \sup\{h(x) : h \text{ is an affine function, } h \leq f\}.$$

(The inequality $h \leq f$ stands for $h(x) \leq f(x)$ for all x.) The supremum above is attained at a point x if and only if the *subdifferential* of f at x is nonempty. Since affine functions are defined by means of linear functions, one can say that convexity is *linearity + envelope representation.*

As it turns out, the contribution of *envelope representation* to the convexity is fairly large. This observation stimulated the development of a rich theory of *convexity without linearity*, also known as *abstract convexity.* (See [137, 118, 175] and references therein.) In particular, functions which can be represented as upper envelopes of subsets of a set H of sufficiently simple (*elementary*) functions are studied in this theory.

By analogy with the classical situation, these functions are called *abstract convex* with respect to H or H-convex functions. The set $\text{supp}(f, H) = \{h \in H, h \leq f\}$ of all H-minorants of a function f is called the *support set* (or H-support set) of this function. The support set accumulates global information about the function f in terms of the set of elementary functions H. A set $U \subset H$ is called *abstract convex* or (H, X)-convex if this set coincides with the support set of an abstract convex function.

Support sets can be considered as an abstract analogue of closed convex sets. (See Example 1.6.)

One of the main notions of convex analysis which plays a key role in various applications is that of the *subdifferential.* There are two equivalent definitions of the subdifferential for a convex function. The first one is based on the global behaviour of the function. A linear function l is called a *subgradient* (i.e., a member of the subdifferential) of the function f at a point y if the affine function $h(x) = l(x) - (l(y) - f(y))$ is a support function with respect to f, that is, $h(x) \leq f(x)$ for all x. The second definition has a local nature and is connected with a local approximation of the function: the subdifferential is a closed convex set of linear functions such that the *directional derivative* $f'_x(u) := \lim_{\alpha \to +0} \alpha^{-1}(f(x + \alpha u) - f(x))$ is the upper envelope of this set. For a differentiable convex function these two definitions represent the support and tangent sides of the gradient, respectively.

Various generalizations of the second definition have led to the development of a rich theory of *nonsmooth analysis.* The natural field for generalizations of the first definition is abstract convexity.

In this preliminary small chapter we present formal definitions of some key notions related to abstract convexity: abstract convex functions and sets, the Minkowski duality, subdifferentials and conjugation. We also give some examples and present some simple statements, which will be used throughout the book. Some properties of abstract convex functions and sets in a general setting will be discussed in Chapter 7. The interested reader can find the detailed historical survey in [118, 175].

1.2. PRELIMINARIES

We begin with some technical remarks. Consider the extended real line $\bar{\mathbb{R}} = \mathbb{R} \cup \{+\infty\} \cup \{-\infty\}$ where $\mathbb{R}$ is the set of real numbers. We shall study functions $f : X \to V$ where X is a set and $V \subset \bar{\mathbb{R}}$ is a segment. As a rule V will coincide with one of the following segments:

$$\mathbb{R} = (-\infty, +\infty), \quad \bar{\mathbb{R}} = [-\infty, +\infty], \quad \mathbb{R}_{+\infty} = (-\infty, +\infty] \tag{1.2.1}$$

and

$$\mathbb{R}_{+} = [0, +\infty), \quad \bar{\mathbb{R}}_{+} = [0, +\infty], \quad \mathbb{R}_{++} = (0, +\infty), \quad \bar{\mathbb{R}}_{++} = (0, +\infty]. \tag{1.2.2}$$

Let $V \subset \bar{\mathbb{R}}$ be a segment. We shall consider the supremum and the infimum of each subset of V, in particular, the supremum and the infimum of the empty subset $\emptyset$ of V. It is convenient to accept the following convention:

$$\inf \emptyset = \sup V, \qquad \sup \emptyset = \inf V. \tag{1.2.3}$$

In particular, $\inf \emptyset = +\infty$ for every V from (1.2.1) and from (1.2.2). If V coincides with one of the sets (1.2.1) then $\sup \emptyset = -\infty$; if V is one of the sets (1.2.2), then $\sup \emptyset = 0$.

Let $V \subset \bar{\mathbb{R}}$ be a segment and let H be a set of functions $h : X \to V$. Assume that

$$\bigcup_{h \in H} \{h(x) : x \in X\} = V, \tag{1.2.4}$$

We shall consider the pointwise supremum and the pointwise infimum of each subset of H. In particular we shall consider the pointwise supremum and infimum of the empty subset, $\emptyset$, of H. The following convention is consistent with (1.2.3):

$$\sup_{h \in \emptyset} h(x) = \inf V, \qquad \inf_{h \in \emptyset} h(x) = \sup V, \qquad (x \in X) \tag{1.2.5}$$

Thus, the infimum of the empty subset of H is the constant function equal to $\sup V$ and the supremum of the empty subset of H is the constant function equal to $\inf V$.

Let $V \subset \mathbb{R}$ be a segment and let $\bar{V}$ be a closure of V, that is,

$$\bar{V} = V \cup \{\sup V\} \cup \{\inf V\}. \tag{1.2.6}$$

Consider a set H of functions $h : X \to V$. Let $U \subset H$ and $f_U(x) = \sup_{h \in U} h(x)$. If U is nonempty, then the function f_U maps into $V \cup \{\sup V\}$. If U is empty then $f(x) = \inf V$ for all $x \in X$. Thus $f_U : X \to \bar{V}$ for each $U \subset H$.

1.3. ABSTRACT CONVEX FUNCTIONS AND SETS

Let us present the main definitions related to abstract convexity.

DEFINITION 1.1 Let $V \subset \bar{\mathbb{R}}$ and let H be a nonempty set of functions $h : X \to V$. A function $f : X \to \bar{V}$ is called *abstract convex* with respect to H (or H-convex) if there exists a set $U \subset H$ such that f is the upper envelope of this set:

$$f(x) = \sup\{h(x) : h \in U\} \text{ for all } x \in X.$$

The set H from Definition 1.1 will be referred to as a set of *elementary functions* .

EXAMPLE 1.1 Let L be the set of all linear functions defined on n-dimensional space $\mathbb{R}^n$ and H be the set of all affine functions defined on $\mathbb{R}^n$. A function h is affine if and only if $h(x) = l(x) - c$, where $l \in L$ and $c \in \mathbb{R}$. Let $V = \mathbb{R}_{+\infty}$. Then a function $p : \mathbb{R}^n \to \mathbb{R}_{+\infty}$ is L-convex if and only if p is lower semicontinuous sublinear function; a function $f : \mathbb{R}^n \to \mathbb{R}_{+\infty}$ is H-convex if and only if f is a lower semicontinuous convex function. (See [134], Theorem 12.1.)

It follows directly from Definition 1.1 that f is H-convex if and only if

$$f(x) = \sup\{h(x) : h \in H, h \leq f\} \qquad (x \in X),$$

where $h \leq f \iff (h(x) \leq f(x) \text{ for all } x \in X)$.

DEFINITION 1.2 Let $f : X \to \bar{V}$. The set

$$\text{supp}(f, H) = \{h \in H, \, h \leq f\} \tag{1.3.1}$$

of all H-minorants of f is called the *support set* of the function f with respect to the set of elementary functions H. Occasionally we shall use the term *lower support set* instead of *support set.*

EXAMPLE 1.2 If L is the set of all linear functions and p is a lower semicontinuous sublinear function such that $p(0) < +\infty$. Then $p(0) = 0$ and

$$\begin{aligned} \text{supp}(p, L) &= \{l \in L : (\forall x \in \mathbb{R}^n) \, l(x) \leq p(x)\} \\ &= \{l \in L : (\forall x \in \mathbb{R}^n) \, l(x) - l(0) \leq p(x) - p(0)\} = \partial p(0), \end{aligned}$$

where $p(0)$ is the subdifferential of the convex function p at the origin.

Let H be a set of elementary functions and assume that (1.2.4) holds. Let

$$i_V(x) \equiv \inf V, \qquad s_V(x) \equiv \sup V, \qquad (x \in X) \tag{1.3.2}$$

It follows directly from Definition 1.1, Definition 1.2, and from (1.2.5) that both functions i_V and s_V are H-convex functions and

$$\text{supp}(i_V, H) = \emptyset, \qquad \text{supp}(s_V, H) = H.$$

If $V = \mathbb{R}$, then $i_V(x) = -\infty$ for all $x \in X$. In such a case each H-convex function f either maps into $\mathbb{R}_{+\infty}$ or is equal to $-\infty$, where $-\infty(x) = -\infty$ for all $x \in X$. If $V = \mathbb{R}_{++}$, then $i_V(x) = 0$ for all $x \in X$. In this situation each H-convex function either maps into $\bar{\mathbb{R}}_{++}$ or is identically equal to zero.

DEFINITION 1.3 Let $f : X \to \bar{V}$. The function $\text{co}_H f$ defined by

$$\text{co}_H f(x) = \sup\{h(x) : h \in \text{supp}(f, H)\}, \qquad (x \in X)$$

is called the *H-convex hull* of the function f.

Clearly f is H-convex if and only if $f = \mathrm{co}_H f$.

EXAMPLE 1.3 If H is the set of all affine functions defined on $\mathbb{R}^n$ then $\mathrm{co}_H f$ is the so-called convex regularization of the function f, i.e., the greatest lower semicontinuous convex function, which does not exceed f, see Subsection 7.3.3 for details.

Denote the set of all H-convex functions defined on X by $P(H, X)$.

DEFINITION 1.4 A set $U \subset H$ is called *abstract convex* with respect to X (or (H, X)-*convex*) if there exists a function $f : X \to \bar{\mathbb{R}}$ such that $U = \mathrm{supp}(f, H)$.

EXAMPLE 1.4 If L is the set of all linear functions defined on $\mathbb{R}^n$ then a set $U \subset L$ is $(L, \mathbb{R}^n)$-convex if and only if this set is closed and convex.

Denote the set of all (H, X)-convex sets by $S(H, X)$. Note that $\emptyset \in S(H, X)$ and $H \in S(H, X)$.

REMARK 1.1 (H, X)-convex sets are called X-convex in [55] and H-convex in [99].

The mapping $f \mapsto \mathrm{supp}(f, H)$ is called the *Minkowski duality*. Some properties of this mapping will be studied in Section 7.4. In particular, it will be shown (see Proposition 7.47) that the Minkowski duality is a one-to-one correspondence. The Minkowski duality extends the well-known dual relationship between sublinear functions and support sets (subdifferentials at zero).

An H-convex function f is called *proper* if $f \neq i_V$ and $f \neq s_V$. An (H, X)-convex set U is called *proper* if $U \neq \emptyset$ and $U \neq H$. It follows from these definitions that the Minkowski duality maps the set of proper H-convex functions onto the set of proper (H, X)-convex sets.

DEFINITION 1.5 The intersection of all (H, X)-convex sets containing a set $U \subset H$ is called the *abstract convex hull* or (H, X)-convex hull of the set U. We shall denote the (H, X)-convex hull of a set U by $\mathrm{co}_{H,X}U$ or, if it does not cause misunderstanding, by $\mathrm{co}_H U$.

REMARK 1.2 The notation $\mathrm{co}_X U$ is more preferable in some instances however, we prefer to keep the same notation co_H for the hulls of both functions and sets.

EXAMPLE 1.5 If L is the set of linear functions defined on $\mathbb{R}^n$ and $U \subset L$, then $\mathrm{co}_H U$ coincides with the closed convex hull of the set U.

If H is a set of elementary functions and U is an (H,X)-convex set, then

$$U = \mathrm{co}_H\, U = \{h \in H : h(x) \leq p_U(x) \text{ for all } x \in X\},$$

where p_U is the upper envelope of the set U:

$$p_U(x) = \sup\{h(x) : h \in U\} \qquad (x \in X). \tag{1.3.3}$$

PROPOSITION 1.1 *Let $U \subset H$ and $p_U(x) = \sup_{h \in U} h(x)$. Then*

$$\mathrm{co}_H\, U = \mathrm{supp}(p_U, H) = \bigcap\{U' : U' \supset U, U' \in S(H,X)\}.$$

Proof: Since $\mathrm{supp}(p_U, H)$ is an (H,X)-convex set and $\mathrm{supp}(p_U, H) \supset U$, it follows that $\mathrm{supp}(p_U, H) \supset \mathrm{co}_H\, U$. Let U' be an (H,X)-convex set, $U' \supset U$. Then $p_{U'} \geq p_U$, hence $U' = \mathrm{supp}(p_{U'}, H) \supset \mathrm{supp}(p_U, H)$. Thus

$$\mathrm{co}_H\, U = \bigcap\{U' : U' \supset U, U' \in S(H,X)\} \supset \mathrm{supp}(p_U, H). \qquad \triangle$$

COROLLARY 1.1 *For each $U \subset H$, $\mathrm{co}_H\, U$ is an (H,X)-convex set.*

REMARK 1.3 Let $V \subset \bar{\mathbb{R}}$. Consider two sets X and H and a coupling function $\phi : X \times H \to V$. Then each element $h \in H$ can be identified with the function defined on X by $x \mapsto \phi(x,h)$ and each element $x \in X$ can be identified with the function defined on H by $h \mapsto \phi(x,h)$. Thus we can consider the set H as a set of functions defined on X and the set X as a set of functions defined on H. This construction allows one to consider (X,H)-convex subsets of X and (H,X)-convex subsets of H. Such a representation of elementary functions by means of a coupling function is especially useful in the study of conjugacy operations.

There are two definitions of convexity for closed sets. One of them is the *inner definition* based on the notion of convex combination. The other (*outer definition*) is based on the separation property. Abstract convexity leads to a generalization of the outer definition. The following lemma is a reformulation of the definition of abstract convex sets in terms of the separation property.

LEMMA 1.1 (Separation property) *Let H be a set of elementary functions defined on a set X. Let U be a proper subset of H, that is $U \neq \emptyset, U \neq H$. Then the set U is (H,X)-convex if and only if for each $h \notin U$ there exists $x \in X$ such that*

$$h(x) > \sup_{h' \in U} h'(x). \tag{1.3.4}$$

Proof: Let U be a proper (H, X)-convex set. Let $p_U(x) = \sup\{h'(x) : h' \in U\}$. It follows from the definition of abstract convex sets that $U = \{h \in H : h(x) \leq p_U(x) \text{ for all } x \in X\}$. Thus if $h \notin U$ there exists $x \in X$ such that $h(x) > p_U(x)$. Assume now that for each $h \notin U$ one can find $x \in X$ such that (1.3.4) holds. Consider the set $\text{supp}(p_U, H)$. Clearly $\text{supp}(p_U, H) \supset U$. If there exists $h \in \text{supp}(p_U, H)$ such that $h \notin U$ then applying (1.3.4) we can find x' such that $h(x') > p_U(x')$, which is impossible since $h(x) \leq p_U(x)$ for all $x \in X$. △

REMARK 1.4 Assume that $U = H$. Then (1.3.4) also holds since there is no $h \notin U$. Assume that $i_V(x) \equiv \inf_{h' \in H} h'(x) < h(x)$ for each $h \in H$ and $x \in X$. Then (1.3.4) also holds for the empty set U.

REMARK 1.5 In contrast with the classical convex case, the separation of a point from a set does not imply the separation property for two sets. However, in some special nonconvex cases the separation property (in a certain sense) is also valid for sets (see Section 2.4 and Section 5.4).

The notion of *abstract convex set* naturally leads to abstract quasiconvexity. We do not examine abstract quasiconvexity in this book. However, it will be useful to present the definition of abstract quasiconvex functions .

DEFINITION 1.6 Let $V \subset \bar{\mathbb{R}}$, let X, H be sets and $\phi : X \times H$ be a coupling function. A function $f : X \to V$ is called abstract quasiconvex (or H-quasiconvex) if its level sets $S_c(f) = \{x \in X : f(x) \leq c\}$ are (X, H)-convex for all $c \in \bar{V}$. (Here $\bar{V}$ is defined by (1.2.6)).

Recall that the function f defined on a vector space X and mapping into $\bar{\mathbb{R}}$ is called *quasiconvex* if its level sets $S_c(f) = \{x : f(x) \leq c\}$ are convex for all $c \in \bar{\mathbb{R}}$.

For applications of abstract convexity, we need to describe H-convex functions and H-convex sets for various specific H. As a rule, describing abstract convex functions is much easier than describing abstract convex sets. We now give some examples, beginning with two classical ones, which will be discussed in detail in Section 7.3. We have already discussed some assertions from these examples.

EXAMPLE 1.6 Let $X = \mathbb{R}^n$ and L be the set of all linear functions defined on $\mathbb{R}^n$. Note that $l \in L$ if and only if there exists $v \in \mathbb{R}^n$ such that $l(x) = [v, x]$. (Here $[v, x]$ stands for the inner product $\sum_i v_i x_i$ of vectors v and x. Hence we can identify L and $\mathbb{R}^n$.) We have:

1) Each L-convex function either maps into $\mathbb{R}_{+\infty}$ or is equal to $-\infty$.

2) A function $p : X \to \mathbb{R}_{+\infty}$ is L-convex if and only if p is lower semicontinuous and sublinear, that is p is positively homogeneous of degree one ($p(\lambda x) = \lambda p(x)$ for $\lambda > 0$) and convex.

3) If $p : X \to \mathbb{R}_{+\infty}$ is a lower semicontinuous sublinear function then $\operatorname{supp}(p, L)$ coincides with the subdifferential $\partial p(0)$ of p at the origin.

4) A set $U \subset L$ is (L, X)-convex if and only if U is closed and convex.

5) A function $f : \mathbb{R}^n \to \bar{\mathbb{R}}$ is L-quasiconvex if and only if f is lower semicontinuous and quasiconvex.

EXAMPLE 1.7 Again, let $X = \mathbb{R}^n$ and let H be the set of all affine functions defined on X. In other words, $h \in H$ if and only if $h(x) = l(x) - c$ with $l \in L$ and $c \in \mathbb{R}$. We identify h with the pair (l, c) and H with $L \times \mathbb{R}$. We have the following:

1) Each H-convex function either maps into $\mathbb{R}_{+\infty}$ or is equal to $-\infty$.

2) A function $f : X \to R_{+\infty}$ is H-convex if and only if f is lower semicontinuous and convex.

3) If $f : X \to \mathbb{R}_{+\infty}$ is a lower semicontinuous convex function, then $\operatorname{supp}(f, H)$ coincides with the epigraph $\operatorname{epi} f^* := \{h = (l, c) : c \geq f^*(l)\}$ of the Fenchel conjugate function f^* to the function f.

4) A set $U \subset H \equiv L \times \mathbb{R}$ is (H, X)-convex if and only if U is closed convex and the following properties hold: a) $((l, c) \in U,\ c' \geq c) \implies (l, c') \in U$ and b) U does not contain vertical lines, that is, lines $\{(x, c) : c \in \mathbb{R}\}$.

5) A function $f : \mathbb{R}^n \to \bar{\mathbb{R}}$ is H-quasiconvex if and only if this function is L-quasiconvex, that is, lower semicontinuous and quasiconvex.

The following example contains results which will be proved in the next chapter.

EXAMPLE 1.8 Let $X = \mathbb{R}^n_{++} := \{x = (x_1, \ldots, x_n) \in \mathbb{R}^n : x_1 > 0, \ldots, x_n > 0\}$ and L be the set of all functions $\phi_l : \mathbb{R}^n_{++} \to \mathbb{R}_{++}$ with $l \in \mathbb{R}^n_{++}$, where

$$\phi_l(x) = \min_{i=1,\ldots,n} l_i x_i \qquad (x \in \mathbb{R}^n_{++}). \tag{1.3.5}$$

Then

1) each L-convex function either maps into $\mathbb{R}_{++}$ or is identically equal to zero or is identically equal to $+\infty$;

2) a function $f : X \to \mathbb{R}_{++}$ is L-convex if and only if this function is IPH (increasing positively homogeneous of degree one);

3) a set $U \subset L \equiv \mathbb{R}^n_{++}$ is (L, X)-convex if and only if it is closed (in the topological space $\mathbb{R}^n_{++}$) and normal: $(l \in U, l' \in L, l' \leq l) \implies l' \in U$; here $\leq$ stands for the coordinate-wise order relation in $\mathbb{R}^n_{++}$.

The following example will be discussed in detail in Section 6.2.

EXAMPLE 1.9 Let $C(Q)$ be the space of continuous functions defined on a compact subset Z of the Euclidean space $\mathbb{R}^n$ and let H be the subspace of $C(Q)$ spanned by the following $n + 2$ functions:

$$h_0(x) = \|x\|^2,\ h_1(x) = x_1,\ \dots, h_n(x) = x_n,\ h_{n+1}(x) = 1.$$

Then a function f defined on Z is H-convex if and only if this function is lower semicontinuous. Abstract convex sets in this case are very large and their description is unknown.

1.4. SUBDIFFERENTIABILITY

As it was mentioned in Section 1.1 there are two different approaches to subdifferentiability of convex functions - local and global. The global definition can be naturally extended to abstract convex setting.

We now give the definition of the abstract subdifferential for abstract convex functions. In this section, we shall consider only finite elementary functions. (A more general situation can be studied by means of *conjugation*.)

DEFINITION 1.7 Let L be a set of finite elementary functions defined on a set X. A function $l \in L$ is called an abstract subgradient (or L-subgradient) of a proper function $f : X \to \mathbb{R}_{+\infty}$ at a point y if $f(x) \geq l(x) - (l(y) - f(y))$ for all $x \in X$. The set $\partial_L f(y)$ of all abstract subgradients of f at y is referred to as the *abstract subdifferential* (or *L-subdifferential*) of the function f at the point y.

It the subdifferential $\partial_L f(y)$ is not empty then $y \in \operatorname{dom} f = \{x \in X : f(x) < +\infty\}$. This assertion follows directly from Definition 1.7.

Let H_L be the closure of the set L under vertical shifts, that is $H_L = \{h' : h'(x) = h(x) - c,\ h \in H,\ c \in \mathbb{R}\}$. Clearly $l \in \partial_L f(y)$ if and only if

$$h_l \in \operatorname{supp}(f, H_L),$$

where the function

$$h_l(x) = l(x) - (l(y) - f(y)) \tag{1.4.1}$$

is an element of H_L. The following assertion holds:

PROPOSITION 1.2 *Let $f : X \to \mathbb{R}_{+\infty}$ be a proper function. Then the subdifferential $\partial_L f(y)$ is not empty if and only if*

$$f(y) = \max\{h(y) : h \leq f, h \in H_L\}. \tag{1.4.2}$$

Proof: Let $l \in \partial_L f(y)$ and let h_l be a function defined by (1.4.1). Then $h_l \in H_L$ and $h_l \leq f$. We also have $h_l(y) = f(y)$. Hence (1.4.2) holds. Assume now that there exists a function $h \in \text{supp}(f, H_L)$ such that $f(y) = h(y)$. Let $h(x) = l(x) - c$ with $l \in L$ and $c \in \mathbb{R}$. Then $c = l(y) - f(y)$, hence the inequality $h(x) - (l(y) - f(y) \leq f(x)$ holds. $\triangle$

COROLLARY 1.2 *If a set L is closed under vertical shifts, that is $L = H_L$ then*

$$\partial_L f(y) = \{l \in \text{supp}(f, L) : l(y) = f(y)\}. \tag{1.4.3}$$

If L is the set of linear functions defined on a vector space X then the set H_L consists of affine functions. In view of this observation we shall consider the initial set L as a set of *abstract linear* functions. Note that we do not define an individual abstract linear function; we consider a *set of abstract linear functions.* Vertical shifts of functions $l \in L$, that is, functions of the form

$$h(x) = l(x) - c, \text{ for all } x \in X \tag{1.4.4}$$

with $l \in L$ and $c \in \mathbb{R}$ are called *L-affine* functions or *abstract affine functions* with respect to L. Thus the set H_L is referred to as the set of L-affine functions. Due to (1.4.4) we shall identify h and the pair $(l, c) \in L \times \mathbb{R}$. It is possible that there are different pairs (l_1, c_1) and (l_2, c_2) such that $h(x) = l_1(x) - c_1 = l_2(x) - c_2$. For example, if L is already closed under vertical shifts and $h(x) = l(x) - c$ then also $h(x) = l'(x) - c'$ with $l' = l + d\mathbf{1} \in L$, $c' = c + d \in \mathbb{R}$, where $\mathbf{1}(x) = 1$ for all $x \in X$ and $d \in \mathbb{R}$. Assume that the following condition (A) holds:

$$(A) \qquad l - c\mathbf{1} \notin L \text{ for all } l \in L \text{ and } c \neq 0. \tag{1.4.5}$$

It follows immediately from (A) that it is possible to represent an L-affine function h in the form (1.4.4) in a unique way. In such a case we can identify the set H_L of all L-affine functions with $L \times \mathbb{R}$.

Let L be a set of abstract linear functions defined on X and let $f : X \to \mathbb{R}_{+\infty}$ be a proper function. Assume that the subdifferential $\partial_L f(x) \neq \emptyset$ for all $x \in X$. Then, as it follows from Proposition 1.2, the function f is abstract convex with respect to the set H_L of abstract affine functions and the supremum $\sup\{h(x) : h \in \text{supp}(f, H_L)\}$ is attained for each $x \in X$. This observation demonstrates that the L-subdifferential is a good tool in the study of H_L-convex functions.

DEFINITION 1.8 A subset X of a vector space is called a *conic set* if $\lambda x \in X$ for each $x \in X$ and $\lambda > 0$.

Sometimes we shall use the term *cone* instead of *conic set*. The definition of a conic set in a more general situation will be given in Section 7.4, see Definition 7.25.

In the subsequent chapters we shall mainly consider sets of abstract linear functions L, which consist of positively homogeneous of degree one functions defined on a conic subset X of a vector space. Obviously such a set L enjoys the property (A). If L consists of positively homogeneous of degree one functions then each L-convex function p also possesses this property. Indeed, we have for $x \in X$ and $\lambda > 0$:

$$\begin{aligned} p(\lambda x) &= \sup\{l(\lambda x) : l \in \text{supp}(p, L)\} \\ &= \lambda \sup\{l(x) : l \in \text{supp}(p, L)\} = \lambda p(x). \end{aligned}$$

The following assertion presents a very simple description of the L-subdifferential for a set L consisting of positively homogeneous functions.

PROPOSITION 1.3 *Let L be a set of abstract linear functions defined on a conic set X and let each $l \in L$ be positively homogeneous of degree one ($l(\lambda x) = \lambda l(x)$ for all $\lambda > 0$.) Then for an L-convex function p and for $y \in X$ we have:*

$$\partial_L p(y) = \{l \in \text{supp}(p, L) : l(y) = p(y)\}. \qquad (1.4.6)$$

Proof: Let $l \in \partial_L(y)$, that is, $p(x) \geq l(x) - (l(y) - p(y))$ for every $x \in X$. Let $x \in X$. Consider vectors λx with $\lambda > 0$. Then

$$\lambda p(x) = p(\lambda x) \geq \lambda l(x) - (l(y) - p(y)), \qquad (1.4.7)$$

so,

$$p(x) \geq l(x) - \frac{l(y) - p(y)}{\lambda}.$$

Passing to the limit as $\lambda \to +\infty$, we obtain $p(x) \geq l(x)$, that is, $l \in \text{supp}(p, L)$. In particular, we have $l(y) \leq p(y)$. On the other hand, passing to the limit in (1.4.7) as $\lambda \to +0$, we have $l(y) \geq p(y)$. Thus $l(y) = p(y)$ so $\partial_L p(y)$ is contained in the set on the right-hand side in (1.4.6). Assume now that $p(x) \geq l(x)$ for all $x \in X$ and $p(y) - l(y) = 0$. Then

$$p(x) \geq l(x) - (l(y) - p(y)) \text{ for all } x \in X.$$

Thus the result holds. △

REMARK 1.6 A similar result is valid if L is closed under vertical shifts (see Corollary 1.2).

The presence of two definitions of the subdifferential (local and global) for the classical convex case is very convenient. In particular, the local definition plays a crucial role in the calculation of the subdifferential and in the numerical applications of convexity.

For applications of abstract convexity we need to have classes of abstract convex functions for which a local characterization of the abstract subdifferential can be given. Some of these classes will be studied in Chapter 3.

1.5. CONJUGATION

Let X and L be two sets. Consider a function $\phi : L \times X \to R$. The function ϕ will be referred to as a *coupling function* and we shall use the notation $\phi(x, l) = \langle l|x \rangle$. As it noted in Remark 1.3 this coupling function allows us to consider an element $l \in L$ as a function defined on X, namely $l(x) = \langle l|x \rangle$ for all $x \in X$ and, similarly, an element $x \in X$ as a function on L.

Let $f : X \to \bar{\mathbb{R}}$. The Fenchel-Moreau L-conjugate, f_L^*, of a function f is defined as follows:

$$f_L^*(l) = \sup_{x \in X}(\langle l|x \rangle - f(x)), \qquad l \in L.$$

The Fenchel-Moreau second L-conjugate, f_L^{**}, is the function defined by:

$$f_L^{**}(x) = \sup_{l \in L}(\langle l|x \rangle - f_L^*(l)), \qquad x \in X.$$

These definitions can be extended for coupling functions mapping into $\bar{\mathbb{R}} = [-\infty, +\infty]$ (we should put $\langle l|x \rangle \dagger (-f(x))$ instead of $\langle l|x \rangle - f(x)$

where $\dagger$ is the lower addition, that is, an extension of usual $+$ by the convention $+\infty\dagger(-\infty) = -\infty\dagger(+\infty) = -\infty$). Consider L as a set of abstract linear functions, defined on X and let H_L be the set of L-affine functions: $H_L = \{h : h(x) = l(x) - c,\ l \in L,\ c \in \mathbb{R}\}$. A classical result of abstract convex analysis, the Fenchel-Moreau theorem (see, for example, [115, 175]), asserts that for a function $f : X \to \bar{\mathbb{R}}$, the equality $f_L^{**} = \mathrm{co}\,_{H_L} f$ holds, where $\mathrm{co}\,_{H_L} f$ is the H_L-convex hull of the function f (see Definition 1.3).

Support sets can be described in terms of conjugate functions. Let $l \in L$ and $c \in \mathbb{R}$. If L possesses the property (A) then we can identify a function $h(x) = \langle l|x\rangle - c$ with the pair (l, c), and H_L with $L \times \mathbb{R}$. It follows easily from the definitions of the Fenchel-Moreau conjugate function and the support set that $\mathrm{supp}(f, H_L)$ coincides with the epigraph, $\mathrm{epi}\, f_L^*$, of the function f_L^*. (We shall prove this assertion in Chapter 7, see Proposition 7.8.) Recall that for a function g defined on a set X, the epigraph, $\mathrm{epi}\, g$, is defined by $\mathrm{epi}\, g := \{(x, c) \in X \times \mathbb{R} : c \geq g(x)\}$.

1.6. ABSTRACT CONCAVE FUNCTIONS AND INFIMA OF ABSTRACT CONVEX FUNCTIONS

Together with abstract convex functions and sets we can define, in a symmetric manner, the concepts of H-concave functions and sets. We consider only H-concave functions. Let $V \subset \bar{\mathbb{R}}$ and let X be a set. Consider a set H of elementary functions $h : X \to V$. A function $f : X \to \bar{\mathbb{R}}$ is said to be *abstract concave* (or H-concave) if there exists a set $U \subset H$ such that $f(x) = \inf\{h(x) : h \in U\}$ for all $x \in X$. (It is supposed here that the infimum over the empty set is defined by (1.2.5).)

DEFINITION 1.9 Let $f : X \to \bar{V}$, where $\bar{V}$ is defined by (1.2.6). The set

$$\mathrm{supp}^+(f, H) = \{h \in H : h \geq f\}$$

is called the *upper support set* of the function f with respect to the set of elementary functions H.

DEFINITION 1.10 Let H be a set of finite elementary functions. A function $h \in H$ is called an *abstract supergradient* (or H-supergradient) of an H-concave function f at a point y if $f(x) \leq h(x) - (h(y) - f(y))$ for all x. The superdifferential $\bar{\partial}_H f(y)$ of an H-concave function f at a point y is defined as the set of all H-supergradients of f at y.

PROPOSITION 1.4 *Let L be a set of abstract linear functions defined on a conic set X and let each $l \in L$ be positively homogeneous of degree* *one. Then for an L-concave function p and for $y \in X$ we have*

$$\bar{\partial}_L(y) = \{l \in \operatorname{supp}^+(p, L) : l(y) = p(y)\}.$$

The proof is similar to that of Proposition 1.3 and we omit it.

Let H be a set of functions defined on a set X and $P(H, X)$ be the set of all H-convex functions. A $P(H, X)$-concave function $f : X \to \bar{\mathbb{R}}$ is called inf-H-convex.

EXAMPLE 1.10 [172] Let H be the set of all affine functions defined on $\mathbb{R}^n$. Then each function $f : \mathbb{R}^n \to \mathbb{R}_{+\infty}$ is inf-H-convex. Indeed, $f(x) = \inf_{z \in \operatorname{dom} f} f_z(x)$, where

$$f_z(x) = \begin{cases} f(z) & \text{if} x = z; \\ +\infty & \text{if} x \neq z. \end{cases}$$

The epigraph, epi f_z, of the function f_z coincides with the ray $\{(z, \lambda) : \lambda \geq f(z)\}$. Since this ray is convex and closed it follows that f_z is a convex and lower semicontinuous function.

The class of inf-convex functions (i.e., inf-H-convex with respect to the set of all affine functions H) is extremely broad. However, for some sets of elementary functions H, the classes of H-convex and inf-H-convex functions coincide, that is the pointwise infimum of H-convex functions is again an H-convex function.

EXAMPLE 1.11 Consider the situation, which was described in Example 1.8, that is, consider the set L of functions defined on $\mathbb{R}^n_{++}$ by (1.3.5). As it was mentioned in Example 1.8 a function p is L-convex if and only if p is IPH (increasing and positively homogeneous of degree one) function. Since the infimum of a family of IPH functions is again IPH, it follows that p is inf-L-convex if and only if this function is L-convex.

Chapter 2

ELEMENTS OF MONOTONIC ANALYSIS: IPH FUNCTIONS AND NORMAL SETS

2.1. INTRODUCTION

In this and next chapters we examine some classes of monotonic functions defined on either the cone $\mathbb{R}^n_+ = \{x \in \mathbb{R}^n : x_i \geq 0 \text{ for all } i = 1, \dots, n\}$ or the cone $\mathbb{R}^n_{++} = \{x \in \mathbb{R}^n_+ : x_i > 0 \text{ for all } i = 1, \dots, n\}$ and the so-called normal subsets of these cones. We shall consider these classes in the framework of abstract convexity. The simplest and most elegant theory can be obtained for functions defined on the cone $\mathbb{R}^n_{++}$ and subsets of this cone. A nonempty set $U \in \mathbb{R}^n_+$ is said to be *normal* if

$$(x \in U,\ x' \in \mathbb{R}^n_+,\ x' \leq x) \implies x' \in U. \tag{2.1.1}$$

Throughout this book we assume that the empty set is a normal subset of $\mathbb{R}^n_+$. Note that the ordering here is the usual coordinate-wise order relation in $\mathbb{R}^n_+$. Normal subsets of the cone $\mathbb{R}^n_{++}$ can be defined in the same way.

Normal sets are closely connected to the so-called IPH (increasing positively homogeneous of degree one) functions.

There is a clear analogy between the class of IPH functions and the class of sublinear functions, and between normal sets and convex sets. (See Subsection 2.2.11 for details.) One of the main tools for the study of sublinearity of functions and the convexity of sets are linear functions. For example, a function p defined on $\mathbb{R}^n$ is lower semicontinuous and sublinear if and only if this function is abstract convex with respect to the set of all linear functions, in other words, there exists a set of linear functions U such that $p(x) = \sup\{u(x) : u \in U\}$. Consider now, for the sake of definiteness, subsets of the cone $\mathbb{R}^n_{++}$ and functions defined on this cone. We shall be interested in the class of functions of the form

$$x \mapsto \min\{l_i x_i : i = 1, 2, \dots, n\} \qquad (x \in \mathbb{R}^n_{++}) \tag{2.1.2}$$

to replace the linear functions $x \mapsto \sum_i l_i x_i$. We shall consider the class L of all functions of the form (2.1.2) with $l \in \mathbb{R}^n_{++}$ as a set of ***abstract linear functions*** (see Section 1.4). The class of linear functions leads to the development of sublinear functions and convex sets whereas the class of functions of the form (2.1.2) leads (in the framework of abstract convexity) to the development of IPH functions and normal sets. Note that in this structure increasing functions will play the role of quasiconvex functions in the sense that a function is quasiconvex if and only if it has convex level sets and a function is increasing if and only if it has normal level sets.

In order to get an analogue of convex functions in this structure we need to consider abstract convex functions with respect to the set H_L of all abstract affine functions (see Section 1.4) generated by the set L. As it turns out, a function f is abstract convex if and only if f is increasing and convex-along-rays (ICAR). The latter means that the restriction of f to each ray starting from zero is a convex function of one variable.

IPH functions can also be considered as L-concave functions (in contrast with the classical convex theory, where a sublinear function is superlinear if and only if it is linear). A set $V \in \mathbb{R}^n_{++}$ is abstract concave if and only if it is *co-normal*, that is, $(x \in V, x' \geq x) \implies x' \in V$. A function f is abstract concave with respect to H_L if and only if f is increasing and concave-along-rays. Decreasing functions can be considered in this structure as analogues of quasiconcave functions. In this book we shall not consider these abstract concave analogues of abstract convex objects.

Another approach in the study of decreasing functions is based on the direct application of IPH functions and normal sets. Namely, the positive part of the hypograph

$$\text{hyp}^+ g = \{(\alpha, y) \in \mathbb{R} \times \mathbb{R}^n_{++} : 0 < \alpha \leq g(y) \text{ for all } y \in \text{dom}\, g\}$$

of a decreasing function g defined on $\mathbb{R}^n_{++}$ is a normal subset of $\mathbb{R}_{++} \times \mathbb{R}^n_{++}$. Hence $\text{hyp}^+ g$ is a support set with respect to L of an IPH function defined on $\mathbb{R}_{++} \times \mathbb{R}^n_{++} \equiv \mathbb{R}^{n+1}_{++}$. These relations are very useful in the study of decreasing functions and their applications.

The corresponding approaches can be applied and corresponding results can be obtained for functions defined on the cone $\mathbb{R}^n_+$ and subsets of this cone. However this case is much more complicated. The reason for this is that functions defined in (2.1.2) are equal to zero for all boundary points of the cone $\mathbb{R}^n_+$. Thus we should substitute functions defined in (2.1.2) by other functions, which coincide with (2.1.2) on int $\mathbb{R}^n_+ \equiv \mathbb{R}^n_{++}$ and not necessarily vanish on the boundary of $\mathbb{R}^n_+$. Unfortunately, it is impossible to find a continuous function of such a sort and the discontinuity causes some difficulties.

The totality of proposed approaches and results obtained in this direction can be considered as certain elements of *monotonic analysis.*

Note that various classes of monotonic functions and normal (conormal) sets have found many applications in several branches of mathematical analysis and its applications.

Consider, for example, system of inequalities $f_t(x) \leq 0\,(t \in T)$, where $(f_t)_{t\in T}$ is a family of increasing functions defined on $\mathbb{R}^n_+$ and T is an arbitrary set of indices. The solution set of this system is normal. Thus the theory of normal sets plays a crucial role in the study of such systems.

The perturbation function

$$\beta(y_1, \dots, y_n) = \inf\{f_0(x) : x \in X,\ f_1(x) \leq y_1, \dots, f_n(x) \leq y_n)$$

of an optimization problem

$$f_0(x) \longrightarrow \min \text{ subject to } x \in X,\ f_i(x) \leq 0,\ i = 1, \dots, n \qquad (2.1.3)$$

is a decreasing function of the vector $y = (y_1, \dots, y_n) \in \mathbb{R}^n_{++}$. The theory of decreasing functions has found interesting applications in the study of various kinds of penalization for problem (2.1.3) with $n = 1$. (See Chapter 4.)

Property (2.1.1) is usually called the *free disposal* property in economic theory ([106]). Thus normal sets are sets with free disposal. In the study of various models of economic equilibrium and dynamics we often consider convex normal sets (see, for example [105, 106]). However there are significant problems which arise in economics for which nonconvex normal sets are required. For example, in the study of the asymptotic behaviour of paths for von Neumann type models of economic dynamics, it is necessary to consider specially structured nonconvex normal sets ([140, 48]). IPH functions also play an important role in the study of models in mathematical economics. For example, they are used as production functions under the assumption of *constant returns to scale.* Thus monotonic analysis can find many applications in mathematical economics.

2.2. INCREASING POSITIVELY HOMOGENEOUS FUNCTIONS DEFINED ON POSITIVE ORTHANT

2.2.1. Overview

In this section we shall study normal subsets of the cone $\mathbb{R}^n_{++}$ and IPH functions defined on this cone. There are two approaches to the relationship between normal sets and IPH functions. Firstly it is often convenient to study normal sets and IPH functions within the framework of abstract convexity. The second approach involves the study of IPH functions as the Minkowski gauge of appropriate normal sets. We shall unite both approaches in order to study a special kind of nonlinear duality for IPH functions defined on $\mathbb{R}^n_{++}$ and normal subsets of this cone. In particular, we show that abstract convexity with respect to the set of all min-type functions defined on $\mathbb{R}^n_{++}$ leads to IPH functions and normal sets. We also discussed a polarity for both IPH functions and normal sets. Some results in this direction easily follow from the general theory, presented in [175, 109]. However the specific structure of the set of min-type functions allows one to obtain much more advanced results.

The normal subsets of the cone $\mathbb{R}^n_+$ and IPH functions defined on this cone will be studied in Section 2.3. The main results of this section were obtained by A.M. Rubinov and B. M. Glover [150]. An interesting approach in the study of IPH functions was suggested by V.D. Matveenko in [110, 111]

2.2.2. Preliminaries

1. Let I be a finite set of indices. We shall use the following notation:

- $\mathbb{R}^I$ is the space of all vectors $(x_i)_{i\in I}$;
- x_i is the i-th coordinate of a vector $x \in \mathbb{R}^I$;
- if $x, y \in \mathbb{R}^I$ then $x \geq y \iff x_i \geq y_i$ for all $i \in I$;
- if $x, y \in \mathbb{R}^I$ then $x \gg y \iff x_i > y_i$ for all $i \in I$;
- $\mathbb{R}^I_+ = \{x = (x_i)_{i\in I} \in \mathbb{R}^I : x \geq 0\}$;
- $\mathbb{R}^I_{++} = \{x = (x_i)_{i\in I} \in \mathbb{R}^I : x \gg 0\}$.

If $I = \{1, \dots, n\}$ we shall also use the notation $\mathbb{R}^n$, $\mathbb{R}^n_+$ and $\mathbb{R}^n_{++}$ instead of $\mathbb{R}^I$, $\mathbb{R}^I_+$ and $\mathbb{R}^I_{++}$, respectively. Thus

$$\mathbb{R}^n_+ = \{x = (x_i)_{i\in I} \in \mathbb{R}^n : x_i \geq 0 \text{ for all } i \in I\},$$

$$\mathbb{R}^n_{++} = \{x = (x_i)_{i \in I} \in \mathbb{R}^n : x_i > 0 \text{ for all } i \in I\}.$$

2. Recall some definitions from the geometry of vector spaces. Let X be a vector space and $x \in X$. The set $\tilde{R}_x = \{\lambda x : \lambda > 0\}$ ($R_x = \{\lambda x : \lambda \geq 0\}$) is called the open ray (closed ray) starting from zero and passing through x. A set $Q \subset X$ is called *conic* if (see Definition 1.8) $x \in Q \implies \tilde{R}_x \subset Q$. Occasionally we shall use a term *cone* instead of a conic set. A convex conic set is called a *convex cone*.It is easy to see that a conic set Q is a convex cone if and only if the following holds: $(x_1, x_2 \in Q) \implies x_1 + x_2 \in Q$.

We shall use the following notation:

$$U + V = \{u + v : u \in U, \, v \in V\} \qquad (U, \, V \subset X); \tag{2.2.1}$$

$$\lambda \cdot U \equiv \lambda U = \{\lambda u : u \in U\} \qquad (U \subset X, \, \lambda > 0); \tag{2.2.2}$$

$$0 \cdot U = \bigcap_{\lambda > 0} \lambda U \qquad (U \subset X). \tag{2.2.3}$$

Using (2.2.2) we can restate the definition of a conic set as follows: Q is called conic if $\lambda Q \subset Q$ for all $\lambda > 0$. It follows from (2.2.3) that $0 \cdot Q = Q$ for a conic set Q. A conic set Q is a convex cone if and only if $Q + Q = Q$.

2.2.3. IPH functions

Let Q be either $\mathbb{R}^n_+$ or $\mathbb{R}^n_{++}$. A function f defined on Q is called *increasing* if $x \geq y$ implies $f(x) \geq f(y)$. Sometimes the term *isotonic* is used instead of increasing. The function f is positively homogeneous of degree one if $f(\lambda x) = \lambda f(x)$ for all $x \in Q$ and $\lambda > 0$. Combining these definitions we get the following definition of an IPH function.

DEFINITION 2.1 Let Q be either $\mathbb{R}^n_+$ or $\mathbb{R}^n_{++}$. A function $f : Q \to \mathbb{R}_{+\infty}$ is an *IPH function* if the following conditions are satisfied:

1 $x \geq y$ implies $f(x) \geq f(y)$;

2 $f(\lambda x) = \lambda f(x)$ for $\lambda > 0$ and $x \in Q$.

We now give some examples of IPH functions.

EXAMPLE 2.1 Let Q be either $\mathbb{R}^n_+$ or $\mathbb{R}^n_{++}$. The following functions defined on Q are IPH:

1) $a(x) = \sum_{i \in I} a_i x_i$ with $a_i \geq 0$;

2) $p_k(x) = (\sum_{i \in I} x_i^k)^{\frac{1}{k}}$ $(k > 0)$;

3) $f(x) = \sqrt{[Ax, x]}$, where A is a matrix with nonnegative entries;

4) $f(x) = \prod_{j \in J} x_j^{t_j}$, where $J \subset I$, $t_j > 0, \sum_{j \in J} t_j = 1$.

It is easy to check that the set of all IPH functions defined on Q enjoys the following properties:

1) it is a convex cone: the sum of two IPH functions is also an IPH function; if f is IPH, then the function γf is IPH for all $\gamma > 0$;

2) if T is an arbitrary index set and $(f_t)_{t \in T}$ is a family of IPH functions, then the functions $f_{\inf}(x) = \inf_{t \in T} f_t(x)$ and $f_{\sup}(x) = \sup_{t \in T} f_t(x)$ are IPH;

3) the pointwise limit of a sequence (and more general, of a directed family) of IPH functions is IPH.

These properties allow us to give two more examples of IPH functions defined on Q.

EXAMPLE 2.2 The following functions are IPH:

1) $f(x) = \max_{k \in K} \min_{j \in J} \sum_{i \in I} a_i^{jk} x_i$, where $a_i^{jk} \geq 0$, $k \in K, j \in J, i \in I$. Here J and K are finite sets of indices;

2)
$$f(x) = \max_{k \in K} \min_{j \in J_k} \sum_{i \in I} a_i^j x_i, \tag{2.2.4}$$

where $a_i^j \geq 0$, $j \in J_k$, $k \in K$. Here K and J_k are finite sets of indices.

Note (see [18]) that an arbitrary piecewise linear function f generated by a collection of linear functions $a^1, \ldots, a^m$ can be represented as (2.2.4); hence an arbitrary piecewise linear function defined on Q and generated by nonnegative vectors, is IPH.

We now describe some properties of IPH functions defined on $\mathbb{R}^n_{++}$. Let f be an IPH function, then

1 $f(x) \geq 0$ for all $x \in \mathbb{R}^n_{++}$. Indeed, consider a vector $x \in \mathbb{R}^n_{++}$. Since $(1/2)x \leq x$, it follows that $(1/2)f(x) = f((1/2)x) \leq f(x)$. Hence $f(x) \geq 0$.

2 If f is IPH and there exists a point $\bar{x} \in \mathbb{R}^n_{++}$ such that $f(\bar{x}) = +\infty$, then $f(x) = +\infty$ for all $x \in \mathbb{R}^n_{++}$. Indeed, if $x \in \mathbb{R}^n_{++}$, then there

exists $\lambda > 0$ such that $x \geq \lambda \bar{x}$. Therefore $f(x) \geq f(\lambda \bar{x}) = \lambda f(\bar{x}) = +\infty$.

3 If there exists a point $\bar{x} \in \mathbb{R}^n_{++}$ such that $f(\bar{x}) = 0$, then $f(x) = 0$ for all $x \in \mathbb{R}^n_{++}$. In fact, for each $x \in \mathbb{R}^n_{++}$ there exists $\lambda > 0$ such that $x \leq \lambda \bar{x}$. Hence $0 \leq f(x) \leq \lambda f(\bar{x}) = 0$.

Thus for an IPH function $f : \mathbb{R}^n_{++} \to \mathbb{R}_{+\infty}$ there are three possibilities:

(i) f maps $\mathbb{R}^n_{++}$ into $(0, +\infty)$;

(ii) $f(x) = +\infty$ for all $x \in \mathbb{R}^n_{++}$;

(iii) $f(x) = 0$ for all $x \in \mathbb{R}^n_{++}$.

4 Each IPH function f is continuous on $\mathbb{R}^n_{++}$.

To see this, assume that f maps $\mathbb{R}^n_{++}$ into $(0, +\infty)$. Let $x \in \mathbb{R}^n_{++}$, and $x_n \to x$. Take $\varepsilon > 0$. For sufficiently large n, we have

$$(1 - \varepsilon)x \leq x_n \leq (1 + \varepsilon)x.$$

Hence

$$(1 - \varepsilon)f(x) \leq f(x_n) \leq (1 + \varepsilon)f(x).$$

Thus $f(x_n) \to f(x)$.

Consider IPH functions defined on $\mathbb{R}^n_+$. Let f be such a function. Then

1) either $f(0) = 0$ or $f(0) = +\infty$. This assertion follows from the equality $f(0) = f(2 \cdot 0) = 2f(0)$;

2) $f(x) \geq 0$ for all $x \in \mathbb{R}^n_+$;

3) if $f(0) \neq 0$, then $f(x) = +\infty$ for all $x \in \mathbb{R}^n_+$. Both 2) and 3) follow from 1) and the isotonicity of f;

4) if there exists a point $\bar{x} \in \mathbb{R}^n_+$ such that $f(\bar{x}) = +\infty$, then $f(x) = +\infty$ for all $x \in \mathbb{R}^n_{++}$. Indeed, let $x' \in \mathbb{R}^n_{++}$, $x' \geq \bar{x}$, then $f(x') = +\infty$. Hence $f(x) = +\infty$ for all $x \in \mathbb{R}^n_{++}$ (see properties of IPH functions defined on $\mathbb{R}^n_{++}$);

5) if there exists a point $\bar{x} \in \mathbb{R}^n_{++}$ such that $f(\bar{x}) = 0$, then $f(x) = 0$ for all $x \in \mathbb{R}^n_+$. In fact, it was shown that $f(x) = 0$ for all $x \in \mathbb{R}^n_{++}$. Since each point from $\mathbb{R}^n_+$ is majorized by a certain point from $\mathbb{R}^n_{++}$, it follows that $f = 0$.

The following example demonstrates that there exist IPH functions defined on $\mathbb{R}^n_+$ which take only two values 0 and $+\infty$.

EXAMPLE 2.3 The function f defined on $\mathbb{R}^n_+$ by

$$f(x) = \begin{cases} +\infty & \text{if } x \in \mathbb{R}^n_{++}; \\ 0 & \text{if } x \notin \mathbb{R}^n_{++} \end{cases}$$

is IPH.

It will be proved later (see Proposition 2.17) that each finite IPH function f is lower semicontinuous. There exist discontinuous finite IPH functions. For example, the function p defined on $\mathbb{R}^2_+$ by

$$p(x) = \begin{cases} x_1 + x_2 & \text{if } x_1 x_2 > 0; \\ 0 & \text{if } x_1 x_2 = 0 \end{cases}$$

is discontinuous on the boundary of $\mathbb{R}^2_+$.

Let p be an IPH function defined on $\mathbb{R}^n_+$. Then its restriction to $\mathbb{R}^n_{++}$ is again an IPH function. We now show that each IPH function defined on $\mathbb{R}^n_{++}$ can be extended to either a lower semicontinuous or an upper semicontinuous IPH function defined on $\mathbb{R}^n_+$. Firstly we recall the well-known notions of *lower regularization* and *upper regularization* of a function (see, for example [168], Subsection 6.3).

DEFINITION 2.2 Consider a function $f : X \to \bar{\mathbb{R}}$, where X is a topological space. The function $\underline{f} : X \to \bar{\mathbb{R}}$ defined by

$$\underline{f}(x) = \min(f(x), \liminf_{x' \to x} f(x')),$$

is called the *lower regularization* of the function f. The function $\bar{f}$: $X \to \bar{\mathbb{R}}$ defined by

$$\bar{f}(x) = \max(f(x), \limsup_{x' \to x} f(x'))$$

is called the *upper regularization* of the function f.

It is well known that $\underline{f}$ is lower semicontinuous and $\bar{f}$ is upper semicontinuous. Moreover, $\underline{f}$ is the greatest lower semicontinuous function, which minorizes f and $\bar{f}$ is the lowest upper semicontinuous function which majorizes f. Due to this fact we can assert that the epigraph, epi $\underline{f}$, is a closure of the epigraph, epi f, of the function f; the hypograph, hyp $\bar{f}$, is a closure of the hypograph, hyp f, of the function f. Recall that

$$\operatorname{epi} f = \{(x, \lambda) \in X \times \mathbb{R} : \lambda \geq f(x)\},$$

$$\operatorname{hyp} f = \{(x, \lambda) \in X \times \mathbb{R} : \lambda \leq f(x)\}.$$

PROPOSITION 2.1 *Let p be an IPH function defined on $\mathbb{R}^n_{++}$. Let p_1 and p_2 be functions defined on $\mathbb{R}^n_+$ by*

$$p_1(x) = \liminf_{x' \in \mathbb{R}^n_{++}, x' \to x} p(x'), \quad p_2(x) = \limsup_{x' \in \mathbb{R}^n_{++}, x' \to x} p(x').$$

Then both p_1 and p_2 are IPH extensions of p, and p_1 is lower semicontinuous, p_2 is upper semicontinuous.

Proof: Positive homogeneity and isotonicity of functions p_1 and p_2 follows immediately from their definitions. Since p is continuous on $\mathbb{R}^n_{++}$, both p_1 and p_2 coincide with p on the cone $\mathbb{R}^n_{++}$. Note that the function p_1 is the lower regularization of the function $\bar{p}$ defined on $\mathbb{R}^n_+$ by

$$\bar{p} = \begin{cases} p(x) & \text{if } x \in \mathbb{R}^n_{++}; \\ +\infty & \text{otherwise} . \end{cases}$$

Hence p_1 is lower semicontinuous. A similar argument shows that p_2 is upper semicontinuous. △

REMARK 2.1 It follows from the properties of the lower (upper) regularization that $\operatorname{epi} p_1 = \operatorname{cl} \operatorname{epi} p$ and $\operatorname{hyp} p_2 = \operatorname{cl} \operatorname{hyp} p$.

2.2.4. Min-type functions and IPH functions

In this subsection we study min-type functions and IPH functions defined on the cone $\mathbb{R}^n_{++}$. We consider $\mathbb{R}^n_{++}$ as a topological space. Thus when we discuss closed subsets of this cone or the closure of subsets, we mean the closure in the topological space $\mathbb{R}^n_{++}$.

The following simple example clarifies connections between the closure in $\mathbb{R}^n$ and the closure in $\mathbb{R}^n_{++}$.

EXAMPLE 2.4 Let $S = \{x = (x_1, x_2) \in \mathbb{R}^2_{++} : x_1 + x_2 \le 1\}$. Then S is a closed subset of $\mathbb{R}^2_{++}$ but S is not closed in $\mathbb{R}^2$. The set S coincides with the closure (in $\mathbb{R}^2_{++}$) of the set $\tilde{S} = \{x = (x_1, x_2) \in \mathbb{R}^2_{++} : x_1 + x_2 < 1\}$.

Consider the set L of all functions l defined on $\mathbb{R}^n_{++}$ by the formula

$$l(x) = \langle l, x \rangle, \tag{2.2.5}$$

where the coupling function $\langle \cdot, \cdot \rangle$ is defined as follows

$$\langle l, x \rangle = \min_{i=1,\dots,n} l_i x_i \tag{2.2.6}$$

with $l = (l_1, \dots, l_n) \in \mathbb{R}^n_{++}$. We will identify the vector $l \in \mathbb{R}^n_{++}$ and the function $l \in L$ which is generated by this vector using (2.2.5).

Clearly the function l is concave, positively homogeneous of degree one (i.e., superlinear) and increasing on the cone $\mathbb{R}^n_{++}$.

The upper level sets $S^c(l) = \{x : l(x) \geq c\}$ of the function l have the following form

$$S^c(l) = \{x : x_i \geq \frac{c}{l_i}, i \in I\},$$

where $I = \{1, \dots, n\}$. Let $l \in \mathbb{R}^n_{++}$. We shall use the following notation in this book:

$$\frac{1}{l} \equiv l^{-1} = \left(\frac{1}{l_1}, \dots, \frac{1}{l_n}\right), \qquad \frac{c}{l} = c\frac{1}{l}, \quad (c \geq 0). \tag{2.2.7}$$

Thus

$$S^c(l) = \frac{c}{l} + \mathbb{R}^n_+.$$

The set L of all min-type functions defined in (2.2.6) is a conic subset of the vector space $C(\mathbb{R}^n_{++})$ of all continuous functions defined on $\mathbb{R}^n_{++}$. The mapping $l \to \langle l, \cdot \rangle$ agrees with the conic structure of the sets $\mathbb{R}^n_{++}$ and L; however, this mapping is not linear. Note that L is not a convex cone.

Let X and Y be conic sets. A mapping $\phi : X \to Y$ is called a *conic isomorphism* between X and Y if ϕ is a one-to-one correspondence, ϕ maps onto Y and $\phi(\lambda x) = \lambda\phi(x)$ for all $x \in X$ and $\lambda > 0$. Thus the following assertion holds.

PROPOSITION 2.2 *The mapping $l \mapsto \langle l, \cdot \rangle$ is a conic isomorphism between $\mathbb{R}^n_{++}$ and L.*

We can introduce two natural order relations on the set L.

1) $l_1 \succeq l_2$ if $l_1(x) \geq l_2(x)$ for all $x \in \mathbb{R}^n_{++}$ (the functional order relation);

2) $l_1 \geq l_2$ if $l_{1i} \geq l_{2i}$ for all $i \in I$, where $l_1 = (l_{11}, l_{12}, \dots, l_{1n})$, $l_2 = (l_{21}, l_{22}, \dots, l_{2n})$ (the vector order relation).

Considering L as a subset of the space $C(\mathbb{R}^n_{++})$, we assume that L is endowed with the functional order relation. The next proposition demonstrates that $\mathbb{R}^n_{++}$ and L are isomorphic ordered spaces.

PROPOSITION 2.3 *Order relations $\succeq$ and $\geq$ coincide.*

Proof: The proof is straightforward. △

We also can consider two types of convergence defined on the set L.

1) the functional convergence

$$l^k \to^1 l \text{ if } \langle l^k, x \rangle \to \langle l, x \rangle \text{ for all } x \in \mathbb{R}^n_{++} \text{ and}$$

2) the vector convergence:

$$l^k \to^2 l \text{ if } l_i^k \to l_i \text{ for each } i \in I.$$

Considering L as a subset of $C(\mathbb{R}^n_{++})$ we assume that L is equipped with the functional convergence. Proposition 2.4 shows that L homeomorphic to $\mathbb{R}^n_{++}$.

PROPOSITION 2.4 *Let l^k be a sequence of elements of $\mathbb{R}^n_{++}$. Then $l^k \to^1 l$ if and only if $l^k \to^2 l$.*

Proof: Clearly $l^k \to^2 l$ implies $l^k \to^1 l$. Assume now that $l^k \to^1 l$, that is, $\langle l^k, x\rangle \to \langle l, x\rangle$ for all $x \in \mathbb{R}^n_{++}$. Then there exists $\varepsilon > 0$ such that $\inf_k l_i^k \geq \varepsilon$ for all i. Otherwise we can find an index m such that $\liminf_k l_m^k = 0$, hence $\liminf_k \langle l^k, x\rangle = 0$ for all $x \in \mathbb{R}^n_{++}$. Clearly there exists j such that $l_j^k \not\to +\infty$ (otherwise $\langle l^k, x\rangle \to +\infty$ for all x). Assume there exists m such that the sequence l_m^k is unbounded. We can assume without loss of generality that $l_m^k \to +\infty$. Then for each x we have

$$\langle l, x\rangle = \min_{i\in I} l_i x_i = \lim_k \min_{i\in I} l_i^k x_i = \lim_k \min_{i\in I, i\neq m} l_i^k x_i. \tag{2.2.8}$$

It follows from (2.2.8) that $\lim_k \min_{i\in I, i\neq m} l_i^k x_i > 0$ for each $x \in \mathbb{R}^n_{++}$. Hence there exists a vector $\bar{x} \in \mathbb{R}^n_{++}$ such that

$$\bar{x}_m < \frac{l_i}{l_m}\bar{x}_i \quad (i \neq m), \qquad \bar{x}_m < \frac{1}{l_m} \lim_k \min_{i\neq m} l_i^k \bar{x}_i. \tag{2.2.9}$$

Now (2.2.9) implies that

$$\langle l, \bar{x}\rangle = \min_{i\in I} l_i \bar{x}_i = l_m \bar{x}_m < \lim_k \min_{i\neq m} l_i^k \bar{x}_i$$

which contradicts (2.2.8). Thus sequences $(l_1^k), \dots, (l_n^k)$ are bounded.

Let $j \in I$ and $x = (x_i)_{i=1}^n$ with $x_j = 1$, $x_i = v$, $i \neq j$, where v is a fairly large number. Then

$$\langle l^k, x\rangle = \min_i l_i^k x_i = \min(l_j^k, v \min_{i\neq j} l_i^k),$$

$$\langle l, x\rangle = \min(l_j, v \min_{i\neq j} l_i).$$

Since $\inf_k l_j^k > 0$ and the sequences l_i^k are bounded for all i, there exists v such that $\langle l^k, x\rangle = l_j^k$ and $\langle l, x\rangle = l_j$. Thus $l_j^k \to l_j$. △

THEOREM 2.1 *The mapping $l \to \langle l, \cdot\rangle$ is a conic and ordered isomorphism and a homeomorphism between $\mathbb{R}^n_{++}$ and L.*

Proof: The proof is a direct consequence of Proposition 2.2, Proposition 2.3 and Proposition 2.4. △

Theorem 2.1 implies that we can identify $\mathbb{R}^n_{++}$ and L as conic and ordered sets and as topological spaces.

2.2.5. Abstract convexity with respect to the set of min-type functions

Consider L as a set of elementary functions. We now describe L-convex functions that are functions of the form

$$f(x) = \sup_{l \in U} \langle l, x \rangle, \tag{2.2.10}$$

where U is a subset of L. Note that a min-type function $x \mapsto \langle l, x \rangle$ is IPH. We shall show that a function f is L-convex if and only if f is IPH.

PROPOSITION 2.5 *A function $f : \mathbb{R}^n_{++} \to \mathbb{R}_{+\infty}$ is abstract convex with respect to L if and only if f is IPH.*

Proof: Since functions $l \mapsto \langle l, x \rangle$ with $l \in \mathbb{R}^n_{++}$ are increasing and positively homogeneous, any L-convex function is IPH. Conversely, suppose that p is IPH and $y \in \mathbb{R}^n_{++}$. Let $l_y = p(y)/y$, where $1/y$ is defined by (2.2.7). We have $\langle l_y, y \rangle = p(y)$. Let $x \in \mathbb{R}^n_{++}$ and $\lambda = \min_i (x_i/y_i)$. Then $x \geq \lambda y$. Therefore,

$$p(x) \geq p(\lambda y) = \lambda p(y) = \min_i \frac{p(y) x_i}{y_i} = \min_i (l_y)_i x_i = \langle l_y, x \rangle.$$

Let $U = \{l_x : x \in \mathbb{R}^n_{++}\}$. Then for each $y \in \mathbb{R}^n_{++}$

$$p(y) = \langle l_y, y \rangle = \max_{l \in U} \langle l, y \rangle.$$

△

We now give a definition of a normal subset of $\mathbb{R}^n_{++}$.

DEFINITION 2.3 1) A subset U of the cone $\mathbb{R}^n_{++}$ is called *normal* if

$$(x \in U,\ x' \in \mathbb{R}^n_{++},\ x' \leq x) \implies x' \in U.$$

In particular the empty set and the set $U = \mathbb{R}^n_{++}$ are normal.
2) A normal set U is called *proper* if $U \neq \emptyset$ and $U \neq \mathbb{R}^n_{++}$.

Let us describe $(L, \mathbb{R}^n_{++})$-convex subsets of the set L (see Definition 1.4). We need the following definition.

DEFINITION 2.4 Let X be a conic subset of a vector space. A set $U \subset X$ is called *closed-along-rays* if

$$(x \in X, \lambda_k > 0, \lambda_k x \in U, k = 1, \dots, \lambda_k \to \lambda > 0) \implies \lambda x \in U.$$

Proposition 2.6 *A subset U of the cone L is $(L, \mathbb{R}^n_{++})$-convex if and only if this set is closed-along-rays and normal.*

Proof: It is easy to check that an $(L, \mathbb{R}^n_{++})$-convex set is closed-along-rays and normal. Let $p(x) = \sup_{l \in U} \langle l, x \rangle$ for all $x \in \mathbb{R}^n_{++}$. Then $U = \{l \in \mathbb{R}^n_{++} : \langle l, x \rangle \leq p(x) \text{ for all } x \in \mathbb{R}^n_{++}\}$. Let $l \in U$ and $l' \leq l$. Then

$$\langle l', x \rangle \leq \langle l, x \rangle \leq p(x) \text{ for all } x \in \mathbb{R}^n_{++},$$

hence $l' \in U$. Thus U is normal. Let

$$l \in L, \quad \lambda_k > 0, \lambda_k \to \lambda > 0, \quad \lambda_k l \in U, \ k = 1, \dots .$$

Since $\langle \lambda_k l, x \rangle \leq p(x)$ for all $x \in \mathbb{R}^n_{++}$, it follows that $\langle \lambda l, x \rangle \leq p(x)$ for all $x \in \mathbb{R}^n_{++}$, hence $\lambda l \in U$. Thus U is closed-along-rays.

Now let U be a closed-along-rays and normal subset of the conic set L. Since the empty set and L are $(L, \mathbb{R}^n_{++})$-convex, we can assume without loss of generality that U is a proper set. Due to Lemma 1.1 we need to show that if $l \in L$ and $l \notin U$, then there exists $y \in \mathbb{R}^n_{++}$ such that $\langle l, y \rangle > \sup_{l' \in U} \langle l', y \rangle$. Consider a function $l \notin U$. Since U is closed-along-rays, there exists an $\varepsilon > 0$ such that $(1 - \varepsilon) l \notin U$. Let $y = (y_1, \dots, y_n)$ be a vector with

$$y_i = \frac{1}{(1 - \varepsilon) l_i}, \qquad i \in I.$$

We have $\langle l, y \rangle = \min_i l_i y_i = 1/(1 - \varepsilon) > 1$. Now let $l' \in U$. Since U is normal, the inequality $l' > (1 - \varepsilon) l$ is not true. Therefore there exists an index i_0 such that $l'_{i_0} < (1 - \varepsilon) l_{i_0}$. Hence

$$\langle l', y \rangle = \min_i l'_i y_i \leq l'_{i_0} y_{i_0} < (1 - \varepsilon) l_{i_0} y_{i_0} = 1.$$

Thus we have constructed a vector y with the property $\langle l, y \rangle > 1 \geq \sup_{l' \in U} \langle l', y \rangle$. △

Remark 2.2 It is easy to check that $(L, \mathbb{R}^n_{++})$-convex set is closed. Thus a normal set is closed if and only if this set is closed-along-rays.

Remark 2.3 Consider the pair of sets $(\mathbb{R}^n_{++}, L)$ with the coupling function

$$\langle l, x \rangle \mapsto \min_{i \in I} l_i x_i.$$

This function allows us to consider an element $l \in L$ as a function defined on $\mathbb{R}^n_{++}$ and an element $x \in \mathbb{R}^n_{++}$ as a function defined on L. (See

Remark 1.3 for details.) Using this function we can consider not only $(L, \mathbb{R}^n_{++})$-convex subsets of L but also $(\mathbb{R}^n_{++}, L)$-convex subsets of $\mathbb{R}^n_{++}$. Note that we have used only the conic and ordered structure of sets L and $\mathbb{R}^n_{++}$ and its topology for description of $(L, \mathbb{R}^n_{++})$-convex sets. Due to Theorem 2.1, L and $\mathbb{R}^n_{++}$ are isomorphic as conic and ordered sets and they are homeomorphic. Hence a set $U \subset \mathbb{R}^n_{++}$ is $(\mathbb{R}^n_{++}, L)$-convex if and only if this set is closed-along-rays and normal.

According to Definition 1.6, a function $f : \mathbb{R}^n_{++} \to \bar{\mathbb{R}}$ is called L-quasiconvex if its level sets $S_c(f) = \{x \in \mathbb{R}^n_{++} : f(x) \le c\}$ are $(\mathbb{R}^n_{++}, L)$-convex for all $c \in \bar{\mathbb{R}}$.

PROPOSITION 2.7 *A function $f : \mathbb{R}^n_{++} \to \bar{\mathbb{R}}$ is L-quasiconvex if and only if f is lower semicontinuous and increasing.*

Proof: In view of Remark 2.3, f is L-quasiconvex if and only if each level set $S_c(f)$ of f is closed and normal. Note that f is lower semicontinuous if and only if S_c are closed for all $c \in \bar{\mathbb{R}}$. It follows directly from corresponding definitions that the normality of S_c for all $c \in \bar{\mathbb{R}}$ is equivalent to the isotonicity of the function f. △

2.2.6. Level sets of IPH functions

In this section we shall show that the level sets of an IPH function are normal and that this function can be expressed as the Minkowski gauge of its level set. The following property of IPH functions will be essential in this section.

PROPOSITION 2.8 *For an IPH function f the level set $S_1(f) = \{x \in \mathbb{R}^n_{++} : f(x) \le 1\}$ is normal and closed (in the topological space $\mathbb{R}^n_{++}$).*

Proof: Since f is increasing, $S_1(f)$ is normal. The closure follows since f is lower semicontinuous. △

REMARK 2.4 Recall (see Subsection 2.2.3) that for an IPH function f defined on $\mathbb{R}^n_{++}$ there are three possibilities

(i) $0 < f(x) < +\infty$ for all $x \in \mathbb{R}^n_{++}$;

(ii) $f(x) = 0$ for all $x \in \mathbb{R}^n_{++}$;

(iii) $f(x) = +\infty$ for all $x \in \mathbb{R}^n_{++}$.

Note that $S_1(f) = \mathbb{R}^n_{++}$ if $f = 0$ and $S_1(f) = \emptyset$ if $f = +\infty$.

For a nonnegative positively homogeneous function of degree one, say f, we have

$$S_c(f) = \{x : f(x) \le c\} = c \cdot S_1(f) \text{ for all } c > 0.$$

Since

$$S_0(f) := \{x : f(x) \leq 0\} = \bigcap_{c>0} S_c(f) = \bigcap_{c>0} c \cdot S_1(f),$$

then using (2.2.3) we have that $S_0(f) = 0 \cdot S_1(f)$. So if f_1, f_2 are IPH, then $f_1 = f_2$ if and only if $S_1(f_1) = S_1(f_2)$.

Let U be a closed normal subset of the space $\mathbb{R}^n_{++}$ and let μ_U be the *Minkowski gauge* of the set U:

$$\mu_U(x) = \inf\{\lambda > 0 : x \in \lambda U\} \quad (x \in \mathbb{R}^n_{++}).$$

If $U = \emptyset$ then $\mu_U(x) = +\infty$ for all x (it follows from (1.2.3)). If $U = \mathbb{R}^n_{++}$ then $\mu_U(x) = 0$ for all x.

REMARK 2.5 A normal set is radiant, i.e., if $x \in U$, then $\lambda x \in U$ for all $\lambda \in [0,1]$. Radiant sets and their Minkowski gauges are discussed in detail in Chapter 5. Note (see (5.2.3)) that

$$\{x : \mu_U(x) < 1\} \subset U \subset \{x : \mu_U(x) \leq 1\}$$

and (see Proposition 5.1) $U = \{x : \mu_U(x) \leq 1\} := S_1(\mu_U)$ for closed-along-rays sets U.

PROPOSITION 2.9 *Let U be a closed normal subset of the space $\mathbb{R}^n_{++}$. Then μ_U is IPH and $U = S_1(\mu_U)$.*

Proof: For $U = \emptyset$ and $U = \mathbb{R}^n_{++}$ the proposition holds. Assume that U is proper. Clearly μ_U is positively homogeneous. Since U is closed, it follows that $U = S_1(\mu_U)$. So we only need to verify that μ_U is increasing. Let $x \leq y$ and $\mu_U(y) = c$. Then $y \in \lambda U$ for all $\lambda > c$. Hence

$$\mu_U(x) = \inf\{\lambda > 0 : x \in \lambda U\} \leq c = \mu_U(y).$$

Thus the result follows. △

COROLLARY 2.1 *The mapping $U \mapsto \mu_U$ is a one-to-one correspondence between the collection of all closed normal sets and the set of all IPH functions.*

DEFINITION 2.5 Let U be a subset of $\mathbb{R}^n_{++}$. The set

$$\mathrm{Nh}\, U = \{x \in \mathbb{R}^n_{++} : \text{ there exists } x' \in U \text{ such that } x \leq x'\}$$

is called the *normal hull* of U (with respect to the cone $\mathbb{R}^n_{++}$).

It is easy to check that the closure (in $\mathbb{R}^n_{++}$) of the normal hull $\mathrm{Nh}\, U$ is also a normal set. (In particular, the closure of a normal set is normal.)

Indeed, let $x \in \text{cl}\,\text{Nh}\,U$ and $0 \ll x' \leq x$. There exists a sequence $x_k \in U$ such that $x_k \to x$. Let $x'_k = \min(x_k, x')$, where min stands for the coordinate-wise minimum. Then $x'_k \in U$ and $x'_k \to x'$. Thus $x' \in \text{cl}\,\text{Nh}\,U$.

PROPOSITION 2.10 *If U' is a closed normal subset of $\mathbb{R}^n_{++}$ and $U' \supset U$ then $U' \supset \text{cl}\,\text{Nh}\,U$.*

Proof: The proof is an immediate consequence of Definition 2.5. △

2.2.7. Polarity for normal sets and IPH functions

We now introduce the following polar operations for both normal sets and IPH functions.

DEFINITION 2.6 Let U be a nonempty subset of $\mathbb{R}^n_{++}$. The set

$$U^\circ = \{l \in \mathbb{R}^n_{++} : \langle l, x \rangle \leq 1 \text{ for all } x \in U\}$$

is called the *polar set with respect to U.*

Clearly we can consider set U° as analogous to the polar set in convex analysis. The following properties of polar sets are consequences of Definition 2.6.

1 U° is closed and normal.

2 $U^\circ = (\text{cl}\,\text{Nh}\,U)^\circ$.

3 $(\mathbb{R}^n_{++})^\circ = \emptyset$. We also accept by definition that $\emptyset^\circ = \mathbb{R}^n_{++}$.

4 Let T be an arbitrary index set and $U_t \subset \mathbb{R}^n_{++}$ for each $t \in T$. Then

$$\left(\bigcup_{t \in T} U_t\right)^\circ = \bigcap_{t \in T} U_t^\circ. \tag{2.2.11}$$

We now define the polar function to an IPH function.

DEFINITION 2.7 Let p be a proper IPH function defined on $\mathbb{R}^n_{++}$. Then the function p° defined on $\mathbb{R}^n_{++}$ by

$$p^\circ(l) = \sup_{x \in \mathbb{R}^n_{++}} \frac{\langle l, x \rangle}{p(x)} \tag{2.2.12}$$

is called the *polar function of p.*

REMARK 2.6 We accept by definition that $p^\circ = +\infty$ if $p = 0$, $p^\circ = 0$ if $p = +\infty$.

Let p be an IPH function and let $U = S_1(p) = \{x : p(x) \leq 1\}$. Since both functions $x \mapsto \langle l, x\rangle$ and p are positively homogeneous, we have

$$p^\circ(l) = \sup_{x \in \mathbb{R}^n_{++}} \frac{\langle l, x\rangle}{p(x)} = \sup_{x \in U} \langle l, x\rangle.$$

Proposition 2.5 shows that p° is an IPH function. Since p is IPH, $U = S_1(p)$ is a closed normal set. Proposition 2.6 also shows that U is an $(L, \mathbb{R}^n_{++})$-convex set.

From Theorem 2.1, we can identify $\mathbb{R}^n_{++}$ and L as conic and ordered sets. Hence the definitions of the polar set and polar function can also be applied to subsets of L and functions defined on L. This observation allows us to define the second polar set to a set $U \subset \mathbb{R}^n_{++}$, and the second polar function to a function p defined on $\mathbb{R}^n_{++}$. By definition

$$U^{**} = (U^\circ)^\circ, \qquad p^{\circ\circ} = (p^\circ)^\circ. \tag{2.2.13}$$

Now let us describe the level set $S_1(p^\circ)$ of the polar function p°. We have that

$$\begin{aligned} S_1(p^\circ) &= \{l \in \mathbb{R}^n_{++} : p^\circ(l) \leq 1\} \\ &= \{l \in \mathbb{R}^n_{++} : (\forall x \in U)\ \langle l, x\rangle \leq 1\} = U^\circ. \end{aligned}$$

Thus the following assertion is valid.

PROPOSITION 2.11 *For an IPH function p the level set $S_1(p^\circ)$ of the polar function p° is the polar set to the level set, $S_1(p)$, of the given function p.*

It is possible to give an explicit description of the polar function.

THEOREM 2.2 *For an IPH function p we have:*

$$p^\circ(l) = \frac{1}{p(\frac{1}{l})}.$$

where $1/l$ is defined by (2.2.7).

Proof: If $p = 0$ or $p = +\infty$ then the result follows from Remark 2.6. So we shall only consider a proper IPH function p. Let $l \in \mathbb{R}^n_{++}$. Then $l_i x_i \geq \langle l, x\rangle$ for each $x \in \mathbb{R}^n_{++}$ and each $i = 1, \dots, n$, hence $x \geq \langle l, x\rangle / l$. Since p is increasing, we have

$$p(x) \geq \langle l, x\rangle p\left(\frac{1}{l}\right),$$

so

$$p^\circ(l) = \sup_x \frac{\langle l, x\rangle}{p(x)} \leq \frac{1}{p\left(\frac{1}{l}\right)}.$$

Let $\bar{x} = 1/l$. Then

$$\frac{\langle l, \bar{x}\rangle}{p(\bar{x})} = \frac{1}{p(\bar{x})} = \frac{1}{p\left(\frac{1}{l}\right)}.$$

This completes the proof. △

The next three corollaries follow immediately from Theorem 2.2.

COROLLARY 2.2 *Let U be a closed normal subset of $\mathbb{R}^n_{++}$ and let p be an IPH function such that $U = S_1(p)$. Then*

$$U^\circ = \{l : p^\circ(l) \leq 1\} = \left\{l : p\left(\frac{1}{l}\right) \geq 1\right\}.$$

COROLLARY 2.3 *For an IPH function p we have $p^{**} = p$, and for a closed normal set U we have $U^{**} = U$.*

COROLLARY 2.4 *If $U \subset \mathbb{R}^n_{++}$ then $U^{**} = \operatorname{cl} \operatorname{Nh} U$.*

We now present some examples of IPH functions and their polar functions.

EXAMPLE 2.5 We define the following family of IPH functions

$$p_k(x) = (\sum_i x_i^k)^{1/k} \qquad (x \in \mathbb{R}^n_{++}),$$

where $0 < k < +\infty$. Theorem 2.2 implies that

$$p_k^\circ(l) = \frac{1}{p_k(1/l)} = \frac{1}{(\sum_i (\frac{1}{l_i})^k)^{1/k}}.$$

Consider in particular, the ℓ_1-norm p_1, in which case the polar function is

$$p_1^\circ(l) = \frac{1}{\frac{1}{l_1} + \cdot + \frac{1}{l_n}}.$$

Note that $np_1^\circ(l)$ is the harmonic mean of the sequence $(l_1, \ldots, l_n)$.

EXAMPLE 2.6 Consider the IPH function defined as follows:

$$p(x) = Cx^{\alpha_1} x^{\alpha_2} \cdots x^{\alpha_n}, \ \sum_i \alpha_i = 1, \ \ \alpha_i \geq 0.$$

In this case the polar function is defined by

$$p^\circ(l) = \frac{1}{C(\frac{1}{l_1})^{\alpha_1}\cdots(\frac{1}{l_n})^{\alpha_n}} = \frac{1}{C} l_1^{\alpha_1}\cdots l_n^{\alpha_n}.$$

Thus if $C = 1$ we have $p = p^\circ$. Note that functions such as p are concave IPH functions and they arise as the so-called Cobb-Douglas production functions in economics (see, for example, [83]).

EXAMPLE 2.7 We next consider the function $p_{+\infty}$ defined by

$$p_{+\infty}(x) = \max_i x_i \qquad (x \in \mathbb{R}^n_{++}).$$

Note that $p_{+\infty}$ is the $\ell_{+\infty}$-norm on $\mathbb{R}^n$ restricted to the cone $\mathbb{R}^n_{++}$. Then the polar function is given as follows:

$$p^\circ(l) = \frac{1}{\max \frac{1}{l_i}} = \frac{1}{\frac{1}{\min l_i}} = \min l_i.$$

2.2.8. Support sets

Recall that the support set $\mathrm{supp}(p, L)$ of an IPH function $p : \mathbb{R}^n_{++} \to \mathbb{R}_+$ with respect to the set L is defined by

$$\mathrm{supp}(p, L) = \{l \in \mathbb{R}^n_{++} : \langle l, x\rangle \le p(x) \text{ for all } x \in \mathbb{R}^n_{++}\}. \qquad (2.2.14)$$

Note that the set $U \subset L$ is $(L, \mathbb{R}^n_{++})$-convex if and only if there exists an IPH function p such that $U = \mathrm{supp}(p, L)$. So (see Proposition 2.6) the following holds:

PROPOSITION 2.12 *Let $U \subset \mathbb{R}^n_{++}$. Then there exists an IPH function p such that U is the support set of p if and only if U is closed and normal.*

It will be shown (see Proposition 7.47) that the mapping $p \mapsto \mathrm{supp}(p, L)$ is a one-to-one correspondence between the set of IPH functions defined on $\mathbb{R}^n_{++}$ and the set of closed normal subsets of $\mathbb{R}^n_{++}$. This mapping is called the *Minkowski duality.*

It follows from the definition that $\mathrm{supp}(p, L) = \emptyset$ if $p = 0$ and $\mathrm{supp}(p, L) = \mathbb{R}^n_{++}$ if $p = +\infty$.

REMARK 2.7 Since we identify a vector l and the function $x \mapsto \langle l, x\rangle$, we shall sometimes present (2.2.14) as

$$\mathrm{supp}(p, L) = \{l \in L : \langle l, x\rangle \le p(x) \text{ for all } x \in X\}.$$

Theorem 2.2 allows us to easily describe support sets.

THEOREM 2.3 *Let p be an IPH function defined on $\mathbb{R}^n_{++}$. Then*

$$\operatorname{supp}(p, L) = \left\{ l \in \mathbb{R}^n_{++} : p\left(\frac{1}{l}\right) \geq 1 \right\}.$$

Proof: We have

$$\begin{aligned}
\operatorname{supp}(p, L) &= \{l : \langle l, x\rangle \leq p(x) \text{ for all } x \in \mathbb{R}^n_{++}\} \\
&= \{l : \max_{x : p(x) \leq 1} \langle l, x\rangle \leq 1\} = \{l : p^\circ(l) \leq 1\} \\
&= \left\{ l : \frac{1}{p(\frac{1}{l})} \leq 1 \right\} = \left\{ l : p(\frac{1}{l}) \geq 1 \right\}.
\end{aligned}$$

This completes the proof. △

We now describe some properties of support sets. Let p_1, p_2 be IPH functions. It follows immediately from the definition of the support set that

$$p_1 \leq p_2 \iff \operatorname{supp}(p_1, L) \subset \operatorname{supp}(p_2, L).$$

PROPOSITION 2.13 *Consider a family $(p_\alpha)_{\alpha \in A}$ of IPH functions and let $p(x) = \inf_{\alpha \in A} p_\alpha(x)$. Then $\operatorname{supp}(p, L) = \bigcap_{\alpha \in A} \operatorname{supp}(p_\alpha, L)$.*

Proof: Since the sets $\operatorname{supp}(p_\alpha, L)$ are normal and closed, it follows that their intersection is also normal and closed. Proposition 2.6 demonstrates that this intersection is an $(L, \mathbb{R}^n_{++})$-convex set, hence there exists an IPH function $\tilde{p}$ such that $\operatorname{supp}(\tilde{p}, L) = \bigcap_{\alpha \in A} \operatorname{supp}(p_\alpha, L)$. Since $\tilde{p} \leq p_\alpha$ for all $\alpha \in A$, it follows that $\tilde{p} \leq \inf_{\alpha \in A} p_\alpha = p$. The function $p(x) = \inf_{\alpha \in A} p_\alpha(x)$ is IPH. Since $p \leq p_\alpha$ for all α, $\operatorname{supp}(p, L) \subset \bigcap_{\alpha \in A} \operatorname{supp}(p_\alpha, L) = \operatorname{supp}(\tilde{p}, L)$ so $p \leq \tilde{p}$. △

REMARK 2.8 A more general result will be proved in the framework of the Minkowski duality (see Subsection 7.4.2)

Recall that for $x \in \mathbb{R}^n_{++}$:

$$\frac{1}{x} \equiv x^{-1} = \left(\frac{1}{x_i}\right)_{i \in I}.$$

For $a, x \in \mathbb{R}^n_{++}$, $U \subset \mathbb{R}^n_{++}$ we shall require the following notational convention:

$$a \cdot x = (a_i x_i)_{i \in I}, \qquad a \cdot U = \{a \cdot u : u \in U\}, \qquad \frac{a}{x} = \left(\frac{a_i}{x_i}\right)_{i \in I}. \tag{2.2.15}$$

Let $a = (a_i)_{i\in I} \in \mathbb{R}^n_{++}$ and p be an IPH function. In the remaining part of this subsection we shall consider the function p^a, where $p^a(y) = p(a \cdot y)$ and $a \cdot y$ is defined by (2.2.15).

PROPOSITION 2.14 *Let $a \gg 0$. Then* $\mathrm{supp}(p^a, L) = a \cdot \mathrm{supp}(p, L) \equiv \{a \cdot l : l \in \mathrm{supp}(p, L)\}$.

Proof: Let $l \in \mathrm{supp}(p^a, L)$, $y = (y_i)_{i\in I} \in \mathbb{R}^n_+$ and $z = (z_i)_{i\in I} = a \cdot y$. Then

$$p(z) = p(a \cdot y) = p^a(y) \geq \langle l, y\rangle = \min_{i\in I} l_i y_i = \min_{i\in I} \frac{l_i}{a_i} z_i.$$

Thus the vector $l' = l/a$ belongs to $\mathrm{supp}(p, L)$. Since $l = a \cdot l'$, we have $l \in a \cdot \mathrm{supp}(p, L)$. We have checked that $\mathrm{supp}(p^a, L) \subset a \cdot \mathrm{supp}(p, L)$. The similar argument shows that the reverse inclusion holds. △

LEMMA 2.1 *If U is a normal set and $a \geq a' \gg 0$ then $a \cdot U \supset a' \cdot U$.*

Proof: Let $l \in a' \cdot U$. Then there exists $u \in U$ such that $l = a' \cdot u$. Thus $(l/a') \in U$. Since U is normal, it follows that $(l/a) \in U$, so $l \in a \cdot U$. △

We now describe support sets for some IPH functions.

EXAMPLE 2.8 Let $I = \{1, \dots, n\}$ and let $p(y) = \max_{i\in I} a_i y_i$ with $a = (a_i)_{i\in I}$ and $a_i > 0$. Clearly p is IPH. Applying Theorem 2.3 we conclude that the support set, $\mathrm{supp}(p, L)$, coincides with the set

$$V_a = \bigcup_{i\in I} \{l = (l_1, \dots, l_n) \in \mathbb{R}^n_{++} : l_i \leq a_i\} = \{l \in \mathbb{R}^n_{++} : \min_{i\in I} \frac{l_i}{a_i} \leq 1\}. \tag{2.2.16}$$

Indeed,

$$\begin{aligned} \mathrm{supp}(p, L) &= \left\{l : p\left(\frac{1}{l}\right) \geq 1\right\} = \left\{l : \max_{i\in I} \frac{a_i}{l_i} \geq 1\right\} \\ &= \left\{l : \min \frac{l_i}{a_i} \leq 1\right\} = V_a. \end{aligned}$$

Assume, more generally, that $a_i \geq 0$ for all $i \in I$ and that the set $I_a = \{i \in I : a_i > 0\}$ is nonempty. It is easy to check that

$$\mathrm{supp}(p, L) = \left\{l \in \mathbb{R}^n_{++} : \min_{i\in I_a} \frac{l_i}{a_i} \leq 1\right\}.$$

EXAMPLE 2.9 Let $0 < k < +\infty$ and

$$p_k(x) = (\sum_{i\in I} x_i^k)^{\frac{1}{k}} \text{ for all } x \in \mathbb{R}^n_{++}.$$

Clearly p_k is IPH. Applying Theorem 2.3, we obtain the following:

$$\mathrm{supp}(p_k, L) = \left\{ l \in \mathbb{R}^n_{++} : \sum_{i \in I} \frac{1}{l_i^k} \geq 1 \right\}. \qquad (2.2.17)$$

2.2.9. Subdifferential

In this subsection we present a description of the L-subdifferential of a proper (that is finite and non-zero) IPH function p defined on $\mathbb{R}^n_{++}$. Recall (see Definition 1.7) that the L-subdifferential $\partial_L p(y)$ of the function p at a point $y \in \mathbb{R}^n_{++}$ is given by

$$\partial_L p(y) = \{l \in L : p(x) \geq \langle l, x\rangle - (\langle l, y\rangle - p(y)) \text{ for all } x \in \mathbb{R}^n_{++}\}.$$

Since the set L consists of positively homogeneous functions of degree one, it follows from Proposition 1.3 that

$$\partial_L p(y) = \{l \in \mathrm{supp}(p, L) : \langle l, y\rangle = p(y)\}. \qquad (2.2.18)$$

THEOREM 2.4 *Let p be a proper IPH function defined on $\mathbb{R}^n_{++}$ and $y \in \mathbb{R}^n_{++}$. Then*

$$\partial_L p(y) = \left\{ l : l \geq \frac{p(y)}{y}, p\left(\frac{1}{l}\right) = 1 \right\}. \qquad (2.2.19)$$

Proof: Let $l \in \partial_L p(y)$. Applying (2.2.18) we conclude that

$$l \in \mathrm{supp}(p, L) \qquad (2.2.20)$$

and

$$\langle l, y\rangle = p(y). \qquad (2.2.21)$$

It follows from (2.2.21) that $l_i y_i \geq p(y)$ for all $i \in I$. Hence $l \geq p(y)/y$. This inequality is equivalent to $1/l \leq y/p(y)$. Since p is IPH, we have

$$p\left(\frac{1}{l}\right) \leq p\left(\frac{y}{p(y)}\right) = \frac{p(y)}{p(y)} = 1.$$

On the other hand, Theorem 2.3 demonstrates that (2.2.20) is equivalent to $p(l^{-1}) \geq 1$. Thus $p(l^{-1}) = 1$ and hence $\partial_L p(y)$ is contained in the set on the right hand side in (2.2.19). Assume now that

$$p\left(\frac{1}{l}\right) = 1 \qquad (2.2.22)$$

and

$$l \geq \frac{p(y)}{y}. \tag{2.2.23}$$

Inequality (2.2.23) implies that $l_i y_i \geq p(y)$ for all i, hence $\langle l, y\rangle \geq p(y)$. Combining Theorem 2.3 and (2.2.22) we deduce that $l \in \text{supp}(p, L)$. In particular, $\langle l, y\rangle \leq p(y)$. Hence $\langle l, y\rangle = p(y)$. △

REMARK 2.9 It follows from Theorem 2.4 that $p(y)/y \in \partial_L p(y)$ for all $y \in \mathbb{R}^n_{++}$.

2.2.10. Concavity of the polar function

In this subsection we shall establish the following statement:

THEOREM 2.5 *If p is a proper IPH convex function then p° is concave.*

The proof of this theorem is based on Theorem 2.2 and the following lemma:

LEMMA 2.2 *Let p be a lower semicontinuous sublinear increasing function $\mathbb{R}^n_+ \to \mathbb{R}_{+\infty}$. Then there exists a set $U \subset \mathbb{R}^n_+$ such that*

$$p(x) = \sup_{l \in U}[l, x] \text{ for all } x \in \mathbb{R}^n_+,$$

where $[l, x] = \sum_i l_i x_i$ is the inner product of vectors l and x.

Proof of Lemma 2.2: Consider the function $p_+(x)$ defined on $\mathbb{R}^n$ by

$$p_+(x) = p(x^+),$$

where $x^+ = (\max(x_1, 0), \ldots, \max(x_n, 0))$ is the positive part of a vector x. Since p is increasing, p_+ is a sublinear function. Clearly this function is lower semicontinuous, hence (see Example 1.6) p_+ is abstract convex with respect to the set of all linear functions L_1 defined on $\mathbb{R}^n$. Since each linear function has a form $x \mapsto [l, x]$ with a vector $l \in \mathbb{R}^n$, we can consider the support set $U := \text{supp}(p_+, L_1)$ of the function p_+ as a subset of $\mathbb{R}^n$. Recall that by the definition

$$U = \{l \in \mathbb{R}^n : [l, x] \leq p_+(x) \text{ for all } x \in \mathbb{R}^n\}.$$

Abstract convexity of p^+ implies

$$p_+(x) = \sup_{l \in U}[l, x].$$

Since $p_+(x) = p(x)$ for $x \in \mathbb{R}^n_+$, we have $p(x) = \sup_{l \in U}[l, x]$. Note that $p_+(-x) = 0$ for $x \in \mathbb{R}^n_+$. Let $l \in U$. Then we have $[l, -x] \le p(-x) = 0$ for all $x \in \mathbb{R}^n_+$, hence $l \in \mathbb{R}^n_+$. $\triangle$

Proof of Theorem 2.5:

1) Let $l = (l_1, \dots, l_n) \in \mathbb{R}^n_+$. Denote the linear function $x \mapsto [l, x]$ by $\tilde{l}$. Clearly $\tilde{l}$ is an IPH function. We have, for $h = (h_1, h_2, \dots, h_n) \in \mathbb{R}^n_{++}$,

$$\tilde{l}^\circ(h) = \frac{1}{\tilde{l}(\frac{1}{h})} = \frac{1}{\frac{l_1}{h_1} + \cdots + \frac{l_n}{h_n}} \quad .$$

We now check that $\tilde{l}^\circ(h)$ is a concave function. For $x = (x_1, x_2, \dots, x_n) \in \mathbb{R}^n_{++}$ set

$$H_n(x) = \frac{1}{\frac{1}{x_1} + \cdots + \frac{1}{x_n}} \quad .$$

We first verify that H_n is concave on $\mathbb{R}^n_{++}$. We shall use induction on n. Calculating the Hessian of H_2 it is easy to check that H_2 is concave on $\mathbb{R}^2_{++}$. Clearly H_2 is also increasing. So the equality

$$H_{n+1}(x_1, x_2, \dots, x_n, x_{n+1}) = H_2(H_n(x_1, x_2, \dots, x_n), x_{n+1})$$

shows that the concavity of H_n implies the concavity of H_{n+1}. Thus H_n is concave on $\mathbb{R}^n_{++}$ for all n. For $h \in \mathbb{R}^n_{++}$ we have that

$$\tilde{l}^\circ(h) = H_\kappa(Ah),$$

where $Ah = (h_i/l_i)_{i \in I}$, $I = \{i : l_i > 0\}$, $\kappa = |I|$. Since A is a linear operator, it follows that $\tilde{l}^\circ$ is a concave function as required.

2) Now let $p \in$ IPH, $p \neq 0$ and assume that p is convex. Thus p is sublinear and increasing. Consider the function p_1 defined on $\mathbb{R}^n_+$ by

$$p_1(x) = \liminf_{x' \in \mathbb{R}^n_{++}, x' \to x} p(x').$$

In the light of Proposition 2.1 we can assert that p_1 is a lower semicontinuous extension of the function p. Remark 2.1 shows that $\operatorname{epi} p_1 = \operatorname{cl} \operatorname{epi} p$, hence p_1 is a sublinear function. Due to Lemma 2.2 we can find a set $U \in \mathbb{R}^n_+$ such that

$$p(x) = p_1(x) = \sup\{\tilde{l}(x) : l \in U\} \qquad (x \in \mathbb{R}^n_{++}).$$

We have

$$p^\circ(h) = \frac{1}{p(\frac{1}{h})} = \frac{1}{\sup_{l \in U} \tilde{l}(\frac{1}{h})} = \inf_{l \in U} \frac{1}{\tilde{l}(\frac{1}{h})} = \inf_{l \in U} \tilde{l}^\circ(h).$$

Since l° is concave it follows that p° is also concave. △

REMARK 2.10 Example 2.6 above shows that the converse to Theorem 2.5 is not valid. It is possible that both p and p° are concave functions.

2.2.11. Comparison with convex analysis

We now present a summary of some results obtained in the previous subsections.

We compare these results with corresponding results from the theory of sublinear functions and convex sets. Sublinear functions are considered as being abstract convex with respect to the set L_1 of all linear functions.

First we recall some results from convex analysis. The detailed presentation of these results can be found in Subsection 7.3.2. Let SUB be the set of all proper lower semicontinuous nonnegative sublinear functions $\mathbb{R}^n \to \mathbb{R}_{+\infty}$ and CON be the totality of all convex closed subsets of $\mathbb{R}^n$ containing zero. If $p \in SUB$, then the subdifferential $\partial p(y)$ of p at a point $y \in \operatorname{dom} p$ is the set

$$\partial p(y) := \partial_{L_1} p(y) = \{l \in \mathbb{R}^n : p(x) \geq [l, x] - ([l, y] - p(y))\}.$$

Here $[l, x] = \sum_{i \in I} l_i x_i$ is the inner product of vectors l and x. In particular,

$$\partial p(0) = \{l : [l, x] \leq p(x) \text{ for all } x \in \mathbb{R}^n\},$$

that is, $\partial p(0)$ coincides with the support set $\operatorname{supp}(p, L_1)$. Due to Proposition 1.3 we have $\partial p(y) = \{l \in \partial p(0) : [l, y] = p(y)\}$.

A lower semicontinuous sublinear function p is nonnegative (that is, $p \in SUB$) if and only if $0 \in \partial p(0)$ (that is, $\partial p(0) \in CON$.) Let $p \in SUB$. The function $p^0 : \mathbb{R}^n \to \mathbb{R}_{+\infty}$ defined by

$$p^0(l) = \sup\{[l, x] : \ p(x) \leq 1\}$$

is usually called the *polar function* to p (see for example [134], Section 15). The function $p^{00} = (p^0)^0$ is called the *bipolar function.*

Let $U \in CON$. The set

$$U^0 = \{l \in R_n : [l, x] \leq 1 \text{ for all } x \in U\}$$

is called the *polar set* to U. The set $U^{00} = (U^0)^0$ is called the *bipolar set.* Let $p \in SUB$ and

$$U = \{x : p(x) \leq 1\}. \tag{2.2.24}$$

Then $U^0 = \{l \in \mathbb{R}^n : p^0(l) \leq 1\}$. Each set $U \in CON$ can be presented in the form (2.2.24), where $p = \mu_U$ is the Minkowski gauge of the set U. Recall that the Minkowski gauge μ_U of a set $U \subset \mathbb{R}^n$ is defined by

$$\mu_U(x) = \inf\{\lambda > 0 : x \in \lambda U\}.$$

It is well known that $U \in CON$ if and only if $\mu_U \in SUB$. We also have $U = \{x : \mu_U(x) \leq 1\}$ for $U \in CON$.

In this subsection we denote the set of all min-type functions $x \mapsto \langle l, x\rangle$ with $l \in \mathbb{R}^n_{++}$ by L_2. In the rest of this section we shall use the symbol **1** for items related to abstract convexity with respect to L_1 and the symbol **2** for items related to L_2.

We now compare the main definitions and results related to L_1 and L_2.

Abstract linear functions:

1. L_1 is the set of "usual" linear functions $x \mapsto [l, x]$ with $l \in \mathbb{R}^n$.

2. L_2 is the set of min-type functions $x \mapsto \langle l, x\rangle$ with $l \in \mathbb{R}^n_{++}$.

Abstract convex functions (see Definition 1.1):

1. Lower semicontinuous sublinear functions $p : \mathbb{R}^n \to \mathbb{R}_{+\infty}$ and the function $-\infty : x \mapsto -\infty$ for all $x \in \mathbb{R}^n$.

2. IPH functions $p : \mathbb{R}^n_{++} \to [0, +\infty]$.

Abstract convex sets (see Definition 1.5):

1. Closed convex subsets of $\mathbb{R}^n$.

2. Closed normal subsets of the cone $\mathbb{R}^n_{++}$.

Abstract quasiconvex functions (see Definition 1.6):

1. Lower semicontinuous quasiconvex functions.

2. Lower semicontinuous increasing functions.

The subdifferential (see Definition 1.7):

1. Let $p : \mathbb{R}^n \to \mathbb{R}_{+\infty}$ be a sublinear function and $x \in \operatorname{int} \operatorname{dom} p$. Then the subdifferential $\partial p(x) := \partial_{L_1} p(x)$ is the support set, $\operatorname{supp}(p'_x, L_1)$, of the directional derivative $p'_x(u) = \lim_{\alpha \to +0} \alpha^{-1}(p(x + \alpha u) - p(x))$.

2. There is an explicit description of the subdifferential $\partial_{L_2}p(x)$ for a proper IPH function p (see Theorem 2.4):

$$\partial_{L_2}p(x) = \left\{l \in \mathbb{R}^n_{++} : l \geq \frac{p(x)}{x}, p\left(\frac{1}{l}\right) = 1\right\}.$$

Note that $p(x)/x \in \partial_{L_2}p(x)$ for all x.

Polar functions:

1. Let p be a sublinear function. The polar p^0 of p is the function defined by $p^0(l) = \sup\{[l,x] : p(x) \leq 1\}$. The following result holds: (see, for example, [134], Theorem 15.1): $p^{00} = p$ for each $p \in SUB$.

2. Let $p \in IPH$. The polar p° of p is the function defined by $p^\circ(l) = \sup\{\langle l,x\rangle : p(x) \leq 1\}$. There is an explicit description of the polar function:
$$p^\circ(l) = \frac{1}{p(\frac{1}{l})}. \tag{2.2.25}$$
It follows from (2.2.25) that $p^{**} = p$ for $p \in IPH$.

Polar sets:

1. Let U be a closed convex set, $0 \in U$. The polar set U^0 of U is defined by $U^0 = \{l \in \mathbb{R}^n : [l,x] \leq 1 \text{ for all } x \in U\}$. The following result is well-known (see, for example [134], Theorem 14.5): $U^{00} = U$ for each $U \in CON$. If U is a closed convex set then $\mu_U \in SUB$ and $U = \{x : \mu_U(x) \leq 1\}$.

2. Let U be a closed normal subset of $\mathbb{R}^n_+$. The set $U^\circ = \{l \in \mathbb{R}^n_{++} : \langle l,x\rangle \leq 1 \text{ for all } x \in U\}$ is called the polar set to U. If $U = \{x : p(x) \leq 1\}$, where p is IPH, then
$$U^\circ = \{l : p^\circ(l) \leq 1\} = \left\{l : p\left(\frac{1}{l}\right) \geq 1\right\}. \tag{2.2.26}$$
If U is a closed normal set then $\mu_U \in IPH$ and $U = \{x : \mu_U(x) \leq 1\}$. Thus $U = U^{**}$ for all closed normal sets U.

Support sets (see Definition 1.2):

1. Let $p \in SUB$. Then $\text{supp}(p, L_1) = \partial p(0)$.

2. Let p be an IPH function and $U = \{x \in \mathbb{R}^n_{++} : p(x) \leq 1\}$. Then

$$\text{supp}(p, L_2) = \left\{l \in \mathbb{R}^n_{++} : p\left(\frac{1}{l}\right) \geq 1\right\} = \{l : p^\circ(l) \leq 1\} = U^\circ.$$

2.3. INCREASING POSITIVELY HOMOGENEOUS FUNCTIONS DEFINED ON THE NON-NEGATIVE ORTHANT

2.3.1. Overview

In this section we shall consider IPH functions defined on the non-negative orthant $\mathbb{R}^n_+$. The theory of such functions, based on abstract convexity, is much more complicated than that for functions defined on $\mathbb{R}^n_{++}$. Let us explain this situation. Consider an IPH function p defined on $\mathbb{R}^n_+$. Let $x \in \mathbb{R}^n_+$ be such that the coordinate x_i is equal to zero for some i and $p(x) > 0$. Let $I = \{1, \dots, n\}$. Since $\min_{i \in I} l_i x_i = 0$ for all $l \in \mathbb{R}^n_{++}$ it follows that min-type functions $x \to \min_{i\in I} l_i x_i$ with $l \in \mathbb{R}^n_{++}$ cannot help to restore the IPH function p. If $l \in \mathbb{R}^n_+$ but there exists $i \in I$ such that $l_i = 0$ then $\min_{i\in I} l_i x_i = 0$ for all $x \in \mathbb{R}^n_+$. In order to take into account vectors $l \in \mathbb{R}^n_+$, with some zero coordinates, we shall change the definition of the min-type function generated by such a vector.

For $y \in \mathbb{R}^n_+$ we define

$$I_+(y) = \{i \in I : y_i > 0\}. \tag{2.3.1}$$

There are two natural ways to define the expression $\langle l, x\rangle$ if a vector l has zero coordinates. We can define $\langle l, x\rangle$ either as $\min_{i\in I_+(l)} l_i x_i$ or as $\min_{i \in I_+(x)} l_i x_i$. In terms of coupling functions both of these expressions are similar. However if we wish to consider the function $l(x)$ defined by $l(x) = \langle l, x\rangle$, the first expression is more preferable since the index set $\{i : l_i > 0\}$ depends on the function itself and does not depend on its argument x. Thus we will define the coupling function in the case under consideration by

$$\langle l, x\rangle = \begin{cases} \min_{i\in I_+(l)} l_i x_i & \text{if } I \neq \emptyset; \\ 0 & \text{if } I = \emptyset. \end{cases} \tag{2.3.2}$$

Note that this coupling function is not symmetrical, that is, $\langle l, x\rangle \neq \langle x, l\rangle$. Unfortunately the function $l \mapsto \langle l, x\rangle$ is not continuous even for $x \in \mathbb{R}^n_{++}$. This fact impedes the work with IPH functions (in particular, min-type functions) defined on $\mathbb{R}^n_+$. In this section we show that abstract convexity with respect to the set L of all min-type functions $x \mapsto \langle l, x\rangle$, where $\langle l, x\rangle$ is defined by (2.3.2) leads to IPH functions defined on $\mathbb{R}^n_+$, we describe also abstract convex sets and abstract subdifferentials of IPH functions. In contrast with the previous section we will not consider polar functions and sets. The corresponding theory can be developed in the framework of the approach presented by J.E. Martinez Legaz and I. Singer in [109]. Duality between Minkowski gauges and support sets is discussed in Subsection 7.2.9.

The main results presented here were published in [143, 147, 156, 161, 162].

2.3.2. Preliminaries

Denote by L the set of all min-type functions of the form $x \mapsto \langle l, x\rangle$, where $\langle l, x\rangle$ is defined by (2.3.2). (The same letter has been used for the class of all min-type functions defined on $\mathbb{R}^n_{++}$ with $l \in \mathbb{R}^n_{++}$. However, this will not lead to any misunderstanding.)

Let $l \in \mathbb{R}^n_+$. We define the vector $\frac{1}{l} \equiv l^{-1}$ by

$$\left(\frac{1}{l}\right)_i = \begin{cases} \frac{1}{l_i} & \text{if } i \in I_+(l); \\ 0 & \text{if } i \notin I_+(l). \end{cases} \tag{2.3.3}$$

In particular, if $l = 0$, then $1/l = 0$. Note that for a vector $l \in \mathbb{R}^n_{++}$ this definition coincides with that given in (2.2.7). Throughout this section we shall use the following notation:

$$\frac{a}{l} = a\frac{1}{l} \text{ for all } a \in \mathbb{R}_+ \text{ and } l \in \mathbb{R}^n_+.$$

In particular, if $l = 0$ then $a/l = 0$ for all $a \geq 0$.

We now discuss some properties of the set L of all min-type functions $x \mapsto \langle l, x\rangle$ defined on $\mathbb{R}^n_+$ with $l \in \mathbb{R}^n_+$. (Compare with the corresponding discussion in Subsection 2.2.4 for functions defined on $\mathbb{R}^n_{++}$.) Of course we can identify the set L with the cone $\mathbb{R}^n_+$. However, we have to distinguish between the algebraic, ordering and topological properties of the set L of vectors $l \in \mathbb{R}^n_+$ and the set of min-type functions, which are generated by vectors $l \in L$ using (2.3.2).

Note that the conic structure of the set $\mathbb{R}^n_+$ is isomorphic to the conic structure of the set L. In other words, the function $x \mapsto \langle \lambda l, x\rangle$ which is generated by the vector λl with $\lambda > 0$ is equal to the function λl, where $l(x) = \langle l, x\rangle$. (Recall that we use the same notation for both a vector and the function generated by this vector). So we can identify L and $\mathbb{R}^n_+$ as conic sets. Clearly the set L is not convex.

Subsequently we assume that L is equipped with the 'functional' order relation $\succeq$.

DEFINITION 2.8 For $l^1, l^2 \in L$

$$l^1 \succeq l^2 \iff \langle l^1, x\rangle \geq \langle l^2, x\rangle \text{ for all } x \in \mathbb{R}^n_+.$$

PROPOSITION 2.15 *For $l^1, l^2 \in L$, $l^1 \succeq l^2$ if and only if*

$$I_+(l^1) \subset I_+(l^2) \quad \text{and} \quad l_i^1 \geq l_i^2 \text{ for all } i \in I_+(l^1). \tag{2.3.4}$$

Proof: 1) Let $l^1 \succeq l^2$ and assume that $I_+(l^1) \not\subset I_+(l^2)$. Then there exists $j \in I_+(l^1)$ such that $j \notin I_+(l^2)$. Take a vector $x \in \mathbb{R}^n_+$ such that $x_i = 1$ for $i \in I_+(l^2)$ and $x_j = 0$. Then $\langle l^1, x\rangle = 0$ and $\langle l^2, x\rangle = \min_{i\in I_+(l^2)} l_i^2 > 0$. Since $\langle l^1, x\rangle < \langle l^2, x\rangle$ it follows that the inequality $l^1 \succeq l^2$ is not valid. Thus we have a contradiction and hence $I_+(l^1) \subset I_+(l^2)$. Now assume that there exists $k \in I_+(l^1)$ such that $l_k^1 < l_k^2$. Take a vector y such that $y_k = 1$ and $y_i > \frac{l_k^1}{l_i^1}$ for all $i \in I_+(l^1)$, $i \neq k$, and $y_i > \frac{l_k^2}{l_i^2}$ for all $i \in I_+(l^2)$. Then

$$l^1(y) = l_k^1 < l_k^2 = l^2(y)$$

and we have a contradiction. Thus (2.3.4) holds.

2) Now assume that (2.3.4) is valid for vectors l^1 and l^2. For $x \in \mathbb{R}^n_+$ we have

$$\langle l^1, x\rangle = \min_{i\in I_+(l^1)} l_i^1 x_i \geq \min_{i\in I_+(l^2)} l_i^1 x_i \geq \min_{i\in I_+(l^2)} l_i^2 x_i = \langle l^2, x\rangle.$$

So, $l^1 \succeq l^2$. △

REMARK 2.11 It follows from Proposition 2.15 that the functional order relation $\succeq$ differs from the vector order relation $\geq$. Consider for example the three-dimensional vectors $l^1 = (1,1,0)$ and $l^2 = (1,1,1)$. Then $l^2 \geq l^1$. Since $I_+(l^1) = \{1,2\} \subset \{1,2,3\} = I_+(l^2)$ and $l_1^1 = l_2^1$, $l_2^1 = l_2^2$, Proposition 2.15 implies that $l^1 \succeq l^2$.

2.3.3. L-convex function

We shall examine abstract convex functions and abstract convex sets with respect to the class L of functions defined by the coupling function (2.3.2). Note that $0 \in L$, so the support set is not empty for each abstract convex function. First we show that finite abstract L-convex functions are IPH (for the definition of IPH functions see Subsection 2.2.3).

PROPOSITION 2.16 *Let L be as above. Then a function $p : \mathbb{R}^n_+ \to \mathbb{R}$ is L-convex if and only if p is IPH.*

Proof: Since a function $l : x \mapsto \langle l, x\rangle$ is IPH for all $l \in L$, it follows that an L-convex function is also IPH. Conversely, suppose p is an IPH function. Let $\bar{x} \in \mathbb{R}^n_+ \setminus \{0\}$ and $l^{\bar{x}} = p(\bar{x})/\bar{x}$, i.e. $l^{\bar{x}} = (l_1^{\bar{x}}, \dots, l_n^{\bar{x}})$, where

$$l_i^{\bar{x}} = \begin{cases} p(\bar{x})/\bar{x}_i & \text{if } i \in I_+(\bar{x}); \\ 0 & \text{if } i \notin I_+(\bar{x}). \end{cases}$$

For the function $l^{\bar{x}} : x \mapsto \langle l^{\bar{x}}, x\rangle$ we have

$$l^{\bar{x}}(\bar{x}) = \min_{i \in I_+(\bar{x})} l_i^{\bar{x}} \bar{x}_i = p(\bar{x}).$$

On the other hand, if $x \in \mathbb{R}^n_+$ and $\lambda = \min_{i \in I_+(\bar{x})}(x_i/\bar{x}_i)$, then $x \geq \lambda \bar{x}$. Therefore

$$p(x) \geq p(\lambda \bar{x}) = \lambda p(\bar{x}) = \min_{i \in I_+(\bar{x})} \frac{p(\bar{x}) x_i}{\bar{x}_i} = \min_{i \in I_+(\bar{x})} l_i^{\bar{x}} x_i = l^{\bar{x}}(x).$$

Let $U_p = \{l^x \in L : x \in \mathbb{R}^n_+ \setminus 0\}$. For $x \in \mathbb{R}^n_+ \setminus 0$ we have

$$p(x) = \max_{l \in U_p} l(x).$$

If $x = 0$, then the equality $p(x) = \max_{l \in U_p} \langle l, x\rangle$ trivially holds. Thus p is an L-convex function. △

REMARK 2.12 We have verified that the subdifferential $\partial_L p(x)$ of a finite IPH function p is nonempty at each point $x \in \mathbb{R}^n_+$ and $p(x)/x \in \partial_L p(x)$. In particular, if $x = 0$ then $p(x)/x = 0 \in \partial_L p(x)$. A full description of L-subdifferential at a point x such that $p(x) > 0$ will be given later (see Proposition 2.30).

Thus a function defined on $\mathbb{R}^n_+$ is abstract convex with respect to the class of min-type functions if and only if this function is IPH. Note that the same result holds for functions defined on $\mathbb{R}^n_{++}$.

PROPOSITION 2.17 *Each finite IPH function p is lower semicontinuous.*

Proof: It follows from Proposition 2.16 that p is a pointwise supremum of a family of continuous functions $(l^{\bar{x}})_{\bar{x} \neq 0}$. Hence p is lower semicontinuous. △

REMARK 2.13 IPH functions mapping into $\mathbb{R}_{+\infty}$ will be examined in the framework of a more general theory of ICAR functions (see Subsection 3.3.3).

2.3.4. Support sets

The description of support sets in the case under consideration is much more complicated than that in the case of min-type functions defined on $\mathbb{R}^n_{++}$. As an illustration, we now provide a geometric interpretation of the support set in the case of min-type functions defined on $\mathbb{R}^2_+$. Let p be an IPH function. Then $\text{supp}(p, L)$ is the union of zero and three sets: two of them are semi-open segments on the coordinate axes and the third is a subset of the open cone $\mathbb{R}^2_{++}$ bounded by the graph of a decreasing function $\psi : (0, +\infty) \to \mathbb{R}_{+\infty}$. Applying Theorem 2.3, we can show that $\psi(y) = p(1/y_1, 1/y_2)$.

We begin with the definition of the normal subset of the cone $\mathbb{R}^n_+$.

DEFINITION 2.9 A subset U of the cone $\mathbb{R}^n_+$ is called *normal* if $x \in U$ and $0 \leq y \leq x$ imply $y \in U$.

Normal sets play a very important role in the study of increasing functions defined on $\mathbb{R}^n_+$. This follows from the fact that for an increasing function f, the level sets $S_c(f) = \{x \in \mathbb{R}^n_+ : f(x) \leq c\}$ are normal.

Since the order relation introduced in the set L does not coincide with the order relation in $\mathbb{R}^n_+$ (see Remark 2.11), the normality with respect to L differs from that with respect to $\mathbb{R}^n_+$.

DEFINITION 2.10 A subset U of the ordered set L is *normal* if

$$l^1 \in U,\ l^2 \in L,\ l^1 \succeq l^2 \implies l^2 \in U.$$

Recall (see Definition 2.4) that a subset U of the conic set L is called closed-along-rays if

$$\lambda_n > 0,\ \ \lambda_n l \in U\ (n = 1, 2, \dots) \text{ and } \lambda_n \to \lambda > 0 \implies \lambda l \in U.$$

This definition is consistent with the conic structure of the set L which is isomorphic to the conic structure of the set $\mathbb{R}^n_+$. The next assertion is an analogue of Proposition 2.6.

PROPOSITION 2.18 *A subset U of the ordered set L is $(L, \mathbb{R}^n_+)$-convex if and only if U is closed-along-rays and normal.*

Proof: It follows immediately from the definition of normality that an $(L, \mathbb{R}^n_+)$-convex set is closed-along-rays and normal. Now let U be a closed-along-rays normal subset of the cone L. Due to Lemma 1.1, we need to show that for each $l \in L \setminus U$ there is $x \in \mathbb{R}^n_+$ such that

$\langle l, x\rangle > \sup_{l' \in U}\langle l', x\rangle$. Consider a vector $l \notin U$. Since U is closed-along-rays, there exists $\varepsilon > 0$ such that $(1-\varepsilon)l \notin U$. Let $\bar{x} = (\bar{x}_1, \dots, \bar{x}_n) = \dfrac{1}{(1-\varepsilon)l}$, that is,

$$\bar{x}_i = \frac{1}{(1-\varepsilon)l_i} \text{ for all } i \in I_+(l), \qquad \bar{x}_i = 0 \text{ for all } i \notin I_+(l).$$

We have $l(\bar{x}) = \min_{i \in I_+(l)} l_i \bar{x}_i = 1/(1-\varepsilon) > 1$. Now let $l' \in U$. Since U is normal, the inequality $l' \succeq (1-\varepsilon)l$ is not true. Applying Proposition 2.15 we can conclude that either

$$I_+(l') \not\subset I_+((1-\varepsilon)l) = I_+(l) \tag{2.3.5}$$

or

$$I_+(l') \subset I_+(l) \;\text{ but } \exists\, i_0 \in I_+(l) \;\text{ such that } l'_{i_0} < (1-\varepsilon)l_{i_0}. \tag{2.3.6}$$

Assume (2.3.5) holds. Then we can find an index $i' \in I_+(l')$ such that $i' \notin I_+(l)$. Since $\bar{x}_{i'} = 0$, we have $\langle l', \bar{x}\rangle = 0 < \langle l, \bar{x}\rangle$. Now assume that (2.3.6) is valid. Then $\bar{x}_{i_0} > 0$. Hence

$$l'(\bar{x}) = \min_{i \in I_+(l')} l'_i \bar{x}_i \le l'_{i_0}\bar{x}_{i_0} < (1-\varepsilon)l_{i_0}\bar{x}_{i_0} = 1.$$

Thus we have constructed a vector $\bar{x}$ with the property

$$l(\bar{x}) > 1 \ge \sup_{l' \in U} l'(\bar{x}). \qquad \triangle$$

REMARK 2.14 We say that a subset U of the set L is pointwise closed if

$$(l^k \in U \ (k = 1, 2, \dots), \quad \langle l^k, x\rangle \to \langle l, x\rangle \text{ for all } x \in \mathbb{R}^n_+) \implies l \in U.$$

It follows directly from the definition of abstract convex sets that an $(L, \mathbb{R}^n_+)$-convex set is pointwise closed. So Proposition 2.18 shows that a normal closed-along-rays subset of L is pointwise closed.

2.3.5. Two kinds of normality

Consider a pair of sets $\mathbb{R}^n_+$ and L with the coupling function

$$\langle l, x\rangle = \min_{i \in I_+(l)} l_i x_i. \tag{2.3.7}$$

This coupling function allows us to consider an element $l \in L$ as a function defined on $\mathbb{R}^n_+$ and an element $x \in \mathbb{R}^n_+$ as a function defined on L. (See Remark 1.3 for details.) The latter has a form

$$x(l) = \min_{i \in I_+(l)} l_i x_i.$$

The coupling function (2.3.7) allows us to consider two kinds of sets, namely $(L, \mathbb{R}^n_+)$-convex subsets of L and $(\mathbb{R}^n_+, L)$-convex subsets of $\mathbb{R}^n_+$. Recall that a set $U \subset L$ is $(L, \mathbb{R}^n_+)$-convex if there exists a function $f : \mathbb{R}^n_+ \to \bar{\mathbb{R}}$ such that $U = \operatorname{supp}(f, L)$. It follows from Proposition 2.18 that U is $(L, \mathbb{R}^n_+)$-convex if and only if U is normal (in L) and closed-along-rays. In order to describe $(\mathbb{R}^n_+, L)$-convex subsets of $\mathbb{R}^n_+$ we will use Lemma 1.1. According to this lemma a set $U \subset \mathbb{R}^n_+$ is $(\mathbb{R}^n_+, L)$-convex if and only if for each $x \in \mathbb{R}^n_+ \setminus U$ there exists $l \in L$ such that $\langle l, x \rangle > \sup_{y \in U} \langle l, y \rangle$.

PROPOSITION 2.19 *Let $U \subset \mathbb{R}^n_+$. The following assertions are equivalent:*

1) U is $(\mathbb{R}^n_+, L)$-convex.

2) U is closed-along-rays and normal (in $\mathbb{R}^n_+$).

Proof: We only prove 2) $\Longrightarrow$ 1). Let $x \notin U$. Since U is closed-along-rays there exists $\varepsilon > 0$ such that $(1-\varepsilon)x \notin U$. Let

$$l = \frac{1}{(1-\varepsilon)x}. \tag{2.3.8}$$

Then

$$\langle l, x \rangle = \min_{i \in I_+(x)} l_i x_i = \frac{1}{1-\varepsilon} > 1.$$

Let $y \in U$. Since U is normal and $(1-\varepsilon)x \notin U$, the inequality $y \geq (1-\varepsilon)x$ does not hold. Hence there exists an index i_0 such that $y_{i_0} < (1-\varepsilon)x_{i_0}$. Then

$$\langle l, y \rangle = \min_{i \in I_+(x)} \frac{y_i}{(1-\varepsilon)x_i} \leq \frac{y_{i_0}}{(1-\varepsilon)x_{i_0}} < 1.$$

Thus

$$\langle l, x \rangle > \sup_{y \in U} \langle l, y \rangle.$$

△

REMARK 2.15 Let U be a closed-along-rays and normal set. It follows from the proof of Proposition 2.19 that for each $x \notin U$ there exists $l' \in \mathbb{R}^n_{++}$ such that $\langle l', x \rangle > \sup_{y \in U} \langle l', y \rangle$. Indeed, let l be a vector defined by (2.3.8) and $l' \in \mathbb{R}^n_{++}$ be a vector such that $l_i = l'_i$ for $i \in I_+(x)$. Then $\langle l', x \rangle = \langle l, x \rangle$ and $\langle l', x' \rangle \leq \langle l, x' \rangle$ for all $x' \in \mathbb{R}^n_+$. Since $\langle l, x \rangle > \sup_{y \in U} \langle l, y \rangle$ it follows that l' is required vector.

REMARK 2.16 It follows from Proposition 2.19 that a normal closed-along-rays subset U of $\mathbb{R}^n_+$ is closed.

PROPOSITION 2.20 *Let $U \subset \mathbb{R}^n_+$. The following assertions are equivalent:*

1) U is normal;

2) for each $x \notin U$ there exists $l \in \mathbb{R}^n_+$ such that

$$\langle l, u \rangle < 1 = \langle l, x \rangle \text{ for all } u \in U. \tag{2.3.9}$$

Proof: 1) $\implies$ 2). Let $x \notin U$ and $l = 1/x$. Obviously, $\langle l, x \rangle = 1$. Consider an element $y \in \mathbb{R}^n_+$ such that $\langle l, y \rangle \geq 1$. Then $y \geq x$, so $y \notin U$. Thus $\langle l, u \rangle < 1$ for all $u \in U$.

2) $\implies$ 1). Assume that U is not normal, so there exists $u \in U$ and $x \notin U$ with $x \leq u$. Let $l \in \mathbb{R}^n_+$ be a vector such that (2.3.9) holds. Then $\langle l, x \rangle \leq \langle l, u \rangle < 1 = \langle l, x \rangle$, which is impossible. △

REMARK 2.17 Recall that a convex set $U \subset \mathbb{R}^n$ is called evenly convex, if U is the intersection of open half-spaces. There exist convex but not evenly convex sets. If we consider a min-type function $x \mapsto \langle l, x \rangle$ as an analogue of a linear function, then the set $\{x \in \mathbb{R}^n_+ : \langle l, x \rangle < 1\}$ plays the role of an open half-space. We can consider this set as an "open min-type lower half-space". Proposition 2.20 shows that a set $U \subset \mathbb{R}^n_+$ is normal if and only if it is "evenly normal", i.e, the intersection of open min type lower half-spaces.

PROPOSITION 2.21 *The closure* $\mathrm{cl}\, U$ *of a normal set $U \subset \mathbb{R}^n_+$ is normal.*

Proof: Let $x \in \mathrm{cl}\, U$ and $0 \leq x' \leq x$. Then there exists a sequence (x_k) such that $x_k \in U$, $x_k \to x$. Set $x'_k = \min(x', x_k)$. Then $x'_k \in U$ (due to the normality of U) and $x'_k \to x'$. Hence $x' \in \mathrm{cl}\, U$. △

Recall (see Definition 1.6) that a function $f : \mathbb{R}^n_+ \to \bar{\mathbb{R}}$ is called L-quasiconvex if its level sets $S_c(f) = \{x \in \mathbb{R}^n_+ : f(x) \leq c\}$ are $(\mathbb{R}^n_+, L)$-convex.

PROPOSITION 2.22 *A function $f : \mathbb{R}^n_+ \to \bar{\mathbb{R}}$ is L-quasiconvex if and only if f is lower semicontinuous and increasing.*

Proof: The proof is similar to that of Proposition 2.7 and we omit it. △

DEFINITION 2.11 Let $U \subset \mathbb{R}^n_+$. The set

$$\mathrm{Nh}\, U = \{x \in \mathbb{R}^n_+ : \exists x' \in U : x \leq x'\}$$

is called the *normal hull* of the set U with respect to the cone $\mathbb{R}^n_+$.

This definition is similar to the definition of the normal hull with respect to the cone $\mathbb{R}^n_{++}$ (see Definition 2.5). Definition 2.11 can be presented in the following form:

$$\mathrm{Nh}\, U = (U - \mathbb{R}^n_+) \cap \mathbb{R}^n_+. \tag{2.3.10}$$

PROPOSITION 2.23 *Let $U \subset \mathbb{R}^n_+$. Then the set* Nh U *is normal. If U is closed then* Nh U *is closed as well.*

Proof: The normality of Nh U is obvious. Assume that U is closed. Let us check that the set Nh U is closed as well. Let $x_k \in \mathrm{Nh}\, U$, $k = 1, \dots$ and $x_k \to x$. Then there exist elements $x'_k \in U$ such that $x'_k \geq x_k$, $k = 1, \dots$. The sequence x_k is bounded therefore there exists $y \in \mathbb{R}^n_+$ such that $x_k \leq y$ for all k and $x \leq y$. Let $\bar{x}_k = \min(x'_k, y)$. Then the sequence $\bar{x}_k$ is bounded, so we can assume without loss of generality that there exists $\bar{x} := \lim \bar{x}_k$. Since $\bar{x}_k \leq x'_k \in U$ it follows that $\bar{x}_k \in U$ for all k, hence $\bar{x} \in U$. We also have $\bar{x}_k = \min(x'_k, y) \geq \min(x_k, x)$, therefore $\bar{x} \geq \lim_k \min(x_k, x) = x$. The inclusion $\bar{x} \in U$ and the inequality $\bar{x} \geq x$ imply $x \in \mathrm{Nh}$. △

The following assertion holds:

PROPOSITION 2.24 *Let $U \subset \mathbb{R}^n_+$. The set* Nh U *coincides with the intersection V of all normal sets $U' \supset U$. If U is closed then* Nh U *coincides with the intersection $\bar{V}$ of all closed normal sets $U' \supset U$.*

Proof: Let U is an arbitrary subset of $\mathbb{R}^n_+$. The inclusion $V \supset \mathrm{Nh}\, U$ is obvious. Since Nh U is normal it follows that $V \subset \mathrm{Nh}\, U$, hence $V = \mathrm{Nh}\, U$. The same argument and Proposition 2.23 show that $\bar{V} = \mathrm{Nh}\, U$ for a closed $U \subset \mathbb{R}^n$. △

COROLLARY 2.5 *Let U be a closed subset of $\mathbb{R}^n_+$. Then $(\mathbb{R}^n_+, L)$-convex hull of U coincides with the normal hull of U. Thus*

$$\mathrm{Nh}\, U = (U - \mathbb{R}^n_+) \cap U.$$

Indeed, it follows directly from Proposition 2.24.

2.3.6. Properties of the support sets

We now describe some properties of the support sets, that is, $(L, \mathbb{R}^n_+)$-convex sets. It follows immediately from the definition of the support set that $p_1 \le p_2$ if and only if $\mathrm{supp}(p_1, L) \subset \mathrm{supp}(p_2, L)$ for IPH functions p_1 and p_2.

PROPOSITION 2.25 *Consider a family $(p_\alpha)_{\alpha\in A}$ of IPH functions and let $p(x) = \inf_{\alpha\in A} p_\alpha(x)$. Then* $\mathrm{supp}(p, L) = \bigcap_{\alpha\in A} \mathrm{supp}(p_\alpha, L)$.

Proof: Since each set $\mathrm{supp}(p_\alpha)$ is normal and closed-along-rays, it follows that the intersection of these sets is also normal and closed-along-rays. Proposition 2.18 shows that this intersection is $(L, \mathbb{R}^n_+)$-convex, so there exists an IPH function $\tilde{p}$ such that $\mathrm{supp}(\tilde{p}, L) = \bigcap_{\alpha\in A} \mathrm{supp}(p_\alpha, L)$. Since $\tilde{p} \le p_\alpha$ for all $\alpha \in A$, it follows that $\tilde{p} \le \inf_{\alpha\in A} p_\alpha$. The function $p(x) = \inf_{\alpha\in A} p_\alpha(x)$ is IPH. Since $p \le p_\alpha$ for all α we have

$$\mathrm{supp}(p, L) \subset \bigcap_{\alpha\in A} \mathrm{supp}(p_\alpha, L) = \mathrm{supp}(\tilde{p}, L)$$

so $p \le \tilde{p}$. $\triangle$

REMARK 2.18 A more general statement is formulated in the framework of the Minkowski duality (see Subsection 7.4.2)

Let J be a nonempty subset of $I = \{1, 2, \dots, n\}$. Denote $\bar{J} = I \setminus J$. We shall use the following notation:

$$\mathbb{R}^J_+ = \{y = (y_1, \dots, y_n) \in \mathbb{R}^n_+ : y_i = 0 \text{ for all } i \in \bar{J}\}, \qquad (2.3.11)$$

$$\mathbb{R}^J_{++} = \{y \in \mathbb{R}^J_+ : y_i > 0 \text{ for all } i \in J\}. \qquad (2.3.12)$$

In particular, $\mathbb{R}^I_+ = \mathbb{R}^n_+$, $\mathbb{R}^I_{++} = \mathbb{R}^n_{++}$. Set $\mathbb{R}^\emptyset_+ = \{0\}, \mathbb{R}^\emptyset_{++} = \emptyset$. For $y \in \mathbb{R}^n_+$ we denote by y_J the projection of y onto $\mathbb{R}^J_+$, that is, $y_J = (y'_1, \dots, y'_n)$ with

$$y'_i = \begin{cases} y_i & \text{if } i \in J; \\ 0 & \text{if } i \in \bar{J}. \end{cases}$$

Let p be an IPH function and $J \subset I$. Let p_J and P_J be defined on $\mathbb{R}^I_+$ and $\mathbb{R}^J_+$ respectively by

$$p_J(y) = p(y_J) \quad (y \in \mathbb{R}^I_+), \qquad P_J(y) = p(y) \quad (y \in \mathbb{R}^J_+).$$

It is clear that p_J and P_J are IPH functions and $P_J(y) = p_J(y)$ for $y \in \mathbb{R}^J_+$. Denote by L_J the set of all min-type functions $x \to \langle l, x\rangle := \min_{i\in J_+(l)} l_i x_i$ defined on $\mathbb{R}^J_+$ with $l \in \mathbb{R}^J_+$. Here $J_+(l) = \{i \in J : l_i > 0\}$.

PROPOSITION 2.26 *Let* $U = \mathrm{supp}(P_J, L_J)$. *Then* $\mathrm{supp}(p_J, L) = U \times \mathbb{R}^{\bar{J}}_+$.

Proof: Let $l \in U \times \mathbb{R}^{\bar{J}}_+$ and $y \in \mathbb{R}^I_+$. Then

$$\begin{aligned} \langle l, y\rangle &= \min_{i \in I_+(l)} l_i y_i = \min\left(\min_{i\in I_+(l)\cap J} l_i y_i, \min_{i \in I_+(l)\cap \bar{J}} l_i y_i\right) \\ &\leq \min_{i\in I_+(l)\cap J} l_i y_i = \langle l_J, y_J\rangle \\ &\leq P_J(y_J) = p_J(y). \end{aligned}$$

Thus

$$l \in \mathrm{supp}(p_J, L). \tag{2.3.13}$$

Assume now that $l \in \mathrm{supp}(p_J)$. Then for $y \in \mathbb{R}^J_+$ we have

$$P_J(y) = p_J(y) \geq \langle l_J, y\rangle = \min_{i\in I_+(l_J)} l_i y_i = \min_{i \in I_+(l_J)\cap J} l_i y_i = \langle l_J, y\rangle.$$

Thus $l_J \in \mathrm{supp}(P_J, L_J) = U$. △

Let $a = (a_i)_{i \in I} \in \mathbb{R}^n_{++}$ and $p \in \mathrm{IPH}$. We shall consider the function $p^a(y) = p(a \cdot y)$, where $a \cdot y = (a_1 y_1, \dots, a_n y_n)$.

PROPOSITION 2.27 *Let* $a \gg 0$. *Then* $\mathrm{supp}(p^a, L) = a \cdot \mathrm{supp}(p, L) \equiv \{a \cdot l : l \in \mathrm{supp}(p, L)\}$.

Proof: The proof is similar to that of Proposition 2.14. △

COROLLARY 2.6 *Let* $a = (a_i)_{i\in I} \in \mathbb{R}^n_+$ *and* $J = \{i : a_i > 0\}$. *Let* $P^a_J(y) = p(a \cdot y)$. *Then* $\mathrm{supp}(P^a_J, L_J) = a_J \cdot \mathrm{supp}(P_J, L_J)$.

EXAMPLE 2.10 (Compare with Example 2.8) Let p be an IPH function given by $p(y) = \max_{i\in I} a_i y_i$, where a_i ($i \in I$) are positive numbers. First we consider special sets which will be useful in describing the support set of max-type functions. For $J \subset I$ and $a = (a_1, \dots, a_n) \in \mathbb{R}^J_{++}$, define

$$V_{a_J} = \bigcup_{i \in J}\{l = (l_1, \dots, l_m) \in \mathbb{R}^J_{++} : l_i \leq a_i\} = \{l \in \mathbb{R}^J_{++} : \min_{i\in J} \frac{l_i}{a_i} \leq 1\}. \tag{2.3.14}$$

Clearly,

$$\mathbb{R}^J_{++} \backslash V_{a_J} = \{l \in \mathbb{R}^J_{++} : l_i > a_i \text{ for all } i \in J\}.$$

PROPOSITION 2.28 *If* $p(x) = \max_{i\in I} a_i x_i$ *with* $a = (a_i)_{i\in I} \in \mathbb{R}^I_{++}$ *then* $\mathrm{supp}(p, L) = \{0\} \cup \bigcup_{J \subset I, J\neq\emptyset} V_{a_J}$.

Proof: Let $l \in \mathrm{supp}(p, L)$, $l \neq 0$ and let $J := I_+(l)$. Then $l \in \mathrm{supp}(P_J, L_J)$, where P_J is defined on $\mathbb{R}^J_+$ by $P_J(y) = \max_{i\in J} a_i y_i$. Since

$l \in \mathbb{R}^J_{++}$ and $\langle l, y\rangle \leq P_J(y)$ for all $y \in \mathbb{R}^J_{++}$, it follows from Example 2.8 that

$$1 \leq P_J\left(\frac{1}{l}\right) = \max_{i\in J}\frac{a_i}{l_i} = \frac{1}{\min_{i\in J}\frac{l_i}{a_i}}.$$

So, $\min_{i\in J}(l_i/a_i) \leq 1$ and by (2.3.14) $l \in V_{a_J}$. Thus $\operatorname{supp}(p, L) \subset \bigcup_{J\subset I} V_{a_J}$. In the same manner we can prove the reverse inclusion. △

Let p be an IPH function defined on $\mathbb{R}^n_+$. For each nonempty $K \subset I = \{1, \dots, n\}$, consider the restriction p_K of the function p to the cone R^K_{++}, defined by (2.3.12). Let $L_K = \{l \in L : I_+(l) = K\}$.

PROPOSITION 2.29 *Let p be an IPH function defined on $\mathbb{R}^n_+$. Then*

$$\operatorname{supp}(p, L) = \bigcup_{K\subset I, K\neq\emptyset} \{x \in \mathbb{R}^K_{++} : p_K\left(\frac{1}{x}\right) \geq 1\} \cup \{0\}.$$

Proof: Let $U = \operatorname{supp}(p, L)$. We apply Theorem 2.3 to obtain

$$\begin{aligned}
U &= \{l \in L : (\forall x \geq 0)\, \langle l, x\rangle \leq p(x)\}\\
&= \bigcup_{\emptyset\neq K\subset I} \{l \in L,\ I_+(l) = K : \ (\forall x \in \mathbb{R}^K_{++})\, \langle l, x\rangle \leq p(x)\} \cup \{0\}\\
&= \bigcup_{\emptyset\neq K\subset I} \{l \in L_K : l \in \operatorname{supp}(p_K, L_K)\} \cup \{0\}\\
&= \bigcup_{\emptyset\neq K\subset I} \{x \in \mathbb{R}^K_{++} : p_K\left(\frac{1}{x}\right) \geq 1\} \cup \{0\}.
\end{aligned}$$

The proof is completed. △

Let us give an example.

EXAMPLE 2.11 Let $p(x) = \sum_{i\in I} a_i x_i$ with $a_i > 0$ for all $i \in I$. We have for nonempty $K \subset I$, $p_K(x) = \sum_{i\in K} a_i x_i$. Thus

$$\operatorname{supp}(p_K, L_K) = \{x \in \mathbb{R}^I_{++} : \sum_{i\in K}\frac{a_i}{x_i} \geq 1\},$$

and

$$\operatorname{supp}(p, L) = \bigcup_{K\subset I, K\neq\emptyset} \{x \in \mathbb{R}^K_{++} : \sum_{i\in K}\frac{a_i}{x_i} \geq 1\} \cup \{0\}.$$

In particular, if $n = 2$, then $\operatorname{supp}(p, L)$ is the union of zero and three sets. Two of these sets are segments on coordinate axes:

$$\operatorname{supp}(p_{\{1\}}, L_{\{1\}}) = \{x = (x_1, x_2) : 0 < x \leq a_1,\ x_2 = 0\},$$

$$\operatorname{supp}(p_{\{2\}}, L_{\{2\}}) = \{x = (x_1, x_2) : x_1 = 0,\ 0 < x_2 \le a_2\},$$

and the third set is

$$\operatorname{supp}(p_{\{1,2\}}, L_{\{1,2\}}) = \{x = (x_1, x_2) \in \mathbb{R}^2_{++} : \frac{a_1}{x_1} + \frac{a_2}{x_2} \ge 1\}.$$

Let p be an IPH function defined on $\mathbb{R}^n_+$. Sometimes it is more convenient to define the support set of p in a different way, namely we can consider the support set of the restriction of p to $\mathbb{R}^n_{++}$ as the support set of the function p itself. Let us give the corresponding definition.

DEFINITION 2.12 Let $\bar{p}$ be an IPH function defined on $\mathbb{R}^n_+$ and let p be the restriction of the function $\bar{p}$ to the cone $\mathbb{R}^n_{++}$. Then the support set of the function p, defined by (2.2.14), is called the strictly positive support set of the function $\bar{p}$ and denoted by $\operatorname{supp}_1(p, L)$.

Theorem 2.3 and the above definition imply that

$$\operatorname{supp}_1(p, L) = \left\{x \in \mathbb{R}^I_{++} : p\left(\frac{1}{x}\right) \ge 1\right\}.$$

It is clear that $\operatorname{supp}_1(p, L)$ coincides with the *main part* of the support set $\operatorname{supp}(p, L)$ which corresponds to the cone $\mathbb{R}^n_{++}$.

2.3.7. Subdifferentials of IPH functions

It has been shown (see Proposition 2.16 and Remark 2.12) that for an arbitrary IPH function p and a point $y \in \mathbb{R}^n_+$, the subdifferential $\partial_L p(y)$ is nonempty and contains the element $p(y)/y$. Here we give a complete description of this subdifferential. It will be convenient to denote the subdifferential $\partial_L p(y)$ by $\partial^- p(y)$. Recall that by the definition (see Definition 1.7)

$$\partial^- p(y) = \{l \in \mathbb{R}^n_+ : p(x) \ge \langle l, x\rangle - (\langle l, y\rangle - p(y))\}.$$

Proposition 1.3 shows that

$$\partial^- p(y) = \{l \in \mathbb{R}^n_+ : \langle l, x\rangle \le p(x) \text{ for all } x \in \mathbb{R}^n_+,\ \langle l, y\rangle = p(y)\}.$$

Recall that for $l \in \mathbb{R}^n_+$ we define the vector $l^{-1} = \dfrac{1}{l}$ by the formula:

$$\left(\frac{1}{l}\right)_i = \begin{cases} \dfrac{1}{l_i} & \text{if } i \in I_+(l); \\ 0 & \text{if } i \notin I_+(l). \end{cases}$$

Clearly, $\langle l, \frac{1}{l}\rangle = 1$ for $l \ne 0$. The following assertion has been proved in [161].

PROPOSITION 2.30 *Let $p : \mathbb{R}^n_+ \to \mathbb{R}$ be an IPH function. Then the set $\partial p^-(y)$ is not empty for all $y \geq 0$. If $p(y) > 0$ then*

$$\partial^- p(y) \equiv \partial_L p(y) = \{l \in \mathbb{R}^n_+ : \langle l, y \rangle = p(y), p\left(\frac{1}{l}\right) = 1\}. \qquad (2.3.15)$$

In particular

$$\frac{p(y)}{y} \in \partial^- p(y).$$

Proof: If $p(y) = 0$, then $0 \in \partial^- p(y)$ and therefore $\partial^- p(y)$ is nonempty. Consider now a point y with $p(y) > 0$.

Let $l \in \partial^- p(y)$. Since $\langle l, y \rangle = \min_{i \in I_+(l)} l_i y_i = p(y) > 0$ it follows that $y_i > 0$ for $i \in I_+(l)$, hence $I_+(y) \supset I_+(l)$. Define a vector y_l by

$$y^l_i = \begin{cases} y_i & \text{if } i \in I_+(l); \\ 0 & \text{if } i \notin I_+(l). \end{cases} \qquad (2.3.16)$$

Since $\langle l, y \rangle = \min_{i \in I_+(l)} l_i y_i = p(y)$ it follows that $y_i \geq (1/l_i)p(y)$ for all $i \in I_+(l)$. The inclusion $I_+(y) \supset I_+(l)$ implies $y \geq y^l \geq p(y)/l$. Since p is IPH, we obtain

$$p(y) \geq p(y^l) \geq p\left(\frac{p(y)}{l}\right) = p(y) p\left(\frac{1}{l}\right),$$

hence $p(1/l) \leq 1$. On the other hand $p(1/l) \geq \langle l, 1/l \rangle = 1$, so $p(1/l) = 1$. Let

$$A = \left\{l \in \mathbb{R}^n_+ : \langle l, y \rangle = p(y), p\left(\frac{1}{l}\right) = 1\right\}.$$

We have proved that $l \in A$, hence $\partial_L p(y) \subset A$.

We now prove the reverse inclusion. Let $l \in A$. First, we check that $\langle l, x \rangle \leq p(x)$ for all $x \in \mathbb{R}^n_+$. Assume, on the contrary, that there exists $z \in \mathbb{R}^n_+$ such that $\langle l, z \rangle = \min_{i \in I_+(l)} l_i z_i > p(z)$. Then $l_i z_i > p(z)$ for all $i \in I_+(l)$, hence $I_+(l) \subset I_+(z)$, and there exists $\varepsilon > 0$ such that

$$z_i \geq \frac{p(z)}{l_i} + \frac{\varepsilon}{l_i} \text{ for all } i \in I_+(l).$$

Thus $z \geq (p(z) + \varepsilon)(1/l)$. Since p is increasing, it follows that

$$p(z) \geq p\left((p(z) + \varepsilon)\frac{1}{l}\right) = (p(z) + \varepsilon) p\left(\frac{1}{l}\right) = p(z) + \varepsilon.$$

We have a contradiction, hence $\langle l, x \rangle \leq p(x)$ for all $x \in \mathbb{R}^n_+$. Since $l \in A$, it follows that $\langle l, y \rangle = p(y)$, hence $l \in \partial_L p(y)$. Finally, let $l = p(y)/y$. Then

$$\langle l, y \rangle \left\langle \frac{p(y)}{y}, y \right\rangle = p(y) \text{ and } p\left(\frac{1}{l}\right) = p\left(\frac{y}{p(y)}\right) = 1,$$

hence $l \in A = \partial_L p(y)$. $\triangle$

REMARK 2.19 In Subsection 3.2 we shall examine finite IPH functions p of $n+1$ variables defined on the cone

$$\mathbb{R}_*^{n+1} = \{(x, \lambda) : x \in \mathbb{R}_+^n, \lambda > 0\} \cup \{0, 0\}.$$

Consider the set $\hat{L}$ of all min-type functions $x \to \langle l, x \rangle$ $(x \in \mathbb{R}_*^{n+1})$ with $l \in \mathbb{R}_*^{n+1}$. If $l \in \hat{L}$, then $1/l \in \hat{L}$ as well. We can identify $\mathbb{R}_*^{n+1}$ and $\hat{L}$ as conic sets. For an IPH function p defined on the cone $\mathbb{R}_*^{n+1}$ and a point $y \in \mathbb{R}_*^{n+1}$ with $p(y) > 0$, consider the set

$$B(p, y) = \left\{ l \in \mathbb{R}_*^{n+1} : \langle l, y \rangle = p(y),\ p\left(\frac{1}{l}\right) = 1 \right\}.$$

The same argument as in the proof of Proposition 2.30 demonstrates that this set is nonempty and coincides with the set

$$\partial_{\hat{L}} p(y) = \{l \in \mathbb{R}_*^{n+1} : \langle l, x \rangle \le p(x) \text{ for all } x \in \mathbb{R}_*^{n+1},\ \langle l, y \rangle = p(y)\}.$$

If $p(y) = 0$ then $0 \in B(p, y)$. It follows from these assertions that a finite IPH function p defined on $\mathbb{R}_*^{n+1}$ is $\hat{L}$-convex. Note that $p(y)/y \in \partial_{\hat{L}} p(y)$ for each $y \in \mathbb{R}_*^{n+1}$.

A function p is called *strictly increasing at a point* $y \in \mathbb{R}_+^n$ if for each $x \in \mathbb{R}_+^n$ with $x < y$ the inequality $p(x) < p(y)$ holds. It follows from this definition that $p(y) > 0$ for an IPH function p, which is strictly increasing at a point y.

PROPOSITION 2.31 *If an IPH function p is strictly increasing at a point y then*

$$\partial^- p(y) = \left\{ \frac{p(y)}{y} \right\}.$$

Proof: In view of Proposition 2.30, we have only to check that the relations

$$\langle l, y \rangle = p(y), \qquad p\left(\frac{1}{l}\right) = 1 \tag{2.3.17}$$

imply $l = p(y)/y$. Take l such that (2.3.17) holds. Since $\min_{i \in I_+(l)} l_i y_i \ge p(y) > 0$, it follows that $I_+(y) \subset I_+(l)$ and $y_i \ge p(y)/l_i$ for $i \in I_+(l)$. Hence $y \ge p(y)/l$. Assume $y \ne p(y)/l$. Since p is strictly increasing at the point y, we have

$$p(y) > p\left(\frac{p(y)}{l}\right) = p(y) p\left(\frac{1}{l}\right).$$

Thus $p(1/l) < 1$, which is a contradiction. $\triangle$

REMARK 2.20 A more general result will be proved later. (See Proposition 3.10.)

Now let us give some examples:

EXAMPLE 2.12 Let $p(x) = \max_{i \in I} x_i$ and $\mathbf{1} = (1, 1, \dots, 1)$. Proposition 2.30 shows that

$$\partial^- p(\mathbf{1}) = \left\{ l : \min_{i \in I_+(l)} l_i = 1, \max_{i \in I_+(l)} \frac{1}{l_i} = 1 \right\} = \{ l : \min_{i \in I_+(l)} l_i = 1 \}.$$

The subdifferential $\partial^- p(\mathbf{1})$ is a very large set. We can present this set in the following form:

$$\partial^- p(\mathbf{1}) = \bigcup_{J \in I, J \neq \emptyset} B_J,$$

where B_J is the boundary of the shifted cone $\mathbf{1}^J + \mathbb{R}^J_+$ in the space $\mathbb{R}^J$ and where $\mathbf{1}^J$ is a vector such that its ith coordinate is equal to 1 for $i \in J$ and equal to zero for $i \notin J$.

EXAMPLE 2.13 Let $p(x) = \min_{i \in I} x_i$. We have

$$\partial^- p(\mathbf{1}) = \left\{ l : \min_{i \in I_+(l)} = 1, \min_{i \in I_+(l)} \frac{1}{l_i} = 1 \right\} = \{ l : \min_{i \in I_+(l)} = 1, \max_{i \in I_+(l)} l_i = 1 \}.$$

Thus $\partial^- p(\mathbf{1}) = \{ \mathbf{1}^J : J \subset I \}$, where

$$(\mathbf{1}^J)_i = \begin{cases} 1 & \text{if } i \in J; \\ 0 & \text{if } i \notin J. \end{cases}$$

In other words, the subdifferential $\partial^- p(\mathbf{1})$ consists of all non-zero vertices of the n-dimensional cube $\{ x : 0 \leq x \leq 1 \}$.

EXAMPLE 2.14 Let $p(x) = (x_1 \cdot x_2 \cdots x_n)^{\frac{1}{n}}$. It is easy to check that p is an IPH function and p is strictly increasing at the point $\mathbf{1}$. It follows from Proposition 2.31 that $\partial^- p(\mathbf{1}) = \{\mathbf{1}\}$.

2.3.8. Abstract concavity and superdifferentials

Along with min-type functions, we also consider max-type functions of the form $l(x) = \langle l, x \rangle^+$ where the coupling function $\langle \cdot, \cdot \rangle^+ : \mathbb{R}^n_+ \times \mathbb{R}^n_+$ is defined by

$$\langle l, x \rangle^+ = \max_{i \in I} l_i x_i.$$

(As usual, $I = \{1, \dots, n\}$.)

Note that $\max_{i \in I} l_i x_i = \max_{i \in I_+(l)} l_i x_i$ for all $x \in \mathbb{R}^n_+$. Denote the set of all functions $x \mapsto \langle l, x \rangle^+$ with $l \in \mathbb{R}^n_+$ by L^+. We shall examine the abstract concavity with respect to L^+, in other words, we wish to describe functions which can be represented as the pointwise infimum of a subset of L^+. Since each $l \in L^+$ is IPH it follows that every L^+-concave function is IPH as well. Since each $l \in L^+$ is continuous, L^+ concave function is upper semicontinuous. Note that each IPH function is lower semicontinuous (see Proposition 2.17), hence every L^+-concave function is continuous IPH. It was mentioned in Subsection 2.2.3 that there exist discontinuous IPH functions, hence the set of L^+-concave functions does not coincide with the set of all L-convex functions (that is, with the set of all IPH functions).

We shall study L^+-concavity of an IPH function p with the help of the superdifferential $\bar{\partial}_{L^+} p(y)$ (see Definition 1.10). For the sake of simplicity we denote this superdifferential by $\partial^+ p(y)$. Due to Proposition 1.4 we have

$$\partial^+ p(y) = \{l \in \mathbb{R}^n_+ : \langle l, x \rangle^+ \geq p(x) \text{ for all } x \in \mathbb{R}^n_+, \ \langle l, y \rangle^+ = p(y)\}.$$

The following example demonstrates that the superdifferential $\partial^+ p(y)$ can be empty even if an IPH function p is continuous on $\mathbb{R}^n_+$.

EXAMPLE 2.15 Let p be an IPH function such that

1) there exists a non-zero element y with the property $p(y) = 0$ (in this case y is a boundary point of the cone $\mathbb{R}^n_+$);

2) if $x^k = x_{(1)} + x^k_{(2)}$ where $x_{(1)}$ is a fixed vector from $\mathbb{R}^n_{++}$, $I_+(x^k_{(2)}) \subset I_+(y)$ and $\|x^k_{(2)}\| \to +\infty$ as $k \to \infty$, then $p(x^k) \to +\infty$.

Assume $l \in \partial^+ p(y)$. Then $\langle l, y \rangle^+ = \max_{i \in I} l_i y_i = p(y) = 0$, hence $l_i = 0$ for $i \in I_+(y)$. Thus $I_+(l) \subset I \setminus I_+(y)$. The inequality $\langle l, x \rangle^+ \geq p(x_k)$ shows that $l \neq 0$. Let $x \gg 0$ be a vector such that $l_i x_i = 1$ for $i \in I_+(l)$ and x_i be large enough for $i \notin I_+(l)$, so that $p(x) > 1$. It follows from 2) that such a vector exists. Since $\langle l, x \rangle^+ = 1$, the inequality $\langle l, x \rangle^+ \geq p(x)$ does not hold. Thus $\partial^+ p(y)$ is empty.

Consider, in particular, the function $p(x) = (x_1 \cdot x_2 \cdots x_n)^{\frac{1}{n}}$ and a boundary point y of the cone $\mathbb{R}^n_+$. Since both 1) and 2) hold for the function p at the point y, the superdifferential $\partial^+ p(y)$ is empty.

Nevertheless, if $y \gg 0$ then the superdifferential $\partial^+ p(y)$ is nonempty and it contains interior points of the cone $\mathbb{R}^n_+$.

PROPOSITION 2.32 *Let p be an IPH function and $y \gg 0$. Then $\partial^+ p(y) \cap \mathbb{R}^n_{++}$ is nonempty and*

$$\partial^+ p(y) \cap \mathbb{R}^n_{++} = \left\{ l \gg 0 : l \leq \frac{p(y)}{y}, p\left(\frac{1}{l}\right) = 1 \right\}. \tag{2.3.18}$$

In particular, $p(y)/y \in \partial^+ p(y)$.

Proof: First we show that the set on the right-hand side of (2.3.18) coincides with the following set B:

$$B = \left\{ l \gg 0 : l \leq \frac{p(y)}{y}, p\left(\frac{1}{l}\right) \leq 1 \right\}.$$

Indeed, if $l \in B$ then $y \leq p(y)/l$, hence

$$p(y) \leq p\left(\frac{p(y)}{l}\right) = p(y) p\left(\frac{1}{l}\right).$$

Since $p(y) > 0$, we have $p(1/l) \geq 1$. On the other hand, $p(1/l) \leq 1$. Thus the required equality has been proved. Let $l \in B$. Since $l \leq p(y)/y$, it follows that

$$\max_{i \in I} l_i y_i = \langle l, y \rangle^+ \leq p(y). \tag{2.3.19}$$

We now check that $\langle l, x \rangle^+ \geq p(x)$ for all $x \in \mathbb{R}^n_+$. Assume, on the contrary, that there exists z such that $\langle l, z \rangle^+ < p(z)$. Then $l_i z_i < p(z)$ for all $i \in I$. Therefore there exists $\varepsilon > 0$ such that $(1+\varepsilon) z_i < p(z)/l_i$ for $i \in I_+(l)$. So $(1+\varepsilon)z < p(z)/l$. We have

$$(1+\varepsilon) p(z) = p((1+\varepsilon) z) \leq p\left(\frac{p(z)}{l}\right) = p(z) p\left(\frac{1}{l}\right).$$

Since $p(z) > 0$, it follows that $p(1/l) > 1 + \varepsilon$. We have a contradiction which shows that $\langle l, x \rangle^+ \geq p(x)$ for all x. In particular, $\langle l, y \rangle^+ \geq p(y)$. Combining this inequality with (2.3.19), we conclude that $\langle l, y \rangle^+ = p(y)$. Thus $l \in \partial^+ p(y)$. Clearly, the element $l = p(y)/y$ belongs to the set B. Thus the superdifferential $\partial^+ p(y)$ is not empty.

Assume now that an element $l \gg 0$ belongs to $\partial^+ p(y)$. Then $1 = \langle l, 1/l \rangle^+ \geq p(1/l)$. It follows from the equality $\langle l, y \rangle^+ = p(y)$ that $l \leq p(y)/y$. Thus $l \in B$. △

REMARK 2.21 The equality (2.3.18) can be represented as

$$\partial^+ p(y) = \left\{ l \gg 0 : \langle l, y \rangle^+ = p(y), p\left(\frac{1}{l}\right) = 1 \right\}.$$

Consider the *upper support set*, $\text{supp}^+(p, L^+)$, of p with respect to L^+ (see Definition 1.9):

$$\text{supp}^+(p, L^+) = \{l \in \mathbb{R}^n_+ : \langle l, x\rangle^+ \geq p(x) \text{ for all } x \in \mathbb{R}^n_+\}.$$

Clearly,

$$\partial^+ p(y) = \{l \in \text{supp}^+(p, L^+) : \langle l, y\rangle^+ = p(y)\}.$$

THEOREM 2.6 *Let p be an IPH function defined on $\mathbb{R}^n_+$. Then p is abstract concave with respect to L^+ if and only if p is continuous.*

Proof: In the beginning of this subsection it was shown that each L^+-concave function is continuous IPH. We now prove the reverse assertion. Let p be a continuous IPH function and let $y \in \text{int}\,\mathbb{R}^n_+ = \mathbb{R}^n_{++}$. It follows from Proposition 2.32 that

$$p(y) = \min\{\langle l, y\rangle^+ : l \in \text{supp}^+(p, L^+)\}.$$

Consider now a boundary point y of the cone $\mathbb{R}^n_+$. Let $y_k \to y$, $y_k \in \mathbb{R}^n_{++}$ and $l_k \in \partial^+ p(y_k)$. Then $\langle l_k, x\rangle^+ \geq p(x)$ for all $x \in \mathbb{R}^n_+$ and $\langle l_k, y_k\rangle^+ = p(y_k) \to p(y)$. Therefore,

$$p(y) = \inf\{\langle l, y\rangle^+ : l \in \text{supp}^+(p, L^+)\}. \qquad \triangle$$

2.4. BEST APPROXIMATION BY NORMAL SETS

2.4.1. Overview

Let X be a normed linear space, $G \subset X$ and let $x^0 \in X$. Then the number

$$\text{dist}\,(x^0, G) := \inf_{g\in G} \left\|x^0 - g\right\|, \tag{2.4.1}$$

is called the *distance* of x^0 to G. An element $g^0 \in G$ such that

$$\left\|x^0 - g^0\right\| = \min_{g\in G} \left\|x^0 - g\right\| \tag{2.4.2}$$

is called an *element of best approximation* of x^0 by the set G. The set of all such elements g^0 will be denoted by $P_G(x^0)$. It is well known that if the dimension of X is finite and G is closed, then $P_G(x^0) \neq \emptyset$.

The theory of best approximation by elements of convex sets and reverse convex sets, i.e., complements of convex sets, in normed linear spaces is well developed (see for example [170], [176] and the references therein). However, convexity or reverse convexity is sometimes a very restrictive assumption, so the problem arises to find other classes of sets, for which a theory of best approximation can be developed.

The aim of the present section is to develop a theory of best approximation by elements of normal sets. We shall consider a normed finite-dimensional space $\mathbb{R}^I$, where $I = \{1, \dots, n\}$, and we shall examine the best approximation of elements $x^0 \in \mathbb{R}^I_+$ by normal subsets of the cone $\mathbb{R}^I_+$.

The most suitable norm to develop a theory of best approximation in $\mathbb{R}^I_+$ by normal sets will be the ℓ_∞-norm:

$$\|x\|_\infty = \max_{i \in I} |x_i| \qquad (x = (x_i)_{i \in I} \in \mathbb{R}^I). \tag{2.4.3}$$

The main tool used in the theory of best approximation of an element x^0 in a normed linear space X by a closed convex set G is the separation of G and the ball $B(x^0, r)$ centred at x^0 with radius $r = \operatorname{dist}(x^0, G)$ by continuous linear functions. It is well known that the separability of G and the ball B is a consequence of the separability of G and any outside point. Since a closed normal set $G \subset \mathbb{R}^I_+$ is abstract convex with respect to the set L of all min-type functions, each point $x \notin G$ can be separated from this set by a min-type function. However, the separability of G and the above ball B is no longer a consequence of this fact.

It will turn out that there are also a number of other differences as compared to the convex case, due to some special situations which may occur.

On the other hand, the best approximation by normal sets is in certain aspects better than that by convex sets. Namely, while in general it is difficult even to estimate the distance to the intersection of two convex sets (see, for example, [78]), we shall show that the distance to the intersection of an arbitrary collection of normal sets is exactly equal to the supremum of the distances to the sets of the intersection. This result has no analogue in the classical convex theory of best approximation and can be applied in the study of linear regularity of a collection of normal sets.

Results from this section can be found in [161, 162].

2.4.2. Preliminaries

A subset G of the cone $\mathbb{R}^I_+$ is called radiant if $(g \in G, 0 < \lambda \leq 1) \implies \lambda g \in G$. Clearly a normal set $G \subset \mathbb{R}^I_+$ is radiant. Let $G \subset \mathbb{R}^I_+$ be a closed radiant set. The function μ_G defined on $\mathbb{R}^I_+$ by

$$\mu_G(x) = \inf\{\lambda > 0 : x \in \lambda G\}$$

is called the Minkowski gauge of the set G. Radiant sets and their Minkowski gauges will be studied in detail in Chapter 5. Let us indicate some properties of the Minkowski gauge of a closed radiant set G (see Subsection 5.2.2).

1) $0 \le \mu_G(x) \le +\infty$; $\mu_G(x) = +\infty$ if and only if $\{\alpha x : \alpha > 0\} \cap G = \emptyset$, $\mu_G(x) = 0$ if and only if $\{\alpha x : \alpha > 0\} \subset G$;

2) μ_G is positively homogeneous of degree one;

3) if $\mu_G(x) < +\infty$ then $x \in \mu_G(x)G$;

4) $G = \{x \in \mathbb{R}^I_+ : \mu_G(x) \le 1\}$;

5) μ_G is lower semicontinuous.

It is easy to check that the Minkowski gauge μ_G is an increasing function if and only if the set G is normal. Each increasing function s with properties 1),2) and 5) is the Minkowski gauge of the closed normal set $G = \{x : s(x) \le 1\}$.

Let $G \subset \mathbb{R}^I_+$. We can consider G as a subset of the topological space $\mathbb{R}^I$ and as a subset of the topological space $\mathbb{R}^I_+$. Correspondingly, we can consider the interior, closure and boundary of the set G with respect to $\mathbb{R}^I$ and with respect to $\mathbb{R}^I_+$. We will use notation $\operatorname{int} G$, $\operatorname{cl} G$ and $\operatorname{bd} G$ if G is considered as a subset of $\mathbb{R}^I$ and notation $\operatorname{int}_0 G$, $\operatorname{cl}_0 G$ and $\operatorname{bd}_0 G$ if G is considered as a subset of $\mathbb{R}^I_+$.

By x^+ we shall denote the positive part of a vector $x \in \mathbb{R}^I$. Namely $(x^+)_i = x_i^+$ for all $i \in I$, where $x_i^+ = \max(x_i, 0)$.

PROPOSITION 2.33 *Let $G \subset \mathbb{R}^I_+$ be a normal set and $G \cap \mathbb{R}^I_{++}$ is not empty. Then $\operatorname{int}_0 G$ is a normal set as well.*

Proof: Let $u \gg 0$ and $\varepsilon > 0$. Consider the neighbourhood $V(\varepsilon, g) = \{y : g - \varepsilon u \le y \le g + \varepsilon u\}$ of a point $g \in \mathbb{R}^I_+$. We have

$$V^+(\varepsilon, g) := V(\varepsilon, g) \cap \mathbb{R}^I_+ = \{y : (g - \varepsilon u)^+ \le y \le g + \varepsilon u\}.$$

If $g \in \operatorname{int}_0 G$ then there exist $u \gg 0$ and $\varepsilon > 0$ such that $V^+(\varepsilon, g) \subset G$. Let $0 \le g' \le g$ and $y' \in V(\varepsilon, g')$, that is, $(g' - \varepsilon u)^+ \le y' \le g' + \varepsilon u$. Then $0 \le y' \le g + \varepsilon u$. Since G is normal, it follows that $y' \in G$. Hence $V(\varepsilon, g') \subset G$ and $g' \in \operatorname{int}_0 G$. △

We need the following well-known definition.

DEFINITION 2.13 Let G be a closed normal set. A point $g \in G$ is called a weak Pareto (shortly w.P.) point of G if $(1 + \varepsilon)g \notin G$ for all $\varepsilon > 0$.

PROPOSITION 2.34 *Let G be a closed normal set and $g \in G$. Then g is a w.P. point of G if and only if $\mu_G(g) = 1$.*

Proof: Since $g \in G$ it follows that $\mu_G(g) \leq 1$. If $\mu_G(g) < 1$, then there exists $\lambda > 1$ such that $\lambda\mu_G(g) = \mu_G(\lambda g) \leq 1$, so $\lambda g \in G$. If $\lambda g \in G$ with $\lambda > 1$, then $\lambda\mu_G(g) \leq 1$, so $\mu_G(g) < 1$. △

COROLLARY 2.7 *Each w.P. point of G belongs to $\mathrm{bd}_0\, G$.*

A closed normal set $G \subset \mathbb{R}^I_+$ is called *radiative* if $G \cap \mathrm{int}\, \mathbb{R}^I_+ \neq \emptyset$ and each ray $R_x := \{\alpha x : \alpha \geq 0\}$ with $x \neq 0$ does not intersect the boundary $\mathrm{bd}_0\, G$ of the set G (in the topological space $\mathbb{R}^I_+$) more than once. Radiative sets in a more general situation will be examined in Chapter 5.

PROPOSITION 2.35 *Let G be a normal closed set and $G \cap \mathrm{int}\, \mathbb{R}^I_+ \neq \emptyset$. Then the set G is radiative if and only if each boundary point (with respect to $\mathbb{R}^I_+$) of the set G is a w.P. point of this set.*

Proof: 1) Let G be radiative and $g \in \mathrm{bd}_0\, G$. Assume $\lambda g \in G$ with $\lambda > 1$. Since G is radiative, $\lambda g \notin \mathrm{bd}_0\, G$, hence $\lambda g \in \mathrm{int}_0\, G$. It follows from Proposition 2.33 that the set $\mathrm{int}_0\, G$ is normal, so $g \in \mathrm{int}_0\, G$ and we arrive at a contradiction.

2) Let $x \in \mathbb{R}^I_+$ be a point such that the intersection $R_x \cap \mathrm{bd}_0\, G$ is not empty, where $R_x = \{\lambda x : \lambda \geq 0\}$. Let $\lambda x \in R_x \cap \mathrm{bd}_0\, G$. Then λx is a w.P. point of G, so $\lambda' x \notin G$ for $\lambda' > \lambda$. Assume there exists nonnegative $\lambda_0 < \lambda$ such that $\lambda_0 x \in \mathrm{bd}_0\, G$. Since $G \cap \mathrm{int}\, \mathbb{R}^n_+ \neq \emptyset$, we conclude that $\lambda_0 \neq 0$. Then $\lambda_0 x$ is again a w.P. point, hence $\lambda x = (\lambda/\lambda_0)(\lambda_0 x) \notin G$. We have arrived at a contradiction, so $R_x \cap \mathrm{bd}_0\, G = \{\lambda x\}$. △

Let $I' \subset I$. We can identify the cone $\mathbb{R}^{I'}_+$ with the face $\{x \in \mathbb{R}^I_+ : x_i = 0 \text{ for } i \notin I'\}$ of the cone $\mathbb{R}^I_+$. Let $\mathrm{Pr} \equiv \mathrm{Pr}_{I'}$ be the operator of projection on $\mathbb{R}^{I'}_+$:

$$\mathrm{Pr}\, x = \begin{cases} x_i & \text{if } i \in I'; \\ 0 & \text{if } i \notin I'. \end{cases}$$

Note that $\mathrm{Pr}\, x \leq x$ for $x \in \mathbb{R}^I_+$. It easily follows from this observation that we can identify the projection $\mathrm{Pr}\, G$ of a normal set G with the intersection $G \cap \mathbb{R}^{I'}_+$.

2.4.3. Distance to a normal set

The subset F of the space $\mathbb{R}^I$ is called *downward* if $(x \in F, x' \leq x) \implies x' \in F$. Let G be a normal closed subset of the cone $\mathbb{R}^I_+$. The intersection G_* of all downward sets containing G is called the *downward hull* of the set G.

Let G be a normal closed subset of $\mathbb{R}^I_+$. Let us indicate some properties of the downward hull G_* of the set G.

PROPOSITION 2.36 *Let G_* be a downward hull of a normal closed set G. Then*

1) $G_* = G - \mathbb{R}^I_+$;

2) if $g \in G_*$ then $g^+ \in G$;

3) if $g \in G_*, g \geq 0$ then $g \in G$;

4) G_* is closed.

Proof: 1) If F is a downward set and $F \supset G$ then $F \supset G - \mathbb{R}^I_+$. Clearly $G - \mathbb{R}^I_+$ is a downward set, hence $G_* = G - \mathbb{R}^I_+$.

2) If $g \in G_*$ then there exists $g' \in G$ such that $g' \geq g$. Since $g' \geq 0$ it follows that $g' \geq \max(g, 0) = g^+$. Since G is normal, we deduce that $g^+ \in G$.

3) This assertion follows from 2), since $g = g^+$ for $g \geq 0$.

4) Let $g_k \in G_*$, $g_k \to g$. Then $g_k^+ \to g^+$. It follows from 1) that $g_k^+ \in G$. Since G is closed we have $g^+ \in G$. The inequality $g \leq g^+$ implies that $g \in G - \mathbb{R}^I_+ = G_*$. △

We need the following assertion.

PROPOSITION 2.37 *Let $x \in \mathbb{R}^I_+$, $g \in \mathbb{R}^I$. Then $\|x - g\| \geq \|x - g^+\|$.*

Proof: Let $I_+ = \{i : g_i \geq 0\}$, $I_- = \{i : g_i < 0\}$. We have

$$|x_i - g_i| = |x_i - g_i^+| \ (i \in I_+), \quad |x_i - g_i| = x_i + g_i \geq x_i = |x_i - g_i^+| \ (i \in I_-).$$

Thus $\|x - g\| \geq \|x - g^+\|$. △

COROLLARY 2.8 Let G be a closed normal set. Then $\text{dist}(x, G) = \text{dist}(x, G_*)$ for each $x \in \mathbb{R}^I_+$.

Proof: Since $G_* \supset G$ it follows that $\text{dist}\,(x, G) \geq \text{dist}\,(x, G_*)$. Conversely, for each $g \in G_*$ there exists $g' = g^+ \in G$ such that $\|x - g\| \geq \|x - g'\|$. Hence, $\text{dist}\,(x, G) \leq \text{dist}\,(x, G_*)$. △

Let $Z \subset \mathbb{R}^n$ be a closed set. For each $x^0 \in \mathbb{R}^n$ consider the set

$$P_Z(x_0) = \{x \in Z : \|x - x^0\| = \text{dist}\,(x^0, Z)\}.$$

An element $x \in Z$ is called an *element of best approximation* of x^0 by the set Z. It is well known that $P_Z(x^0)$ is not empty for each closed set Z and each $x^0 \in \mathbb{R}^n$.

PROPOSITION 2.38 *Let F be a downward set and $x^0 \notin F$. Then there exists the least element f^0 of the set $P_F(x^0)$ and*

$$f^0 = x^0 - r\mathbf{1}, \tag{2.4.4}$$

where $r = \operatorname{dist}(x^0, F)$.

Proof: Since $B(x^0, r) = \{x : x^0 - r\mathbf{1} \le x \le x^0 + r\mathbf{1}\}$ it follows that the element f^0 defined by (2.4.4) is the least element of the ball $B(x^0, r)$. Let $f \in P_F(x^0)$. Then $\|f - x^0\| = r$, that is, $f \in B(x^0, r)$, so $f^0 \le f$. Since the set F is downward it follows that $f^0 \in F$. Hence $f^0 \in P_F(x^0)$ and f^0 is the least element of $P_F(x^0)$. △

THEOREM 2.7 *Let G be a closed normal set and $x^0 \in \mathbb{R}^I_+$, $x^0 \notin G$. Then there exists the least element g^0 of the set $P_G(x^0)$ and $g^0 = (g^0_*)^+$ where g^0_* is the least element of the set $P_{G_*}(x^0)$.*

Proof: Proposition 2.38 and Corollary 2.8 demonstrate that $g^0_* = x^0 - r\mathbf{1}$, where $r = \operatorname{dist}(x, G_*) = \operatorname{dist}(x, G)$. Hence $(g^0_*)^+ = \max(x^0 - r\mathbf{1}, 0)$. Let $I_+ = \{i : (g^0_*)_i > 0\} = \{i : x^0_i > r\}$. Then

$$((g^0_*)^+)_i = \begin{cases} x^0_i - r & \text{if } i \in I_+; \\ 0 & \text{if } i \notin I_+. \end{cases} \tag{2.4.5}$$

It follows from Proposition 2.37 that $r = \|x^0 - g^0_*\| \ge \|x^0 - (g^0_*)^+\|$. Since $(g^0_*)^+ \in G$ (See Proposition 2.36 (2)) we have $\|x^0 - (g^0_*)^+\| \ge r$. Thus

$$\|x^0 - g^0_*\| = \|x^0 - (g^0_*)^+\| = r$$

that is $(g^0_*)^+ \in P_G(x)$. Let $g \in P_G(x^0)$. Then $\|g - x^0\| = r$. Since $g \in G \subset G_*$ it follows that $g \in P_{G_*}(x^0)$, so $g \ge g^0_*$. We also have $g \ge 0$, hence $g \ge (g^0_*)^+$. △

COROLLARY 2.9 *Let g^0 is the least element of the set $P_G(x^0)$ and $r =$* $\operatorname{dist}(x^0, G)$. *Then $g^0 = (x^0 - r\mathbf{1})^+$, that is,*

$$g^0_i = \begin{cases} x^0_i - r & \textit{if } i \in I_+; \\ 0 & \textit{if } i \notin I_+. \end{cases}$$

where $I_+ = \{i \in I : x^0_i > r\}$.

COROLLARY 2.10 *Let G be a closed normal set and g^0 be the least element of $P_G(x^0)$. Then $g^0 \le x^0$.*

Let $B_0(x^0, r) = B(x^0, r) \cap \mathbb{R}^I_+$ be the ball in the metric space $\mathbb{R}^I_+$ with centre x^0 and radius r:

$$B_0(x^0, r) = \{x \in \mathbb{R}^I_+ : \|x^0 - x\| \leq r\}.$$

PROPOSITION 2.39 *The least element g^0 of the set $P_G(x^0)$ is the least element of the ball $B_0(x^0, r)$.*

Proof: Since $g^0 \geq 0$ and $g^0 \in B(x^0, r)$ it follows that $g^0 \in B_0(x^0, r)$. Let $g \in B_0(x^0, r)$. Then $g \geq 0$ and $g \in B(x^0, r)$, that is, $g \geq x^0 - r\mathbf{1}$. Thus $g \geq (x^0 - r\mathbf{1})^+ = g^0$. △

COROLLARY 2.11 $B_0(x^0, r) \subset \{y \in \mathbb{R}^I_+ : y \geq g^0\}$.

Later on we shall use the notation g^0 for the least element of $P_G(x^0)$ without any special mention.

2.4.4. Separation

Let G be a normal closed subset of $\mathbb{R}^I_+$. It follows from the abstract convexity of G that each point $x \in \mathbb{R}^I_+ \setminus G$ can be separated from G by a min-type function l. In this subsection we shall study "min-type" separation of G and the ball $B_0(x^0, r')$, where $x^0 \notin G$ and $r' = \|x^0 - g'\|$ for some $g' \in G$. In other words we will examine the following problem: under which conditions there exists $l \in \mathbb{R}^I_+ \setminus \{0\}$ such that

$$\langle l, g \rangle \leq \langle l, y \rangle \text{ for all } g \in G \text{ and } y \in B_0(x^0, r')? \qquad (2.4.6)$$

We can present (2.4.6) in the following form: there exists a number $\gamma \geq 0$ such that

$$\langle l, g \rangle \leq \gamma \leq \langle l, y \rangle \text{ for all } g \in G \text{ and } y \in B_0(x^0, r').$$

If $\gamma > 0$ we can consider a vector l/γ instead of l so we can assume without loss of generality that $\gamma = 1$. Thus we can consider the following two kinds of separation:

1) there exists $l \in \mathbb{R}^I_+$ such that

$$\langle l, g \rangle \leq 1 \leq \langle l, y \rangle \qquad (g \in G, y \in B_0(x^0, r')); \qquad (2.4.7)$$

2) there exists $l \neq 0$ such that

$$\langle l, g \rangle = 0 \leq \langle l, y \rangle \qquad (g \in G, y \in B_0(x^0, r')). \qquad (2.4.8)$$

We now show that (2.4.8) holds for some $l \in \mathbb{R}^I_+, l \neq 0$ if and only if $G \cap \mathbb{R}^I_{++} = \emptyset$. Indeed, if $g \gg 0$ then $\langle l, g \rangle > 0$ for all $l \neq 0$ so (2.4.8) implies that $G \cap \mathbb{R}^I_{++}$ is empty. On the other hand if $G \cap \mathbb{R}^I_{++} = \emptyset$ then for each $g \in G$ there exists $i \in I$ such that $g_i = 0$. Let $l \gg 0$. Then $\langle l, g \rangle = 0$ for all $g \in G$. We shall study in this subsection only the separation of the form (2.4.7).

Let $R_x = \{\alpha x : \alpha \geq 0\}$ be a ray starting from zero and passing through an element $x \in \mathbb{R}^I_+$. First, we present simple necessary conditions for separation.

PROPOSITION 2.40 *Let G be a closed normal set, $x^0 \in \mathbb{R}^I_+ \setminus G$ and (2.4.7) holds for some r'. Then*

1) $R_g \not\subset G$ for each $g \in G$ with $\langle l, g \rangle > 0$;

2) $R_g \not\subset G$ for $g \in G \cap \mathbb{R}^I_{++}$;

3) $\langle l, g \rangle = 1$ for $g \in P_G(x^0)$;

4) $R_g \not\subset G$ for $g \in P_G(x^0)$.

Proof: 1) If $\langle l, g \rangle > 0$ and $R_g \subset G$ then

$$1 \geq \sup_{g' \in G} \langle l, g' \rangle \geq \sup_{\alpha > 0} \langle l, \alpha g \rangle = +\infty.$$

and we arrive at a contradiction.

2) This assertion follows from 1) since $\langle l, g \rangle > 0$ for $g \gg 0$.

3) If $g \in P_G(x_0)$ then $g \in G$ and $g \in B_0(x^0, r')$. Hence $\langle l, g \rangle = 1$.

4) This assertion follows from 1) and 3). △

We now give an example of a set G and a point x^0 such that (2.4.7) does not hold.

EXAMPLE 2.16 Let $I = \{1, 2\}$ and $G = \{(0, g_2) : g_2 \geq 0\}$. Clearly G is normal and closed and $G \supset R_g$ for each $g \in G$, $g \neq 0$. Let $x^0 \in \mathbb{R}^I_+ \setminus G$ and let $g^0 \in P_G(x^0)$. Then $R_{g^0} \subset G$, so (see Proposition 2.40 (4)), there exists no $l \in \mathbb{R}^I_+$ satisfying (2.4.7).

We now give necessary and sufficient conditions for separability in terms of the least element g^0 of the set $P_G(x^0)$.

THEOREM 2.8 *Let G be a closed normal set. Let $x^0 \in \mathbb{R}^I_+ \setminus G$, let g^0 be the least element of the set $P_G(x^0)$ and $r = \min_{g \in G} \|x^0 - g\| = \|x^0 - g^0\|$. Then the following assertions are equivalent:*

1) there exists $l \in \mathbb{R}^I_+$ such that (2.4.7) holds with $r' = r$;

2) g^0 is a weak Pareto point of the set G.

Proof: 1) $\Longrightarrow$ 2). Let (2.4.7) holds. Then (see Proposition 2.40 (3)) $\langle l, g^0 \rangle = 1$. If g^0 is not a w.P. point then there exists $\lambda > 1$ such that $\lambda g^0 \in G$. Then $\langle l, \lambda g^0 \rangle = \lambda > 1$ which contradicts (2.4.7).

2) $\Longrightarrow$ 1). Let g^0 be a w.P. point of the set G and $l = 1/g^0$. Let $y \in B_0(x, r)$. Then (see Corollary 2.11) $y \geq g^0$, so $\langle l, y \rangle \geq \langle l, g^0 \rangle = 1$. Since g^0 is a w.P. point of G, it follows that $\mu_G(g^0) = 1$. Applying Proposition 2.30 we conclude that $1/g = l \in \partial \mu_G(g^0)$. Hence for each $g \in G$ we have

$$\langle l, g \rangle \leq \mu_G(g) \leq 1 = \langle l, g \rangle. \qquad \triangle$$

REMARK 2.22 If g^0 is a w.P. point of G and $r' = r$, then (2.4.7) holds with $l = 1/g^0$.

We now show that separability holds for normal radiative sets. We need the following assertion.

PROPOSITION 2.41 *Let G be a normal closed set such that $G \cap \operatorname{int} \mathbb{R}^I_+ \neq \emptyset$, let $x^0 \in \mathbb{R}^I_+ \setminus G$ and g^0 be the least element of $P_G(x^0)$. Then $g^0 \in \operatorname{bd}_0 G$.*

Proof: Since $G \cap \operatorname{int} \mathbb{R}^I_+$ is not empty it follows that $\operatorname{int}_0 G$ is not empty. Assume $g^0 \in \operatorname{int}_0 G$. Since $x^0 \geq g^0$ (see Corollary 2.10) and $g^0 \in \operatorname{int}_0 G$ it follows that there exists $\lambda > 0$ such that $g^0 + \lambda(x^0 - g^0) \in G$. We have

$$\|x^0 - g^0\| \leq \|x^0 - g^0 - \lambda(x^0 - g^0)\| = \|(1-\lambda)(x^0 - g^0)\| < \|x^0 - g^0\|$$

and we arrive at a contradiction. Hence $g^0 \in \operatorname{bd}_0 G$. $\triangle$

THEOREM 2.9 *Let G be a normal closed radiative set. Then for each $x^0 \notin G$ there exists an element $l \in \mathbb{R}^I_+$ such that (2.4.7) holds with $r' = r$.*

Proof: It follows from Theorem 2.8, Proposition 2.41 and Proposition 2.35. $\triangle$

We now discuss some sufficient conditions for the separability of non-radiative sets. As a matter of fact we shall describe a situation, where $\mu_G(g^0) = 1$.

Let G be a closed normal set. Let $x^0 \in \mathbb{R}^I_+ \setminus G$ and $r = \operatorname{dist}(x^0, G)$. Consider the following sets:

$$I_+ = \{i : x_i^0 > r\}, \; I_0 = \{i : x_i^0 = r\}, \; I_- = \{i : x_i^0 < r\}. \tag{2.4.9}$$

Corollary 2.9 implies that

$$I_+ = \{i \in I : g_i^0 > 0\}, \quad I_0 \cup I_- = \{i \in I : g_i^0 = 0\}$$

and

$$g_i^0 = \begin{cases} x_i^0 - r & i \in I_+; \\ 0 & i \in I_0 \cup I_-. \end{cases} \tag{2.4.10}$$

THEOREM 2.10 *Let G be a closed normal set and let $x^0 \in \mathbb{R}_+^I \setminus G$. Furthermore, let g^0 be the least element of the set $P_G(x^0)$. Assume that*

$$R_{g^0} \not\subset G \quad \text{and} \quad I_0 = \emptyset.$$

Then there exists a vector l such that (2.4.7) holds:

$$\langle l, g\rangle \le 1 \le \langle l, y\rangle \text{ for all } g \in G, \ \ y \in B_0(x, r).$$

Proof: Let Pr be the projection on the space $\mathbb{R}^{I_+}$ (see Preliminaries for the definition of projection). Denote the projection $\Pr x$ of a vector $x \in \mathbb{R}_+^I$ by x_+ and the norm in the space $\mathbb{R}^{I_+}$ by $\|\cdot\|_+$. We shall identify g^0 and g_+^0. Let $G_+ = \{g_+ : g \in G\}$. Since G is normal it follows that we can identify G_+ with the intersection $G \cap \mathbb{R}_+^{I_+}$. Hence, G_+ is normal and closed, $g^0 \in G_+$ and $R_{g^0} \not\subset G_+$. It follows from (2.4.10) that

$$\|(x^0)_+ - g^0\|_+ = \max_{i \in I_+} x_i^0 - g_i^0 = r.$$

Assume that there exists a vector $g \in G$ such that $\max_{i \in I_+} |x_i^0 - g_i| < r$. The set I_0 is empty, so we can define the vector $\bar{g} \in \mathbb{R}_+^I$, where

$$\bar{g}_i = \begin{cases} g_i & \text{if } i \in I_+; \\ 0 & \text{if } i \in I_-. \end{cases}$$

Since $\bar{g} \le g$ and G is normal, $\bar{g} \in G$.

Since $x_i^0 < r$ and $g_i^0 = 0$ for $i \in I_-$ (see 2.4.9 and and (2.4.10)) we have

$$\max_{i \in I} |x_i^0 - \bar{g}_i| = \max(\max_{i \in I_+} |x_i^0 - g_i|, \max_{i \in I_-} x_i^0) < r,$$

which is impossible. Hence $\|x_+^0 - g_+\|_+ \ge r$ for all $g \in G$ and

$$\|x_+^0 - g^0\| = r = \min_{g_+ \in G_+} \|g_+ - x_+^0\|_+.$$

Thus $g^0 \in P_{G_+}(x_+^0)$. Consider now the Minkowski gauge μ_{G_+} of the set G_+.

Note that g^0 is strictly positive in $\mathbb{R}^{I+}$, that is $g^0 \in \operatorname{int} \mathbb{R}^{I+}_+$. We now verify that $\mu_{G_+}(g^0) = 1$. Let $\mu_{G_+}(g^0) = c$. Clearly $0 \le c \le 1$. Since G_+ does not contain the ray R_{g^0} it follows that $c > 0$. Assume that $c < 1$. Since μ_{G_+} is lower semicontinuous at the point g^0 and $\mu_{G_+}(g^0) < 1$, there exists a neighbourhood V of g^0 such that $\mu_{G_+}(g) < 1$ for all $g \in V \cap \operatorname{int} \mathbb{R}^{I+}_+$. Since $g^0 \in \operatorname{int} \mathbb{R}^{I+}_+$, we can find $\lambda > 0$ such that $x'_\lambda := \lambda x^0_+ + (1-\lambda) g^0 \in G_+$ and

$$\|x^0_+ - x_\lambda\|_+ = (1-\lambda)\|x^0_+ - g^0\| = (1-\lambda) r < r,$$

which is impossible. Thus $c = \mu_{G_+}(g^0) = 1$. It follows from Theorem 2.8 that the element $l_+ = 1/(g^0)$ separates the set G_+ and the ball $B_+ := \{x_+ \in \mathbb{R}^{I+}_+ : \|x^0_+ - x_+\|_+ \le r\}$ that is

$$\langle l_+, g_+ \rangle \le 1 \le \langle l_+, y_+ \rangle \text{ for all } g_+ \in G_+ \text{ and } y_+ \in B_+.$$

Let

$$l = \begin{cases} (l_+)_i & \text{if } i \in I_+; \\ 0 & \text{if } i \in I_-. \end{cases}$$

Then $\langle l, g \rangle = \langle l_+, \operatorname{Pr} g \rangle \le 1$ for all $g \in G$. Note that $B_+ = \operatorname{Pr} B_0(x^0, r)$, so $\langle l, y \rangle = \langle l', \operatorname{Pr} y \rangle \ge 1$ for all $y \in B_+(x^0, r)$. Thus the result follows. $\triangle$

REMARK 2.23 By Proposition 2.40, $\mathbb{R}_{g_0} \not\subset G$ is a necessary condition for separability (2.4.7) with $r' = r$. However it is not a sufficient condition (see Example 2.16). Thus condition $I_0 = \emptyset$ cannot be deleted from Theorem 2.10.

A more detailed analysis of separability has been carried out in the paper [161]. The following statement was proved in this paper.

THEOREM 2.11 *Let G be a closed normal set, $x^0 \in \mathbb{R}^I_+ \setminus G, g' \in G$ and $r' := \|x^0 - g'\|$. For $l \in \mathbb{R}^I_+$ consider the following statements:*

1) l separates G and $B_0(x^0, r')$ in the sense (2.4.7);

2) $I_+(l) \subset I_+ = \{x^0_i > r\}$;

3) $\langle l, g' \rangle = 1$;

4) $\frac{1}{l}$ is a weak Pareto point of G.

Then 1) $\implies$ 2). If g' coincides with the least element g^0 of the set $P_G(x^0)$, then the converse is also true, i.e., the following statements are equivalent:

1') l separates G and $B_0(x^0, r)$ in the sense of (2.4.7), where $r := \|x^0 - g^0\|$.

2') $I_+(l) \subset I_+$, $\langle l, g^0 \rangle = 1$ and $\frac{1}{l}$ is a w.P. point of G.

The separability allows us to obtain necessary and sufficient conditions for a given $g' \in G$ to be a nearest point to x^0. We present two statements from [161] without proofs.

Let G be a closed normal set, $x^0 \in \mathbb{R}^I_+ \setminus G$. Let $g' \in G$ and $r' = \|x^0 - g'\|$.

THEOREM 2.12 (sufficient condition) *If there exists $l \in \mathbb{R}^I_+$ which separates G and $B_0(x^0, r')$ in the sense (2.4.7), then $g' \in P_G(x^0)$. Moreover, if (2.4.7) holds with $l = 1/g'$, then g' coincides with the least element g^0 of the set $P_G(x^0)$.*

THEOREM 2.13 (necessary and sufficient conditions) *The following statements are equivalent:*

1) $g' \in P_G(x^0)$ and the least element g^0 of $P_G(x^0)$ is a weak Pareto point of G.

2) There exists $l \in \mathbb{R}^I_+$ which separates G and $B_0(x^0, r')$ with $r' = \|x^0 - g'\|$, in the sense (2.4.7).

2.4.5. Distance to the union and the intersection of normal sets

Consider the totality $\mathcal{G}$ of all closed normal subsets of the cone $\mathbb{R}^I_+$. Let $(G_t)_{t \in T} \subset \mathcal{G}$ be a family of closed normal sets. It follows directly from the definition of normality and Proposition 2.21 that both sets $\bigcap_{t \in T} G_t$ and $\operatorname{cl} \bigcup_{t \in T} G_t$ are elements of $\mathcal{G}$.

We shall now calculate the distance to the sets $\operatorname{cl} \bigcup_{t \in T} G_t$ and $\bigcap_{t \in T} G_t$.

PROPOSITION 2.42 *Let T be an arbitrary index set and $G_t \in \mathcal{G}$ for all $t \in T$. Let $G = \operatorname{cl} \bigcup_{t \in T} G_t$. Then*

$$\operatorname{dist}(x, G) = \inf_{t \in T} \operatorname{dist}(x, G_t) \qquad (x \in \mathbb{R}^I_+). \tag{2.4.11}$$

Proof: We have for $x \in \mathbb{R}^I_+$:

$$\begin{aligned} \operatorname{dist}(x, G) &= \min_{g \in G} \|x - g\| = \inf_{g \in \bigcup_{t \in T} G_t} \|x - g\| \\ &= \inf_{t \in T} \min_{g \in G_t} \|x - g\| = \inf_{t \in T} \operatorname{dist}(x, G_t). \end{aligned}$$

Thus the result holds. △

Note that we did not use the properties of the norm $\|\cdot\|_\infty$ when we proved Proposition 2.42. We shall now prove that the distance to the intersection of closed normal is equal to the supremum of distances. This result certainly does not hold for an arbitrary norm, so here the properties of the norm $\|\cdot\|_\infty$ are essential.

THEOREM 2.14 *Let T be an arbitry index set and $G_t \in \mathcal{G}$ for all $t \in T$, and let $G = \bigcap_{t\in T} G_t$. Then*

$$\operatorname{dist}(x, G) = \sup_{t\in T} \operatorname{dist}(x, G_t) \qquad (x \in \mathbb{R}^I_+). \tag{2.4.12}$$

Proof: Let $x \in \mathbb{R}^I_+$ and $r_t = \operatorname{dist}(x, G_t)$ for all $t \in T$. Let $r = \sup_{t\in T} r_t$. Since $G_t \supset G$ it follows that $r_t \leq \operatorname{dist}(x, G)$ for all $t \in T$, hence $r \leq \operatorname{dist}(x, G)$. We now prove the opposite inequality. It follows from Corollary 2.9 that $(x - r_t\mathbf{1})^+ \in G_t$ for all $t \in T$. Since $r \geq r_t$ we have $x - r\mathbf{1} \leq x - r_t\mathbf{1}$, hence $(x - r\mathbf{1})^+ \leq (x - r_t\mathbf{1})^+$. The normality of G_t ($t \in T$) implies $(x - r\mathbf{1})^+ \in G_t$, hence $(x - r\mathbf{1})^+ \in \bigcap_{t\in T} G_t = G$. Since

$$(x - (x - r\mathbf{1})^+)_i = \begin{cases} r & \text{if } x_i \geq r; \\ x_i & \text{if } x_i \leq r, \end{cases}$$

we conclude, that $\|x - (x - r\mathbf{1})^+\| \leq r$. Since $(x - r\mathbf{1})^+ \in G$, it follows that $\operatorname{dist}(x, G) \leq r$. △

COROLLARY 2.12 *Let T be an arbitrary index set and $G_t \in \mathcal{G}$ for $t \in T$, let $G = \bigcap_{t\in T} G_t$. Consider $x^0 \in \mathbb{R}^I_+$ and let g^0 be the least element of the set $P_G(x^0)$ and g^0_t be the least element of $P_{G_t}(x^0)$. Then*

$$g^0 = \inf_{t\in T} g^0_t \tag{2.4.13}$$

where inf stands for the coordinate-wise infimum.

Proof: Let $r_t = \operatorname{dist}(x, G_t)$ ($t \in T$) and $r = \operatorname{dist}(x, G)$. It follows from Corollary 2.9, that $g^0_t = (x - r_t\mathbf{1})^+$ and $g^0 = (x - r\mathbf{1})^+$. Since by Theorem 2.14 we have $r = \sup_t r_t$, it follows that

$$\inf_{t\in T} g^0_t = \inf_{t\in T}(x - r_t\mathbf{1})^+ = (x - (\sup_{t\in T} r_t)\mathbf{1})^+ = (x - r\mathbf{1})^+ = g^0.$$

△

We now apply Theorem 2.14 in the study of an analogue of linear regularity for normal sets. First, we recall that a collection $\{G_t\}_{t\in T}$ of subsets of a normed linear space X is said to be *linearly regular* (see, for example, [17]), if there exists a constant $C > 0$ such that

$$\operatorname{dist}(x^0, \cap_{\alpha \in A} G_\alpha) \leq C \sup_{\alpha \in A} \operatorname{dist}(x^0, G_\alpha) \qquad (x^0 \in X). \tag{2.4.14}$$

This concept plays an important role in the theory of error bounds for convex inequality systems. Among other results, it is known (see [17], Corollary 5.26) that if a collection $\{G_\alpha\}_{\alpha \in A}$ is finite and each G_α is a convex polyhedron, then $\{G_\alpha\}_{\alpha \in A}$ is linearly regular.

For closed normal sets the situation is completely different. Namely, we have

PROPOSITION 2.43 *Every collection $\{G_\alpha\}_{\alpha \in A}$ of closed normal subsets of $\mathbb{R}^I_+$ is linearly regular.*

Proof: By Theorem 2.14, for $C = 1$ we have even equality in (2.4.14).$\triangle$

Chapter 3

ELEMENTS OF MONOTONIC ANALYSIS: MONOTONIC FUNCTIONS

3.1. INTRODUCTION

The theory of IPH functions defined on either the cone $\mathbb{R}^n_{++}$ or the cone $\mathbb{R}^n_+$ can be applied in the study of various classes of monotonic functions. One of possible approaches in this direction is to use the hypographs of decreasing functions and the epigraphs of increasing functions. Consider, for example, a decreasing upper semicontinuous function g defined on the cone $\mathbb{R}^n_{++}$. The positive part $\text{hyp}^+ g = \{(x, \lambda) : x \in \mathbb{R}^n_{++}, 0 < \lambda < g(x)\}$ of the hypograph of this function is a closed normal subset of the cone $\mathbb{R}^{n+1}_{++}$, hence there exists an IPH function p defined on $\mathbb{R}^{n+1}_{++}$ such that $\text{hyp}^+ g$ is the support set of p with respect to the set L of all min-type functions. This observation allows us to examine decreasing functions with the help of IPH functions (See Section 3.4).

If f is a lower semicontinuous increasing function defined on $\mathbb{R}^n_{++}$ then its epigraph $\text{epi}\, f = \{(x, \lambda) : x \in \mathbb{R}^n_{++} : \lambda > 0\}$ is a closed conormal subset of the cone $\mathbb{R}^{n+1}_{++}$, so we can use ***abstract concave*** with respect to L (that is, closed conormal) subsets of $\mathbb{R}^n_{++}$ and corresponding them IPH functions in the study of increasing functions. This approach leads to the theory, which is similar to the corresponding theory of decreasing functions, so we do not consider it here.

Various classes of lower semicontinuous increasing functions can be studied in the framework of abstract convexity, corresponding to certain sets of elementary functions. We shall examine two of these classes in this chapter. First, we consider the so-called ICR (increasing coradiant) functions, that is, increasing functions f defined on $\mathbb{R}^n_+$ such that $f(\alpha x) \geq \alpha f(x)$ for all $x \in \mathbb{R}^n_+$ and $\alpha \in [0, 1]$. We study lower semicontinuous ICR functions as abstract convex functions with respect to the set of min-type elementary functions of the form

$$x \mapsto \min(\min_{i \in I_+(l)} l_i x_i, c) \qquad (x \in \mathbb{R}^n_+),$$

where $l \in \mathbb{R}^n_+$, $c > 0$. The construction of the positively homogeneous extension allows us to reduce the theory of ICR functions defined on $\mathbb{R}^n_+$ to the theory of IPH functions defined on $\mathbb{R}^n_+ \times \mathbb{R}_{++}$. (See Section 3.2.)

We also examine the class of ICAR (increasing convex-along-rays) functions, that is, increasing functions f such that the restriction of f to each ray starting at the origin is a convex function of one variable. ICAR functions arise as abstract convex functions with respect to the class H_L of abstract affine elementary functions, where L is the class of all min-type functions. In the framework of the comparison between abstract convexity based on linear functions and abstract convexity based on min-type functions (see Subsection 2.2.11), ICAR functions play the role of lower semicontinuous convex functions.

We shall study in detail ICR and ICAR functions and their subdifferentials with respect to the corresponding set of elementary functions. In particular, we shall show that the abstract subdifferentials of finite ICR and ICAR functions are nonempty at each point and that there exists an abstract subgradient which depends only on a local behaviour of the function at a given point.

Both classes of ICR and ICAR functions are very broad. However, they are quite different. For example, consider a family of functions

$$f(x) = x_1^{\alpha_1} \cdots x_n^{\alpha_n} \qquad (x = (x_1, \dots, x_n) \in \mathbb{R}^n_+),$$

where $\alpha_i > 0$, $i = 1, \dots, n$. Then f is an ICR function if $\sum_i \alpha_i \leq 1$, f is an ICAR function, if $\sum_i \alpha_i \geq 1$ and f is an IPH function, if $\sum_i \alpha_i = 1$.

Note that the unit simplex $S = \{x \in \mathbb{R}^n_+ : \sum_i x_i = 1\}$ is completely unordered in the sense of the coordinate-wise order relation $\geq$: for each two points $x, x' \in S$ neither $x \geq x'$ nor $x' \geq x$. So every function defined on S can be considered as increasing with respect to $\geq$. We shall show that each positive Lipschitz function defined on S is the restriction of an ICAR function defined on $\mathbb{R}^n_+$, so ICAR functions can be used in the study of Lipschitz functions. Some applications of this result to numerical methods can be found in Chapter 9.

3.2. INCREASING CO-RADIANT FUNCTIONS

3.2.1. Overview

The theory of IPH functions defined on $\mathbb{R}^n_+$ can be applied in the study of the much more general class of increasing co-radiant (ICR) functions. In this section we define and examine these functions. We demonstrate that the class of such functions is very broad and describe some properties of an individual ICR function and the class of all ICR functions. We show that a function f defined on $\mathbb{R}^n_+$ is ICR if and only if its positively homogeneous extension $\hat{f}(x, \lambda) = \lambda f(x/\lambda)$ is an IPH function defined on the cone $\mathbb{R}^n_+ \times (0, +\infty)$. This result allows us to establish abstract convexity of ICR functions with respect to a certain class of min-type functions and describe the abstract subdifferentials of this functions.

The main results of this section can be found in [2, 147].

3.2.2. Definition and properties of ICR functions

Let $Q \subset \mathbb{R}^n$ be a conic set. We shall use the following definitions: A set $U \subset Q$ is called *radiant (star-shaped with respect to zero)* if

$$(x \in U, 0 < \lambda \leq 1) \implies \lambda x \in U.$$

A set $V \subset Q$ is called *co-radiant (co-star-shaped with respect to zero or star-shaped with respect to* ∞*)* if

$$(x \in V, \lambda \geq 1) \implies \lambda x \in V.$$

Radiant and co-radiant sets are examined in detail in Chapter 5.

A function $f : Q \to \mathbb{R}_{+\infty}$ defined on a cone $Q \subset \mathbb{R}^n$ is called star-shaped with respect to zero (or radiant) if $f(\lambda x) \leq \lambda f(x)$ for all $x \in Q$ and all nonnegative $\lambda \leq 1$. This term was introduced in a slightly different situation by Pallaschke and Rolewicz [118]. If f is star-shaped with respect to zero then its lower level sets $\{x : f(x) \leq c\}$ are radiant for all c.

Keeping this terminology in mind, we introduce the following definition: A function $f : Q \to \mathbb{R}_{+\infty}$ is called *co-radiant* (or *star-shaped with respect to infinity*) if

$$f(\lambda x) \geq \lambda f(x) \text{ for all } x \in Q,\ \lambda \in [0, 1]. \tag{3.2.1}$$

If $0 \in Q$, then $f(0) \geq 0$ for a co-radiant function f. Indeed, $f(0) = f(\lambda 0) \geq \lambda f(0)$ for all $\lambda \in (0,1)$, which is impossible if $f(0) < 0$.

It is easy to check that f is co-radiant if and only if

$$f(\mu x) \leq \mu f(x) \text{ for all } x \in Q, \ \mu \geq 1. \tag{3.2.2}$$

Indeed, let f be co-radiant. Let $\mu \geq 1$, $x \in Q$ and $\mu x = x'$. Since $x = \lambda x'$ with $\lambda = 1/\mu \leq 1$, it follows that $f(x) \geq \lambda f(x')$ so $\mu f(x) \geq f(\mu x)$. We have proved that (3.2.1) implies (3.2.2). The same reasoning shows that (3.2.2) implies (3.2.1).

We shall mainly consider increasing co-radiant functions defined on the cone $\mathbb{R}^n_+$. We shall use notation ICR for *increasing co-radiant.* These functions were introduced and studied in [147] as increasing star-shaped with respect to infinity (abbreviated ISSI) functions.

Since an ICR function f is increasing and $f(0) \geq 0$ it follows that $f(x) \geq 0$ for all $x \in \mathbb{R}^n_+$. Let us check that a finite ICR function f is continuous on the cone $\mathbb{R}^n_{++}$. In fact, let $x \gg 0$ and $x_k \to x$. Let $\varepsilon > 0$. Then for sufficiently large k, we have $(1-\varepsilon)x \leq x_k \leq (1+\varepsilon)x$. Using the properties of f, (3.2.1) and (3.2.2) we have

$$(1-\varepsilon)f(x) \leq f((1-\varepsilon)x) \leq f(x_k) \leq f((1+\varepsilon)x) \leq (1+\varepsilon)f(x).$$

These inequalities imply the continuity.

Let $\mathcal{F}(ICR)$ denote the set of all ICR functions $f : \mathbb{R}^n_+ \to \mathbb{R}_{+\infty}$. It is easy to show that $\mathcal{F}(ICR)$ possesses the following useful properties:

1) $\mathcal{F}(ICR)$ is a convex cone: if f and g are ICR and λ, μ are positive numbers then also $\lambda f + \mu g$ is also ICR.

2) $\mathcal{F}(ICR)$ is a conditionally complete lattice. More precisely, for an arbitrary family $(f_\alpha)_{\alpha \in A}$ of ICR functions the pointwise supremum, $\sup_{\alpha \in A} f_\alpha$, is ICR as well. If the family $(f_\alpha)_{\alpha \in A}$ is bounded below, that is, there exists an ICR function f such that $f_\alpha \geq f$ for all $\alpha \in A$, then the pointwise infimum $\inf_{\alpha \in A} f_\alpha$ is also ICR.

3) If $f_k \in \mathcal{F}(ICR)$ and $f_k(x) \to f(x)$ for all $x \in \mathbb{R}^n_+$, then f is an ICR function.

4) Let f be a finite ICR function defined on $\mathbb{R}^n_+$ and let ϕ be an ICR function defined on $\mathbb{R}_+$. Then the composition $g = \phi \circ f$ is ICR.

We only prove the last property. For $x \in \mathbb{R}^n_+$ and $\lambda \in (0,1)$, we have

$$g(\lambda x) = \phi(f(\lambda x)) \geq \phi(\lambda f(x)) \geq \lambda \phi(f(x)) = \lambda g(x).$$

We need the following definitions.

DEFINITION 3.1 An increasing function $f : \mathbb{R}^n_+ \to \mathbb{R}$ is called *convex-along-rays* (abbreviated *ICAR*) if for each $x \in \mathbb{R}^n_+$, the function $f_x =: f(\alpha x)$ is convex on $(0, +\infty)$. An increasing function f is called *concave-along-rays* (abbreviated *ICAR(E)*) function if the function $f_x(\alpha)$ is concave on $(0, +\infty)$ for each $x \in \mathbb{R}^n_+$.

ICAR functions are studied in detail in Subsection 3.3. It is easy to check that an ICAR(E) function f such that $f(0) \geq 0$ is ICR. In fact, we have for each $x \in \mathbb{R}^n_+$ and $\lambda \in [0, 1]$:

$$f(\lambda x) = f_x(\lambda) = f_x(\lambda + (1-\lambda)0) \geq \lambda f_x(1) + (1-\lambda)f(0) \geq \lambda f(x).$$

Let us give some examples of ICR functions.

EXAMPLE 3.1 An increasing positively homogeneous function f of degree δ, where $0 < \delta \leq 1$, is ICR. Indeed such a function is ICAR(E). (If $\delta \geq 1$ then f is an ICAR function). In particular a Cobb-Douglas function

$$f(x) = Cx_1^{\alpha_1} \cdots x_n^{\alpha_n} \text{ with } \alpha_i \geq 0, \sum_i \alpha_i \leq 1 \tag{3.2.3}$$

is ICR. Note that if $\sum_i \alpha_i \geq 1$, then the Cobb-Douglas function is ICAR. The function $f_p(x) = (\sum_i x_i^p)^{\frac{1}{p}}$ $(x \in \mathbb{R}^n_+)$ with $p > 0$ is also ICR. Note that this function is convex if $p \geq 1$ and concave if $p \leq 1$.

EXAMPLE 3.2 The pointwise supremum f of a family of ICAR(E) functions $(f_\alpha)_{\alpha \in A}$ is an ICR function. Note that this function is not necessarily ICAR(E).

EXAMPLE 3.3 Let f_j, $j = 1, \ldots, k$ be an increasing positively homogeneous function of degree $0 < \delta_j \leq 1$. Then the sum, minimum and maximum of a family $f_1, \ldots, f_k$ are ICR functions.

EXAMPLE 3.4 Let P be a polynomial of degree m with nonnegative coefficients. Then the function $f(x) = (P(x))^{1/m}$ is ICR. Indeed, let $P(x) = \sum_k a_k g_k(x)$, where g_k is a monomial of degree m_k with $0 \leq m_k \leq m$ and $a_k \geq 0$. Let $\lambda \in (0, 1)$. Then

$$P(\lambda x) = \sum_k a_k g_k(\lambda x) = \sum_k a_k \lambda^{m_k} g_k(x) \geq \sum_k a_k \lambda^m g_k(x) = \lambda^m P(x).$$

Hence $(P(\lambda x))^{1/m} \geq \lambda (P(x))^{1/m}$.

3.2.3. ICR functions and IPH functions

ICR functions and IPH functions are closely related. In order to establish this relation we need the following definition: Let f be a function defined on the cone $\mathbb{R}^n_+$. A function $\hat{f}$ defined on the cone

$$\mathbb{R}^{n+1}_* = \{(x,\lambda) : x \in \mathbb{R}^n_+,\ \lambda > 0\} \cup \{0,0\} \tag{3.2.4}$$

by the formula

$$\hat{f}(x,\lambda) = \lambda f\left(\frac{x}{\lambda}\right), \quad (x \in \mathbb{R}^n_+,\ \lambda > 0), \quad \hat{f}(0,0) = 0 \tag{3.2.5}$$

is called the *positively homogeneous extension* of the function f. We shall study the positively homogeneous extension in Subsection 7.2.8. The following result has been proved in [2].

THEOREM 3.1 *A function f defined on $\mathbb{R}^n_+$ is ICR if and only if its positively homogeneous extension $\hat{f}(x,\lambda)$ is increasing in both variables x, λ.*

Proof: Let f be an ICR function. Since $f(x) \geq 0$ for all $x \in \mathbb{R}^n_+$, it follows that $\hat{f}(x,\lambda) = \lambda f(x/\lambda) \geq 0$. Consider now two points (x_1,λ_1) and (x_2,λ_2) with $x_1, x_2 \in \mathbb{R}^n_+$, $x_1 \geq x_2$ and $\lambda_1 \geq \lambda_2 > 0$. We have

$$(x_1,\lambda_1) \geq (x_2,\lambda_1) \geq (x_2,\lambda_2).$$

Therefore

$$\begin{aligned} \hat{f}(x_1,\lambda_1) &= \lambda_1 f\left(\frac{x_1}{\lambda_1}\right) \geq \lambda_1 f\left(\frac{x_2}{\lambda_1}\right) = \lambda_1 f\left(\frac{\lambda_2}{\lambda_1}\frac{x_2}{\lambda_2}\right) \\ &\geq \lambda_2 f\left(\frac{x_2}{\lambda_2}\right) = \hat{f}(x_2,\lambda_2). \end{aligned}$$

Thus $\hat{f}$ is an increasing function. We now assume that $\hat{f}$ is increasing. Then $\hat{f}(x,\lambda) \geq \hat{f}(0,0) = 0$, in particular, $f(0) = \hat{f}(0,1) \geq 0$. If $x_1 \geq x_2$, then $f(x_1) = \hat{f}(x_1,1) \geq \hat{f}(x_2,1) = f(x_2)$. Thus f is increasing. Let $\lambda \in (0,1)$. Then $(\lambda x, \lambda) \leq (\lambda x, 1)$, hence

$$\lambda f(x) = \hat{f}(\lambda x, \lambda) \leq \hat{f}(\lambda x, 1) = f(\lambda x).$$

If $\lambda = 0$, then $0 = \lambda f(x) \leq f(0) = f(\lambda x)$. Thus f is an ICR function. $\triangle$

REMARK 3.1 The same reasoning shows that the following assertion holds: a function f defined on $\mathbb{R}^n_+$ is radiant and decreasing if and only if the positively homogeneous extension $\hat{f}$ is decreasing in both variables.

3.2.4. Abstract convexity of ICR functions

Theorem 3.1 allows us to study ICR functions by means of IPH functions, which are simpler. In particular, we can examine abstract convexity of ICR functions with respect to a certain class of elementary functions.

Let f be an ICR function. Consider its positively homogeneous extension $\hat{f}$ defined on the cone $\mathbb{R}^{n+1}_*$. It follows from Theorem 3.1 that $\hat{f}$ is an IPH function defined on the cone $\mathbb{R}^{n+1}_*$, so (see Remark 2.19) this function is abstract convex with respect to the set $\hat{L}^-$ of all functions $\hat{\ell}$ defined on $\mathbb{R}^{n+1}_+$ by $\hat{\ell}(x) = \langle \hat{\ell}, x\rangle$, where $\hat{\ell} \in \mathbb{R}^{n+1}_*$. We represent a vector $\hat{\ell} \in \mathbb{R}^{n+1}_*$ in the form $\hat{\ell} = (l, c)$ where $l \in \mathbb{R}^n_+$, $c > 0$. Then for $\hat{x} = (x, 1)$ we have

$$\langle \hat{\ell}, \hat{x}\rangle = \min_{i \in I_+(\hat{\ell})} \hat{\ell}_i \hat{x}_i = \min(\min_{i \in I_+(l)} l_i x_i, c).$$

So

$$\begin{aligned} f(x) &= \hat{f}(\hat{x}) = \sup\{\min_{i \in I_+(\hat{\ell})} \hat{\ell}_i \hat{x}_i : \hat{\ell} \in \hat{L}^-, \hat{\ell} \leq \hat{f}\} \\ &= \sup\{\min(\langle l, x\rangle, c) : \hat{\ell} = (l, c) \leq f, l \in \mathbb{R}^n_+, c > 0\}. \quad (3.2.6) \end{aligned}$$

It follows from (3.2.6) that f is abstract convex with respect to the set $\mathcal{H}^-$ of functions $h : \mathbb{R}^n_+ \to \mathbb{R}_+$ of the form $h(x) = \min(\langle l, x\rangle, c)$, where $l \in \mathbb{R}^n_+$ and $c > 0$.

Let f be an ICR function and $y \in \mathbb{R}^n_+$. Denote by $\bar{\partial}^- f(y)$ the subdifferential of the function f at the point y with respect to the set $\mathcal{H}^-$. Assume that $f(y) > 0$. Then $\hat{f}(\hat{y}) = f(y) > 0$ and therefore (see Remark 2.19)

$$\frac{\hat{f}(\hat{y})}{\hat{y}} = \frac{f(y)}{(y, 1)} \in \partial_{\hat{L}} \hat{f}(y). \qquad (3.2.7)$$

Let

$$h(x) = \min\left(\left\langle \frac{f(y)}{y}, x\right\rangle, f(y)\right), \qquad x \in \mathbb{R}^n_+.$$

Clearly, $h \in \mathcal{H}^-$. We have

$$h(x) = \min\left(\min_{i \in I_+(y)} \frac{f(y) x_i}{y_i}, \frac{f(y)}{1}\right) = \left\langle \frac{f(y)}{(y, 1)}, (x, 1)\right\rangle, \qquad x \in \mathbb{R}^n_+.$$

It follows from (3.2.7) that $h(x) \leq \hat{f}(x, 1) = f(x)$ for all $x \in \mathbb{R}^n_+$ and $h(y) = \hat{f}(\hat{y}) = f(y)$. Thus h is an element of the subdifferential $\bar{\partial}^- f(y)$

of the function f at the point y with respect to the set $\mathcal{H}^-$, so $\bar{\partial}^- f(y)$ is not empty.

We now give a description of the subdifferential $\bar{\partial}^- f(y)$ by assuming that $f(y) > 0$. Applying Remark 2.19 we can conclude that

$$\partial_{\hat{L}} \hat{f}(\hat{y}) = \left\{ \hat{\ell} = (l, \mu) \in \mathbb{R}_*^{n+1} : \langle \hat{\ell}, \hat{y} \rangle = \hat{f}(\hat{y}),\ \hat{f}\left(\frac{1}{\hat{\ell}}\right) = 1 \right\}. \tag{3.2.8}$$

Since

$$\hat{f}\left(\frac{1}{(l, \mu)}\right) = \frac{1}{\mu}\hat{f}\left(\frac{\mu}{l}, 1\right) = \frac{1}{\mu} f\left(\frac{\mu}{l}\right)$$

we can represent (3.2.8) in the following form:

$$\partial^- \hat{f}(y) = \left\{ (l, \mu) \in \mathbb{R}_+^n \times \mathbb{R}_{++} : \min(\langle l, y \rangle, \mu) = f(y),\ f\left(\frac{\mu}{l}\right) = \mu \right\}. \tag{3.2.9}$$

Consider the set $\mathcal{H}^+ = \{h_{l,c} : l \in \mathbb{R}_+^n, c \in \mathbb{R}_+\}$, where $h_{l,c}$ is a max-type function given by

$$h(x) = \max(\langle l, x \rangle^+, c), \qquad x \in \mathbb{R}_+^n.$$

Using the same argument as above and Theorem 2.6 we can demonstrate that an ICR function f is abstract concave with respect to $\mathcal{H}^+$ if and only if f is continuous. We can easily describe the superdifferential of f at a point $y \gg 0$ with respect to the set $\mathcal{H}^+$. In particular, the function

$$h(x) = \max\left(\left\langle \frac{f(y)}{y}, x \right\rangle^+, f(y)\right)$$

belongs to the superdifferential.

3.3. INCREASING CONVEX-ALONG-RAYS FUNCTIONS

3.3.1. Overview

Let L be the set of all min-type functions defined either on the cone $\mathbb{R}_{++}^n$ or on the cone $\mathbb{R}_+^n$. We treated L as the set of abstract linear functions, that is, as an analogue of the set of linear functions, and then we considered the abstract convexity based on this set. We got IPH functions, which play the role of sublinear functions, and normal sets, which play the role of convex sets. In the current section we make the next step in the implementation of this scheme, namely we examine a corresponding analogue of convex functions. For this purpose we consider the set H_L of abstract affine functions and then the upper envelopes of subsets of this set. As the result we get the so-called ICAR (increasing-convex-along-ray) functions.

We shall focus on the L-subdifferentiability of ICAR functions. As it turns out, the L-subdifferential of a finite ICAR function f is not empty at each point $x \in \mathbb{R}^n_+$. Moreover, we can give an explicit description of an L-subgradient for an arbitrary ICAR function and completely describe the L-subdifferential for a strictly increasing CAR function. We also show that the class of ICAR functions is very broad. In particular, each Lipschitz function defined on the unit simplex $S = \{x \in \mathbb{R}^n_+ : \sum_i x_i = 1\}$ can be extended to a finite ICAR function defined on $\mathbb{R}^n_+$. The main results of this section can be found in [1, 2, 143, 148, 152].

3.3.2. ICAR functions: definition, examples and some properties

We begin with the following definition.

DEFINITION 3.2 Let $Q \subset \mathbb{R}^n$ be a conic set. A function $f : Q \to \mathbb{R}_{+\infty}$ is called *convex-along-rays* (briefly *CAR*) if, for each $x \in Q$, the function of one variable

$$f_x(t) = f(tx) \qquad t \in [0, +\infty) \tag{3.3.1}$$

is convex.

Denote by R_x the closed ray $\{tx : 0 \leq t < +\infty\}$ starting at the origin and passing through a point x. We can rephrase the definition of convexity-along-rays in the following form. A function $f : Q \to \mathbb{R}_{+\infty}$ is convex-along-rays if its restriction to each ray R_x with $x \in Q$ is a convex function.

In this section we consider increasing convex-along-rays functions defined on either $Q = \mathbb{R}^n_+$ or $Q = \mathbb{R}^n_{++}$. Thus we get the following definition of an ICAR (increasing convex-along-rays) function.

DEFINITION 3.3 Let Q be either $\mathbb{R}^n_+$ or $\mathbb{R}^n_{++}$. A function $f : Q \to \mathbb{R}_{+\infty}$ is an ICAR function if the following are satisfied:

1) f is increasing: $x \geq y$ implies $f(x) \geq f(y)$;

2) for each $x \in Q$, the function f_x defined by (3.3.1) is convex.

We now present some examples of ICAR functions:

EXAMPLE 3.5 An increasing convex function defined on $\mathbb{R}^n_+$ is ICAR.

EXAMPLE 3.6 An increasing positively homogeneous function of degree $m \geq 1$ defined on $\mathbb{R}^n_+$ is ICAR. In particular,

1) an IPH function defined on $\mathbb{R}^n_+$ is ICAR.

2) A monomial

$$f(x) = x_1^{m_1} x_2^{m_2} \dots x_n^{m_n}, \qquad x = (x_1, x_2, \dots, x_n) \in \mathbb{R}^n_+ \tag{3.3.2}$$

with $m_1 + \dots + m_n \geq 1$ is ICAR.

Note that an ICAR function

$$f_k(x) = \left(\sum_{i \in I} x_i^k\right)^{\frac{1}{k}}, \qquad k > 0 \tag{3.3.3}$$

is IPH and concave if $0 < k < 1$, hence there exist concave IPH functions.

PROPOSITION 3.1 *The set $\mathcal{F}(ICAR)$ of all ICAR functions defined on either $\mathbb{R}^n_+$ or $\mathbb{R}^n_{++}$ possesses the following properties:*

1) $\mathcal{F}(ICAR)$ is a convex cone: if f_1, f_2 are ICAR functions and $\alpha_1, \alpha_2 > 0$ then the function $\alpha_1 f_1 + \alpha_2 f_2$ is ICAR;

2) $\mathcal{F}(ICAR)$ is a complete upper semilattice: if $(f_s)_{s \in S}$ is a family of ICAR functions, where S is an arbitrary index set, then the function $x \mapsto \sup_{s \in S} f_s(x)$ is ICAR;

3) The set $\mathcal{F}(ICAR)$ is closed in the topology of pointwise convergence.

Proof: The proof follows directly from the definition of ICAR functions. $\triangle$

Proposition 3.1 allows us to construct more examples of ICAR functions.

EXAMPLE 3.7 1) A polynomial with nonnegative coefficients is ICAR.

2) A function of the form $f(x) = \sup_{s \in S} p_s(x)$, where S is an arbitrary index set and p_s is a polynomial with nonnegative coefficients, is ICAR.

3) A series $f(x) = \sum_{j=1}^{+\infty} p_j(x)$, where p_j is a polynomial with nonnegative coefficients, is ICAR.

We now describe the level sets of increasing (in particular ICAR) functions. Recall that a set $U \subset \mathbb{R}^n_+$ is called normal if $(x \in U,\ 0 \leq x' \leq x) \implies x' \in U$. We also need the following definition.

DEFINITION 3.4 A set $V \in \mathbb{R}^n_+$ is called *conormal* if $(x \in V,\ x' \geq x) \implies x' \in V$.

It follows directly from the definition that U is normal if and only if its complement $V = \mathbb{R}^n_+ \setminus U$ is conormal. Indeed, assume that U is normal and $V = \mathbb{R}^n_+ \setminus U$. Let $x \in V$ and $x' \geq x$. If $x' \notin V$ then $x \in V$; since U is normal and $x \leq x'$ it follows that $x \in V$, that is, $x \notin V$. We have a contradiction, which shows that V is conormal. The normality of the complement to conormal sets can be shown by the same argument.

PROPOSITION 3.2 *Let f be an increasing function defined on $\mathbb{R}^n_+$. Then the level sets $\{x \in \mathbb{R}^n_+ : f(x) \leq c\}$ are normal and the level sets $\{x \in \mathbb{R}^n_+ : f(x) \geq c\}$ are conormal. In particular, the set* $\operatorname{dom} f = \{x \in \mathbb{R}^n_+ : f(x) < +\infty\}$ *is normal and the set $\{x : f(x) = +\infty\}$ is conormal.*

Proof: The result follows directly from the definition. △

PROPOSITION 3.3 *Let f be an ICAR function defined on $\mathbb{R}^n_+$ and $x \in \mathbb{R}^n_{++}$. If there exists $\lambda > 1$ such that $\lambda x \in \operatorname{dom} f$, then the function f is continuous at the point x.*

Proof: Let $x_k \to x$. Take a positive number ε such that $1 + \varepsilon \leq \lambda$. For large enough k the inequality $(1 - \varepsilon)x \leq x_k \leq (1 + \varepsilon)x$ holds. Since the function f is increasing, we have

$$f((1-\varepsilon)x) \leq f(x_k) \leq f((1+\varepsilon)x); \qquad f((1-\varepsilon)x) \leq f(x) \leq f((1+\varepsilon)x).$$

Since the convex function $f_x : \alpha \mapsto f(\alpha x)$ is continuous on the segment $[0, \lambda)$, it follows that $f((1 + \varepsilon)x) - f((1 - \varepsilon)x) \to 0$ as $\varepsilon \to 0$. △

REMARK 3.2 If f is a lower semicontinuous ICAR function then the set $\operatorname{dom} f$ is closed and normal. We can express the condition "there exists $\lambda > 1$ such that $\lambda x \in \operatorname{dom} f$" in the following form: "x is not a weak Pareto point of the closed normal set $\operatorname{dom} g$".

REMARK 3.3 A finite ICAR function defined on $\mathbb{R}^n_+$ can be discontinuous at a boundary point of the cone $\mathbb{R}^n_+$. For example, the function

$$g_1(x) = \begin{cases} \sum_i x_i & \text{if } x \gg 0; \\ 0 & \text{otherwise} \end{cases}$$

is ICAR and discontinuous at each boundary point of $\mathbb{R}^n_+$ excluding the origin.

REMARK 3.4 A finite ICAR function defined on $\mathbb{R}^n_{++}$ is continuous.

In the rest of this section we shall study ICAR functions defined on $\mathbb{R}^n_+$.

PROPOSITION 3.4 *Let f be an ICAR function defined on $\mathbb{R}^n_+$ and let $x \neq 0$ be a point such that $\lambda x \in \operatorname{dom} f$ for some $\lambda > 1$. Then*

$$f(x) = \liminf_{x' \to x} f(x').$$

Proof: Let $\varepsilon > 0$ be a fairly small number. The set

$$N = \{x' \in \mathbb{R}^n_+ : x'_i > (1-e)x_i, \ (i \in I_+(x))\}$$

is open (in $\mathbb{R}^n_+$) and contains the point x, hence N is a neighbourhood of x. Since f is increasing, $f(x') \geq f((1-\varepsilon)x)$ for all $x' \in N$. This implies $\liminf_{x' \to x} f(x') \geq f((1-\varepsilon)x)$. The increasing function $f_x(t) = f(tx)$ is convex on $[0, +\infty)$ and finite at the point $t = \lambda > 1$, hence $1 \in \operatorname{int} \operatorname{dom} f_x$. Therefore f_x is continuous at the point $t = 1$, so

$$\liminf_{x' \to x} f(x') \geq \sup_{\varepsilon > 0} f((1-\varepsilon)x) = \lim_{\varepsilon \to 0} f((1-\varepsilon)x) = f(x).$$

On the other hand,

$$\liminf_{x' \to x} f(x') \leq \lim_{t \to 1} f(tx) = f(x).$$

Thus the result follows. △

COROLLARY 3.1 *A finite ICAR function is lower semicontinuous.*

REMARK 3.5 There exist ICAR functions $\mathbb{R}^n_+ \to \mathbb{R}_{+\infty}$ which are not lower semicontinuous. For example the ICAR function

$$g_2(x) = \begin{cases} \sum_i x_i & \text{if } \sum_i x_i < 1 \\ +\infty & \text{otherwise} \end{cases}$$

is not lower semicontinuous.

Let L be the set of all min-type functions defined on the cone $\mathbb{R}^n_+$ by (2.3.2). Recall that $l \in L$ if and only if $l(x) = \langle l, x \rangle$ $(x \in \mathbb{R}^n_+)$, where

$$\langle l, x \rangle = \min_{i \in I_+(l)} l_i x_i \qquad (l, x \in \mathbb{R}^n_+),$$

and $I_+(l) = \{i : l_i > 0\}$. We shall consider L as the set of abstract linear functions (see Section 1.4 for details). Let H_L be the set of all abstract affine functions with respect to the set L. Thus $h \in H_L$ if and only if there exist $l \in \mathbb{R}^n_+$ and $c \in \mathbb{R}$ such that

$$h(x) = \langle l, x \rangle - c \qquad (x \in \mathbb{R}^n_+). \tag{3.3.4}$$

We shall identify the function h of (3.3.4) with the pair (l, c). The set L possesses the property (A) (see (1.4.5)), hence we can identify the set H_L with $L \times \mathbb{R}$. It is easy to check that the following assertion holds.

PROPOSITION 3.5 *Any H_L-convex function $f : \mathbb{R}^n_+ \to \mathbb{R}_{+\infty}$ is lower semicontinuous and ICAR.*

Proof: Let f be an H_L-convex function, that is, $f(x) = \sup\{h(x) : h \in \text{supp}(f, H_L)\}$. Then for each $x \in \mathbb{R}^n_+$ we have $f_x(t) = \sup\{h_x(t) : h \in \text{supp}(f, H_L)\}$, where $f_x(t)$ and $h_x(t)$ are defined as in (3.3.1). Note that the function h_x is affine for each $h \in H_L$ and $x \in \mathbb{R}^n_+$, hence the function f_x is convex. Thus f is convex-along-rays. Since each function $h \in H_L$ is increasing, it follows that f is increasing as well. Since each $h \in H_L$ is a continuous function, we conclude that the function f, which is the upper envelope of the set $\text{supp}(f, H_L)$ of continuous functions, is lower semicontinuous. △

The following statement shows that the class of ICAR functions is very large.

PROPOSITION 3.6 *Let f be a lower semicontinuous function defined on the unit simplex $S = \{x \in \mathbb{R}^n_+ : \sum_i x_i = 1\}$. Then there exists an ICAR extension of f, that is, an ICAR function $\tilde{f} : \mathbb{R}^n_+ \to \mathbb{R}_{+\infty}$ such that $\tilde{f}(x) = f(x)$ for all $x \in S$.*

We shall prove this proposition in Section 6.2 (see Proposition 6.5).

REMARK 3.6 Proposition 3.3 shows that the following assertion is valid: if a finite lower semicontinuous function f is discontinuous at a point $x \in S$, then $\tilde{f}(\lambda x) = +\infty$ for any extension $\tilde{f}$ of this function and for any $\lambda > 1$. Thus $\tilde{f}(y) = +\infty$ for all $y \gg x$.

It can be shown (see Subsection 3.3.6) that each positive Lipschitz function defined on S has a locally Lipschitz (hence finite) extension $\tilde{f}$.

3.3.3. ICAR functions as abstract convex functions

Proposition 3.5 shows that each H_L-convex function $f : \mathbb{R}^n_+ \to \mathbb{R}_{+\infty}$ is lower semicontinuous and ICAR. In this section we shall prove that the reverse statement is also valid. We need the following simple assertion.

LEMMA 3.1 *Let $f : \mathbb{R}_+ \to \mathbb{R}_{+\infty}$ be an increasing lower semicontinuous convex function. Then there exists a closed convex set $V \subset \mathbb{R}^2$ such that $f(t) = \sup_{v \in V}(v_1 t - v_2)$ and $v_1 \geq 0$ for each $v \in V$.*

Proof: Let $\tilde{f}$ be a function defined on $\mathbb{R}$ by

$$\tilde{f}(t) = \begin{cases} f(t) & \text{if } t \geq 0; \\ f(0) & \text{if } t < 0. \end{cases} \tag{3.3.5}$$

It follows from the definition that $\tilde{f}$ is a lower semicontinuous convex function, hence $\tilde{f}$ is abstract convex with respect to the set A of all affine functions defined on $\mathbb{R}$. (See, for example, [134], Theorem 12.1.) Let V be the support set of f with respect to A. Then

$$\tilde{f}(t) = \sup_{v \in V}(v_1 t - v_2) \quad \text{for all } t \in \mathbb{R}. \tag{3.3.6}$$

It follows from (3.3.6) that $f(t) = \sup_{v \in V}(v_1 t - v_2)$ for all $t \geq 0$. Let $v \in V$. Then $\tilde{f}(t) \geq v_1 t - v_2$. Dividing by $t < 0$ we have

$$\frac{f(0)}{t} = \frac{\tilde{f}(t)}{t} \leq v_1 - \frac{v_2}{t}.$$

Passing to the limit as $t \to -\infty$ we obtain the desired inequality $v_1 \geq 0$. $\triangle$

PROPOSITION 3.7 *Let $f : \mathbb{R}^n_+ \to \mathbb{R}_{+\infty}$ be an ICAR function such that for each $x \in \mathbb{R}^n_+$ the function f_x defined by (3.3.1) is lower semicontinuous on $\mathbb{R}_+$. Then the function f is abstract convex with respect to the set H_L of all functions h defined by (3.3.4).*

Proof: If $f(x) = +\infty$ for all $x \in \mathbb{R}^n_+$, then the function f is H_L-convex. So we consider only proper functions. Let f be such a function. Consider the support set

$$\text{supp}(f, H_L) = \{(l, c) : \langle l, x \rangle - c \leq f(x) \text{ for all } x \in \mathbb{R}^n_+\} \tag{3.3.7}$$

of f. We need to show that $f(x) = \sup_{(l,c) \in \text{supp}(f,H_L)}(\langle l, x \rangle - c)$ for all $x \in \mathbb{R}^n_+$.

Let $y \in \mathbb{R}^n_+$ be a fixed vector. A function f_y defined on $[0, +\infty)$ by $f_y(t) = f(ty)$ is lower semicontinuous and convex. Applying Lemma 3.1, we conclude that there exists a set V of affine functions

$$t \mapsto v_1 t - v_2 \ (t \geq 0) \tag{3.3.8}$$

such that $f_y(t) = \sup_{v \in V}(v_1 t - v_2)$ for $t \geq 0$ and $v_1 \geq 0$ for all $v \in V$. Denote the function (3.3.8) by (v_1, v_2). For $v = (v_1, v_2) \in V$ set

$$h_v(x) = v_1 \left\langle \frac{1}{y}, x \right\rangle - v_2.$$

(For the definition of the vector $\frac{1}{y}$ see (2.3.3).)

Let us check that for all $x \in \mathbb{R}^n_+$ and for each $v = (v_1, v_2) \in V$

$$f(x) \geq h_v(x). \tag{3.3.9}$$

First assume that $v_1 = 0$. Then $h_v(x) = -v_2$. Note that $f(0) = f_y(0) \geq -v_2$. Since f is increasing, it follows that

$$f(x) \geq f(0) \geq -v_2 = h_v(x) \text{ for all } x \in \mathbb{R}^n_+,$$

hence (3.3.9) holds. The same argument demonstrates that (3.3.9) holds if $y = 0$.

Now assume that $v_1 > 0$ and $y \neq 0$. If $f(x) = +\infty$, then the inequality (3.3.9) holds. Assume now that $f(x) < +\infty$ and (3.3.9) does not hold for the vector x. Let $\beta = h_v(x)$. Since

$$\beta = h_v(x) = v_1 \left\langle \frac{1}{y}, x \right\rangle - v_2 = v_1 \min_{i \in I_+(y)} \frac{x_i}{y_i} - v_2, \tag{3.3.10}$$

it follows that

$$v_1 \frac{x_i}{y_i} \geq \beta + v_2 \text{ for all } i \in I_+(y).$$

Therefore,

$$x \geq \frac{1}{v_1}(\beta + v_2)y \geq 0.$$

Since the function f is increasing and (3.3.9) does not hold at the point x, we have

$$\begin{aligned} \beta &> f(x) \geq f\left(\frac{1}{v_1}(\beta + v_2)y\right) = f_y\left(\frac{1}{v_1}(\beta + v_2)\right) \\ &\geq v_1 \cdot \frac{1}{v_1}(\beta + v_2) - v_2 = \beta. \end{aligned}$$

Thus we have a contradiction which shows that (3.3.9) holds for all $x \in \mathbb{R}^n_+$. Let $(v_1, v_2) \in V$ and $l = v_1/y$ and $c = v_2$. It follows from (3.3.9) that $(l, c) \subset \text{supp}(f, H_L)$. We also have

$$f(y) = f_y(1) = \sup_{(v_1,v_2)\in V} (v_1 - v_2) = \sup_{(v_1,v_2)\in V} v_1 \left\langle \frac{1}{y}, y \right\rangle - v_2.$$

Since

$$\{(l, c) : l = \frac{v_1}{y}, c = v_2 \text{ with } v \in V\} \subset \text{supp}(f, H_L)\}$$

it follows that $f(y) = \sup\{\langle l, y\rangle - c : (l, c) \in \text{supp}(f, H_L)\}$. This completes the proof. △

COROLLARY 3.2 *Each finite ICAR function is abstract convex with respect to H_L.*

Corollary 3.2 follows from Proposition 3.7 and Corollary 3.1.

REMARK 3.7 The proof of Proposition 3.7 consists of three parts. Let f be an ICAR function and $y \in \mathbb{R}^n_+$. First we construct a set of affine functions V with the required properties. Then, we verify that for an arbitrary $(v_1, v_2) \in V$, the pair (l, c) with $l = v_1/y$ and $c = v_2$ belongs to $\text{supp}(f, H_L)$. Finally, we show that f is the upper envelope of the set $\text{supp}(f, H_L)$. The second part plays the main role in the proof. We shall use this part in the proof of Theorem 3.3 below.

THEOREM 3.2 *Let H_L be the class of all functions h defined by (3.3.4). A function $f : \mathbb{R}^n_+ \to \mathbb{R}_{+\infty}$ is H_L-convex if and only if f is lower semicontinuous and ICAR.*

Proof: Let f be a lower semicontinuous ICAR function. Then the function f_y is lower semicontinuous for each y. Thus Proposition 3.7 demonstrates that f is H_L-convex. The reverse assertion follows from Proposition 3.5. △

REMARK 3.8 Proposition 3.7 and Theorem 3.2 imply that ICAR function $f : \mathbb{R}^n_+ \to \mathbb{R}_{+\infty}$ is lower semicontinuous if and only if its restriction to each ray $\{ty : t \geq 0\}$ is lower semicontinuous.

REMARK 3.9 Let L be a set of functions defined on $\mathbb{R}^n_{++}$ by $l(x) = \min_{i=1,\dots,n} l_i x_i$ with $l \in \mathbb{R}^n_{++}$. It follows from Theorem 3.2 that a function $f : \mathbb{R}^n_{++} \to \mathbb{R}_{+\infty}$ is H_L-convex if and only if this function is lower semicontinuous and ICAR.

Theorem 3.2 asserts that for each lower semicontinuous ICAR function f and each $x \in \mathbb{R}^n_+$, we have

$$f(x) = \sup\{h(x) : h \in \text{supp}(f, H_L)\}. \tag{3.3.11}$$

It is important to describe points x such that the supremum in (3.3.11) is attained. We shall study this question. First we give some remarks on the directional derivative of an ICAR function and its restriction to rays starting from zero.

REMARK 3.10 Let $f : \mathbb{R}^n_+ \to \mathbb{R}$ be a convex-along-rays function. For $y \in \mathbb{R}^n_+$ consider the function f_y defined by (3.3.1). The function f_y is convex on the half line $\{t : t \geq 0\}$. Assume that $t \in \text{int dom} f_y$, that is there exists $\varepsilon > 0$ such that $(t + \varepsilon)y \in \text{dom} f$. Then the subdifferential $\partial f_y(t)$ at the point t is not empty. It is well known (and easy to check) that $\partial f_y(t)$ coincides with the segment $[(f_y)'_-(t), (f_y)'_+(t)]$, where $(f_y)'_-(t)$ and $(f_y)'_+(t)$ are the left and right derivative of the function f_y at the point t respectively. An easy calculation shows that $(f_y)'_-(1) = -f'(y, -y)$ and $(f_y)'_+(1) = f'(y, y)$, where

$$f'(y, u) := \lim_{\alpha \to +0} \frac{1}{\alpha}(f(y + \alpha u) - f(y))$$

is the directional derivative of the function f at the point y in the direction u. Thus $\partial f_y(1) = [f'(y, -y), f'(y, y)]$. It follows from this remark that the directional derivatives $f'(y, -y)$ and $f'(y, y)$ do exist.

THEOREM 3.3 *Let f be an ICAR function and $y \in \mathbb{R}^n_+ \setminus \{0\}$ be a point such that $(1 + \varepsilon)y \in \operatorname{dom} f$ for some $\varepsilon > 0$. Then there exists $h \in \operatorname{supp}(f, H_L)$ such that $h(y) = f(y)$. In particular, $x \mapsto \langle l, x \rangle - c$, where $l = v_1/y$ with $v_1 \in \partial f_y(1)$ and $c = v_1 - f(y)$ possesses this property.*

Proof: Consider a convex function f_y defined on $\mathbb{R}_+$ and let $\tilde{f}_y$ be the extension of this function to the real line defined by (3.3.5). Then $f_y(1 + \varepsilon) = f((1 + \varepsilon)y) < +\infty$. Since f is increasing it follows that $\tilde{f}_y$ is finite on the segment $(-\infty, 1 + \varepsilon]$, so the subdifferential $\partial \tilde{f}_y(1)$ of the convex function $\tilde{f}_y$ at $t = 1$ is nonempty. Let $v_1 \in \partial \tilde{f}_y(1)$. Then

$$v_1 t - v_1 \le \tilde{f}_y(t) - \tilde{f}_y(1) \text{ for all } t \in \mathbb{R}. \qquad (3.3.12)$$

In particular $v_1 t - v_2 \le f_y(t) - f_y(1)$ for all $t \ge 0$. Hence $v_1 \in \partial f_y(1)$. Let $v_2 = v_1 - \tilde{f}_y(1) = v_1 - f(y)$, then $v_1 t - v_2 \le \tilde{f}_y(t)$. Therefore $v = (v_1, v_2) \in V$, where V is the support set of the function $\tilde{f}_y$ with respect to the set of all affine functions. Let $l = v_1/y, c = v_2$. It has been shown in the proof of Proposition 3.7 (see also Remark 3.7) that $(l, c) \in \operatorname{supp}(f, H_L)$, that is, $\langle l, x \rangle - c \le f(x)$ for all x. Since $v \in \partial f_y(1)$ we also have

$$\langle l, y \rangle - c = v_1 \left\langle \frac{1}{y}, y \right\rangle - v_2 = v_1 - (v_1 - f_y(1)) = f_y(1) = f(y).$$

Thus the result follows. △

The following proposition will be useful in the next section.

PROPOSITION 3.8 *Let f be an ICAR function and $y \in \mathbb{R}^n_+ \setminus \{0\}$ be a vector, such that $(1 + \varepsilon)y \in \operatorname{dom} f$ for some $\varepsilon > 0$. Then there exists $l \in \mathbb{R}^n_+$ with the property*

$$\langle l, x \rangle - \langle l, y \rangle \le f(x) - f(y) \textit{ for all } x \in \mathbb{R}^n_+.$$

In particular the vector $l = u/y$ with $u \in \partial f_y(1)$ enjoys this property.

Proof: Let $h(x) = \langle l, x\rangle - c$ where $l = u/y$ with $u \in \partial f_y(1)$ and $c = u - f(y)$. Theorem 3.3 implies that $h(x) \leq f(x)$ for all x and $h(y) = f(y)$. In other words,

$$\langle l, x\rangle - c \leq f(x) \text{ for all } x \in \mathbb{R}^n_+ \quad \text{and } \langle l, y\rangle - c = f(y). \tag{3.3.13}$$

The result immediately follows from (3.3.13). △

3.3.4. Subdifferentiability of ICAR functions

Recall that we consider abstract convexity with respect to the set H_L of all L-affine functions, where L is the set of all min-type functions defined on $\mathbb{R}^n_+$. It has been proved that a function $f : \mathbb{R}^n_+ \to \mathbb{R}_{+\infty}$ is abstract convex with respect to H_L if and only if f is lower semicontinuous and ICAR. In this subsection we shall study the L-subdifferential $\partial_L f(x)$ of an ICAR function f at a point $x \in \operatorname{dom} f$. Recall (see Definition 1.7) that

$$\partial_L f(x) = \{l \in L : \langle l, y\rangle - \langle l, x\rangle \leq f(y) - f(x)\}.$$

THEOREM 3.4 *Let f be an ICAR function and $x \in \mathbb{R}^n_+ \setminus \{0\}$ be a point such that $(1+\varepsilon)x \in \operatorname{dom} f$ for some $\varepsilon > 0$. Then the subdifferential $\partial_L f(x)$ is not empty and*

$$\partial_L f(x) \supset \left\{\frac{u}{x} : u \in \partial f_x(1)\right\}, \tag{3.3.14}$$

where $f_x(t) = f(tx)$.

Proof: The result follows directly from Proposition 3.8. △

Since f is increasing it follows that $0 \in \partial_L f(0)$ therefore, $\partial_L f(0)$ is also nonempty.

Let f be a finite ICAR function. For $x \in \mathbb{R}^n_+$ we consider the following vector

$$f^{\nabla}(x) = \frac{f'(x,x)}{x}. \tag{3.3.15}$$

It follows immediately from Theorem 3.4 that

$$f^{\nabla}(x) \in \partial_L f(x) \text{ for all } x \in \mathbb{R}^n_+. \tag{3.3.16}$$

In the remaining part of this subsection we describe the L-subdifferential of ICAR function f at the point $y = 0$.

PROPOSITION 3.9 *Let f be an ICAR function and $0 \in \operatorname{int} \operatorname{dom} f$ (here int stands for the interior with respect to the topological space $\mathbb{R}^n_+$). Then $\partial_L f(0)$ is not empty and*

$$\partial_L f(0) = \{l \in \mathbb{R}^n_+ : \langle l, x\rangle \leq (f_x)'_+(0) \textit{ for all } x \in \mathbb{R}^n_+\},$$

where $(f_x)'_+$ is the right-hand derivative of the function f_x defined by (3.3.1).

Proof: Since f is increasing it follows that $0 \in \partial_L f(0)$, so $\partial_L f(0) \neq \emptyset$. Applying the equality $\langle l, 0\rangle = 0$, we deduce that $l \in \partial_L f(0)$ if and only if $\langle l, x\rangle \leq f(x) - f(0)$ for all $x \in \mathbb{R}^n_+$, hence $l \in \partial_L f(0)$ if and only if

$$\alpha \cdot \langle l, x\rangle - 0 \cdot \langle l, x\rangle \leq f(\alpha x) - f(0) = f_x(\alpha) - f_x(0)$$

for all $x \in \mathbb{R}^n_+$ and all $\alpha > 0$. Thus

$$\begin{aligned} l \in \partial_L f(0) &\iff (\forall x \in \mathbb{R}^n_+)\ \langle l, x\rangle \in \partial f_x(0) \\ &\iff (\forall x \in \mathbb{R}^n_+)\ \langle l, x\rangle \leq (f_x)'_+(0). \end{aligned}$$

△

COROLLARY 3.3 *Assume there exists a vector $x \in \operatorname{int} \mathbb{R}^n_+$ such that $(f_x)'_+(0) = 0$. Then $\partial_L f(0) = \{0\}$.*

Proof: For $l \in \partial_L f(0)$ with nonempty $I_+(l)$ we have

$$\langle l, x\rangle = \min_{i \in I_+(l)} l_i x_i \leq 0.$$

Since $x_i > 0$, for all i, it follows that there exists $i_0 \in I_+(l)$ such that $l_{i_0} = 0$. We get a contradiction with the definition of $I_+(l)$. Thus $I_+(l) = \emptyset$ and $l = 0$. △

3.3.5. Subdifferentiability of strictly ICAR functions

It has been proved (see Theorem 3.4) that the L-subdifferential $\partial_L f(y)$ of an ICAR function f at a point y contains the set $A(f, y) = \{l : l = u/y, u \subset \partial f_y(1)\}$. For some functions f the set $\partial_L f(y)$ can be very large, much larger than the set $A(f, y)$.

EXAMPLE 3.8 Let $p(x) = \max_{i \in I} x_i$ and $y \equiv \mathbf{1} = (1, 1, \ldots, 1)$. Then (see Example 2.12)

$$\partial_L p(y) \equiv \partial^- p(y) = \{l \in \mathbb{R}^n_+ : \min_{i \in I_+(l)} l_i = 1\}.$$

Thus $\partial_L p(y)$ contains all vectors $l \geq \mathbf{1}$ with $\min_{i \in I} l_i = 1$. We also have $p_y(t) = t$, hence $\partial p_y(1) = \{1\}$ and the set $A(p, y)$ consists of the unique element $1/\mathbf{1} = \mathbf{1}$.

We now demonstrate that for a strictly increasing ICAR function the L-subdifferential, $\partial_L f(y)$, coincides with the set $A(f, y)$ (compare with Proposition 2.31, where this result has been proved for IPH functions).

DEFINITION 3.5 A function f defined on $\mathbb{R}^n_+$ is *strictly increasing* at a point $y \in \mathbb{R}^n_+$ if the inequality $x < y$ (that is, $x \le y$ and $x \ne y$) implies $f(x) < f(y)$.

PROPOSITION 3.10 *Let f be an ICAR function defined on $\mathbb{R}^n_+$. Assume f is strictly increasing at the point y such that $\lambda y \in \operatorname{dom} F$ for some $\lambda > 1$. Then*

$$\partial_L f(y) = \left\{ \frac{u}{y} : u \in \partial f_y(1) \right\},$$

where $f_y(t) = f(ty)$ for $t > 0$.

Proof: The inclusion $(u/y) \in \partial_L f(y)$ with $u \in \partial f_y(1)$ has already been proved (see Theorem 3.4).

Conversely, take $l \in \partial_L f(y)$ and let $u = \langle l, y \rangle$. For $x = \alpha y$ with $\alpha > 0$, we have

$$f_y(\alpha) - f_y(1) = f(x) - f(y) \ge \langle l, x \rangle - \langle l, y \rangle = \langle l, y \rangle(\alpha - 1).$$

Thus $u = \langle l, y \rangle \in \partial f_y(1) = [(f_y)'_-(1), (f_y)'_+(1)]$. Since f is strictly increasing at the point y, it follows that the convex function f_y is strictly increasing at the point $t = 1$. Hence the left-hand derivative $(f_y)'_-(1)$ of this function is positive and $u \ge (f_y)'_-(1) > 0$. We now check that

$$I_+(l) \supset I_+(y). \tag{3.3.17}$$

If (3.3.17) does not hold, then there is an index $j \notin I_+(l)$ such that $y_j > 0$. Take a positive $\varepsilon < y_j$ and consider the vector z:

$$z = \begin{cases} y_i & \text{if } i \ne j; \\ y_j - \varepsilon & \text{if } i = j. \end{cases}$$

The function f is strictly increasing at the point y, so the inequality $z < y$ implies that $f(z) - f(y) < 0$. Since $j \notin I_+(l)$, we have:

$$\langle l, z \rangle = \min_{i \in I_+(l)} l_i z_i = \min_{i \in I_+(l)} l_i y_i = \langle l, y \rangle,$$

and since $l \in \partial_L f(y)$,

$$0 = \langle l, z \rangle - \langle l, y \rangle \le f(z) - f(y).$$

Thus we have a contradiction which shows that (3.3.17) is valid.

Now we check that $I_+(l) = I_+(y)$. If this equality does not hold then by (3.3.17) we have that $\langle l, y \rangle = \langle l, \alpha y \rangle = 0$ for all $\alpha > 0$. Thus for all $\alpha > 0$

$$0 = \langle l, \alpha y \rangle - \langle l, y \rangle \le f(\alpha y) - f(y).$$

Hence $f(\alpha y) \geq f(y)$. Since f is strictly increasing at the point y, it follows that $f(\alpha y) < f(y)$ for $\alpha < 1$. Thus equality must hold in (3.3.17).

We now check that $l_j = u/y_j$ for all $j \in I_+(y)$. The inequality $l_j < u/y_j$ for an index $j \in I_+(y)$ leads to $\langle l, y\rangle < u$, which contradicts the definition of u. Thus $l_j \geq u/y_j$ for all $j \in I_+(y)$. Assume there exists an index $j \in I_+(y)$ such that $l_j > u/y_j$. Then $l_j y_j > u = \langle l, y\rangle$. Let $0 < \varepsilon < l_j - (u/y_j)$. Take a vector z such that

$$z_i = \begin{cases} y_i & \text{if } i \neq j; \\ y_j - \varepsilon & \text{if } i = j. \end{cases}$$

Since f is strictly increasing, $f(z) < f(y)$. Also, $\langle l, y\rangle = \langle l, z\rangle$ and since $l \in \partial_L f(y)$, we have

$$f(z) \geq \langle l, z\rangle - (\langle l, y\rangle - f(y)) = f(y),$$

which contradicts the inequality $f(z) < f(y)$. Thus $l_j = u/y_j$ for all $j \in I_+(l)$. $\triangle$

DEFINITION 3.6 A function $f : \mathbb{R}^n_+ \to \mathbb{R}$ is called *differentiable-along-rays* if the function f_y defined by (3.3.1) is differentiable for each $y \in \mathbb{R}^n_+ \setminus \{0\}$.

PROPOSITION 3.11 *Let f be a strictly ICAR differentiable-along-rays function. Then for each $y \neq 0$ the L-subdifferential $\partial_L f(y)$ consists of the unique element $f^{\nabla}(y)$ defined by (3.3.15).*

Proof: The result follows directly from Proposition 3.10.

REMARK 3.11 We now compare Proposition 3.11 with the corresponding results for "usual" convex functions. Let L_1 be a set of all linear functions defined on $\mathbb{R}^n$ and f be a convex function (that is, abstract convex with respect to the set of all affine functions). The subdifferential $\partial f(x)$ consists of the unique point $l \in L_0$ if and only if f is differentiable at the point x and $l = \nabla f(x)$. From this point of view the unique element $f^{\nabla}(x)$ of the L-subdifferential $\partial_L f(x)$ of a strictly ICAR differentiable-along-rays function is similar to the unique element $\nabla f(x)$ of the subdifferential of a differentiable convex function. Thus we can consider strictly ICAR differentiable-along-rays functions as a certain analogue of differentiable convex functions.

3.3.6. Lipschitz function and ICAR functions

As it was mentioned in Subsection 3.3.2 the class of ICAR functions is very broad. It follows from Proposition 3.6 that each lower semicontinuous function f defined on the unit simplex can be extended to an

ICAR function $g : \mathbb{R}^n_+ \to \mathbb{R}_{+\infty}$. In this subsection we show that each positive Lipschitz function defined on the unit simplex can be extended to a finite ICAR function, more precisely to an increasing positively homogeneous function of degree $p \geq 1$. Recall that a function f defined on a subset X of a Banach space is called Lipschitz if there exists a constant K' such that $|f(x) - f(y)| \leq K'\|x - y\|$ for all $x, y \in X$. Consider the least number K which possesses this property. Clearly

$$K = \sup_{x,y\in X, x\neq y} \frac{f(x) - f(y)}{\|x - y\|}.$$

Let $S = \{y \in \mathbb{R}^n_+ : \sum_{i=1}^n y_i = 1\}$ be the unit simplex and $y \in S$. We assume that the space $\mathbb{R}^n$ is endowed with the ℓ_1-norm $\|x\| = \sum_{i=1}^n |x_i|$.

Let f be a Lipschitz function defined on S. Consider the function g defined on the cone $\mathbb{R}^n_+$ by

$$g(x) = \begin{cases} f\left(\frac{x}{\sum_{i=1}^n x_i}\right) (\sum_{i=1}^n x_i)^p & \text{if } x \neq 0; \\ 0 & \text{if } x = 0 \end{cases} \tag{3.3.18}$$

where $p \geq 0$. Let us indicate some properties of the function g :

1) $g(y) = f(y)$ for all $y \in S$.

2) $g(\alpha x) = \alpha^p g(x)$ for all $x \in \mathbb{R}^n_+$ and $\alpha > 0$.

Let $p \geq 1$. It follows from 2) that g is convex-along-rays. Thus g is a convex-along-rays extension of the function f.

THEOREM 3.5 *Let f be a Lipschitz positive function, defined on the simplex S and*

$$K = \sup_{\substack{x\neq y \\ x,y\in S}} \frac{|f(x) - f(y)|}{\|x - y\|}$$

be the Lipschitz constant of f. Let $c = \min_{x\in S} f(x) > 0$ and

$$p \geq \frac{2K}{c}.$$

Then the function $g(x)$ is increasing. If

$$p > \frac{2K}{c}$$

then g is strictly increasing for all $x \in \mathbb{R}^n_+ \setminus \{0\}$.

Proof: Let $x, z \in \mathbb{R}^n_+$ and $z > x$, in other words, $z \geq x$ and $z \neq x$.

1) First we assume that $x \in S$, that is, $\|x\| = 1$. Then $\|z\| := \lambda > 1$ so the point $y = (1/\lambda)z$ belongs to S. We have

$$\begin{aligned} f(x) &\leq f(y) + K\|x-y\| = f(y)\left(1 + \frac{K\|x-y\|}{f(y)}\right) \\ &\leq f(y)\left(1 + \frac{K\|x-y\|}{c}\right). \end{aligned} \tag{3.3.19}$$

Note that

$$\|y - z\| = \|y - \lambda y\| = |1-\lambda|\|y\| = \lambda - 1$$

and

$$\|z - x\| = \sum_i |z_i - x_i| = \sum_i (z_i - x_i) = \sum_i z_i - \sum_i x_i = \lambda - 1.$$

Hence

$$\|y - x\| \leq \|y - z\| + \|z - x\| = 2(\lambda - 1). \tag{3.3.20}$$

Also

$$g(x) = f(x), \quad g(z) = g(\lambda y) = \lambda^p g(y) = \lambda^p f(y). \tag{3.3.21}$$

Combining (3.3.19),(3.3.20) and (3.3.21), we conclude that

$$g(x) \leq \frac{1}{\lambda^p} g(z)\left(1 + \frac{2K(\lambda - 1)}{c}\right). \tag{3.3.22}$$

Consider the functions $s(\lambda) = 1 + (2K(\lambda-1))/c$ and $t(\lambda) = \lambda^p$. We have $t(1) = s(1) = 1$. Since $p \geq (2K)/c$, it follows for $\lambda \geq 1$:

$$\frac{dt}{d\lambda} = p\lambda^{p-1} \geq \frac{2K}{c} = \frac{ds}{d\lambda}; \qquad t(1) = s(1) = 1.$$

Hence $t(\lambda) \geq s(\lambda)$ for $\lambda > 1$. Applying (3.3.22) we deduce that

$$g(x) \leq \frac{1}{t(\lambda)} s(\lambda) g(z) \leq g(z).$$

If $p > (2K)/c$ then $t(\lambda) > s(\lambda)$ for all $\lambda > 1$, hence $g(x) < g(z)$.

2) Consider now an arbitrary point $x \in \mathbb{R}^n_+$ and let $z > x$. If $x = 0$ then $g(z) \geq 0 = g(0)$, so we can assume that $x \neq 0$. Let $x' = x/\|x\|$, $z' = z/\|x\|$. Then $x' \in S$ and $x' \leq z'$, hence $g(x') \leq g(z')$. We have

$$g(x) = g(\|x\|x') = \|x\|^p g(x') \leq \|x\|^p g(z') = g(z).$$

If $p > (2K)/c$ then $g(x') < g(z')$, hence $g(x) < g(z)$. The proof is completed. $\triangle$

REMARK 3.12 Theorem 3.5 was established in [148]. The proof presented here was suggested by Hoang Tuy.

COROLLARY 3.4 *If* $p \geq \max\left\{1, \dfrac{2K}{c}\right\}$ *with* $c = \min_{y \in S} f(y)$ *then* g *is ICAR.*

Let $g_y(\alpha) = g(\alpha y)$ for $\alpha > 0$ $(y \in S)$. Since

$$g_y(\alpha) = \alpha^p g_y(1) = \alpha^p g(y) = \alpha^p f(y),$$

g_y is differentiable at the point $\alpha = 1$ and

$$g'_y(1) = g'(y, y) = pf(y).$$

Let $p > 2K/c$. Applying Proposition 3.10 and Theorem 3.5, we can deduce that the L-subdifferential of the function g at a point $y \in S$ consists of the unique element $g^{\nabla}(y) = pf(y)/y$.

3.4. DECREASING FUNCTIONS

3.4.1. Overview

In this section we apply the theory of IPH functions to study decreasing functions defined on the positive orthant $\mathbb{R}^n_{++}$. Our approach is based on the following simple observation. A positive part of the hypograph of a decreasing upper semicontinuous function defined on $\mathbb{R}^n_{++}$ is a closed normal subset of $\mathbb{R}^n_{++} \times \mathbb{R}_{++}$. In turn, each closed normal subset of $\mathbb{R}^n_{++} \times \mathbb{R}_{++}$ is a support set of an IPH function defined on $\mathbb{R}^n_{++} \times \mathbb{R}_{++}$. We also introduce and study an operation of multiplicative inf-convolution. We demonstrate that the set of upper semicontinuous decreasing functions is a natural domain for this operation (like the set of convex functions is a natural domain for the usual inf-convolution). The results presented in this section can be found in [157].

3.4.2. Decreasing functions and IPH functions

Let $I = \{1, \dots, n\}$, $I' = \{0\} \cup I$ and let U be a normal closed subset of $\mathbb{R}^{I'}_{++}$. Consider the function

$$g_U(y) = \sup\{\alpha : \ (\alpha, y) \in U\}, \qquad y \in \mathbb{R}^I_{++}. \tag{3.4.1}$$

This function maps $\mathbb{R}^I_{++}$ into $\mathbb{R}_{+\infty} = \mathbb{R}_+ \cup \{+\infty\}$. Since U is closed, it follows that $g_U(y) = \max\{\alpha : \ (\alpha, y) \in U\}$ whenever $g_U(y) < +\infty$. By the normality of the set U we obtain:

$$\begin{aligned} U &= \{(\alpha, y) : y \in \operatorname{dom} g_U,\ 0 < \alpha \le g_U(y)\} \\ &\cup\ \{(\alpha, y) : y \notin \operatorname{dom} g_U,\ 0 < \alpha < +\infty\} \end{aligned} \tag{3.4.2}$$

where $\operatorname{dom} f = \{y : f(y) < +\infty\}$.

Let g be a function defined on $\mathbb{R}^I_{++}$. Consider the hypograph, $\operatorname{hyp} g$, of the function g:

$$\operatorname{hyp} g = \{(\alpha, y) \in \mathbb{R} \times \mathbb{R}^I_{++} : \alpha \le g(y) \text{ for all } y \in \mathbb{R}^n_{++}\}.$$

DEFINITION 3.7 Let g be a nonnegative function defined on $\mathbb{R}^I_{++}$. The set $\operatorname{hyp}^+ g = \mathbb{R}^{I'}_{++} \cap \operatorname{hyp} g$ is called the *positive part of the hypograph* of g.

It follows from Definition 3.7 that

$$\operatorname{hyp}^+ g = \{(\alpha, y) \in \mathbb{R} \times \mathbb{R}^I_{++} : 0 < \alpha \le g(y) \text{ for all } y \in \operatorname{dom}' g\},$$

where $\operatorname{dom}' g = \{y : g(y) > 0\}$.

Let U be a normal closed subset of $\mathbb{R}^{I'}_{++}$. It is easy to see that U coincides with the positive part $\operatorname{hyp}^+ g_U$ of the function g_U.

PROPOSITION 3.12 *For a closed normal subset U of $\mathbb{R}^{I'}_{++}$, the function g_U is decreasing (that is, $y_1 \ge y_2 \implies g_U(y_1) \le g_U(y_2)$), nonnegative and upper semicontinuous.*

Proof: Let $y_1 \ge y_2$ and $(\alpha, y_1) \in U$. Since U is normal and $(\alpha, y_1) \ge (\alpha, y_2)$ we conclude that $(\alpha, y_2) \in U$. Therefore, $g_U(y_1) \le g_U(y_2)$. Thus g_U is decreasing. It is clear that $g(y) \ge 0$ for all y. Let $y_k \to y$. First assume that there exists a subsequence k_s such that $g_U(y_{k_s}) = +\infty$. Thus $(\alpha, y_{k_s}) \in U$ for all $\alpha > 0$ and therefore, $(\alpha, y) \in U$ for all $\alpha > 0$. So $g_U(y) = +\infty \ge \limsup_k g_U(y_k)$. Assume now that $g_U(y_k) < +\infty$ for all k. Then $(g_U(y_k), y_k) \in U$. If $\lambda := \limsup_k g_U(y_k) < +\infty$, then $(\lambda, y) \in U$ and hence $\lambda \le g_U(y)$. If $\lambda = +\infty$ then $g_U(y) = +\infty$. Thus $\limsup_k g_U(y_k) \le g_U(y)$ in both cases, so g_U is upper semicontinuous. $\triangle$

PROPOSITION 3.13 *Let g be a nonnegative decreasing and upper semicontinuous function. Furthermore, let*

$$U = \{(\alpha, y) :\ y \gg 0,\ 0 < \alpha \le g(y) \text{ for all } y \in \operatorname{dom}' g\}.$$

Then U is a normal closed set and $g = g_U$.

Proof: We first show that U is normal. Let $(\alpha_1, y_1) \in U$, $\alpha_2 > 0$, $y_2 \gg 0$ and $(\alpha_1, y_1) \geq (\alpha_2, y_2)$. Since g is decreasing, we have $\alpha_2 \leq \alpha_1 \leq g(y_1) \leq g(y_2)$. Thus $(\alpha_2, y_2) \in U$. Since g is upper semicontinuous, U is closed. We also have that

$$g_U(y) = \sup\{\alpha : (\alpha, y) \in U\} = \sup\{\alpha : \alpha \leq g(y)\} = g(y) \quad (y \in \operatorname{dom} g).$$

$\triangle$

REMARK 3.13 It follows from Proposition 3.12 and Proposition 3.13 that the mapping $U \mapsto g_U$ is a one-to-one correspondence between the totality of all normal closed subsets of $\mathbb{R}^{I'}_{++}$ and the totality of all decreasing nonnegative upper semicontinuous functions defined on $\mathbb{R}^{I}_{++}$. If U is the empty subset of $\mathbb{R}^{I'}_{++}$, then $g_U = 0$. (This follows from (1.2.3).) If $U = \mathbb{R}^{I}_{++}$ then $g_U(y) = +\infty$ for all $y \in \mathbb{R}^{I}_{++}$.

Recall that $I' = \{0\} \cup I$. Consider an IPH function p defined on the cone $\mathbb{R}^{I'}_{++}$. Let $U = \operatorname{supp}(p, L)$ be the support set of the function p with respect to the set L of all min-type functions. Then the set U generates the function g_U (see (3.4.1) for the definition of g_U).

DEFINITION 3.8 Let p be an IPH function defined on $\mathbb{R}^{I}_{++}$ and $U = \operatorname{supp}(p, L)$. Then the function g_U defined by (3.4.1) is called the ***associated function*** to p. We shall denote the associated function to p by h_p.

Consider the set $\mathcal{F}_{I'}(IPH)$ of all IPH functions defined on $\mathbb{R}^{I'}_{++}$ and the set $\mathcal{N}_{I'}$ of all normal closed subsets of $\mathbb{R}^{I'}_{++}$. Let L be the set of all min-type functions $x \mapsto \langle l, x\rangle$ $(x \in \mathbb{R}^{I'}_{++})$ with $l \in \mathbb{R}^{I'}_{++}$, where

$$\langle l, x\rangle = \min_{i \in I'} l_i x_i.$$

It is known (see Proposition 2.5) that $\mathcal{F}_{I'}(IPH)$ coincides with the set of all L-convex functions, and (see Proposition 2.6 and Remark 2.2) $\mathcal{N}_{I'}$ coincides with the set of all L-convex subsets of $\mathbb{R}^{I'}_{++}$. The mapping $p \mapsto \operatorname{supp}(p, L)$ (the Minkowski duality) is a one-to-one correspondence between $\mathcal{F}_{I'}(IPH)$ and $\mathcal{N}_{I'}$ (see Subsection 7.4.2 for details). As it was mentioned in Remark 3.13, the mapping $U \mapsto g_U$ is a one-to-one correspondence between $\mathcal{N}_{I'}$ and the set D_I of all nonnegative decreasing upper semicontinuous functions defined on $\mathbb{R}^{I}_{++}$. Hence the mapping $p \mapsto h_p$ is a one-to-one correspondence between the set $\mathcal{N}_{I'}$ and D_I.

We now calculate associated functions for some IPH functions.

EXAMPLE 3.9 Let

$$p(\delta, y) = \max\{\alpha\delta, a_1 y_1, \dots, a_n y_n\}$$

with $\alpha > 0$, $a = (a_1, \dots, a_n) \in \mathbb{R}^n_{++}$. Then (see Example 2.8) $U :=$ supp(p, L) coincides with the set $V_{(\alpha,a)}$ defined as follows

$$V_{(\alpha,a)} = \{y \in \mathbb{R}^{I'}_{++} : \min\left(\frac{y_0}{\alpha}, \min_{i \in I} \frac{y_i}{a_i}\right) \leq 1\},$$

where $I_a = \{i \in I : a_i > 0\}$. If $y \in \mathbb{R}^I_{++}$ is a vector such that $y_i \leq a_i$ for an index $i \in I$, then for each $\delta > 0$ we have $\min(\delta/\alpha, \min_{i \in I} y_i/a_i) \leq 1$. Thus $(\delta, y) \in U$ for all $\delta > 0$, so $h_p(y) = g_U(y) = +\infty$. Now let $y_i > a_i$ for all $i \in I$. Since $(\delta, y) \in U$ if and only if

$$\min\left(\frac{\delta}{\alpha}, \min_{i \in I} \frac{y_i}{a_i}\right) \leq 1,$$

it follows that $(\delta, y) \in U$ if and only if $\delta \leq \alpha$. So $h_p(y) = \alpha$. Thus

$$h_p(y) = \begin{cases} \alpha & \text{if } y_i > a_i \text{ for all } i \in I; \\ +\infty & \text{otherwise}. \end{cases}$$

Assume now that $\alpha = 0$. Example 2.8 demonstrates that the set $U :=$ supp(p, L) coincides with the set

$$V_{(0,\alpha)} = \{y \in \mathbb{R}^{I'}_{++} : \min_{i \in I} \frac{l_i}{a_i} \leq 1\}.$$

If there exists $i \in I$ such that $y_i/a_i \leq 1$, then we have again $g_U(y) = +\infty$. If $y_i > a_i$ for all $i \in I$, then the set $\{\delta : (\delta, y) \in V_{(0,\alpha)}\}$ is empty, so $g_U(y) = \sup\{\delta : (\delta, y) \in V_{(0,\alpha)}\} = 0$. Thus g_U coincides with the indicator function δ_Z of the set $Z = \{y : y_i > a_i \ (i \in I_a)\}$.

EXAMPLE 3.10 Let $0 < k < +\infty$ and $p_k(\delta, y) = (\delta^k + \sum_{i \in I} y_i^k)^{\frac{1}{k}}$. Let $U = $ supp(p, L). Furthermore, let

$$u = \left(1 - \sum_{i \in I} \frac{1}{y_i^k}\right)^{\frac{1}{k}}.$$

Then (see Example 2.9) we have:

$$U = \left\{(\alpha, y) \in \mathbb{R}^{I'}_{++} : \frac{1}{\alpha^k} + \sum_{i \in I} \frac{1}{y_i^k} \geq 1\right\}.$$

So,

$$h_p(y) = \sup\left\{\alpha : \frac{1}{\alpha^k} \geq 1 - \sum_{i \in I} \frac{1}{y_i^k}\right\}$$

$$= \begin{cases} \frac{1}{u} & \text{if } 1 > \sum_{i\in I} \frac{1}{y_i^k}; \\ +\infty & \text{otherwise.} \end{cases}$$

Let p be an IPH function defined on $\mathbb{R}^{I'}_{++}$. We now show that

$$\sup_{y\gg 0} p(1,y) = \sup_{y\gg 0} h_p(y).$$

We need the following simple assertion.

LEMMA 3.2 *Let $\psi(\lambda)$ be a continuous decreasing function defined on the segment $(0,+\infty)$ and let*

$$\sup_{\lambda>0} \psi(\lambda) = \lim_{\lambda\to+0} \psi(\lambda) = \bar{M} > 0.$$

Let $\chi_b(\lambda) = \min\{\psi(\lambda), b\lambda\}$ for $b > 0$. Then

1) The equation $\psi(\lambda) = b\lambda$ has a unique solution λ_b;

2) $\chi_b(\lambda_b) = \max_{\lambda>0} \chi_b(\lambda)$ and λ_b is the least maximizer of the function χ_b;

3) $\lambda_b \to 0$ as $b \to +\infty$;

4) $\chi_b(\lambda_b) = \psi(\lambda_b) = b\lambda_b \to \bar{M}$ as $b \to +\infty$.

Proof: Consider the function $\phi(\lambda) = \psi(\lambda)/\lambda$. Since $\lim_{\lambda\to+0} \psi(\lambda) > 0$, it follows that $\lim_{\lambda\to 0} \phi(\lambda) = +\infty$. We also have $\lim_{\lambda\to+\infty} \phi(\lambda) = 0$. Hence for each $b > 0$ there exists a solution λ_b of the equation $\phi(\lambda) = b$. Clearly $\psi(\lambda_b) = \lambda_b b$. We have

$$\lambda < \lambda_b \implies \psi(\lambda) \geq \psi(\lambda_b) = \lambda_b b > \lambda b, \tag{3.4.3}$$

$$\lambda < \lambda_b \implies \psi(\lambda) \leq \psi(\lambda_b) = \lambda_b b < \lambda b. \tag{3.4.4}$$

If $\lambda < \lambda_b$ then applying (3.4.3) we have

$$\chi_b(\lambda) = \min(\psi(\lambda), \lambda b) = \lambda b < \lambda_b b = \chi_b(\lambda_b).$$

If $\lambda > \lambda_b$ then, it follows from (3.4.4) that

$$\chi_b(\lambda) = \psi(\lambda) \leq \lambda_b b = \chi(\lambda_b).$$

Thus λ_b is the least maximizer of the function ψ_b. Since

$$\lim_{b\to+\infty} \lambda_b = \lim_{b\to+\infty} \frac{\psi(\lambda_b)}{b} = 0$$

we have

$$\lim_{b\to+\infty} \chi_b(\lambda_b) = \lim_{b\to+\infty} \psi(\lambda_b) = \lim_{\lambda\to+0} \psi(\lambda) = \bar{M}.$$

△

PROPOSITION 3.14 *Let p be an IPH function defined on $\mathbb{R}^{I'}_{++}$. Then*

$$\sup_{y\gg 0} p(1,y) = \sup_{y\gg 0} h_p(y).$$

Proof: First we shall verify that

$$p(1,y) = \sup_{z\gg 0} \min\{h_p(z), \langle z,y\rangle\} \text{ for all } y \in \mathbb{R}^I_{++}. \tag{3.4.5}$$

Indeed, it follows, from the definition of the associated function h_p, that $\text{supp}(p,L) = \{(\delta,z) : z \gg 0,\ 0 < \delta \le h_p(z)\}$. So for $y \gg 0$ we have:

$$\begin{aligned} p(1,y) &= \sup\{\langle(\delta,z),(1,y)\rangle : (\delta,z) \in \text{supp}(p,L)\} \\ &= \sup_{z\gg 0,\, 0<\delta\le h_p(z)} \min(\delta, \langle z,y\rangle). \end{aligned}$$

Thus (3.4.5) holds. It follows from (3.4.5) that for an arbitrary $y \gg 0$ and $\varepsilon > 0$, there exists a vector $z \gg 0$ such that

$$p(1,y) - \varepsilon \le \min(h_p(z), \langle z,y\rangle) \le h_p(z) \le \sup_{u\gg 0} h_p(u).$$

Thus

$$\sup_{y\gg 0} p(1,y) \le \sup_{u\gg 0} h_p(u).$$

We now verify the reverse inequality. Fix a vector $z \gg 0$ and consider the ray $\{\lambda z : \lambda > 0\}$. Let $\psi(\lambda) \equiv \psi_z(\lambda) = h_p(\lambda z)$. The function ψ is decreasing and

$$\lim_{\lambda\to+0} \psi(\lambda) = \lim_{\lambda\to+0} h_p(\lambda z) = \lim_{y\gg 0, y\to 0} h_p(y) = \sup_{y\gg 0} h_p(y). \tag{3.4.6}$$

For $y \gg 0$ consider the function $\chi_b(\lambda) = \min\{\psi(\lambda), b_y\lambda\}$ where $b_y = \langle z,y\rangle$. Let λ_y be a solution of the equation $\psi(\lambda) = b_y\lambda$. Lemma 3.2 implies that $\max_{\lambda>0} \min\{\psi(\lambda), b_y\lambda\}$ is attained at the point λ_y and is equal to $\psi(\lambda_y)$. It follows from (3.4.5) that

$$\begin{aligned} p(1,y) &= \sup_{u\gg 0} \min(h_p(u), \langle u,y\rangle) \\ &\ge \sup_{\lambda>0} \min(h_p(\lambda z), \langle \lambda z,y\rangle) \\ &= \sup_{\lambda>0} \min(\psi_z(\lambda), \lambda b_y) = \psi(\lambda_y). \end{aligned} \tag{3.4.7}$$

Note that $b_y \to +\infty$ as $\min_i y_i \to +\infty$, so (see Lemma 3.2) $\lambda_y \to 0$ and $\psi(\lambda_y) \to \sup_{\lambda>0} \psi(\lambda)$ as $\min_i y_i \to +\infty$. Applying (3.4.6) and (3.4.7) we have:

$$\sup_{y \gg 0} p(1, y) \geq \sup_{y \gg 0} \psi(\lambda_y) = \lim_{\lambda \to +0} \psi(\lambda) = \lim_{\lambda \to +0} h_p(\lambda z) = \sup_{u \gg 0} h_p(u). \quad \triangle$$

We now show that the associated function h_p can be expressed in terms of the initial function p.

PROPOSITION 3.15 *Assume that p is an IPH function defined on $\mathbb{R}^{I'}_{++}$. Let $z \in \mathbb{R}^I_{++}$. Then the following statements hold:*

1) *If $\lim_{\tau \to +0} p(\tau, z^{-1}) \geq 1$, then $h_p(z) = +\infty$.*

2) *If $\lim_{\tau \to +\infty} p(\tau, z^{-1}) < 1$ or $\lim_{\tau \to +\infty} p(\tau, z^{-1}) = 1 > p(\tau, z^{-1})$ for all τ, then $h_p(z) = 0$.*

3) *Otherwise (if $\lim_{\tau \to +0} p(\tau, z^{-1}) < 1$ and either $\lim_{\tau \to +\infty} p(\tau, z^{-1}) > 1$ or $\lim_{\tau \to +\infty} p(\tau, z^{-1}) = 1$ and there is $\tau > 0$ such that $p(\tau, z^{-1}) = 1$),*

$$h_p(z) = \frac{1}{b(\frac{1}{z})},$$

where $b(z)$ is the least solution of the equation $p(b, z) = 1$.

Proof: It follows from Theorem 2.3 that

$$\text{supp}(p, L) = \{(\alpha, y) \in \mathbb{R}^{I'}_{++} : p\left(\frac{1}{\alpha}, \frac{1}{y}\right) \geq 1\}.$$

Let $z \in \mathbb{R}^I_{++}$ and $y = z^{-1}$. Then

$$\begin{aligned} h_p(z) = h_p\left(\frac{1}{y}\right) &= \sup\{\alpha : \left(\alpha, \frac{1}{y}\right) \in \text{supp}(p, L)\} \\ &= \sup\{\alpha : p\left(\frac{1}{\alpha}, y\right) \geq 1\} \\ &= \sup\{\tau^{-1} : p(\tau, y) \geq 1\} \\ &= \frac{1}{\inf\{\tau : p(\tau, y) \geq 1\}}. \end{aligned}$$

Thus for $\psi_y(\tau) = p(\tau, y)$ we have

$$h_p(z) = \frac{1}{\inf\{\tau : \psi_y(\tau) \geq 1\}}. \tag{3.4.8}$$

It follows, from the properties of the function p, that ψ_y is an increasing continuous function on $(0, +\infty)$. Let

$$\gamma_- = \lim_{\tau\to+0} \psi_y(\tau) \text{ and } \gamma_+ = \lim_{\tau\to+\infty} \psi_y(\tau).$$

If $\gamma_- \geq 1$ then $\inf\{\tau : \psi_y(\tau) \geq 1\} = 0$; if either $\gamma_+ < 1$ or $\gamma_+ = 1$, but there exists no τ such that $\psi_y(\tau) = 1$, then the set $\{\tau : \psi_y(\tau) \geq 1\}$ is empty and so the infimum of this set is defined as $+\infty$. If $\gamma_- < 1 < g_+$ or $\gamma_+ = 1$ and there exists τ such that $\psi(\tau) = 1$, then the equation $\psi_y(\tau) = 1$ has a solution and $\inf\{\tau : \psi_y(\tau) \geq 1\}$ is equal to the smallest root of this equation. Thus $\inf\{\tau : \psi_y(\tau) \geq 1\} = \inf\{\tau : p(\tau, y) = 1\}$. Denote the last infimum by $b(y)$. Recall that $y = z^{-1}$. Applying (3.4.8) we conclude that

$$h_p(z) = \frac{1}{b(y)} = \frac{1}{b\left(\frac{1}{z}\right)}.$$

△

COROLLARY 3.5 *If* $0 < h_p(z) < +\infty$, *then*

$$h_p(z) = \frac{1}{b(y)},$$

where

$$p(b(y), y) = 1 \text{ with } y = \frac{1}{z}. \qquad (3.4.9)$$

REMARK 3.14 Subsequently we shall consider limits of the following form: $\lim_{\|z\|\to+\infty} h(z)$ where h is a function defined on $\mathbb{R}^n_{++}$. Clearly we can represent this limit as $\lim_{\max_i z_i\to+\infty} h(z)$. Assume that $z - 1/y$. Then

$$\lim_{\|z\|\to+\infty} h(z) \equiv \lim_{\max_i z_i\to+\infty} h(z) = \lim_{\min_i y_i\to 0} h\left(\frac{1}{y}\right)$$

and

$$\lim_{\|z\|\to 0} h(z) = \lim_{\max_i z_i\to 0} h(z) = \lim_{\min_i y_i\to+\infty} h\left(\frac{1}{y}\right).$$

Proposition 3.15 allows us to describe some properties of IPH functions in terms of associated functions.

Let p be an IPH function defined on the cone $\mathbb{R}^{I'}_+$. We shall use the strictly positive support set, $\text{supp}_1(p, L)$, of the function p (see Definition 2.12). Recall that $\text{supp}_1(p, L)$ coincides with the support set of the restriction of p to $\mathbb{R}^{I'}_{++}$. We will denote this restriction by the same letter p.

The following propositions will be useful in the next chapter.

Proposition 3.16 *Let p be a continuous IPH function defined on the cone $\mathbb{R}^{I'}_{+} = \mathbb{R}_{+} \times \mathbb{R}^{I}_{+}$. Then $\lim_{\min_i z_i \to +\infty} h_p(z) = 1$ if and only if $p(1,0) = 1$.*

Proof: Let $\lim_{\min_i z_i \to +\infty} h_p(z) = 1$. Then $0 < h_p(z) < +\infty$ for large enough $z \gg 0$. It follows from Corollary 3.5 that $h_p(z) = (b(y))^{-1}$, where $p(b(y), y) = 1$ and $y = z^{-1}$. Since p is continuous we conclude that

$$\begin{aligned} 1 &= \lim_{\max_i y_i \to 0} p(b(y), y) = \lim_{\min_i z_i \to +\infty} p\left(b\left(\frac{1}{z}\right), \frac{1}{z}\right) \\ &= \lim_{\min_i z_i \to +\infty} p\left(\frac{1}{h_p(z)}, \frac{1}{z}\right) = p(1,0). \end{aligned} \tag{3.4.10}$$

Now assume that $p(1,0) = 1$. Since p is continuous, we have for an arbitrary vector $z \gg 0$:

$$\lim_{\tau \to +\infty} p(\tau, z^{-1}) = \lim_{\tau \to +\infty} \tau p(1, z^{-1}/\tau) = +\infty.$$

Assume that $\lim_{\tau \to +0} p(\tau, z^{-1}) \geq 1$ for all $z \gg 0$. Then $p(0, y) \geq 1$ for all $y \gg 0$ which contradicts continuity of p. Hence there exists $\bar{z} \gg 0$ such that $\lim_{\tau \to +0} p(\tau, \bar{z}^{-1}) < 1$. Let Z be the set of points $z \in \mathbb{R}^n_{++}$ such that

$$\lim_{\tau \to +\infty} p(\tau, z^{-1}) = +\infty, \qquad \lim_{\tau \to +0} p(\tau, z^{-1}) < 1. \tag{3.4.11}$$

Then $\bar{z} \in Z$ so Z is nonempty. Since p is an increasing function and the first equality in (3.4.11) holds for all $z \gg 0$, we conclude that Z is conormal, that is, $(z \in Z, z' \geq z) \implies z \in Z$. Applying Proposition 3.15, we conclude that $h_p(z)$ is defined for vectors $z \in Z$ by (3.4.9). Since Z is conormal,

$$\lim_{\min_i z_i \to +\infty} h_p(z) = \lim_{z \in Z, \min_i z_i \to +\infty} h_p(z).$$

Let $z \in Z$ and $t(z) = \{\alpha : p(\alpha, z) = 1\}$. Since p is positively homogeneous and $p(1,0) = 1$, it follows that $t(0) = \{1\}$. By the continuity of p we have:

$$(\alpha \to \alpha', \quad z \to 0, \quad \alpha \in t(z)) \implies \alpha' = 1.$$

Since $b(z^{-1}) \in t(z^{-1})$, we infer that $h_p(z) = (b(z^{-1}))^{-1} \to 1$ as $\min_i z_i \to +\infty$. $\triangle$

PROPOSITION 3.17 *Let p be a continuous IPH function defined on $\mathbb{R}^{I'}_{+}$ with $I' = \{0\} \cup I$ and*

$$\lim_{\|u\| \to +\infty} p(1, u) = +\infty. \qquad (3.4.12)$$

Then $\lim_{\min_i z_i \to 0} h_p(z) = +\infty$.

Proof: It follows from (3.4.12) that the support set $\text{supp}_1(p, L)$ contains vectors (α, u) with $\alpha > 0$, so there exists $\bar{z} \gg 0$ such that $h_p(\bar{z}) > 0$. Since h_p is decreasing, it follows that $h_p(z) > 0$ for all strictly positive $z \leq \bar{z}$. We shall consider only such vectors z. It is sufficient to consider only vectors $z \gg 0$ such that $h_p(z) < +\infty$. It follows from Corollary 3.5 that for such vectors

$$h_p\left(\frac{1}{y}\right) = \frac{1}{b(y)},$$

where $p(b(y), y) = 1$. We now show that

$$(p(b(y), y) = 1, \ \|y\| \to +\infty) \implies b(y) \to 0. \qquad (3.4.13)$$

If $b(y) \geq 1$, then $p(b(y), y) \geq p(1, y) \to +\infty$ (as $\|y\| \to +\infty$), so the inequality $b(y) \geq 1$ is impossible for y with fairly large $\|y\|$. Thus $b(y) < 1$ for such y. Since p is IPH, it follows from $p(b(y), y) = 1$ that $p(1, (y/b(y))) = 1/b(y)$. Since p is an increasing function we can conclude that:

$$\lim_{\|y\| \to +\infty} \frac{1}{b(y)} = \lim_{\|y\| \to +\infty} p\left(1, \frac{y}{b(y)}\right) \geq \lim_{\|y\| \to +\infty} p(1, y) = +\infty.$$

Hence $b(y) \to 0$. It follows from (3.4.13) that

$$\lim_{\min_i z_i \to 0} h_p(z) = \lim_{\|y\| \to +\infty} h_p\left(\frac{1}{y}\right) = \lim_{\|y\| \to \infty} (b(y))^{-1} = +\infty. \qquad \triangle$$

3.4.3. Multiplicative inf-convolution

The inf-convolution $\hat{+}$ is one of the main operations in convex analysis. By definition $(f \hat{+} g)(x) = \inf_y (f(y) + g(x - y))$. It is well known that this operation corresponds to the summation of epigraphs of the functions f and g (see Subsection 7.3.3 for a corresponding discussion). We now consider a multiplicative analogue to the inf-convolution. We shall consider functions f defined on $\mathbb{R}^I_{++}$ and mapping into $\bar{\mathbb{R}}_+ = [0, +\infty]$.

A function f defined on $\mathbb{R}^I_{++}$ is called ***positive*** if f maps into $(0, +\infty]$.

DEFINITION 3.9 Let h and l be positive functions defined on $\mathbb{R}^I_{++}$. The function

$$(h \diamond l)(z) = \inf_{y \gg 0} h(y) l\left(\frac{z}{y}\right), \qquad z \in \mathbb{R}^I_{++} \qquad (3.4.14)$$

is called the *multiplicative inf-convolution* of the functions h and l.

Since

$$\inf_{y \gg 0} h(y) l\left(\frac{z}{y}\right) = \inf_{u \gg 0} h\left(\frac{z}{u}\right) l(u),$$

it follows that the multiplicative inf-convolution is a commutative operation: $h \diamond l = l \diamond h$.

PROPOSITION 3.18 *The multiplicative inf-convolution is an associative operation.*

Proof: Let h, l and m be positive functions defined on $\mathbb{R}^I_{++}$. Then

$$\begin{aligned} h \diamond (l \diamond m)(z) &= \inf_{u \gg 0} h\left(\frac{z}{u}\right) (l \diamond m)(u) \\ &= \inf_{u \gg 0} \inf_{y \gg 0} h\left(\frac{z}{u}\right) l\left(\frac{u}{y}\right) m(y). \end{aligned}$$

On the other hand,

$$(h \diamond l) \diamond m(z) = \inf_{y \gg 0} m(y) \inf_{v \gg 0} h\left(\frac{z}{y \cdot v}\right) l(v).$$

Let $u = y \cdot v$. Then

$$(h \diamond l) \diamond m(z) = \inf_{y \gg 0} \inf_{u \gg 0} m(y) h\left(\frac{z}{u}\right) l\left(\frac{u}{y}\right) = h \diamond (l \diamond m)(z). \qquad \triangle$$

If l is a positive decreasing function then, applying (3.4.14), it is easy to check that the multiplicative inf-convolution, $h \diamond l$, of l and an arbitrary positive function h is also decreasing. Assume now that l is an upper semicontinuous function. Then for an arbitrary function h, the function $z \mapsto h(y) l(z/y)$ is upper semicontinuous for all $y \gg 0$. Therefore $h \diamond l$ is also upper semicontinuous. In particular, the following assertion holds.

PROPOSITION 3.19 *If l is a positive decreasing upper semicontinuous function, then $h \diamond l$ is decreasing and upper semicontinuous for any arbitrary positive function h.*

We now consider some examples:

EXAMPLE 3.11 Let α be a nonnegative number and $a = (a_1, \dots, a_m)$ be a positive vector. Let $p(\delta, y) = \max\{\alpha\delta, a_1 y_1 \dots, a_m y_m\}$. Then (see Example 3.9) $h_p(y) = \alpha$ if $y \gg a$ and $h_p(y) = +\infty$ otherwise. Let l be a continuous decreasing function defined on $\mathbb{R}^I_{++}$. We have

$$(l \diamond h_p)(z) = \inf_{y \gg 0} l(y) h_p\left(\frac{z}{y}\right) = \inf_{u \gg 0} l\left(\frac{z}{u}\right) h_p(u) = \inf_{u \gg a} l\left(\frac{z}{u}\right) \alpha.$$

Since l is continuous and decreasing, we conclude that $\inf_{u \gg a} l(z/u) = l(z/a)$. Thus

$$(l \diamond h_p)(z) = \alpha l\left(\frac{z}{a}\right).$$

In particular, if $\alpha = 1$ and $a = (1, \dots, 1)$, then $l \diamond h_p = l$ for all continuous decreasing functions l.

We now describe the positive part of the hypograph of the multiplicative inf-convolution of decreasing functions.

PROPOSITION 3.20 *Let l and h be decreasing positive functions defined on $\mathbb{R}^I_{++}$. Assume, in addition, that l is a finite function. Let $U = \{(\alpha, y) \in \mathbb{R}^{I'}_{++} : \alpha \le h(y)\}$ be a positive part* $\operatorname{hyp}^+ h$ *of the hypograph* $\operatorname{hyp} h$. *Then*

$$\bigcap_{y \gg 0} (l(y), y) \cdot U = \operatorname{hyp}^+(l \diamond h), \tag{3.4.15}$$

where the product $a \cdot U$ is defined by (2.2.15): $a \cdot U = \{a \cdot u : u \in U\}$ and $a \cdot u = (a_1 u_1, \dots, a_n u_n)$.

Proof: We should check that

$$\bigcap_{y \gg 0} (l(y), y) \cdot U = \{(\delta, z) : 0 < \delta \le (l \diamond h)(z), z \gg 0\}.$$

First we prove that for all $y \gg 0$

$$(l(y), y) \cdot U = \left\{(\delta, z) : \ \delta \le l(y) h\left(\frac{z}{y}\right)\right\}. \tag{3.4.16}$$

Indeed, since $l(y) > 0$, it follows from the definition of the set U that

$$\begin{aligned}
(l(y), y) \cdot U &= \{(l(y) \cdot \gamma, y \cdot u) : (\gamma, u) \in U\} \\
&= \{(l(y) \cdot \gamma, y \cdot u) : \ \gamma \le h(u)\} \\
&= \{(\delta, z) : \ \frac{\delta}{l(y)} \le h\left(\frac{z}{y}\right)\} \\
&= \{(\delta, z) : \ \delta \le l(y) h\left(\frac{z}{y}\right)\}.
\end{aligned}$$

Let V be the set on the left-hand side in (3.4.15). Then

$$\begin{aligned}
(\delta, z) \in V &\iff (\delta, z) \in (l(y), y) \cdot U \quad (\forall y \gg 0) \\
&\iff \delta \le l(y) h\left(\frac{z}{y}\right) \quad (\forall y \gg 0) \\
&\iff \delta \le \inf_{y \gg 0} l(y) h\left(\frac{z}{y}\right).
\end{aligned}$$

So $V = \{(\delta, z) : 0 < \delta \leq (l \diamond h)(z),\ z \gg 0\}$. △

COROLLARY 3.6 *Let l be a positive finite decreasing function defined on $\mathbb{R}^I_{++}$ and p be an IPH function defined on $\mathbb{R}^{I'}_{++}$ with $I' = \{0\} \cup I$. Let h_p be the associated function for p. Then*

$$\bigcap_{y \gg 0} (l(y), y) \cdot \mathrm{supp}(p, L) = \mathrm{hyp}^+(l \diamond h_p). \tag{3.4.17}$$

REMARK 3.15 Let l, h be functions as in Proposition 3.20 and U_l be the positive part of the hypograph of the function l. The set $\{(l(y), y) : y \gg 0\}$ represents *the upper boundary* of the normal set U_l. We can consider the set defined by (3.4.15) as a 'product' of the sets U_l and U. Since the set U is the positive part of the hypograph hyp h and similarly U_l is the positive part of hyp l, it follows that the positive part of the hypograph of the multiplicative inf-convolution of l and h coincides with the 'product' of the positive parts of the hypographs hyp h and hyp l.

Let $h = l \diamond h_p$ be the multiplicative inf-convolution, where l and h_p are as in Corollary 3.6. It follows from Proposition 3.19 that h is a decreasing and upper semicontinuous function and therefore there exists an IPH function r such that $h = h_r$.

PROPOSITION 3.21 *Let l be a decreasing finite positive function defined on $\mathbb{R}^I_{++}$ and let p, $r : \mathbb{R}^{I'}_{++} \to \mathbb{R}_{+\infty}$ be IPH functions. Then $h_r = l \diamond h_p$ if and only if*

$$\mathrm{supp}(r, L) = \bigcap_{y \gg 0} (l(y), y) \cdot \mathrm{supp}(p, L). \tag{3.4.18}$$

Proof: It follows from the definition of the associated function that $\mathrm{supp}(p, L) = \mathrm{hyp}^+ h_p$ and $\mathrm{supp}(r, L) = \mathrm{hyp}^+ h_r$. Proposition 3.20 shows that $\mathrm{hyp}^+(l \diamond h_p) = \bigcap_{y \gg 0}(l(y), y) \cdot \mathrm{supp}(p, L)$. Hence (3.4.18) holds if and only if $h_r = l \diamond h_p$. △

LEMMA 3.3 *Let h be a decreasing function defined on $\mathbb{R}^I_{++}$. Then the following limits exist:*

$$\lim_{z \to 0} h(z) = \sup_{z \gg 0} h(z), \qquad \lim_{\min_i z_i \to +\infty} h(z) = \inf_{z \gg 0} h(z).$$

Proof: The proof is straightforward. △

We now present the main result of this section.

THEOREM 3.6 *Let l and h be decreasing functions defined on $\mathbb{R}^I_{++}$ such that*

1) $0 < \gamma := \lim_{\min_i z_i \to +\infty} l(z)$; $\quad M := \lim_{y\to 0} l(y) < +\infty$;

2) $\operatorname{dom} h = \{y : \ h(y) < +\infty\} \neq \emptyset$ *and* $H := \lim_{\min_i z_i \to +\infty} h(z) > 0$;

3) $\liminf_{\min z_i \to 0} h(z) > \frac{M}{\gamma} H$.

Then

$$\lim_{z\to 0}(h \diamond l)(z) = \lim_{z\to 0} l(z) \times \lim_{\min_i z_i \to +\infty} h(z) = MH. \tag{3.4.19}$$

Proof: First we show that $(h \diamond l)(z) \leq MH$ for all $z \gg 0$. Let $\mathbf{1} = (1, \dots, 1)$. For the functions $u(\lambda) = l(\lambda \mathbf{1})$ and $v_z(\lambda) = h(\frac{1}{\lambda} z)$ with $z \gg 0$, we have

$$u(\lambda) \leq \sup_{\lambda'>0} u(\lambda') = \lim_{\lambda'\to 0} l(\lambda' \mathbf{1}) = \lim_{y\to 0} l(y) = M,$$

$$\inf_{\lambda>0} v_z(\lambda) = \inf_{\mu>0} h(\mu z) = \lim_{\mu\to+\infty} h(\mu z) = \lim_{\min_i y_i \to +\infty} h(y) = H.$$

So

$$\begin{aligned} (h \diamond l)(z) &= \inf_{y\gg 0} l(y) h\left(\frac{z}{y}\right) \leq \inf_{\lambda>0} l(\lambda\mathbf{1}) h\left(\frac{z}{\lambda\mathbf{1}}\right) \\ &= \inf_{\lambda>0} u(\lambda) v_z(\lambda) \leq M \inf_{\lambda>0} v_z(\lambda) = MH. \end{aligned} \tag{3.4.20}$$

Thus

$$\lim_{z\to 0}(h \diamond l)(z) \leq MH. \tag{3.4.21}$$

We now prove the reverse inequality.

It follows from 3) that there exist numbers $\mu > 0$ and $\varepsilon > 0$ such that $h(u) \geq (1+\varepsilon)(1/\gamma)MH$ whenever $\min_i u_i \leq \mu$. Thus, if $\min_i (z/y)_i \leq \mu$, then

$$l(y) h\left(\frac{z}{y}\right) \geq \gamma h\left(\frac{z}{y}\right) \geq (1+\varepsilon) MH.$$

Thus $\inf_{y\gg 0,\ \min(z/y)_i \leq \mu} l(y) h(z/y) > MH$. Applying (3.4.20) we can conclude that

$$\begin{aligned} MH &\geq (h \diamond l)(z) = \inf_{y\gg 0} l(y) h\left(\frac{z}{y}\right) \\ &= \min\left(\inf_{y\gg 0,\ \min(z/y)_i \leq \mu} l(y) h\left(\frac{z}{y}\right), \inf_{y\gg 0, (z/y)\gg \mu\mathbf{1}} l(y) h\left(\frac{z}{y}\right)\right) \\ &= \inf_{y\gg 0, (z/y)\gg \mu\mathbf{1}} l(y) h\left(\frac{z}{y}\right). \end{aligned}$$

Let $z \in \mathbb{R}^I_{++}$ and $z_\mu = (1/\mu)z$. We have

$$(h \diamond l)(z) = \inf_{y \gg 0,\ z/y \gg \mu \mathbf{1}} l(y)h\left(\frac{z}{y}\right) = \inf_{0 \ll y \ll z_\mu} l(y)h\left(\frac{z}{y}\right). \quad (3.4.22)$$

Since l is decreasing, we have $l(y) \geq l(z_\mu)$ for $0 \ll y \ll z_\mu$. So

$$(h \diamond l)(z) \geq \inf_{0 \ll y \ll z_\mu} l(z_\mu)h\left(\frac{z}{y}\right) = l(z_\mu) \inf_{u \gg \mu \mathbf{1}} h(u).$$

Due to the fact that h is decreasing and applying Lemma 3.3, we conclude that

$$\inf_{u \gg \mu \mathbf{1}} h(u) = \inf_{u \gg 0} h(u) = \lim_{\min_i u_i \to +\infty} h(u) = H,$$

so $(h \diamond l)(z) \geq Hl(z_\mu)$. Thus

$$\lim_{z \to 0} (h \diamond l)(z) \geq H \lim_{z \to 0} l(z_\mu) = H \lim_{y \to 0} l(y) = HM. \quad (3.4.23)$$

It follows from (3.4.20) and (3.4.23) that (3.4.19) holds. △

REMARK 3.16 The item 3) in Theorem 3.6 holds if

$$\frac{\liminf_{\min_i z_i \to 0} h(z)}{\lim_{\min_i z_i \to +\infty} h(z)} > \frac{\lim_{z \to 0} l(z)}{\lim_{\min_i z_i \to +\infty} l(z)}$$

and if both denominators are positive. It is clear that this inequality holds if $\lim_{\min_i z_i \to 0} h(z) = +\infty$.

Chapter 4

APPLICATION TO GLOBAL OPTIMIZATION: LAGRANGE AND PENALTY FUNCTIONS

4.1. INTRODUCTION

One of the main approaches to the solution of convex optimization problems involves the exploitation of properties of the traditional linear Lagrange function (Lagrangian) and the penalty function. In particular, the zero duality gap property between the primal convex optimization problem and its Lagrange (penalty) dual problem has enabled important algorithms to be proposed and developed, see for example [21, 57, 113, 136] and references therein. The zero duality gap property guarantees that there exists a sequence of unconstrained problems, such that their solutions tend to a solution of the initial constrained optimization problem. These problems are formed by means of a special choice of Lagrange multipliers (penalty parameters). The situation becomes much simpler if there exists the exact Lagrange multiplier (penalty parameter). In this case it is possible to build an unconstrained problem, which is equivalent to the initial problem. The exact penalty parameter can be found in many instances, however sometimes this parameter is very large, so the unconstrained problem discussed above becomes ill-conditioned.

The Lagrange (penalty) function can also be constructed for nonconvex optimization problems although the zero duality gap property does not necessarily hold in this case.

Thus the question arises of finding a generalization of classical Lagrange and penalty functions, which can be applied for a broad class of nonconvex problems. This question has been considered by many researches. We point out here the so-called augmented Lagrangian (see

[136] and references therein), a very general scheme, based on theorems of the alternative, which was suggested by F. Giannessi ([59]), and nonlinear analogues of Lagrangians and penalty functions, suggested by Yu. Evtushenko and V. Zhadan ([197]). The convolution of the objective function and constraints by the maximum (instead of the sum) have been considered in [68, 197, 198]. Very general results for this type of convolution have been obtained by C.J. Goh and X.Q. Yang [67]. The convolution by p-norm has been studied by D. Li [104].

Consider the following inequality constrained optimization problem:

$$(COP) \quad f_0(x) \longrightarrow \inf \text{ subject to } x \in X, \ f_i(x) \le 0, \ i \in I,$$

where $X \subset \mathbb{R}^n$, $I = \{1, \dots, m\}$, and $f_i : X \to \mathbb{R}$, $i \in \{0\} \cup I$ are finite functions. We shall assume in this chapter that $f_0(x) > 0$ for all $x \in X$. If the lower estimate c of the function f_0 on X is known, then the problem (COP) can be replaced by the problem with the same constraints and the objective function $f_0(x) - c$, which is positive. We shall be interested in some convolutions of functions $f_0, f_1, \dots, f_m$, which enjoy the zero duality gap property. This convolution should be accomplished by a function p defined on either the entire space $\mathbb{R} \times \mathbb{R}^m \equiv \mathbb{R}^{1+m}$ (in the case of the Lagrange function) or on the cone $\mathbb{R}^{1+m}_+$ (in the case of the penalty function). Thus we shall consider functions of the form

$$\mathcal{L}(x, d) = p(d_0 f_0(x), d_1 f_1(x), \dots, d_m f_m(x)) \tag{4.1.1}$$

with a function p defined on $\mathbb{R}^{1+m}$ and

$$\mathcal{L}^+(x, d) = p(d_0 f_0(x), d_1 f_1^+(x), \dots, d_m f_m^+(x)) \tag{4.1.2}$$

with a function p defined on $\mathbb{R}^{1+m}_+$. Here $y^+ = \max(y, 0)$. It is assumed that numbers $d_0, d_1, \dots, d_m$ are positive.

The function defined by (4.1.1) will be called the *extended Lagrange function* or *extended Lagrangian*; the function defined by (4.1.2) will be called the *extended penalty function*. In Section 4.2 we shall show that the convolution by an increasing function with some additional properties very often leads to the zero duality gap property. We need these additional properties in order to exclude linear increasing functions defined on the entire space R^{1+m} (this is not surprising since the zero duality gap property does not hold for the traditional (linear) Lagrangian without convexity assumptions).

For increasing functions defined on the positive orthant, the additional assumptions we need are not very restrictive, in particular, linear functions $x \to \sum c_i x_i$ with positive coefficients c_i will satisfy the assumptions.

It is well known that all constraints f_i can be convoluted into a single constraint (see [197] for a detailed discussion.) For example, we can use a convolution by maximum: $\max_i f_i(x) \leq 0 \iff f_i(x) \leq 0$ for all i. So we shall consider in Section 4.3 only problems with a single constraint, which has the form

$$f_0(x) \longrightarrow \inf \text{ subject to } x \in X, \ f_1(x) \leq 0 \tag{4.1.3}$$

where $f_0 : X \to \mathbb{R}_{++} := \{x \in \mathbb{R} : x > 0\}$ and $f_1 : X \to \mathbb{R}$. Exploiting IPH convolution functions allows us to obtain a more advanced theory of extended penalization for problem (4.1.3). This theory is based on the properties of IPH functions and properies of decreasing functions, which were established in Chapter 2 and Chapter 3.

The perturbation function

$$\tilde{\beta}(y) = \inf\{f_0(x) : \ x \in X, \ f_i(x) \leq y_i, \ i = 1, \dots, m\}, \ y = (y_i)_{i=1}^m \in \mathbb{R}^m \tag{4.1.4}$$

has useful applications in the study of the nonlinear programming problem:

$$f_0(x) \longrightarrow \inf \text{ subject to } x \in X, \ f_i(x) \leq 0, \ i = 1, \dots, m,$$

(see for example [21, 29, 30, 101, 113, 136] and the references therein). The perturbation function of the problem (4.1.3) can be considered on the real line. However we shall assume in Section 4.3 that this function is defined only on $\mathbb{R}_{++}$. The perturbation function is decreasing, that is, $y^1 \geq y^2 \implies \beta(y^1) \leq \beta(y^2)$. Consequently, the study of perturbation functions should be based on a theory of decreasing functions.

The extended penalty function $\mathcal{L}_p^+(x, d)$ of the problem (4.1.3) has the form

$$\mathcal{L}_p^+(x, d) = p(f_0(x), d \max(f_1(x), 0)), \tag{4.1.5}$$

where p is an IPH function defined on $\mathbb{R}^2_+$. Among many questions which arise in connection with such a setting we indicate the following:

1 How do we obtain conditions ensuring the zero duality gap property?

2 How do we obtain conditions ensuring the exact penalization?

3 How do we find convolution functions p such that an exact penalty parameter for the function $\mathcal{L}_p^+$ is substantially smaller than that of the classical function?

The corresponding questions for the classical function have been discussed, for example, in [9, 29, 31, 113, 136].

These three questions are addressed in Section 4.3. We consider the penalty function of the form (4.1.5) involving only an IPH function p with some natural properties.

We completely describe the class $\mathcal{P}_+$ of IPH functions such that for $p \in \mathcal{P}_+$ the zero duality gap property

$$\inf\{f_0(x) : \ x \in X, \ f_1(x) \leq 0\} = \sup_{d>0} \inf_{x \in X} \mathcal{L}_p^+(x, d), \qquad (4.1.6)$$

holds if and only if the perturbation function

$$\beta(y) = \inf\{f_0(x) : \ f_1(x) \leq y\}, \qquad y \geq 0$$

is lower semicontinuous at the origin.

Thus the zero duality gap property depends only on the problem itself and does not depend on an outer convolution function from a very large class of such functions. The proof of this fact is based on the theory of multiplicative inf-convolution developed in Section 3.4. In contrast to this result we show that the exact penalization essentially depends on an outer convolution function. In particular, it is proved that as a rule, the penalization with respect to the function $p_{+\infty}(\alpha, y) = \max\{\alpha, y\}$ is not exact.

We also show that if (4.1.6) holds for all problems with the lower semicontinuous at zero perturbation function, then $p \in \mathcal{P}_+$. The convolution with respect to the family of IPH functions

$$p_k(\delta, y) = (\delta^k + y^k)^{\frac{1}{k}} \quad (0 < k < +\infty)$$

is considered. We study in detail the exact penalization by $p_{\frac{1}{2}}$ and demonstrate that this penalization always can be accomplished with a smaller penalty parameter than that of the classical convolution function p_1. We also obtain an asymptotically sharp estimate of the ratio $\bar{d}_{\frac{1}{2}}/\bar{d}_1$, where $\bar{d}_k$ is the least exact penalization parameter with respect to p_k. This estimate allows us to draw the following conclusion.

If the constrained minimum of the objective function is not very distant from the unconstrained minimum of this function then the penalization by $p_{\frac{1}{2}}$ can be accomplished with a substantially smaller exact penalty parameter $\bar{d}_{\frac{1}{2}}$ than $\bar{d}_1$.

We prove that the class of IPH functions is sufficiently large to provide the exact penalization: an exact modified penalty function can be found for a given problem under some very weak assumptions.

4.2. EXTENDED LAGRANGE AND PENALTY FUNCTIONS

4.2.1. Overview

In this section we examine extended Lagrange and penalty functions, which are formed by means of special increasing functions. We consider problems with a continuous objective function and continuous constraints. If a convolution function is IPH then the dual penalty function corresponding to p is IPH as well. We describe the value of the dual problem in terms of the support set of this IPH function. The main results from this section can be found in [156].

4.2.2. Preliminaries

Let Q be either the space $\mathbb{R} \times \mathbb{R}^m = \mathbb{R}^{1+m}$ or the cone $\mathbb{R}_+^{1+m}$. Consider a continuous increasing function p, which enjoys the following properties:

A) There exist positive numbers $a_0, a_i, \dots a_m$ with $a_0 = 1$ such that for $\bar{y} = (y_0, y_1, \dots, y_m) \in Q$ with $y_0 \in \mathbb{R}_+$ we have

$$p(\bar{y}) \geq \max_{0 \leq i \leq m} a_i y_i. \tag{4.2.1}$$

B) For $y_0 \in \mathbb{R}_+$

$$p(y_0, 0, \dots, 0) = y_0. \tag{4.2.2}$$

Let $Z(Q)$ be the set of all continuous increasing functions which possess properties A) and B). It is easy to see that

1) $Z(Q)$ is a convex set: if $p_1, p_2 \in Z(Q)$ and $0 < \alpha < 1$ then $\alpha p_1 + (1 - \alpha)p_2 \in Z(Q)$.

2) if $p_1, p_2 \in Z(Q)$ then $\max(p_1, p_2) \in Z(Q)$.

3) if $p_1 \in Z(Q)$ and p_2 is an increasing continuous nonnegative function defined on Q and such that $p_2(y_0, 0, \dots, 0) = 0$ for all $y_0 \geq 0$ then $p_1 + p_2 \in Z(Q)$.

We now present some examples. First, we consider functions defined on the entire space $\mathbb{R}^{1+m}$.

EXAMPLE 4.1 a) The function

$$p(y_0, y) = \max\ (y_0, a_1y_1, \dots, a_my_m) \qquad (y_0, y) \in \mathbb{R}^{1+m}$$

with positive $a_1, \dots, a_m$ is an increasing function with properties A) and B).

b) Let $\psi(y_0, y)$ be an arbitrary increasing function defined on $\mathbb{R}^{1+m}$ such that $\psi(y_0, 0) = y_0$ and $\psi(y_0, y) \geq y_0$ for all y. Then the function

$$p(y_0, y) = \max\ (\psi(y_0, y), a_1y_1, \dots, a_my_m)$$

is increasing and possesses properties A) and B). In particular, the function

$$p(y_0, y) = \max\ (y_0 + c_1\overset{+}{y_1} + \dots + c_m\overset{+}{y_m}, a_1y_1, \dots, a_my_m)$$

with nonnegative c_i and positive a_i enjoys these properties.

We now consider an example of increasing functions with properties A) and B) defined on the cone $\mathbb{R}^{1+m}_+$.

EXAMPLE 4.2 Let $\bar{y} = (y_0, y) \in \mathbb{R}^{1+m}_+$ with $y = (y_1, \dots, y_m)$. The following increasing functions defined on $\mathbb{R}^{m+1}_+$ have properties A) and B):

$$p_k(\bar{y}) = \left(\sum_{i=0}^{m} y_i^k\right)^{\frac{1}{k}}, \quad 0 < k < +\infty, \tag{4.2.3}$$

$$p_{+\infty}(\bar{y}) = \max_{0 \leq i \leq m} y_i. \tag{4.2.4}$$

REMARK 4.1 Let us compare the set of conditions (4.2.1) and (4.2.2) for increasing functions defined on $\mathbb{R}^{1+m}$ and on $\mathbb{R}^{1+m}_+$.

Since the zero duality gap property is not valid for the classical Lagrange function defined by means of a linear convolution function, it is necessary to have conditions which exclude linearity. Note that (4.2.1) holds for linear positive functions defined on $\mathbb{R}^{1+m}_+$ and does not hold for linear positive functions defined on $\mathbb{R}^{1+m}$. The inequality (4.2.1) is not restrictive for IPH functions p defined on $\mathbb{R}^{1+m}_+$. Indeed, let $p(e_k) = a_k > 0$. Then for each $y = (y_1, \dots, y_m) \geq 0$ and each $k \in J := \{1, \dots, m\}$ we have $y \geq y_ke_k$, so $p(y) \geq p(y_ke_k) = y_kp(e_k)$. Thus $p(y) \geq \max_{k \in J} a_ky_k$. At the same time (4.2.1) is very restrictive for functions defined on the entire space. We need to have condition A) in order to prove the zero duality gap property for Lagrange and penalty functions. This property formally holds for both Lagrange and penalty functions under the same condition A). However, this condition is very strong for the Lagrange case and very weak for the penalty case.

4.2.3. Extended Lagrange functions

Consider the inequality constrained optimization problem

$$(COP) \quad f_0(x) \longrightarrow \inf \text{ subject to } x \in X,\ f_i(x) \leq 0,\ i \in J, \qquad (4.2.5)$$

where $X \subset \mathbb{R}^n$, $J = \{1, \dots, m\}$, $f_i : X \to \mathbb{R}$, $i \in \{0\} \cup J$ are continuous functions. The following assumption will be needed throughout the section.

ASSUMPTION 4.1 *The function f_0 is positive on the set X.*

Let

$$X_0 = \{x \in X : f_i(x) \leq 0, i \in J\}.$$

be the set of all feasible solutions for (COP). Consider the vector-valued function $F : X \times \mathbb{R}^{1+m}_+ \to \mathbb{R}^{1+m}$ defined as follows

$$F(x, d) = (d_0 f_0(x), d_1 f_1(x), \dots, d_m f_m(x)),$$

where $x \in X$, $d \in \mathbb{R}^{1+m}_+$. Let $p : \mathbb{R}^{1+m} \to \mathbb{R}_{+\infty}$ be an increasing function with properties A) and B). We now define the extended Lagrange function, the dual function and the dual problem for the problem (COP), corresponding to p.

DEFINITION 4.1 The function $\mathcal{L}$ defined by

$$\mathcal{L}(x, d) = p(F(x, d)), \qquad x \in X, \quad d \in \mathbb{R}^{1+m}_+.$$

is called the *extended Lagrange function* for the problem (COP) corresponding to a function p.

DEFINITION 4.2 The dual function for (COP) corresponding to p is defined by

$$q_p(d_0, d) = \inf_{x \in X} \mathcal{L}(x, d_0, d), \quad (d_0, d) \in \mathbb{R}^{1+m}_+.$$

DEFINITION 4.3 The problem:

$$(D_p^*) \qquad q_p(1, d) \longrightarrow \sup \text{ subject to } d \in \mathbb{R}^m_+$$

is called the *dual problem* to the problem (COP) corresponding to the function p.

Let M be the optimal value of the problem (COP) and M_p^* be the optimal value of the dual problem (D_p^*), that is,

$$M = \inf\{f_0(x) : x \in X_0\}, \quad M_p^* = \sup\{q_p(1, d) : d \in \mathbb{R}^m_+\}. \qquad (4.2.6)$$

The properties A) and B) of an increasing function p allow us to establish the following result.

LEMMA 4.1 *Let p be an increasing function with properties A) and B). Then the equality $p(F(x,d)) = f_0(x)$ holds for all $x \in X_0$ and $d \in \mathbb{R}^{1+m}_+$ with $d_0 = 1$.*

Proof: Since $x \in X_0$ it follows that $f_i(x) \leq 0$ $(i \in J)$. Applying the isotonicity of p and property B) we conclude that

$$p(f_0(x), d_1 f_1(x), \ldots, d_m f_m(x)) \leq p(f_0(x), 0, \ldots, 0) = f_0(x).$$

Also property A) shows that for some numbers $a_i > 0$ $(i \in J)$

$$\begin{aligned} p(F(x,d)) &= p(f_0(x), d_1 f_1(x), \ldots, d_m f_m(x)) \\ &\geq \max(f_0(x), a_1 d_1 f_1(x), \ldots, a_m d_m f_m(x)) = f_0(x). \end{aligned}$$

Thus the desired result follows. △

We now establish a zero duality gap result (that is, the equality $M = M^*_p$) applicable to the optimization problem (COP) and its extended Lagrangian dual problem (D^*_p). First we prove the following assertion.

LEMMA 4.2 *Let M and M^*_p be defined by (4.2.6). Then $M \geq M^*_p$.*

Proof: Let $x \in X_0$ and $d \in \mathbb{R}^{1+m}_+$ with $d_0 = 1$. It follows from Lemma 4.1 that $p(F(x,d)) = f_0(x)$. Therefore, $f_0(x) \geq \inf_{x' \in X} p(F(x',d)) = q_p(d)$. Thus

$$M^*_p = \sup_{d \in \mathbb{R}^{1+m}_+, d_0 = 1} q_p(d) \leq \inf_{x \in X_0} f_0(x) = M. \qquad \triangle$$

THEOREM 4.1 *(Zero duality gap). Assume that X_0 is a compact set and (if X is unbounded) the function f_0 enjoys the following property*

$$\lim_{x \to \infty, x \in X} f_0(x) = +\infty.$$

Then $M = M^*_p$.

REMARK 4.2 Recall that due to Assumption 4.1 the function f_0 is positive on the set X. This assumption plays a crucial role in establishing the zero duality gap property.

Proof: It follows from Lemma 4.2 that $M \geq M^*_p$. Assume $M > M^*_p$. Then there exist $\varepsilon > 0$ and $\tilde{x} \in X_0$ such that

$$M - 3\varepsilon > q_p(d) \text{ for all } d \in \mathbb{R}^{1+m}_+,\ d_0 = 1 \qquad (4.2.7)$$

and

$$f_0(\tilde{x}) \geq M > f_0(\tilde{x}) - 2\varepsilon. \tag{4.2.8}$$

Assume that ε is sufficiently small so that $f_0(\tilde{x}) - 2\varepsilon > 0$. Thus (4.2.7) and (4.2.8) imply

$$f_0(\tilde{x}) - 3\varepsilon \geq M - 3\varepsilon > q_p(d) \text{ for all } d \in \mathbb{R}^{1+m}_+,\ d_0 = 1. \tag{4.2.9}$$

Applying Lemma 4.1 and (4.2.8) we deduce that

$$\begin{aligned} q_p(d) &= \min\{\inf_{x\in X_0} p(F(x,d)), \inf_{x\in \mathrm{cl}\,(X\setminus X_0)} p(F(x,d))\} \\ &= \min\{M, \inf_{x\in \mathrm{cl}\,(X\setminus X_0)} p(F(x,d))\} \\ &\geq \min\{f_0(\tilde{x}) - 2\varepsilon, \inf_{x\in \mathrm{cl}\,(X\setminus X_0)} p(F(x,d))\} \\ &\geq \min\{f_0(\tilde{x}) - 3\varepsilon, \inf_{x\in \mathrm{cl}\,(X\setminus X_0)} p(F(x,d))\}, \end{aligned}$$

where $\mathrm{cl}\,(X\setminus X_0)$ is the closure of the set $X\setminus X_0$. Combining the last inequality with (4.2.9), we obtain

$$q_p(d) \geq \inf_{x\in \mathrm{cl}\,(X\setminus X_0)} p(F(x,d)) \text{ for all } d \in \mathbb{R}^{1+m}_+,\ d_0 = 1. \tag{4.2.10}$$

Since f_0 is continuous and X_0 is compact it follows that we can find $\delta > 0$ such that for each $x \in \mathrm{cl}\,(X\setminus X_0)$ with the property $\rho(x, X_0) := \min\{\|x - x_0\| : x_0 \in X_0\} \leq \delta$, the inequality

$$f_0(x') < f_0(x) + \varepsilon$$

holds for some $x' \in X_0$. It follows from (4.2.8) that $f_0(x') \geq M \geq f_0(\tilde{x}) - 2\varepsilon$. So

$$f_0(\tilde{x}) - 3\varepsilon < f_0(x) \text{ for all } x \in \mathbb{R}^{1+m}_+\setminus X_0,\ \rho(x, X_0) \leq \delta. \tag{4.2.11}$$

Let $X^\delta = \{x \in \mathrm{cl}\,(X\setminus X_0) : \rho(x, X_0) \geq \delta\}$ and

$$M_1(d) = \inf_{x\in \mathrm{cl}\,(X\setminus X_0),\ \rho(x,X_0)\leq\delta} p(F(x,d)),$$

$$M_2(d) = \inf_{x\in \mathrm{cl}\,(X\setminus X_0),\ \rho(x,X_0)>\delta} p(F(x,d)) = \inf_{x\in X^\delta} p(F(x,d)).$$

It follows from (4.2.11) that $M_1(d) \geq f_0(\tilde{x}) - 3\varepsilon$. Applying (4.2.10) we have

$$q_p(d) = \min\{M_1(d), M_2(d)\} \geq \min\{f_0(\tilde{x}) - 3\varepsilon, M_2(d)\} \tag{4.2.12}$$

for all $d \in \mathbb{R}_+^{1+m}$ with $d_0 = 1$. Combining (4.2.9) and (4.2.12) we have

$$q_p(d) \geq M_2(d) = \inf_{x \in X^\delta} p(F(x,d)) \text{ for all } d \in \mathbb{R}_+^{1+m},\ d_0 = 1.$$

Using property A) we deduce, that there exist numbers $a_i > 0$, $(i = 0, 1 \ldots, m)$ with $a_0 = 1$ such that for $d \in \mathbb{R}_+^{1+m}$ with $d_0 = 1$ we have

$$p(F(x,d)) = p(f_0(x), d_1 f_1(x), \ldots, d_m f_m(x)) \geq \max_{0 \leq i \leq m} a_i d_i f_i(x) \geq f_0(x).$$

Since $\lim_{x \in X, x \to \infty} f_0(x) = +\infty$, there exists a number Δ such that

$$q_p(d) \geq \inf_{x \in X^\delta} p(F(x,d)) = \inf_{x \in X^\delta, \|x\| \leq \Delta} p(F(x,d)), \quad d \in \mathbb{R}_+^{1+m},\ d_0 = 1. \tag{4.2.13}$$

Let $g(x) = \max_{1 \leq i \leq m} f_i(x)$. It follows directly from the definition of the set X_0 that $X_0 = \{x \in X : g(x) \leq 0\}$. Since the set $\{x \in X^\delta : \|x\| \leq \Delta\}$ is compact and the function g is continuous, we conclude that

$$\min\{g(x) : \ x \in X^\delta, \|x\| \leq \Delta\} = 2\beta > 0.$$

Thus for any $x \in X^\delta$, $\|x\| \leq \Delta$ there exists i such that $x \in B_i$, where

$$B_i = \{x \in X^\delta, \|x\| \leq \Delta : f_i(x) > \beta\}.$$

Let $\tilde{d}_0 = 1$ and let $\tilde{d}_i$ $(i \in J)$ be numbers, defined by

$$\tilde{d}_i = \begin{cases} \sigma_i & \text{if } B_i \neq \emptyset; \\ 1 & \text{if } B_i = \emptyset, \end{cases}$$

where

$$\sigma_i = \max_{x \in B_i} \frac{f_0(\tilde{x}) - 2\varepsilon}{a_i f_i(x)}.$$

Recall that ε has been chosen so small that $f(\tilde{x}) - 2\varepsilon > 0$. So $\tilde{d} = (\tilde{d}_0, \tilde{d}_1 \ldots, \tilde{d}_m) \in \mathbb{R}_+^{1+m}$. Applying (4.2.13) and (4.2.1), we can find a vector $u \in X^\delta$, $\|u\| \leq \Delta$ such that

$$q_p(\tilde{d}) = \inf_{x \in X^\delta,\ \|x\| \leq \Delta} p(F(x,\tilde{d})) \geq p(F(u,\tilde{d})) - \varepsilon.$$

There exists an index i_u such that $u \in B_{i_u}$. Thus by (4.2.1)

$$p(F(u,\tilde{d})) \geq a_{i_u} \tilde{d}_{i_u} f_{i_u}(u).$$

Hence

$$q_p(\tilde{d}) \geq a_{i_u} \tilde{d}_{i_u} f_{i_u}(u) - \varepsilon = a_{i_u} f_{i_u}(u) \max_{x \in B_{i_u}} \frac{f_0(\tilde{x}) - 2\varepsilon}{a_{i_u} f_{i_u}(x)} - \varepsilon \geq f_0(\tilde{x}) - 3\varepsilon.$$

This contradicts (4.2.9) and the proof is complete. $\triangle$

4.2.4. Extended penalty functions

The set X_0 of feasible solutions for the problem (COP) can be represented in the following form:

$$X_0 = \{x \in X : f_i^+(x) = 0, \, i \in J\},$$

where $f_i^+(x) = \max(f_i(x), 0)$. Thus the problem (COP) is equivalent to the following problem:

$$f_0(x) \longrightarrow \inf \text{ subject to } x \in X, \; f_i^+(x) = 0, \quad i \in J. \qquad (4.2.14)$$

Due to Assumption 4.1 we have $f_0^+(x) = f_0(x)$. Since $f_i^+(x) \geq 0$ for all $i = 0, 1, \ldots, m$, dual functions and dual problems can be constructed by using an increasing function p defined only on the cone $\mathbb{R}_+^{1+m}$. Let

$$F^+(x, d) = (d_0 f_0(x), d_1 f_1^+(x), \ldots, d_m f_m^+(x)), \qquad x \in X, \, d \in \mathbb{R}_+^{1+m}.$$

In this subsection we assume that the function p is defined on $\mathbb{R}_+^{1+m}$ and possesses properties A) and B). The dual function $\mathcal{L}^+$ for problem (4.2.14), corresponding to the increasing function p, has the following form:

$$\mathcal{L}^+(x, d) = p(F^+(x, d)), \qquad x \in X, \, d \in \mathbb{R}_+^{1+m}. \qquad (4.2.15)$$

In particular, if $p(y) = \sum_{i=0}^m y_i$, then

$$\mathcal{L}^+(x, d) = d_0 f_0(x) + \sum_{i=1}^m d_i f_i^+(x). \qquad (4.2.16)$$

We can consider the function $\mathcal{L}^+$ defined by (4.2.15) as a penalty function for the problem (COP). Thus the function $\mathcal{L}^+$ for problem (4.2.14) can be considered as an *extended penalty function* for the problem (COP).

The dual function q_p^+ corresponding to p can be represented as follows:

$$q_p^+(d) = \inf_{x \in X} \mathcal{L}^+(x, d) \equiv \inf_{x \in X} p(d_0 f_0(x), d_1 f_1^+(x), \ldots, d_m f_m^+(x)). \qquad (4.2.17)$$

The dual problem (D_p^+) to problem (4.2.14) has the following form:

$$q_p^+(1, d) \longrightarrow \sup \text{ subject to } d = (d_1, \ldots, d_m) \in \mathbb{R}_+^m, \qquad (4.2.18)$$

where q_p^+ is the dual function defined by (4.2.17). It is easy to check that Lemma 4.1, Lemma 4.2 and Theorem 4.1 (Zero duality gap) hold for the dual problem (D_p^+) corresponding to increasing functions p defined on $\mathbb{R}_+^{1+m}$ and having properties A) and B).

4.2.5. Examples

We shall discuss two examples in this subsection.

EXAMPLE 4.3 Consider the increasing function p_k defined on $\mathbb{R}_+^{1+m}$ by (4.2.3) for $0 < k < +\infty$:

$$p_k(\bar{y}) = \left(\sum_{i=0}^{m} y_i^k\right)^{\frac{1}{k}}, \qquad \bar{y} = (y_0, y_1, \ldots, y_m) \in \mathbb{R}_+^{1+m}. \tag{4.2.19}$$

The extended penalty function for the problem (COP) can be specified as follows:

$$\mathcal{L}_k^+(x, d) = \left(d_0 f_0^k(x) + \sum_{i=1}^{m} d_i (f_i^+(x))^k\right)^{\frac{1}{k}}, \quad x \in X,\ d \in \mathbb{R}_+^{1+m}.$$

The dual function for the corresponding problem (4.2.14) is now defined as

$$q_{p_k}^+(d) = \inf_{x \in X} \left(f_0^k(x) + \sum_{i=1}^{m} d_i (f_i^+(x))^k\right)^{\frac{1}{k}}.$$

If the conditions in Theorem 4.1 hold, then

$$\inf_{x \in X_0} f_0(x) = \sup_{d \in \mathbb{R}_+^m} \inf_{x \in X} \left(f_0^k(x) + \sum_{i=1}^{m} d_i (f_i^+(x))^k\right)^{\frac{1}{k}}.$$

EXAMPLE 4.4 Let

$$p_\infty(\bar{y}) = \max_{0 \le i \le m} y_i, \qquad \bar{y} = (y_0, y_1, \ldots, y_m) \in \mathbb{R}^{1+m}. \tag{4.2.20}$$

Then we obtain the following extended Lagrange function:

$$\mathcal{L}_\infty(x, d) = \max_{0 \le i \le m} d_i f_i(x). \tag{4.2.21}$$

Let

$$p_{+\infty}(\bar{y}) = \max_{0 \le i \le m} y_i, \qquad \bar{y} = (y_0, y_1, \ldots, y_m) \in \mathbb{R}_+^{1+m}. \tag{4.2.22}$$

Then we obtain the following extended penalty function:

$$\mathcal{L}_{+\infty}^+(x, d) = \max_{0 \le i \le m} d_i f_i^+(x). \tag{4.2.23}$$

If the conditions of Theorem 4.1 hold, then the zero duality gap property will hold, both for the extended Lagrangian and extended penalty function dualities. The first of them can be represented as follows:

$$\inf_{x\in X_0} f_0(x) = \sup_{d\in \mathbb{R}^m_+} \inf_{x\in X} \max\{f_0(x), \max_{1\le i\le m} d_i f_i(x)\}. \tag{4.2.24}$$

The zero duality gap property for the extended penalty function has the following form:

$$\inf_{x\in X_0} f_0(x) = \sup_{d\in \mathbb{R}^m_+} \inf_{x\in X} \max(f_0(x), \max_{1\le i\le m} d_i f_i^+(x)). \tag{4.2.25}$$

Since $\max_{1\le i\le m} d_i f_i^+(x) = \max(0, \max_{1\le i\le m} d_i f_i(x))$ and $f_0(x) > 0$, (4.2.25) is reduced to (4.2.24). Thus the zero duality gap property for the extended penalty functions coincides with the zero duality gap for the extended Lagrangian function.

Let $g(x) = \max_{i\in J} f_i(x)$. The set X_0 of all feasible solutions of the problem (COP) can be represented by using only one constraint function g, namely $X_0 = \{x \in X : g(x) \le 0\}$. The dual function $\bar{q}(x)$ in this case has the following form:

$$\bar{q}_{p_{+\infty}}(d_0, d_1) = \inf_{x\in X} \max\{d_0 f_0(x), d_1 g(x)\}, \qquad d_0, d_1 \ge 0.$$

We now compare $\bar{q}_{p_{+\infty}}$ with the dual function $q_{p_{+\infty}}$ defined for the problem (COP) with the given constraint functions $f_1, \dots, f_m$. We have for $d_0, d_1, \dots d_m \ge 0$

$$q_{p_{+\infty}}(d_0, d_1, \dots, d_m) = \inf_{x\in X} \max\{d_0 f_0(x), d_1 f_1(x), \dots, d_m f_m(x)\}.$$

Clearly

$$\bar{q}_{p_{+\infty}}(d_0, d) = q_{p_{+\infty}}(d_0, d, \dots, d).$$

Since q is an increasing function, it follows that

$$\sup_{d>0} \bar{q}_{p_{+\infty}}(1, d) = \sup_{d_1,\dots,d_m>0} q_{p_{+\infty}}(1, d_1, \dots, d_m).$$

Thus the value of the dual problem with respect to one constraint g coincides with the value $M^*_{p_{+\infty}}$ of the dual problem with respect to given constraints $f_1, \dots, f_m$. The same conclusion is valid for the extended penalty function.

4.2.6. Support set of the dual function

Let p be an IPH function defined on $\mathbb{R}^{1+m}_+$ and M^*_p be the value of the dual problem (4.2.18) corresponding to p. In this subsection M^*_p will be described in terms of the support set $\text{supp}(q_p^+, L)$ of the dual function q_p^+ with respect to the set of all min-type functions defined on $\mathbb{R}_+^{1+m}$. Recall that for problem (4.2.14) we have

$$q_p^+(\bar{d}) = \inf_{x \in X} p(d_0 f_0(x), d_1 f_1^+(x), \dots, d_m f_m^+(x)), \quad \bar{d} \in \mathbb{R}_+^{1+m}, \quad (4.2.26)$$

where $\bar{d} = (d_0, d)$. It follows immediately from the definition of an IPH function that q_p^+ is also IPH. Since the function q_p^+ is increasing we can consider only the restriction of q_p^+ to $\mathbb{R}_{++}^{1+m}$. We denote this restriction by the same symbol q_p^+.

Consider the support set $\text{supp}(q_p^+, L)$ of the IPH function q_p^+ defined on $\mathbb{R}_{++}^{1+m}$ with respect to the set L of all min-type functions:

$$\begin{aligned} \text{supp}(q_p^+, L) &= \{(l_0, l) \gg 0 : \langle (l_0, l), (d_0, d) \rangle \\ &\leq q_p^+(d_0, d)) \text{ for all } (d_0, d) \in \mathbb{R}_{++}^{m+1}\}, \end{aligned}$$

where

$$\langle (l_0, l), (d_0, d) \rangle = \min(l_0 d_0, \langle l, d \rangle) \text{ and } \langle l, d \rangle = \min_{i \in J} l_i x_i. \quad (4.2.27)$$

Here $J = \{1, \dots, m\}$. The following assertion describes links between the value M_p^* of the dual problem (D_p^+) and the support set $\text{supp}(q_p^+, L)$.

PROPOSITION 4.1 $M_p^* = \sup\{l_0 : (l_0, l) \in \text{supp}(q_p^+, L)\}$.

Proof: Let $(l_0, l) \in \mathbb{R}_{++}^{1+m}$. Then

$$\sup_{d \in \mathbb{R}_{++}^m} \min(l_0, \langle l, d \rangle) = l_0. \quad (4.2.28)$$

Indeed, the inequality $\sup_{d \in \mathbb{R}_+^m} \min(l_0, \langle l, d \rangle) \leq l_0$ is obvious. On the other hand, for each l_0 there exists $d \in \mathbb{R}_{++}^m$ such that $\langle l, d \rangle \geq l_0$, so (4.2.28) holds. Using (4.2.28), we have

$$\begin{aligned} M_p^* &= \sup_{\bar{d} \in \mathbb{R}_{++}^{1+m}, d_0 = 1} q_p(\bar{d}) \\ &= \sup_{\bar{d} \in \mathbb{R}_{++}^{m+1}, d_0 = 1} \sup_{(l_0, l) \in \text{supp}(q_p^+, L)} \langle (l_0, l), (d_0, d) \rangle \\ &= \sup_{d \in \mathbb{R}_{++}^m} \sup_{(l_0, l) \in \text{supp}(q_p^+, L)} \min(l_0, \langle l, d \rangle) \\ &= \sup_{(l_0, l) \in \text{supp}(q_p^+, L)} \sup_{d \in \mathbb{R}_{++}^m} \min(l_0, \langle l, d \rangle) \\ &= \sup_{(l_0, l) \in \text{supp}(q_p^+, L)} l_0. \end{aligned}$$

Thus the result follows. △

Proposition 4.1 shows that the optimal value, M_p^*, of problem (D_p^+) is equal to the supremum of all values l_0, where $(l_0, l) \in \text{supp}(q_p, L)$.

4.2.7. Another approach

The different scheme for investigation of the zero duality gap property for extended Lagrange and penalty functions was proposed by M. Yu. Andramonov [3]. These functions are constructed with the help of increasing convolution functions, however instead of Properties A) and B) a different set is assumed to hold, which guarantees that the zero duality gap property is valid.

A combination of nonlinear penalty functions with classical Lagrange functions is examined in [3] as well.

The combined Lagrange-penalty extended function is defined by

$$U(x, d, r) = G(L(x, d), r_1 f_1(x), \dots, r_m f_m(x)),$$

where $L(x, d) = f_0(x) + \sum_{i=1}^m d_i f_i(x)$ is a classical Lagrange function. Then the auxiliary problem to be solved is

$$\sup_{d \in \mathbb{R}_+^m} \inf_{x \in X} U(x, d, \bar{r}), \tag{4.2.29}$$

where $\bar{r}$ is is a fixed vector. The zero duality gap property can be proved under some assumptions. In particular, it is assumed that the following properties hold:

1) G is an increasing function on $\mathbb{R}^{m+1}$.

2) All the constraints f_i are bounded from below on $\mathbb{R}^n$.

3) $G(y_0, y_1, \dots, y_m) \to +\infty$ whenever $\max_{1 \le i \le m} y_i \to +\infty$ and y_0 is bounded from below.

A similar scheme was considered in [197]. However, the scheme from [3] does not require that a saddle point of the classical Lagrangian should exist. If this point exists then the following result is valid (see [3]).

THEOREM 4.2 *Let $f_i(\bar{x}) \le 0$ for all $1 \le i \le m$ and let $\bar{d}$ be a vector such that the following complementary slackness condition holds:*

$$\bar{d}_i f_i(\bar{x}) = 0 \qquad \textit{for all } 1 \le i \le m.$$

Then $(\bar{x}, \bar{d})$ is a saddle point for the function $U(x, d, \bar{r})$ for any fixed vector $\bar{r}$ of penalty parameters.

Even if no saddle points exist, the optimal value of the auxiliary problem (4.2.29) tends to M under mild assumptions. Consider the set-valued mapping $A_\varepsilon : \mathbb{R}_+ \to X$ where

$$A(\varepsilon) = \{x \in X : f_i(x) \leq \varepsilon, \ i \in J\}.$$

THEOREM 4.3 *Let the mapping $A(\varepsilon)$ be upper semicontinuous. Then for any $\varepsilon > 0$ there exists $Q > 0$ such that whenever $\min_{1 \leq i \leq m} \overline{r}_i > Q$ we have*

$$M - \varepsilon \leq \sup_{d \in \mathbb{R}^m_+} \inf_{x \in X} U(x, d, \overline{r}) \leq M.$$

We now indicate connections between the optimal solutions of the initial constrained problem and the saddle points of the auxiliary problem.

THEOREM 4.4 *Let $(\overline{x}, \overline{d})$ be a saddle point for problem (4.2.29). Then $\overline{x}$ is an optimal solution of the initial problem.*

Note that the scheme has improved numerical efficiency for a number of nonsmooth constrained optimization problems (see [3] for details).

4.3. EXTENDED PENALIZATION FOR PROBLEMS WITH ONE CONSTRAINT

4.3.1. Overview

In this section we shall examine the following constrained optimization problem (*COP*):

$$f_0(x) \longrightarrow \inf \text{ subject to } x \in X, \ f_1(x) \leq 0, \qquad (4.3.1)$$

where X is a metric space and $f_i : X \to \mathbb{R}$, $i = 0, 1$. Note that a more general problem

$$f_0(x) \longrightarrow \inf \text{ subject to } x \in X, \ g_i(x) \leq 0, i \in J \qquad (4.3.2)$$

can be represented by (4.3.1) with $f_1(x) = \sup_{i \in J} g_i(x)$.

We shall consider an exact penalization by means of IPH function defined on $\mathbb{R}^2_+$. The theory of IPH functions and decreasing functions, presented in Chapter 2 allows us to avoid the assumption A) (see (4.2.1)) and to build a complete theory of extended penalization with an IPH convolution function. The main results of this section can be found in [157] and [166].

4.3.2. Preliminaries

It will be more convenient if we denote problem (4.3.1) by $P(f_0, f_1)$ and its value by $M(f_0, f_1)$. Thus

$$M(f_0, f_1) = \inf_{x \in X, f_1(x) \le 0} f_0(x).$$

We will require the following assumption:

ASSUMPTION 4.2 1) $\inf_{x \in X} f_0(x) := \gamma > 0$.

2) There exist $x_1, x_2 \in X$ such that $f_1(x_1) \le 0$, $f_1(x_2) > 0$.

Note that 1) is stronger than Assumption 4.1. Denote by C_X^1 the class of all problems $P(f_0, f_1)$ with finite functions f_0 and f_1 defined on a set X and such that Assumption 4.2 holds.

Let $P(f_0, f_1) \in C_X^1$ and let X_0 be the set of all feasible solutions for $P(f_0, f_1)$:

$$X_0 = \{x \in X : f_1(x) \le 0\}. \tag{4.3.3}$$

Suppose that we have a continuous IPH function p defined on $\mathbb{R}^2_+$. The dual penalty function of the problem $P = P(f_0, f_1)$ corresponding to p has the following form (see (4.2.17)):

$$q_p^+(d_0, d) = \inf_{x \in X} p(d_0 f_0(x), d f_1^+(x)) \equiv \inf_{x \in X} \mathcal{L}_p^+(x, d_0, d), \quad d_0 > 0, d > 0. \tag{4.3.4}$$

The corresponding dual problem is

$$(D_p^*) \qquad q_p^+(1, d) \longrightarrow \sup \text{ subject to } d > 0, \tag{4.3.5}$$

where q_p^+ is defined by (4.3.4).

We will denote by $M_p^*(f_0, f_1)$ the value of the dual problem:

$$M_p^*(f_0, f_1) = \sup\{q_p^+(1, d) : d > 0\}.$$

Note that the function $d \mapsto q_p^+(1, d)$ is increasing and $M_p^*(f_0, f_1) = \lim_{d \to +\infty} q_p^+(1, d)$.

We can express the function $q_p^+(d_0, d)$ defined by (4.3.4) in the following form:

$$q_p^+(d_0, d) = \inf_{x \in X} \mathcal{L}_p^+(x, d_0, d) = \min\{\inf_{x \in X_0} \mathcal{L}_p^+(x, d_0, d), \inf_{x \notin X_0} \mathcal{L}_p^+(x, d_0, d)\}. \tag{4.3.6}$$

Let

$$X_1 = \{x \in X : f_1(x) > 0\} = \{x \in X : x \notin X_0\} \tag{4.3.7}$$

and

$$r_p(d_0, d) = \inf_{x \in X_1} p(d_0 f_0(x), d f_1(x)). \tag{4.3.8}$$

PROPOSITION 4.2 *We have*

$$q_p^+(d_0, d) = \min(\inf_{x \in X_0} \mathcal{L}_p^+(x, d_0, d), r_p(d_0, d)).$$

Proof: The proof follows directly from (4.3.6) and (4.3.8). △

We now study the function r_p defined on $\mathbb{R}^2_{++}$ by (4.3.8).

PROPOSITION 4.3 *r_p is an IPH function.*

Proof: First, we demonstrate that r_p is positively homogeneous of degree one. We have for $\lambda > 0$:

$$\begin{aligned} r_p(\lambda d_0, \lambda d) &= \inf_{x \in X_1} p(\lambda d_0 f_0(x), \lambda d f_1(x)) \\ &= \lambda \inf_{x \in X_1} p(d_0 f_0(x), d f_1(x)) = \lambda r_p(d_0, d). \end{aligned}$$

Next, we show that r_p is increasing. Let $d_0' \geq d_0$ and $d' \geq d$. Since $f_0(x) > 0$ and $f_1(x) > 0$ for all $x \in X_1$ and p is an increasing function, we have

$$\begin{aligned} r_p(d_0', d') &= \inf_{x \in X_1} p(d_0' f_0(x), d' f_1(x)) \\ &\geq \inf_{x \in X_1} p(d_0 f_0(x), d f_1(x)) = r_p(d_0, d). \end{aligned}$$

△

Let s be an IPH function defined on $\mathbb{R}^2_+$. We shall consider the support set, $\text{supp}_1(s, L)$, of the function s, where L is the set of all min-type functions $x \mapsto \langle l, x \rangle \equiv \min(l_1 x_1, l_2 x_2)$ defined on $\mathbb{R}^2_{++}$ with $l = (l_1, l_2) \gg 0$. Recall (see Definition 2.12) that

$$\text{supp}_1(s, L) = \{l \in L : \langle l, x \rangle \leq s(x) \text{ for all } x \in \mathbb{R}^2_{++}\}.$$

In other words, $\text{supp}_1(s, L)$ is the usual support set of the restriction of s to $\mathbb{R}^2_{++}$. We shall denote this restriction by the same letter s, if it does not cause misunderstanding.

Thus we have two IPH functions defined on $\mathbb{R}^2_+$, namely p and r_p. We now describe the support set, $\text{supp}_1(r_p, L)$, of r_p in terms of the set $\text{supp}_1(p, L)$.

PROPOSITION 4.4 *Let p be an IPH function defined on $\mathbb{R}^2_+$. Consider a problem $P(f_0, f_1) \in C^1_X$ and let the set X_1 and the function r_p be defined for this problem by (4.3.7) and (4.3.8) respectively . Then the following holds:*

$$\text{supp}_1(r_p, L) = \bigcap_{x \in X_1} (f_0(x_0), f_1(x)) \cdot \text{supp}_1(p, L), \tag{4.3.9}$$

where the product $a \cdot U$ is defined by (2.2.15): $a \cdot U = \{a \cdot u : u \in U\}$, where $a \cdot u = (au_1, au_2)$.

Proof: We have for $d_0, d > 0$:

$$r_p(d_0, d) = \inf_{x \in X_1} p(d_0 f_0(x), df_1(x)) = \inf_{x \in X_1} q_p^x(d_0, d),$$

where for $x \in X_1$

$$q_p^x(d_0, d) = p((f_0(x), f_1(x)) \cdot (d_0, d)) = p((f_0(x), f_1^+(x)) \cdot (d_0, d)).$$

It follows from Proposition 2.25 and Proposition 2.27 that

$$\text{supp}_1(r_p, L) = \bigcap_{x \in X_1} \text{supp}_1(q_p^x, L) = \bigcap_{x \in X_1} (f_0(x), f_1(x)) \cdot \text{supp}_1(p, L). \quad \triangle$$

4.3.3. Perturbation functions

We now study the perturbation function for the problem $P(f_0, f_1) \in C^1_X$ defined by (4.3.1).

DEFINITION 4.4 The function β defined on $\mathbb{R}_+ = \{y \in \mathbb{R} : y \geq 0\}$ by

$$\beta(y) = \inf\{f_0(x) : x \in X, \ f_1(x) \leq y\} \tag{4.3.10}$$

is called the *perturbation function* of the problem $P(f_0, f_1)$.

The value $\beta(0)$ of the perturbation function at the origin coincides with the value $M(f_0, f_1)$ of the problem $P(f_0, f_1)$. We also have

$$\inf_{y>0} \beta(y) = \inf_{y>0} \inf_{x \in X, \ f_1(x) \leq y} f_0(x) = \inf_{x \in X} f_0(x).$$

Since $\inf_{x \in X} f_0(x) = \gamma > 0$ (by Assumption 4.2), $\inf_{y>0} \beta(y) = \gamma > 0$. It follows directly from the definition that the perturbation function is decreasing, i.e., $y_1 \geq y_2 \implies \beta(y_1) \leq \beta(y_2)$.

ASSUMPTION 4.3 Let X_0 and X_1 be the sets defined by (4.3.3) and (4.3.7) respectively. There exists a sequence $x_k \in X_1$ such that $f_1(x_k) \to$

0 and $f_0(x_k) \to M(f_0, f_1)$, where $M(f_0, f_1) = \inf_{x \in X_0} f_0(x)$ is the value of the problem (4.3.1).

If Assumption 4.3 holds, then $\inf_{x \in X_1, f_1(x) \le y} f_0(x) \le M(f_0, f_1)$ for each $y > 0$. Keeping in mind that

$$\begin{aligned}\beta(y) &= \min(\inf_{x \in X_0} f_0(x), \inf_{x \in X_1, f_1(x) \le y} f_0(x)) \\ &= \min(M(f_0, f_1), \inf_{x \in X_1, f_1(x) \le y} f_0(x)),\end{aligned}$$

we conclude that

$$\beta(y) = \inf_{x \in X_1, f_1(x) \le y} f_0(x), \qquad y > 0. \tag{4.3.11}$$

PROPOSITION 4.5 *Let p be an IPH function defined on $\mathbb{R}^2_{++}$ and let $U = \mathrm{supp}_1(p, L)$. Let Assumptions 4.2 and 4.3 hold. Then for the function r_p defined by (4.3.8) we have*

$$\mathrm{supp}_1(r_p, L) = \bigcap_{y>0} (\beta(y), y) \cdot U.$$

Proof: Let

$$A = \bigcap_{x \in X_1} (f_0(x), f_1(x)) \cdot U \text{ and } B = \bigcap_{y>0} (\beta(y), y) \cdot U.$$

It follows from Proposition 4.4 that $\mathrm{supp}_1(r_p, L) = A$. We now check that $B \subset A$. Let $x \in X_1$ and $y = f_1(x)$. Then (see (4.3.11))

$$\beta(y) = \inf_{x' \in X_1, f_1(x') \le y} f_0(x') \le f_0(x)$$

and therefore $(f_0(x), f_1(x)) \ge (\beta(y), y)$. It follows from Lemma 2.1 that $(f_0(x), f_1(x)) \cdot U \supset (\beta(y), y) \cdot U$ so

$$A = \bigcap_{x \in X_1} (f_0(x), f_1(x)) \cdot U \supset \bigcap_{y = f_1(x),\ x \in X_1} (\beta(y), y) \cdot U \supset B.$$

We now prove that $A \subset B$. Let $y > 0$. We shall check that

$$(\beta(y), y) \cdot U \supset \bigcap_{x \in X_1} (f_0(x), f_1(x)) \cdot U. \tag{4.3.12}$$

Assume to the contrary that (4.3.12) does not hold. Then there exists a vector $v = (v_1, v_2)$ such that

$$v \in \bigcap_{x \in X_1} (f_0(x), f_1(x)) \cdot U \tag{4.3.13}$$

and $v \notin (\beta(y), y) \cdot U$. Since U is closed, $\lambda v \notin (\beta(y), y) \cdot U$ for some positive number $\lambda < 1$. In other words,

$$(v_1, v_2) \notin \left(\frac{\beta(y)}{\lambda}, \frac{y}{\lambda}\right) \cdot U. \tag{4.3.14}$$

Combining the inequality $(\beta(y)/\lambda, y/\lambda) \geq (\beta(y)/\lambda, y)$ and Lemma 2.1, we conclude that

$$\left(\frac{\beta(y)}{\lambda}, \frac{y}{\lambda}\right) \cdot U \supset \left(\frac{\beta(y)}{\lambda}, y\right) \cdot U. \tag{4.3.15}$$

It follows from (4.3.14) and (4.3.15) that

$$(v_1, v_2) \notin \left(\frac{\beta(y)}{\lambda}, y\right) \cdot U. \tag{4.3.16}$$

Let $\varepsilon = (\beta(y)/\lambda) - \beta(y)$. Then $\varepsilon > 0$. It follows from (4.3.11) that there exists a vector $\bar{x} \in X_1$ such that $f_1(\bar{x}) \leq y$ and $f_0(\bar{x}) \leq \beta(y) + \varepsilon$. Combining (4.3.13) and Lemma 2.1 we deduce that

$$v \in (f_0(\bar{x}), f_1(\bar{x})) \cdot U \subset (\beta(y) + \varepsilon, y) \cdot U = \left(\frac{\beta(y)}{\lambda}, y\right) \cdot U$$

which contradicts (4.3.16). Thus we have proved that (4.3.12) holds for arbitrary $y > 0$. It follows from (4.3.12) that

$$B = \bigcap_{y>0} (\beta(y), y) \cdot U \supset \bigcap_{x \in X_1} (f_0(x), f_1(x)) \cdot U = A. \qquad \triangle$$

THEOREM 4.5 *The function h_{r_p} coincides with the multiplicative inf-convolution $\beta \diamond h_p$ of β and h_p.*

Proof: The proof of Theorem 4.5 follows immediately from Proposition 3.21 and Proposition 4.5. $\triangle$

REMARK 4.3 The perturbation function β does not depend on the IPH function p, and the associated function h_p does not depend on the problem $P(f_0, f_1)$. Thus Theorem 4.5 allows one to display the influence of both an outer IPH function p and of the problem $P(f_0, f_1)$ itself to the associated function h_{r_p}.

We need the following simple assertion.

PROPOSITION 4.6 *Let β be the perturbation function of the problem $P(f_0, f_1)$. Then the following are equivalent:*

1) β is lower semicontinuous at the origin;

2) β is continuous at the origin;

3) $\beta(0) = \sup_{y>0} \beta(y)$.

Proof: Since $\beta(y)$ is a decreasing function, there exists

$$\lim_{y\to+0} \beta(y) = \sup_{y>0} \beta(y) \leq \beta(0) \equiv M(f_0, f_1).$$

Hence, the function β is upper semicontinuous at the origin and the equality

$$M(f_0, f_1) = \lim_{y\to+0} \beta(y) \tag{4.3.17}$$

holds if and only if β is lower semicontinuous at the point zero. The proposition directly follows from this assertion. △

We now give simple sufficient conditions for the lower semicontinuity of the perturbation function. More general conditions can be found for example in the book [118], Section 3.1.

PROPOSITION 4.7 *Let f_0 and f_1 be lower semicontinuous functions defined on a metric space X. Assume that there exists $y' > 0$ such that the set $\{x \in X : f_1(x) \leq y'\}$ is compact. Then the function β is lower semicontinuous at the origin.*

Proof: Let $y_k \to 0$ and $x_k \in \arg\min_{x\in X:\, f_1(x)\leq y_k} f_0(x)$, that is $f_1(x_k) \leq y_k$ and $f_0(x_k) = \beta(y_k)$. Without loss of generality, assume that $x_k \to \bar{x} \in X_0$. Since f_1 is lower semicontinuous, $f_1(\bar{x}) \leq \liminf_{k\to+\infty} f_1(x_k) \leq 0$. Since f_0 is lower semicontinuous it follows that

$$\liminf_{k\to+\infty} \beta(y_k) = \liminf_{k\to+\infty} f_0(x_k) \geq f_0(\bar{x}) \geq \inf_{x\in X_0} f_0(x) = \beta(0). \quad △$$

THEOREM 4.6 *Let Assumption 4.2 and Assumption 4.3 hold and let p be a continuous IPH function defined on $\mathbb{R}^2_+$. Let h_p be the associated function for p and assume that*

$$\lim_{z\to+\infty} h_p(z) = 1 \text{ and } \lim_{z\to+0} h_p(z) = +\infty.$$

Then

$$\sup_{z>0} h_{r_p}(z) = \sup_{d>0} r_p(1, d) = \inf_{x\in X_0} f_0(x) \tag{4.3.18}$$

if and only if the perturbation function β is lower semicontinuous at the point zero.

Proof: This follows directly from Theorem 4.5, Theorem 3.6, Remark 3.16 and Proposition 3.14. △

The conditions in Theorem 4.6 are given in terms of the associated function h_p of an IPH function p. Applying Proposition 3.16 and Proposition 3.17 we can represent conditions ensuring the validity of (4.3.18) in terms of the function p itself.

THEOREM 4.7 *Let conditions in Theorem 4.6 hold. Let $p(1,0) = 1$ and $\lim_{u\to+\infty} p(1,u) = +\infty$. Then (4.3.18) holds if and only if the function β is lower semicontinuous at the point zero.*

Let us now consider the dual problem (4.3.5) to the problem $P(f_0, f_1)$. Recall that the value $M_p^*(f_0, f_1)$ of this problem is calculated by

$$M_p^*(f_0, f_1) = \sup_{d>0} q_p^+(1,d) := \sup \mathcal{L}_p^+(x,1,d),$$

where

$$\mathcal{L}_p^+(x,d_0,d) = p(d_0 f_0(x), d f_1^+(x)).$$

LEMMA 4.3 *Let p be an IPH function defined on $\mathbb{R}^2_+$ with $p(1,0) = 1$. Then $M_p^*(f_0, f_1) \le M(f_0, f_1)$ and $\mathcal{L}_p^+(x,d_0) = f_0(x)$ for all $x \in X_0$ and $d > 0$.*

Proof: Let $d > 0$ and $x \in X_0$. Since $f_1^+(x) = 0$ and p is positively homogeneous with $p(1,0) = 1$, we have:

$$f_0(x) = p(f_0(x), 0) = p(f_0(x), d f_1^+(x)) = \mathcal{L}_p^+(x,1,d).$$

Also

$$\mathcal{L}_p^+(x,1,d) \ge \inf_{x'\in X} p(f_0(x'), d f_1^+(x')) = q_p^+(1,d).$$

Thus $M(f_0, f_1) = \inf_{x\in X_0} f_0(x) \ge \sup_{d>0} q_p^+(1,d) = M_p^*(f_0, f_1)$. △

Denote by $\mathcal{P}_+$ the class of all continuous IPH functions defined on $\mathbb{R}^2_+$ and such that

$$p(1,0) = 1, \qquad \lim_{u\to+\infty} p(1,u) = +\infty.$$

LEMMA 4.4 *Let $p \in \mathcal{P}_+$. Then $M_p^*(f_0, f_1) = M := \lim_{y\to+0} \beta(y)$.*

Proof: By Proposition 4.2 and Lemma 4.3 we have

$$q_p^+(1,d) = \min\{\inf_{x\in X_0} \mathcal{L}_p^+(x,1,d), r_p(1,d)\} = \min\{M(f_0, f_1), r_p(1,d)\}. \tag{4.3.19}$$

Proposition 3.16 and Proposition 3.17 show that

$$\lim_{z\to+\infty} h_{r_p}(z) = 1 \text{ and } \lim_{z\to+0} h_{r_p}(z) = +\infty. \tag{4.3.20}$$

Applying Theorem 4.5 and Theorem 3.6 we can conclude that

$$\sup_{d>0} h_(r_p) = \lim_{d\to+0} h_{r_p}(d) = \lim_{y\to+0} \beta(y) = M.$$

It follows from Proposition 3.14 that

$$\lim_{d\to+0} r_p(1,d) = \sup_{d>0} r_p(1,d) = \sup_{d>0} h_{r_p} = M. \tag{4.3.21}$$

Since $M(f_0, f_1) \geq M$ we have, by applying (4.3.19) and (4.3.21), that:

$$\begin{aligned} M_p^*(f_0, f_1) &= \lim_{d\to+\infty} q_p^+(1,d) = \lim_{d\to+\infty} \min\{M(f_0, f_1), r_p(1,d)\} \\ &= \min\{M(f_0, f_1), M\} = M. \end{aligned}$$

△

REMARK 4.4 Let $\lim_{y\to+0} \beta(y) = M(f_0, f_1)$. Then $q_p^+(1,d) = r_p(1,d)$ for all $d > 0$. Indeed, since the function $d \mapsto r_p(1,d)$ is increasing, it follows from (4.3.21) that $r_p(1,d) \leq M(f_0, f_1)$ for all $d > 0$. Applying (4.3.19) we can conclude that $q_p^+(1,d) = r_p(1,d)$ for all $d > 0$.

THEOREM 4.8 *Let $p \in \mathcal{P}_+$. Then $M_p^*(f_0, f_1) = M(f_0, f_1)$ if and only if the perturbation function β is lower semicontinuous at the origin.*

Proof: The proof follows directly from Lemma 4.4. △

Let X be a metric space. Consider the subset C_X of the set C_X^1, which consists of all problems $P(f_0, f_1)$ with f_0 and f_1 defined on X, such that

A_1) $\inf_{x\in X} f_0(x) > 0$;

A_2) there exists a sequence $x_k \in X_1$ such that $f_1(x_k) \to 0$ and $f_0(x_k) \to M(f_0, f_1)$;

A_3) there exists a point $x \in X$ such that $f_1(x) \leq 0$;

A_4) the perturbation function of the problem $P(f_0, f_1)$ is lower semicontinuous at the point $y = 0$.

Note that A_1), A_2) and A_3) hold if and only if both Assumption 4.2 and Assumption 4.3 are valid. Theorem 4.8 demonstrates that each IPH function $p \in \mathcal{P}_+$ provides the penalization with the zero duality gap property for each $P(f_0, f_1) \in C_X$. We now show that a continuous IPH function, which enjoys this property, belongs to $\mathcal{P}_+$.

PROPOSITION 4.8 *Let p be an IPH continuous function defined on $\mathbb{R}^2_+$. If*

$$\sup_{d_1>0} r_p(1, d_1) = M(f_0, f_1) \tag{4.3.22}$$

for all problems $P(f_0, f_1) \in C_X$, then $p \in \mathcal{P}_+$, that is, $p(1,0) = 1$ and $\lim_{u\to+0} p(1,u) = +\infty$.

Proof: Let $p(1,0) = a$. First assume that $a < 1$. Consider an arbitrary problem $P(f_0, f_1) \in C_X$. It follows from A_2) that there exists a sequence $x_k \in X_1$ such that $f_1(x_k) \to 0$ and $f_0(x_k) \to M(f_0, f_1)$. We have for each $d_1 > 0$:

$$\begin{aligned} r_p(1, d_1) &= \inf_{x\in X_1} p(f_0(x), d_1 f_1(x)) = \inf_{x\in X_1} f_0(x) p\left(1, d_1 \frac{f_1(x)}{f_0(x)}\right) \\ &\le \lim_{k\to+\infty} f_0(x_k) p\left(1, d_1 \frac{f_1(x_k)}{f_0(x_k)}\right) \\ &= M(f_0, f_1) \lim_{k\to+\infty} p\left(1, d_1 \frac{f_1(x_k)}{f_0(x_k)}\right) \\ &= M(f_0, f_1) p(1,0) = aM(f_0, f_1). \end{aligned}$$

Thus $\sup_{d_1>0} r(1, d_1) \le M(f_0, f_1)a < M(f_0, f_1)$, which contradicts the equality (4.3.22).

Assume that $a > 1$. Consider a problem $P(f_0, f_1) \in C_X$ such that $\inf_{x\in X} f_0(x) = \inf_{x\in X_1} f_0(x) = \gamma > (M(f_0, f_1)/a)$. We have for each $d_1 > 0$:

$$\begin{aligned} r(1, d_1) &= \inf_{x\in X_1} f_0(x) p\left(1, \frac{d_1 f_1(x)}{f_0(x)}\right) \\ &\ge \gamma p(1,0) > \frac{M(f_0, f_1)}{a} a = M(f_0, f_1). \end{aligned}$$

Thus $\sup_{d_1>0} r_p(1, d_1) > M(f_0, f_1)$ and we have a contradiction. We have proved that $a = p(1,0) = 1$.

We now check that $\lim_{u\to+\infty} p(1,u) = +\infty$. Assume to the contrary that this limit is equal to $c < +\infty$. It is easy to see that there exists a problem $P(f_0, f_1) \in C_X$ and a point $x' \in X_1$ such that

$$f_0(x') < \frac{M(f_0, f_1)}{c}. \tag{4.3.23}$$

In fact, we can consider for example a function f_0, which is defined by a point $\bar{x} \in X$ and two neighbourhoods of this point, N_1 and N_2, such that $N_1 \subset \operatorname{int} N_2$. Assume that f_0 is a continuous function, such that $f_0(x) = u_1$ for $x \in N_1$, $f_0(x) = u_2$ for $x \notin N_2$ and all values of f_0 are placed between numbers u_1 and u_2. Let $f_1(x) = \rho(x, \bar{x})$, where ρ is a metric on X. Then $X_0 = \{\bar{x}\}$ and $M(f_0, f_1) = u_1$. Clearly, $P(f_0, f_1) \in C_X$. If $u_2 < u_1/c$ and $x' \notin N_2$ then (4.3.23) holds. For a problem $P(f_0, f_1)$ such that (4.3.23) is valid we have

$$\begin{aligned}
\sup_{d_1>0} r_p(1, d_1) &= \sup_{d_1>0} \inf_{x \in X_1} f_0(x) p\left(1, \frac{d_1 f_1(x)}{f_0(x)}\right) \\
&\leq \sup_{d_1>0} f_0(x') p\left(1, d_1 \frac{f_1(x')}{f_0(x')}\right) \\
&= f_0(x') \lim_{u \to +\infty} p(1, u) < \frac{M(f_0, f_1)}{c} c \\
&= M(f_0, f_1),
\end{aligned}$$

which is again a contradiction. △

4.3.4. Exact penalization

Consider the optimization problem $P(f_0, f_1)$ defined by (4.3.1). In this section we will discuss the existence of an (exact) penalty parameter, that is a number $d > 0$ such that $M_p^* = q_p^+(1, d)$, for a given IPH function p.

If Assumption 4.2 and Assumption 4.3 hold and the perturbation function β is lower semicontinuous at the origin then (see Remark 4.4) $q_p^+(1, d) = r_p(1, d)$ where r_p is defined by (4.3.8). It has been shown (see Theorem 4.5) that the function h_{r_p} associated to r_p can be represented as the multiplicative inf-convolution of the perturbation function β and the function h_p: $h_{r_p} = \beta \diamond h_p$. We will use this formula in the study of the problem under consideration.

We shall show that in contrast to the zero duality gap property, which depends only on a problem $P(f_0, f_1)$ and does not depend on an outer convolution function p, the exactness of penalization depends on an outer function p. More precisely, we shall demonstrate that for each problem $P(f_0, f_1) \in C_X$, where C_X is a class defined by properties $A_1) - A_4)$ in Subsection 4.3.3, there exists an IPH function $p \in \mathcal{P}_+$ such that the penalization by p is exact. On the other hand there is no $p \in \mathcal{P}_+$, which provides the exact penalization for each problem from C_X, that is for each $p \in \mathcal{P}_+$ there exists $P(f_0, f_1) \in C_X$ such that the penalization of this problem by p is not exact. However, the zero duality gap property holds.

The following propositions will be useful in this section.

PROPOSITION 4.9 *Let $P(f_0, f_1) \in C_X$ and $p \in \mathcal{P}_+$. Then $r_p(1, \bar{d}) = M(f_0, f_1)$ if and only if the associated function h_{r_p} is constant on the segment $[0, M(f_0, f_1)/\bar{d}]$, that is,*

$$h_{r_p}(y) = M(f_0, f_1), \quad 0 \le y \le \frac{M(f_0, f_1)}{\bar{d}}.$$

Proof: Denote $M(f_0, f_1) = M$. Assume that there exists $\bar{d}$ such that $r_p(1, \bar{d}) = M$. The function $r_p(1, d)$ is increasing and $\sup_{d>0} r_p(1, d) = M$, so $r_p(1, d) = M$ for all $d \ge \bar{d}$. Thus for $d/d_0 \ge \bar{d}$ we have

$$r_p(d_0, d) = d_0 r_p(1, d/d_0) = d_0 M.$$

For the support set, $\mathrm{supp}_1(r_p, L)$, of the function r_p the following is valid (see Theorem 2.3)

$$\mathrm{supp}_1(r_p, L) = \left\{ \ell = (l_0, l_1) : r_p\left(\frac{1}{l_0}, \frac{1}{l_1}\right) \ge 1 \right\}.$$

Consider the point $(l_0, l_1) = (M, M/\bar{d})$. We have $(\frac{1}{l_1})(\frac{1}{l_0})^{-1} = \bar{d}$. Thus

$$r_p\left(\frac{1}{l_0}, \frac{1}{l_1}\right) = \frac{1}{l_0} r_p\left(1, \frac{1}{l_1}\left(\frac{1}{l_0}\right)^{-1}\right) = \frac{1}{l_0} r_p(1, \bar{d}) = \frac{1}{M} M = 1.$$

Hence $(M, M/\bar{d}) \in \mathrm{supp}_1(r_p, L)$. The set $\mathrm{supp}_1(r_p, L)$ is normal so $\{(l_0, l_1) \ge 0 : l_0 \le M, l_1 \le M/\bar{d}\} \subset \mathrm{supp}_1(r_p, L)$. By the definition of the associated function we have

$$h_{r_p}(y) = \sup\{\alpha : (\alpha, y) \in \mathrm{supp}_1(r_p, L)\}.$$

So if $0 \le y \le M/\bar{d}$ then $h_{r_p}(y) \ge M$. On the other hand, $h_{r_p}(y)$ is a decreasing function with $h_{r_p}(0) = M$. Thus $h_{r_p}(y) = M$ if $y \le M/\bar{d}$.

Assume now that $h_{r_p}(y) = M$ for $0 \le y \le \bar{y} = M/\bar{d}$. Since

$$h_{r_p}(\bar{y}) = \sup\{\alpha : (\alpha, \bar{y}) \in \mathrm{supp}_1(r_p, L)\},$$

$\mathrm{supp}_1(r_p, L)$ is closed and $h_{r_p}(\bar{y}) < +\infty$, we can deduce that $(M, \bar{y}) \in \mathrm{supp}(r_p, L)$. Thus

$$r_p\left(\frac{1}{M}, \frac{\bar{d}}{M}\right) = \frac{1}{M} r_p(1, \bar{d}) \ge 1,$$

that is, $r_p(1, \bar{d}) \ge M$. On the other hand, $r_p(1, d) \le M$ for all d. Thus $r_p(1, \bar{d}) = M$. △

EXAMPLE 4.5 Let $p(\alpha, y) = \max\{\alpha, y\}$ and β be the perturbation function of the problem $P(f_0, f_1)$. Assume that β is continuous. Then (see Example 3.11) $h_{r_p}(z) = (\beta \diamond h_p)(z) = \beta(z)$. Let $d > 0$. Since the "upper boundary" of the support set $\text{supp}_1(r_p, L)$ coincides with the graph $\{(h_{r_p}(z), z) : z > 0\}$ of the function h_{r_p}, we can conclude that

$$\begin{aligned} r_p(1,d) &= \max\{\langle l, (1,d)\rangle : l \in \text{supp}_1(r_p, L)\} \\ &= \max_{z>0} \langle (h_{r_p}(z), z), (1,d)\rangle \\ &= \max_{z>0} \min(h_{r_p}(z), zd) = \max_{z>0} \min(\beta(z), zd). \end{aligned}$$

Assume that the perturbation function β is strictly decreasing for sufficiently small y, that is, $\inf\{f_0(x) : f_1(x) \le y_1\} > \inf\{f_0(x) : f_1(x) \le y_2\}$ whenever $y_1 < y_2$. Then $M(f_0, f_1) = \lim_{y \to 0} \beta(y) > \beta(z)$ for all $z > 0$. Since $\max_{z>0} \min(\beta(z), zd) = \beta(z_d)$ where z_d is a solution of the equation $\beta(z) = zd$, we have

$$q_p^+(1,d) = r_p(1,d) = \max_{z>0} \min(\beta(z), zd) = \beta(z_d) < M(f_0, f_1).$$

Thus there is no $d > 0$ such that $r_p(1, d) = M(f_0, f_1)$.

PROPOSITION 4.10 *Let X be a metric space with the following property (*): there exists a function $g : X \to \mathbb{R}_+$ such that $g(X) = \mathbb{R}_+$. Let $\beta : \mathbb{R}_+ \to \mathbb{R}_+$ be a decreasing function, which is lower semicontinuous at the origin and such that*

$$\inf\{\beta(y) : y > 0\} > 0. \tag{4.3.24}$$

Then there exists a problem $P(f_0, f_1) \in C_X$ with the perturbation function equal to β.

Proof: Let $f_0(x) = \beta(g(x)), f_1(x) = g(x)$. We have

$$X_0 = \{x \in X : f_1(x) \le 0\} = \{x \in X : g(x) = 0\}.$$

Thus $f_0(x) = \beta(0)$ for all $x \in X_0$. We now check that the conditions $A_1) - A_3)$ from the definition of the class C_X are valid.

1) The inequality (4.3.24) implies $\inf_{x \in X} f_0(x) > 0$, so $A_1)$ holds.

2) Let $y_k \in \mathbb{R}_+, y_k \to 0$ and let $x_k \in X$ be such that $g(x_k) = y_k (k = 1, 2, \cdots)$. Then $f_0(x_k) = \beta(y_k) \to \beta(0)$. Thus $A_2)$ holds.

3) Since g maps onto $\mathbb{R}_+$, there exists $x \in X$ such that $f_1(x) = g(x) = 0$. Hence $A_3)$ is valid.

We now calculate the perturbation function of the problem $P(f_0, f_1)$. We have

$$\inf\{f_0(x) : f_1(x) \leq y\} = \inf\{\beta(g(x)) : g(x) \leq y\}.$$

Since g maps onto $\mathbb{R}_+$, it follows that there exists $x' \in X$ such that $g(x') = y$. Since β is decreasing we have

$$\inf\{\beta(g(x)) : g(x) \leq y\} = \beta(g(x')) = \beta(y).$$

Thus the perturbation function coincides with β. Since β is lower semicontinuous at zero it follows from the definition of the class C_X that the condition A_4) is also valid. △

We now show that there is no a function $p \in \mathcal{P}_+$ such that for each problem $(f_0, f_1) \in C_X$, there exists $\bar{d}$ with the property $r_p(1, \bar{d}) = M(f_0, f_1)$.

THEOREM 4.9 *Let X be a set with the property (*) (see Proposition 4.10). Then for each $p \in \mathcal{P}_+$ there exists a problem $P(f_0, f_1) \in C_X$ such that $r_p(1, d) < M(f_0, f_1)$ for all $d > 0$.*

Proof: Let $p \in \mathcal{P}_+$. Consider the associated function h_p of the function p. Since $p \in \mathcal{P}_+$, Proposition 3.16 and Proposition 3.17 imply that

$$\lim_{y \to +0} h_p(y) = +\infty, \qquad \lim_{y \to +\infty} h_p(y) = 1. \tag{4.3.25}$$

Consider a strictly increasing continuous function l defined on $\mathbb{R}_+$ such that $l(y)/y$ is strictly decreasing and

$$\lim_{y \to +0} \frac{l(y)}{y} = +\infty, \qquad \lim_{y \to +\infty} \frac{l(y)}{y} = c, \tag{4.3.26}$$

where $c > 0$ is a given number. For example (4.3.26) is fulfilled for the following function:

$$l(y) = \begin{cases} 2c\sqrt{y} & \text{if } y \leq 1; \\ c(y+1) & \text{if } y \geq 1. \end{cases}$$

Consider now the function

$$g(y) = \begin{cases} \dfrac{1}{h_p\left(\frac{l(y)}{y}\right)} & \text{if } y > 0; \\ 1 & \text{if } y = 0. \end{cases}$$

Since both $h_p(z)$ and $l(y)/y$ are decreasing, g is decreasing as well. By applying (4.3.25) and (4.3.26), we also have that $\lim_{y \to +0} g(y) = 1$. Thus

g is continuous at the origin. Since the associated function h_p is upper semicontinuous and $y \mapsto l(y)/y$ is continuous, we conclude that g is lower semicontinuous. Assume that the number c in (4.3.26) is chosen so that $h_p(c) < +\infty$. Since h_p is upper semicontinuous we have

$$\lim_{y\to+\infty} h_p\left(\frac{l(y)}{y}\right) \le \limsup_{u\to c} h_p(u) \le h_p(c) < +\infty.$$

Thus

$$\inf_{y>0} g(y) = \lim_{y\to+\infty} g(y) = \frac{1}{\lim_{y\to+\infty} h_p(\frac{l(y)}{y})} \ge \frac{1}{h_p(c)} > 0.$$

Let $\tilde{l}$ be a continuous strictly decreasing function defined on $\mathbb{R}_+$ with $\tilde{l}(0) = 1$ and $\inf_{y\ge c} \tilde{l}(y) > 0$. Set

$$\beta(y) = g(y)\tilde{l}(y), \quad y \ge 0.$$

Then β is a lower semicontinuous decreasing function with $\inf_{y>0} \beta(y) > 0$. Proposition 4.10 implies that there exists a problem $P(f_0, f_1) \in C_X$ such that β coincides with the perturbation function of this problem. The value $M(f_0, f_1)$ of the problem $P(f_0, f_1)$ is $\beta(0) = g(0)\tilde{l}(0) = 1$. Let us calculate the associated function h_{r_p} for the function $r_p(d_0, d_1)$ constructed for the problem $P(f_0, f_1) \in C_X$. We have (see Theorem 4.5) $h_{r_p} = \beta \diamond h_p$, that is,

$$h_{r_p}(z) = \inf_{y>0} \beta(y) h_p\left(\frac{z}{y}\right) \quad (z > 0). \tag{4.3.27}$$

For a given $z > 0$, take $y_z = l^{-1}(z)$. Then $z = l(y_z)$ and

$$\begin{aligned} h_{r_p}(z) &\le \beta(y_z) h_p\left(\frac{z}{y_z}\right) = g(y_z)\tilde{l}(y_z) h_p\left(\frac{l(y_z)}{y_z}\right) \\ &= \tilde{l}(y_z) < 1 = M(f_0, f_1). \end{aligned}$$

Thus $h_{r_p}(z) < M(f_0, f_1)$ for all $z > 0$. It follows from Proposition 4.9 that $r_p(1, d_1) < M(f_0, f_1)$ for all $d_1 > 0$. △

We conclude this subsection by showing that for each $P(f_0, f_1) \in C_X$ it is possible to find an IPH function $p \in \mathcal{P}^+$, which provides the exact penalization of this problem.

THEOREM 4.10 *For each $P(f_0, f_1) \in C_X$ there exists an IPH function $p \in \mathcal{P}_+$ and a number $d > 0$ such that*

$$M(f_0, f_1) = q_p^+(1, d).$$

Proof: Let β be the perturbation function of the problem $P(f_0, f_1)$. Consider the lower regularization $\mathrm{cl}\,\beta$ of the function β (see Definition 2.2) :

$$(\mathrm{cl}\,\beta)(y) = \min(\beta(y), \liminf_{y' \to y, y' \neq y} \beta(y')).$$

Clearly, $\mathrm{cl}\,\beta$ is decreasing and lower semicontinuous. Since β is lower semicontinuous at the origin, $(\mathrm{cl}\,\beta)(0) = \beta(0) = M(f_0, f_1)$. Let

$$g(y) = \begin{cases} +\infty & \text{if } 0 < y \leq 1; \\ \frac{M(f_0,f_1)}{(\mathrm{cl}\,\beta)(y^{-1})} & \text{if } y > 1. \end{cases}$$

Clearly, g is a decreasing function. Since $\mathrm{cl}\,\beta$ is lower semicontinuous it follows that g is upper semicontinuous. We also have

$$\lim_{y \to +\infty} g(y) = \lim_{u \to +0} \frac{M(f_0, f_1)}{(\mathrm{cl}\,\beta)(u)} = 1.$$

Since g is upper semicontinuous we conclude (see Proposition 3.13) that there exists a normal closed set $U \subset \mathbb{R}^2_{++}$ such that $g = g_U$. Consider an IPH function $\bar{p}$ defined on $\mathbb{R}^2_{++}$ by

$$\bar{p}(y) = \sup\{\min(l_1 y_1, l_2 y_2) : l \in U\}.$$

Since the function $y \to \bar{p}(1, y)$ is increasing, $a = \lim_{y \to +0} p(1, y) < +\infty$. Let the function p be defined on $\mathbb{R}^2_+$ by

$$p(y_1, y_2) = \begin{cases} \bar{p}(y_1, y_2) & \text{if } y_1, y_2 > 0; \\ 0 & \text{if } y_1 = 0; \\ a y_1 & \text{if } y_2 = 0. \end{cases}$$

It is easy to check that p is a continuous IPH function and $h_p(y) = g(y)$ for $y > 0$. Since $\lim_{y \to +\infty} h_p(y) = \lim_{y \to +\infty} g(y) = 1$, it follows from Proposition 3.16 that $a = p(0, 1) = 1$. The equality $h_p(y) = +\infty$ for $y \leq 1$ shows that $\lim_{u \to +\infty} p(1, u) = +\infty$. Hence $p \in \mathcal{P}_+$. Theorem 4.7 implies that $M(f_0, f_1) = \sup\{h_{r_p}(z) : z > 0\}$, so $h_{r_p}(z) \leq M(f_0, f_1)$ for all $z > 0$. On the other hand, we have

$$h_{r_p}(z) = \inf_{y>0} \beta(y) h_p\left(\frac{z}{y}\right) = \inf_{y>0} \beta(y) g\left(\frac{z}{y}\right) = \inf_{y<z} \beta(y) \frac{M(f_0, f_1)}{\mathrm{cl}\,\beta\left(\frac{y}{z}\right)}.$$

Let $z = 1$. If the function β is continuous at a point y then $\beta(y) = (\mathrm{cl}\,\beta)(y)$; otherwise $\beta(y) \geq (\mathrm{cl}\,\beta)(y)$. So

$$h_{r_p}(1) = \inf_{y<1} \frac{\beta(y)}{(\mathrm{cl}\,\beta)(y)} M(f_0, f_1) = M(f_0, f_1).$$

Thus $h_{r_p}(z) = M(f_0, f_1)$ for $z \leq 1$. It follows from Proposition 4.9 that there exists a number $\bar{d} > 0$ such that $r_p(1, \bar{d}) = M(f_0, f_1)$. △

4.3.5. Penalization by IPH functions p_k

In this subsection we discuss the penalization by the convolution function p_k ($k > 0$) defined by

$$p_k(\delta, y) = \left(\delta^k + y^k\right)^{\frac{1}{k}} \qquad (\delta > 0,\ y > 0). \tag{4.3.28}$$

For the sake of simplicity, we shall denote the function r_{p_k} by $r_{[k]}$ and its associated function $h_{r_{[k]}}$ by $h_{[k]}$. The following assertion will be useful in the study of both conditions for the exact penalization and of estimations of penalty parameters.

LEMMA 4.5 *Let Assumption 4.2 and Assumption 4.3 hold. Let $k > 0$ and p_k be the function defined by (4.3.28). Let $h_{[k]}$ be the associated function to r_{p_k}. Then $M(f_0, f_1) = h_{[k]}(z)$ if and only if*

$$M(f_0, f_1) \le \beta(y)\frac{z}{\left(z^k - y^k\right)^{\frac{1}{k}}} \text{ for } 0 < y < z. \tag{4.3.29}$$

Proof: It easily follows from Example 3.10 that

$$h_{p_k}(y) = \begin{cases} \frac{y}{(y^k-1)^{1/k}} & \text{if } y > 1; \\ +\infty & \text{if } y \le 1. \end{cases}$$

Since $\lim_{y\to+\infty} h_{p_k}(y) = 1$ and $\lim_{y\to+0} h_{p_k}(y) = +\infty$ we can apply Theorem 4.6 and deduce that

$$M(f_0, f_1) = M^*_{p_k} = \sup_{z>0} h_{[k]}(z) = \lim_{z\to 0} h_{[k]}(z). \tag{4.3.30}$$

By Theorem 4.5 we have

$$h_{[k]}(z) = \inf_{y>0} \beta(y) h_{p_k}\left(\frac{z}{y}\right) = \inf_{0<y<z} \beta(y) h_{p_k}\left(\frac{z}{y}\right) = \inf_{0<y<z} \beta(y)\frac{z}{\left(z^k - y^k\right)^{\frac{1}{k}}}.$$

It follows from (4.3.30) that $M(f_0, f_1) \ge h_{[k]}(z')$ for all z', so $M(f_0, f_1) = h_{[k]}(z)$ if and only if $M(f_0, f_1) \le h_{[k]}(z)$, that is,

$$M(f_0, f_1) \le \inf_{0<y<z} \beta(y)\frac{z}{\left(z^k - y^k\right)^{\frac{1}{k}}}. \tag{4.3.31}$$

Thus the result follows. △

DEFINITION 4.5 A function $f : \mathbb{R}^n \to \mathbb{R}_{+\infty}$ is said to be *calm* of degree $k \in (0, +\infty)$ at a point $x_0 \in \operatorname{dom} f$ if

$$\liminf_{x\to x_0} \frac{f(x) - f(x_0)}{\|x - x_0\|^k} > -\infty. \tag{4.3.32}$$

The notion of calmness (see for example [28, 30, 136] and references therein) has been used to study exact penalization with classical penalty functions generated by the IPH function p_1, where $p_1(y_1, y_2) = y_1 + y_2$. Definition 4.5 can be represented in the following equivalent form: f is calm of degree k at a point x_0 if there exists $C \in \mathbb{R}$ such that

$$f(x) - f(x_0) \geq C\|x - x_0\|^k$$

in a neighbourhood of the point x_0.

THEOREM 4.11 *Let $P(f_0, f_1) \in C_X$. Let $k > 0$ and let $p = p_k$ be defined by (4.3.28). Then there exists a number $\tilde{d} > 0$ such that $M(f_0, f_1) = q_p^+(1, \tilde{d})$ if and only if β is calm of degree k at the origin, that is,*

$$\liminf_{y \to +0} \frac{\beta(y) - \beta(0)}{y^k} > -\infty. \tag{4.3.33}$$

Proof: It follows from Proposition 4.9 that $M(f_0, f_1) = r_{[k]}(1, \tilde{d})$ for some $\tilde{d} > 0$ if and only if there exists $z > 0$ such that $M(f_0, f_1) = h_{[k]}(z)$. According to Lemma 4.5, this equality holds if and only if

$$\frac{\beta(y)}{M(f_0, f_1)} \geq \left(1 - \left(\frac{y}{z}\right)^k\right)^{\frac{1}{k}} \quad \text{for } 0 < y < z. \tag{4.3.34}$$

Since $M(f_0, f_1) = \beta(0)$ we can represent (4.3.34) in the following form:

$$\frac{\beta(y) - \beta(0)}{y^k} \geq \frac{\beta(0)}{y^k}\left(\left(1 - \left(\frac{y}{z}\right)^k\right)^{\frac{1}{k}} - 1\right) \quad \text{for } 0 < y < z.$$

Let $u = y/z$. Then $u \in (0, 1)$ and

$$\frac{\beta(y) - \beta(0)}{y^k} \geq \frac{\beta(0)}{z^k} \frac{(1 - u^k)^{\frac{1}{k}} - 1}{u^k} \quad \text{for } 0 < y < z. \tag{4.3.35}$$

Assume that β is calm of degree k at the origin. Then $\inf_{0<y<y'}(\beta(y) - \beta(0))y^{-k} = v > -\infty$ for a fairly small positive y. Let w be an upper bound of the function $|(1 - u^k)^{\frac{1}{k}} - 1|/u^k$ on the segment $[0, 1]$ and $z^k < \frac{\beta(0)}{|v|} w$. Then

$$|v| \leq \frac{\beta(0)}{z^k} \frac{|(1 - u^k)^{\frac{1}{k}} - 1|}{u^k}.$$

Hence

$$\frac{\beta(y)-\beta(0)}{y^k} \geq v \geq \frac{\beta(0)}{z^k}\frac{(1-u^k)^{\frac{1}{k}}-1}{u^k}.$$

Thus (4.3.35) holds. Assume now that (4.3.35) is valid. Then

$$\frac{\beta(y)-\beta(0)}{y^k} \geq \frac{\beta(0)}{z^k} \inf_{0<u<1} \frac{(1-u^k)^{\frac{1}{k}}-1}{u^k} > -\infty.$$

Hence (4.3.33) holds. △

REMARK 4.5 The family (P_y) of perturbed problems

$$f_0(x) \longrightarrow \min \text{ subject to } x \in X,\ f_1^+(x) := \max(f_1(x), 0) \leq y,\ (y > 0)$$

is said to be *calm* at the point zero if (4.3.33) holds for $k = 1$. Using Theorem 4.11 it is possible to derive the well-known result (see for example [28, 30, 136]) that there exists a number $\tilde{d} > 0$ such that $M(f_0, f_1) = q_{p_1}(1, \tilde{d})$ if and only if the family (P_y) is calm.

For many problems the exact penalization with respect to the classical penalty function p_1 can be accomplished only with very large penalty parameters which leads to ill-conditioned unconstrained optimization problems. Applying penalization with respect to another convolution function p, we can sometimes decrease the penalty parameter, which provides exact penalization. We shall next study this problem for the convolution function $p_{\frac{1}{2}}$. First, we consider an arbitrary $k > 0$ and describe the least penalty parameter, that is, the smallest number for which the equality $M(f_0, f_1) = r_{[k]}(1, d)$ holds.

ASSUMPTION 4.4 Assume that the perturbation function $\beta(y)$ is continuous on $[0, +\infty)$ and $\beta(y) < M(f_0, f_1)$ for $y > 0$.

Let Assumption 4.4 hold and let $k > 0$. Set

$$v_k(y) = \frac{y}{\left(1 - \left(\frac{\beta(y)}{\beta(0)}\right)^k\right)^{\frac{1}{k}}} \qquad (y > 0) \tag{4.3.36}$$

and

$$\varphi_k(z) = \inf_{0<y<z} v_k(y). \tag{4.3.37}$$

Since $v_k(y)$ is continuous for $y > 0$, it follows that $\varphi_k(z) = \inf_{0<y\leq z} v_k(y)$ is continuous for $z > 0$. Obviously, φ_k is a decreasing function.

LEMMA 4.6 *Let $P(f_0, f_1) \in C_X$ and let Assumption 4.4 be valid. Assume that (4.3.33) holds for $k > 0$. Then the least exact penalty parameter, $\bar{d}_k$, of the problem $P(f_0, f_1)$ with respect to the function p_k is equal to $1/\bar{z}_k$, where $\bar{z}_k$ is the solution of the equation $z = \varphi_k(z)$.*

Proof: It follows from Proposition 4.9 that $r_{[k]}(1, d) = M(f_0, f_1)$ if and only if $h_{[k]}(z) = M(f_0, f_1)$ with $z = 1/d$. Lemma 4.5 demonstrates that the equality $M(f_0, f_1) = h_{[k]}(z)$ is valid if and only if (4.3.29) holds, that is,

$$\frac{\beta(0)}{\beta(y)} \leq \frac{z}{(z^k - y^k)^{\frac{1}{k}}} \text{ for } 0 < y < z. \tag{4.3.38}$$

We have

$$\begin{aligned}
\frac{\beta(0)}{\beta(y)} &\leq \frac{z}{(z^k - y^k)^{\frac{1}{k}}} \iff \beta(0)^k(z^k - y^k) \leq z^k \beta(y)^k \\
&\iff z^k(\beta(0)^k - \beta(y)^k) \leq \beta(0)^k y^k \iff z \leq \frac{\beta(0)y}{\beta(0)^k - \beta(y)^k)^{\frac{1}{k}}} \\
&= \frac{y}{\left(1 - \left(\frac{\beta(y)}{\beta(0)}\right)^k\right)^{\frac{1}{k}}} = v_k(y).
\end{aligned}$$

Hence (4.3.38) is equivalent to

$$z \leq v_k(y) \text{ for } 0 < y < z,$$

where v_k is defined by (4.3.36), so $M(f_0, f_1) = h_{[k]}(z)$ if and only if

$$z \leq \inf_{0<y<z} v_k(y) := \varphi_k(z).$$

Since (4.3.33) holds, it follows from Theorem 4.11 that there exists $z > 0$ such that $M(f_0, f_1) = h_{[k]}(z)$, that is, $z \leq \varphi_k(z)$. Thus $r_{[k]}(1, d) = M(f_0, f_1)$ if and only if $z \leq \varphi_k(z)$, so the least exact penalty parameter $\bar{d}_k$ is equal to the inverse of the greatest element $\bar{z}_k$ of the set $\{z > 0 : z \leq \varphi_k(z)\}$. Note that this set is nonempty. Since φ_k is a continuous decreasing function, we can deduce that the equation $z = \varphi_k(z)$ has a unique solution. It is clear that this solution coincides with $\bar{z}_k$. △

REMARK 4.6 It was shown in the proof of Lemma 4.6 that $\bar{z}_k$ is the greatest element of the set $\{z : z \leq \varphi_k(z)\}$.

Let $\beta(y) < M(f_0, f_1)$ for $y > 0$. Set

$$u(y) = \frac{1 + \sqrt{\frac{\beta(y)}{\beta(0)}}}{1 - \sqrt{\frac{\beta(y)}{\beta(0)}}} \qquad (y > 0). \tag{4.3.39}$$

Since β is a decreasing and continuous function, it follows that u is a decreasing and continuous function as well.

LEMMA 4.7 *Let Assumption 4.4 hold and let v_k, φ_k and u be functions defined by (4.3.36),(4.3.37) and (4.3.39) respectively. Then*

$$\varphi_{\frac{1}{2}}(z) \geq \varphi_1(z)u(z) \qquad (z > 0).$$

Proof: We have

$$v_1(y) = \frac{y}{1 - \frac{\beta(y)}{\beta(0)}}, \qquad v_{\frac{1}{2}}(y) = \frac{y}{\left(1 - \sqrt{\frac{\beta(y)}{\beta(0)}}\right)^2}.$$

An easy calculation shows that

$$v_{\frac{1}{2}}(y) = u(y)v_1(y) \qquad (y > 0).$$

Let $z > 0$ and $0 < y < z$. Since $\varphi_1(z) \leq v_1(y)$ and u is a decreasing function, it follows that

$$\varphi_{\frac{1}{2}}(z) = \inf_{0<y<z} v_{\frac{1}{2}}(y) \geq \varphi_1(z) \inf_{0<y<z} u(y) = \varphi_1(z)u(z). \qquad \triangle$$

LEMMA 4.8 *Suppose $P(f_0, f_1) \in C_X$, Assumption 4.4 holds and β is calm of degree one at the origin. Let $\bar{d}_k$ be the least exact penalty parameter with respect to the function p_k and $\bar{z}_k = 1/\bar{d}_k$ $(k = 1, \frac{1}{2})$. Let $\mu \in (0, 1)$ be a number such that*

$$\beta(z) \geq \mu\beta(0) \text{ for } z \in (0, \lambda\bar{z}_1), \tag{4.3.40}$$

where

$$\lambda = \frac{1 + \sqrt{\mu}}{1 - \sqrt{\mu}}. \tag{4.3.41}$$

Then

$$\bar{d}_{\frac{1}{2}} \leq \frac{1 - \sqrt{\mu}}{1 + \sqrt{\mu}} \bar{d}_1.$$

Proof: Let λ be the number defined by (4.3.41). Since the function $a \mapsto (1 + \sqrt{a})(1 - \sqrt{a})^{-1}$ is increasing and $\beta(z)/\beta(0) \geq \mu$ for $z \in \lambda\bar{z}_1$, we have

$$u(z) \geq \frac{1 + \sqrt{\mu}}{1 - \sqrt{\mu}} = \lambda \text{ for } 0 < z < \lambda\bar{z}_1.$$

Hence (see Lemma 4.7)

$$\varphi_{\frac{1}{2}}(z) \geq u(z)\varphi_1(z) \geq \lambda\varphi_1(z) \text{ for } 0 < z \leq \lambda\bar{z}_1. \tag{4.3.42}$$

The equality (4.3.41) shows that $\lambda > 1$, hence $\bar{z}_1 < \lambda\bar{z}_1$. It follows from (4.3.42) that $\varphi_{\frac{1}{2}}(\bar{z}_1) \geq \lambda\varphi_1(\bar{z}_1)$. By the definition of the number $\bar{z}_1$ we have $\bar{z}_1 = \varphi_1(\bar{z}_1)$. Thus

$$\varphi_{\frac{1}{2}}(\bar{z}_1) \geq \lambda\bar{z}_1.$$

Since (see Remark 4.6) $\bar{z}_{\frac{1}{2}}$ is the greatest element of the set $\{z > 0 : z \leq \varphi_{\frac{1}{2}}(z)\}$, we deduce that $\bar{z}_{\frac{1}{2}} \geq \lambda\bar{z}_1$. Thus $\bar{d}_{\frac{1}{2}} \leq \frac{1}{\lambda}\bar{d}_1$. △

THEOREM 4.12 *Assume all conditions in Lemma 4.8 hold. Then*

$$\bar{d}_{\frac{1}{2}} \leq \frac{1 - \sqrt{\frac{\gamma}{M(f_0,f_1)}}}{1 + \sqrt{\frac{\gamma}{M(f_0,f_1)}}}\bar{d}_1, \tag{4.3.43}$$

where $\gamma = \inf_{x\in X} f_0(x) = \inf_{z>0} \beta(z)$.

Proof: Since $\beta(z) \geq \mu\beta(0) = M(f_0, f_1)$ for all z, where $\mu = \gamma/M(f_0, f_1)$, the proof follows from Lemma 4.8. △

Theorem 4.12 allows us to draw the following conclusion:

COROLLARY 4.1 *The exact penalty parameter $\bar{d}_{\frac{1}{2}}$ is always less than $\bar{d}_1$. If the constrained minimum $M(f_0, f_1)$ is not very distant from the unconstrained minimum γ, then the penalization by the convolution function $p_{\frac{1}{2}}$ can be accomplished with a substantially smaller exact penalty parameter than that of the classical penalty function p_1.*

The following simple example confirms this conclusion.

EXAMPLE 4.6 Let $0 < b < c < a$ be real numbers and $X = [0, c]$. Consider the problem $P(f_0, f_1)$ with $f_0(x) = (a - x)^2$, $f_1(x) = x - b$:

$$(a - x)^2 \longrightarrow \min \text{ subject to } x - b \leq 0,\ x \in X. \tag{4.3.44}$$

We have $M(f_0, f_1) = (a - b)^2$, $\beta(y) = (a - b + y)^2$ for $y > 0$. Thus $P(f_0, f_1) \in C_X$ and Assumption 4.4 holds.

Let $k = 1/2$ and $r_{[\frac{1}{2}]} = r_{p_k}$. We have $r_{[\frac{1}{2}]}(1, d) = \min_{b\leq x\leq c} h^2(x)$ where $h(x) = (a - x) + d^{\frac{1}{2}}(x - b)^{\frac{1}{2}}$. Since h is a concave function, h attains its minimum over the segment $[b, c]$ at one of the endpoints of this segment. Since h is positive we can conclude that h^2 also attains its minimum at one of these endpoints. Thus

$$r_{[\frac{1}{2}]}(1,d) = \begin{cases} [(a-c)+d^{\frac{1}{2}}(c-b)^{\frac{1}{2}}]^2 & \text{if } 0 < d \le c-b; \\ (a-b)^2 & \text{if } d > c-b. \end{cases} \tag{4.3.45}$$

An easy calculation shows that $\sup_{d>0} r_{[\frac{1}{2}]}(1,d) = (a-b)^2$ is attained at the point $c-b$. Using (4.3.45) we conclude that $\bar{d}_{\frac{1}{2}} = c-b$. Note that $\bar{d}_{[\frac{1}{2}]}$ does not depend on a.

Consider now the classical penalty function with $k=1$. We have

$$r_{[1]}(1,d) = \min_{b \le x \le c} ((a-x)^2 + d(x-b)).$$

An easy calculation shows that $r_{[1]}(1,d) = -(1/4)d^2 + d(a-b)$ for $d \ge 2(a-c)$, hence $\sup_{d>0} r_{[1]}(1,d) = (a-b)^2$ and $\bar{d}_1 = 2(a-b)$. Thus $\bar{d}_1 \to +\infty$ as $a \to +\infty$.

PROPOSITION 4.11 *The upper bound in (4.3.43) is asymptotically sharp in the following sense. For each $\varepsilon > 0$ there exists a metric space X and a problem $P(f_0, f_1) \in C_X$ such that the difference between expressions in the right-hand side and the left-hand side of (4.3.43) is less than ε.*

Proof: Consider the problem (4.3.44) from Example 4.6. We have

$$M(f_0, f_1) = (a-b)^2, \ \gamma = (a-c)^2, \ \bar{d}_1 = 2(a-b), \ \bar{d}_{\frac{1}{2}} = c-b.$$

Hence inequality (4.3.43) can be represented in the following form:

$$c - b \le 2(a-b)\frac{1-((a-c)/(a-b))}{1+((a-c)/(a-b)}. \tag{4.3.46}$$

Note that the difference between expressions in the right-hand side and left-hand side in (4.3.46) is equal to $(c-b)^2(2a-c-b)^{-1}$, so this difference tends to zero as $c-b \to 0$. △

REMARK 4.7 Consider the following problem $P(f_0 + c, f_1)$:

$$f_0(x) + c \longrightarrow \min \ \text{subject to} \ f_1(x) \le 0,$$

which is equivalent to the problem $P(f_0, f_1)$ in the following sense: both problems $P(f_0 + c, f_1)$ and $P(f_0, f_1)$ have the same set of minimizers. Obviously, both of these problems have the same exact penalty parameter $\bar{d}_1$ with respect to the classical penalty function p_1. On the other hand they have different exact penalty parameters $\bar{d}_{\frac{1}{2}}$ with respect to the convolution function $p_{\frac{1}{2}}$. Let $\bar{d}_{\frac{1}{2}}(c)$ be the least penalty parameter with respect to $p_{\frac{1}{2}}$ for the problem $P(f_0 + c, f_1)$. It follows from (4.3.43) that $\bar{d}_{\frac{1}{2}}(c)$ tends to zero as $c \to +\infty$. Note that the corresponding unconstrained optimization problem can become ill-conditioned for very large c.

Chapter 5

ELEMENTS OF STAR-SHAPED ANALYSIS

5.1. INTRODUCTION

Abstract convexity based on the set of linear functions defined on $\mathbb{R}^n$ (as the set of elementary functions) leads to the classical convex analysis. Abstract convexity based on the set of min-type functions

$$x \mapsto \min_{i=1,\dots,n} l_i x_i \qquad (x \in \mathbb{R}^n_{++}) \tag{5.1.1}$$

leads to monotonic analysis on the cone $\mathbb{R}^n_{++}$. Abstract convexity based on the set of functions

$$x \mapsto \min_{i \in I_+(l)} l_i x_i \qquad (x \in \mathbb{R}^n_+) \tag{5.1.2}$$

leads to monotonic analysis on the cone $\mathbb{R}^n_+$. The function $x \mapsto l_i x_i$ $(x \in \mathbb{R}^n)$ is a very special case of a linear function defined on $\mathbb{R}^n$, hence the function $x \mapsto \min_{i=1,\dots,n} l_i x_i$ is a very special case of the function defined on $\mathbb{R}^n$ by

$$x \mapsto \min_{i=1,\dots,k} [l(i), x], \tag{5.1.3}$$

where $l(i) \in \mathbb{R}^n$, $i = 1, \dots, n$ and $[l, x]$ stands for the inner product of vectors l and x. It is quite interesting to build abstract convexity based on different sets of functions of the form (5.1.3). Note that the monotonicity of abstract convex functions with respect to the set of all functions of the form either (5.1.1) or (5.1.2) arises as a corollary of the monotonicity of the functions $x \mapsto l_i x_i$. Since an arbitrary linear function is not necessarily monotone, it follows that sets of elementary functions of the form (5.1.3) do not necessarily lead to monotonicity. We shall consider two sets of such a kind in this chapter: the set $\mathcal{L}_n$ and the set $\mathcal{L}_{n+1}$, which consist of all functions of the form (5.1.3) with $k = n$ and $k = n + 1$, respectively. We show that a function is $\mathcal{L}_{n+1}$-convex if and only if this function is lower semicontinuous and positively homogeneous; we also show that a nonnegative lower semicontinuous positively homogeneous function is $\mathcal{L}_n$-convex. Abstract convexity with respect to the set of $H_{\mathcal{L}_{n+1}} := \mathcal{H}_{n+1}$ of abstract affine functions leads to convexity-along-rays. As it turns out, a function f with $f(0) < +\infty$ is $\mathcal{H}_{n+1}$-convex if and only if f is lower semicontinuous and convex-along-rays. This is no longer valid for a function with $f(0) = +\infty$. We describe a large class of such functions which are $\mathcal{H}_{n+1}$-convex.

Nonnegative positively homogeneous functions have a very interesting geometric interpretation as the Minkowski gauges of radiant (star-shaped with respect to zero) sets and the Minkowski co-gauges of co-radiant (co-star-shaped with respect to zero) sets. We present a detailed discussion of corresponding links. In particular, we give a description of lower semicontinuous, upper semicontinuous, continuous and Lipschitz continuous functions in terms of corresponding sets. Thus we study nonnegative positively homogeneous functions in two ways: as Minkowski gauges and co-gauges, and as $\mathcal{L}_n$-convex functions. Such a combination is a very useful tool for the theory of nonnegative positively homogeneous functions.

We also show that each radiant set is abstract convex with respect to $\mathcal{L}_n$. This result is based on a special kind of a separation of a point from the radiant set. This kind of a separation was first considered by A. Shveidel [182]. Instead of the separation by a hyperplane in the classical convex analysis, we separate a point from a radiant (or co-radiant) set by a cone. Then we show that the cone-separation can be expressed analytically as a separation by a min-type function of the form (5.1.3) with $k = n$. At the same time the separation of a point from a set by a function of the form (5.1.3) with $k = n+1$ can be geometrically expressed as a separation by an n-dimensional simplex. We also study the cone-separation of a point from a closed co-radiant set. This separation leads to the representation of co-radiant sets as abstract concave with respect to $\mathcal{L}_n$, however we will not present results related to abstract concavity here. As it turns out the separability of a point from a star-shaped (co-star-shaped) set leads to a special separability of two star-shaped (co-star-shaped) sets. We study this kind of separability as well.

The totality of proposed approaches and results obtained in this area can be considered as certain elements of star-shaped analysis. This branch of analysis includes also many interesting questions, related in particular to inverse and harmonic sums (see [165, 126]). We believe that star-shaped analysis will find many applications to optimization and dynamical systems.

5.2. RADIANT AND CO-RADIANT SETS AND THEIR GAUGES

5.2.1. Overview

Positively homogeneous functions of degree one have found many applications in various areas of mathematics. In particular, positively homogeneous functions are used in nonsmooth analysis and nonsmooth optimization, since an approximation of nonsmooth functions can often be accomplished using positively homogeneous functions (see, for example, [43]). There is a close connection between certain classes of positively homogeneous functions and certain classes of sets. For example, a positively homogeneous function is convex if and only if this function is the support function of a convex set (see for example [134]). A nonnegative positively homogeneous function is convex if and only if this function is the Minkowski gauge of a convex set ([134]). A positively homogeneous function p defined on the cone $\mathbb{R}^n_{++}$ of all vectors with positive coordinates is increasing if and only if p is the support function of a *normal* set and if and only if p is the Minkowski gauge of a normal set (see Chapter 2).

In this section we study nonnegative positively homogeneous functions defined on the n-dimensional space $\mathbb{R}^n$. Each of these functions can be considered as the Minkowski gauge of a radiant (star-shaped with respect to zero) set and as the Minkowski co-gauge of a co-radiant (co-star-shaped with respect to zero) set. We discuss links between various classes of nonnegative positively homogeneous functions and the corresponding classes of sets. In particular, we describe radiant sets such that the corresponding Minkowski gauges are Lipschitz functions.

The main results of this section can be found in [182, 160, 145].

5.2.2. Radiant and co-radiant sets

In this section we shall consider some special subsets of a closed cone in a finite dimensional vector space. Recall (see Definition 1.8) that a subset Q of a vector space X is called a *conic set* or a *cone* if $\lambda Q \subset Q$ for all $\lambda > 0$. In this book we shall use both terms "cone" and "conic set". Note that a closed conic set contains zero.

Definition 5.1 Let Q be a conic set in $\mathbb{R}^n$. A nonempty set $U \subset Q$ is called a *radiant* subset of the cone Q if

$$(x \in U,\ 0 < \lambda \leq 1) \implies \lambda x \in U.$$

Remark 5.1 L. Bragard (see [24, 25]) used the term *radiant* in a different sense (for subsets of $Q = \mathbb{R}^n$). J.-P. Penot (see [122]) included convexity in the definition of a radiant set.

Let U be a radiant subset of a conic set Q and let $x \in Q$. Consider the set $\Lambda_x = \{\lambda > 0 : x \in \lambda U\}$ and assume that this set is nonempty. It follows directly from the definition that $(\lambda \in \Lambda_x, \lambda' > \lambda) \implies \lambda' \in \Lambda_x$. Hence Λ_x is either an open ray $\{\lambda : \lambda > a\}$ or a closed ray $\{\lambda : \lambda \geq a\}$ with $a \geq 0$. Let $\mu_U(x)$ be the starting point of the ray Λ_x if Λ_x is nonempty and $+\infty$ if Λ_x is empty, that is

$$\mu_U(x) = \inf \Lambda_x \equiv \inf\{\lambda > 0 : x \in \lambda U\}. \tag{5.2.1}$$

It is assumed that the infimum over the empty set is equal to $+\infty$ (see Section 1.2).

Definition 5.2 Let Q be a conic set and U be a radiant subset of Q. The function μ_U defined on Q by (5.2.1) is called the *Minkowski gauge* of the set U (with respect to the cone Q).

The Minkowski gauge is a nonnegative function mapping into $[0, +\infty] = \mathbb{R}_+ \cup \{+\infty\}$.

Let $x \in \mathbb{R}^n$, $x \neq 0$. We shall denote by R_x ($\tilde{R}_x$) the closed ray $\{\lambda x : \lambda \geq 0\}$ (the open ray $\{\lambda x : \lambda > 0\}$).

Let U be a radiant subset of the cone Q. The function $\mu_U : Q \to [0, +\infty]$ enjoys the following properties:

1) $\mu_U(0) = \begin{cases} 0 & \text{if } 0 \in U; \\ +\infty & \text{if } 0 \notin U. \end{cases}$

2) If $x \in Q, x \neq 0$, then

$$\mu_U(x) = 0 \iff \tilde{R}_x \subset U \text{ and } \mu_U(x) = +\infty \iff \tilde{R}_x \cap U = \emptyset. \tag{5.2.2}$$

3) The function μ_U is positively homogeneous (of degree one), that is, $\mu_U(\lambda x) = \lambda \mu_U(x)$ for all $x \in Q$ and $\lambda > 0$.

4) The following inclusions hold:

$$\{x \in Q : \mu_U(x) < 1\} \subset U \subset \{x \in Q : \mu_U(x) \leq 1\}. \tag{5.2.3}$$

Recall (see Definition 2.4) that a set $U \subset Q$ is called closed-along-rays if $(\lambda_n > 0, \lambda_n \to \lambda > 0, \lambda_n x \in U) \implies \lambda x \in U$.

PROPOSITION 5.1 *Let U be a radiant subset of a conic set Q. Then $U = \{x \in Q : \mu_U(x) \leq 1\}$ if and only if U is closed-along-rays.*

Proof: Let U be a closed-along-rays set. We should only check that $U \supset \{x \in Q : \mu_U(x) = 1\}$. Let $x \in Q$, $\mu_U(x) = 1$ and $1 > \lambda_k > 0$, $\lambda_k \to 1$. Then $\mu_U(\lambda_k x) = \lambda_k < 1$, hence $\lambda_k x \in U$. Since U is closed-along-rays it follows that $x \in U$. Assume now that

$$U = \{x \in Q : \mu_U(x) \leq 1\}. \tag{5.2.4}$$

Let $\lambda_k > 0$, $\lambda_k \to \lambda, \lambda_k x \in U$. Then $\mu_U(\lambda_k x) = \lambda_k \mu_U(x) \to \lambda \mu_U(x)$. Since $\lambda_k x \in U$ it follows that $\mu_U(\lambda_k x) \leq 1$, hence $\mu_U(\lambda x) = \lambda \mu_U(x) \leq 1$. It follows from (5.2.4) that $\lambda x \in U$. △

PROPOSITION 5.2 *Let Q be a closed conic set. Let $p : Q \to [0, +\infty]$ and $U = \{x \in Q : p(x) \leq 1\}$. Then the following statements are equivalent:*
1) p is positively homogeneous, lower semicontinuous and $p(0) = 0$;
2) U is a nonempty closed radiant subset of Q and $p = \mu_U$.

Proof: 1) $\implies$ 2). The set U contains zero, hence nonempty. This set is closed. Since p is nonnegative and positively homogeneous, it follows for $x \in U$ and $\lambda \in (0,1)$ that

$$p(\lambda x) = \lambda p(x) \leq p(x) \leq 1,$$

so $\lambda x \in U$. Hence U is radiant. Let $x \in Q$. Then

$$\begin{aligned} \mu_U(x) &= \inf\left\{\lambda > 0 : \frac{x}{\lambda} \in U\right\} = \inf\left\{\lambda > 0 : p\left(\frac{x}{\lambda}\right) \leq 1\right\} \\ &= \inf\{\lambda > 0 : p(x) \leq \lambda\} = p(x). \end{aligned}$$

2) $\implies$ 1). Obviously $\mu_U(0) = 0$ and μ_U is positively homogeneous. We now verify that μ_U is lower semicontinuous. Since the set U is closed, this set is closed-along-rays, so $U = \{x \in Q : \mu_U(x) \leq 1\}$. Note that $0 \in U$. Consider the level sets, $S_c = \{x \in Q : \mu_U(x) \leq c\}$, of the function μ_U for $c \geq 0$. Since $S_c = cU$ for $c > 0$ it follows that the sets S_c are closed for all $c > 0$. The set $S_0 = \bigcap_{c>0} S_c$ is also closed. Thus μ_U is a lower semicontinuous function. △

REMARK 5.2 Let $Q \subset \mathbb{R}^n$ be a closed conic set and let U be a radiant subset of Q. Then U is a radiant subset of any closed cone $P \supset Q$. In particular U is a radiant subset of $\mathbb{R}^n$. Let μ_U^Q (μ_U^P) be the Minkowski gauge of U with respect to Q (with respect to P). Then

$$\mu_U^P(x) = \begin{cases} \mu_U^Q(x) & \text{if } x \in Q; \\ +\infty & \text{if } x \in P \setminus Q. \end{cases}$$

The class of all closed radiant subsets of a closed conic set Q will be denoted by $\mathcal{U}$. Note that $0 \in U$ for all $U \in \mathcal{U}$. So $\bigcap_{t\in T} U_t$ is not empty for an arbitrary family $(U_t)_{t\in T}$ of elements $\mathcal{U}$. The following proposition is obvious.

PROPOSITION 5.3 *Let $U_t \in \mathcal{U}$ $(t \in T)$, where T is an arbitrary index set. Then $\bigcap_{t\in T} U_t \in \mathcal{U}$ and* $\mathrm{cl} \bigcup_{t\in T} U_t \in \mathcal{U}$.

Subsequently, we shall assume that the set $\mathcal{U}$ is ordered by anti-inclusion: $(U_1 \geq U_2) \iff (U_1 \subset U_2)$. Proposition 5.3 implies that $\mathcal{U}$ is the complete lattice with the greatest element $\{0\}$ and the least element Q.

Consider now the class $\mathcal{P}_*$ of all positively homogeneous, lower semi-continuous and nonnegative functions $p : Q \to [0, +\infty]$. This class is equipped with the natural order relation:

$$p_1 \geq p_2 \iff p_1(x) \geq p_2(x) \text{ for all } x \in Q. \tag{5.2.5}$$

It is easy to check that the following holds:

1) Let $(p_t)_{t\in T}$ be a family of functions, $p_t \in \mathcal{P}_*$, $t \in T$. Let $\bar{p}(x) = \sup_{t\in T} p_t(x)$. Then $\bar{p} \in \mathcal{P}_*$.

If f is an arbitrary nonnegative positively homogeneous function defined on Q, then the function $f_*(x) = \sup\{p(x) : p \in \mathcal{P}_*, p \leq f\}$ is easily seen to be the lower regularization of the function f (see Definition 2.2). We suppose here that $\sup \emptyset = 0$ (see Section 1.2). It follows from 1) that $f_* \in \mathcal{P}$ for each positively homogeneous function f.

2) Let $(p_t)_{t\in T}$ be the same family as in 1) and let p_* be the lower regularization of the function p where $p(x) = \inf_{t\in T} p_t(x)$. Then $p_* \in \mathcal{P}_*$.

Let $\phi : \mathcal{U} \to \mathcal{P}_*$ be the mapping defined by $\phi(U) = \mu_U$. Then ϕ is a one-to-one correspondence onto $\mathcal{P}_*$. Indeed, each $p \in \mathcal{P}_*$ has the form $p = \mu_U$ where $U = \{x \in Q : p(x) \leq 1\}$ and $U_1 = U_2$ if and only if $\mu_{U_1} = \mu_{U_2}$.

PROPOSITION 5.4 *The mapping ϕ is an ordered isomorphism between $\mathcal{U}$ and $\mathcal{P}_*$. The following holds:*

1) If $U = \sup_{t\in T} U_t$ (i.e., $U = \bigcap_{t\in T} U_t$), then $\mu_U(x) = \sup_{t\in T} \mu_t(x)$.

2) If $U = \inf_{t\in T} U_t$ (i.e $U = cl \bigcup_{t\in T} U_t$), then $\mu_U(x)$ coincides with the lower semicontinuous hull of the function $x \mapsto \inf_{t\in T} \mu_t(x)$.

Proof: The result holds since ϕ is a one-to-one correspondence onto $\mathcal{P}_*$ and $U_1 \geq U_2$ is equivalent to $\mu_{U_1} \geq \mu_{U_2}$. △

We shall need the following assertion.

PROPOSITION 5.5 *Let $(U_t)_{t\in T}$ be a family of radiant sets and $U = \bigcup_{t\in T} U_t$. Then $\mu_U(x) = \inf_{t\in T} \mu_{U_t}(x)$ for all $x \in Q$.*

Proof: We have

$$\left\{\lambda > 0 : \frac{x}{\lambda} \in \bigcup_{t\in T} U_t\right\} = \bigcup_{t\in T} \left\{\lambda > 0 : \frac{x}{\lambda} \in U_t\right\}.$$

Hence

$$\begin{aligned} \mu_U(x) &= \inf\left\{\lambda > 0 : \frac{x}{\lambda} \in \bigcup_{t\in T} U_t\right\} = \inf_{t\in T} \inf\left\{\lambda > 0 : \frac{x}{\lambda} \in U_t\right\} \\ &= \inf_{t\in T} \mu_{U_t}(x). \end{aligned}$$

Thus the result holds. △

We now turn to co-radiant sets.

DEFINITION 5.3 Let $Q \subset \mathbb{R}^n$ be a closed cone. A nonempty set $V \subset Q$ is called a *co-radiant* subset of Q if $(x \in V,\ \lambda \geq 1) \implies (\lambda x \in V)$.

REMARK 5.3 Co-radiant sets have been considered in [160] as a special subclass of the class of sets, which are star shaped with respect to infinity. J.-P. Penot (see [122]) used the term *co-starshaped* for co-radiant sets. However, we prefer to keep this term for shifts of co-radiant sets.

Let V be a co-radiant set and let $x \in Q$. Assume that the set $\Lambda^x = \{\lambda > 0 : x \in \lambda V\}$ is nonempty. It follows directly from the definition that $(\lambda \in \Lambda^x, \lambda' < \lambda) \implies \lambda' \in \Lambda^x$. Thus if the set Λ^x is bounded then Λ^x is either the segment $(0, c)$ or the segment $(0, c]$ with $c > 0$. Denote the right end of this segment by $\nu_V(x)$. We now define the function ν_V in a general situation.

DEFINITION 5.4 Let V be a co-radiant subset of a closed cone Q. The function ν_V defined on Q by

$$\nu_V(x) = \sup \Lambda^x \equiv \sup\{\lambda > 0 : x \in \lambda V\} \tag{5.2.6}$$

is called the *Minkowski co-gauge* of V (with respect to the cone Q). It is assumed here that the supremum over the empty set is equal to zero.

The Minkowski co-gauges have been studied for example in [16, 149].

Let V be a co-radiant subset of a closed cone Q. The Minkowski co-gauge ν_V enjoys the following properties:

1) If $0 \in V$, then $\nu_V(0) = +\infty$; if $0 \notin V$, then $\nu_V(0) = 0$;

2) $(x \in Q, x \neq 0, \ \nu_V(x) = 0) \iff \tilde{R}_x \cap V = \emptyset$;

3) $\nu_V(x) = +\infty \iff \tilde{R}_x \subset V$;

4) ν_V is a positively homogeneous function;

5) $\{x \in Q : \nu_V(x) > 1\} \subset V \subset \{x \in Q : \nu_V(x) \geq 1\}$.

PROPOSITION 5.6 *Let V be a co-radiant subset of a closed conic set Q. Then $V = \{x \in Q : \nu_V(x) \geq 1\}$ if and only if V is closed-along-rays in the sense of Definition 2.4.*

Proof: The proof is similar to that of Proposition 5.1. △

Let $X \subset Q$. The complement, $Q \setminus X$, of the set X with respect to Q is denoted by $\mathcal{C}X$.

PROPOSITION 5.7 *Let Q be a closed conic set. The nonempty set $U \subset Q$, $U \neq Q$ is radiant if and only if its complement $V = \mathcal{C}U$ is co-radiant. If U is radiant, then $\mu_U = \nu_V$.*

Proof: Let V be co-radiant and let $y \in U$. Let $0 < \lambda \leq 1$. Since $y \notin V$, it follows that $\lambda y \notin V$, that is, $\lambda y \in U$. Thus U is radiant. The same argument shows that V is co-radiant if U is radiant.

Let $x \in Q$. Consider the sets $\Lambda_x = \{\lambda > 0 : x \in \lambda U\}$ and $\Lambda^x = \{\lambda > 0 : x \in \lambda V\}$. We have $\Lambda^x = \mathbb{R} \setminus \Lambda_x$, hence $\mu_U(x) = \inf \Lambda_x = \sup \Lambda^x = \nu_V(x)$. △

REMARK 5.4 Let V be a co-radiant subset of a closed cone Q. Then V is a co-radiant subset of any closed cone $P \supset Q$, in particular V is a co-radiant subset of $\mathbb{R}^n$. Let ν_V^Q (ν_V^P) be the Minkowski co-gauge of the set V with respect to cone Q (P). Then

$$\nu_V^P(x) = \begin{cases} \nu_V^Q(x) & \text{if } x \in Q; \\ 0 & \text{if } x \in P \setminus Q. \end{cases}$$

PROPOSITION 5.8 *Let Q be a closed conic set, $p : Q \to [0, +\infty]$ and $V = \{x \in Q : p(x) \geq 1\}$. The following statements are equivalent:*

a) $p \neq 0$ *is upper semicontinuous, nonnegative and positively homogeneous;*
b) V *is closed co-radiant and* $p = \nu_V$.

Proof: The proof is similar to that of Proposition 5.2. △

5.2.3. Radiative and co-radiative sets

Let Q be a closed cone. We will consider Q as a topological space equipped with the natural topology of a subspace: a fundamental system of neighbourhoods of a point $x \in Q$ is the family of sets $N_\varepsilon = \{y \in Q : \|y - x\| < \varepsilon\}, \varepsilon > 0$. We denote by $\text{int}_Q X$ and $\text{bd}_Q X$ the interior of a set $X \subset Q$ and its boundary in this topology, respectively .

DEFINITION 5.5 Let $Q \subset \mathbb{R}^n$ be a closed cone. A closed nonempty set $U \subset Q$ is called a *radiative* subset of Q if $0 \in \text{int}_Q U$ and if for each $x \in Q$, the ray R_x does not intersect the boundary $\text{bd}_Q U$ more than once.

REMARK 5.5 Radiative sets are close to "irradiés" sets as defined by L. Bragard (see [24, 25]). These sets were considered in [165] as star-shaped at zero sets.

Let U be a radiative set. Since $0 \in \text{int}_Q U$, it follows that the intersection of the ray R_x with U is nonempty for each $x \in Q$.

REMARK 5.6 Let P and Q be closed cones and $P \supset Q$. Then radiative subsets of Q are not necessarily radiative subsets of P (compare with Remark 5.2). The following simple example illustrates this assertion.

EXAMPLE 5.1 Let $A = \{(x_1, x_2) \in \mathbb{R}^2 : x_1 \geq 0, x_2 \geq 0, x_1 + x_2 \leq 1\}$. Then A is a radiative subset of the cone $\mathbb{R}^2_+$ of all vectors with nonnegative coordinates but A is not a radiative subset of $\mathbb{R}^2$.

PROPOSITION 5.9 *Let* U *be a radiative subset of* Q, *let* $x \in U, x \neq 0$ *and* $\lambda \in [0, 1)$. *Then* $\lambda x \in \text{int}_Q U$.

Proof: If $R_x \cap \text{bd}_Q U = \emptyset$ then $R_x \subset \text{int}_Q U$, hence $\lambda x \in \text{int}_Q U$. Assume that $R_x \cap \text{bd}_Q U \neq \emptyset$. Then there exists $\rho > 0$ such that $R_x \cap \text{bd}_Q U = \{\rho x\}$. Let $x' = \rho x$ and let $\Lambda = \{\lambda \in (0, 1) : \lambda x' \notin U\}$. Assume that Λ is not empty. Since U is closed, it follows that Λ is open, hence Λ is the union of disjoint open segments. Denote by (α, β) one of these segments. Then $\beta x' \in \text{bd}_Q U$. Since U is a radiative subset of Q, we have $\beta x' = x'$ hence $\beta = 1$. This observation demonstrates that Λ coincides with the open segment $(\alpha, 1)$. Since $0 \in \text{int}_Q U$ it follows that

$\alpha > 0$, hence $\alpha x' \in \mathrm{bd}_Q U$. However, this fact contradicts the definition of a radiative subset of Q. Thus Λ is empty and $\lambda x' \in U$ for all $\lambda \in [0,1)$. Since $\mathrm{bd}_Q U \cap R_{x'} = \{x'\}$, we have $\lambda x' \in \mathrm{int}_Q U$ for $\lambda < 1$.

We now show that $\rho \geq 1$. Let $\Lambda' = \{\lambda > 1 : \lambda x' \notin U\}$. Then Λ' is an open set, hence Λ' is the union of disjoint open segments. Let (α, β) be one of such segments. Then $\alpha x \in \mathrm{bd}_Q U$, so $\alpha = 1$. If $\beta < +\infty$, then also $\beta x \in \mathrm{bd}_Q U$, which is impossible. Thus $\Lambda' = (1, +\infty)$. Since $x = (1/\rho)x' \in U$, it follows that $1/\rho \notin \Lambda'$, hence $\rho \geq 1$. Consider a number $\lambda < 1$. Since $\rho \geq 1$, we conclude that $(\lambda/\rho) < 1$ and hence $\lambda x = (\lambda/\rho)x' \in \mathrm{int}_Q U$. △

COROLLARY 5.1 *A radiative subset U of a cone Q is a radiant subset of Q.*

It follows from this corollary that we can consider the Minkowski gauge μ_U of a radiative set U. We need the following simple assertion.

LEMMA 5.1 *Let $p : Q \to [0, +\infty]$ be a positively homogeneous and upper semicontinuous function. Let $U = \{x \in Q : p(x) \leq 1\}$. Then $\nu x \in \mathrm{int}_Q U$ for all $x \in U$ and $\nu \in (0,1)$.*

Proof: Let $0 < \nu < 1$ and $x \in U$. Then $p(\nu x) = \nu p(x) \leq \nu < 1$. Since p is upper semicontinuous it follows that a neighbourhood of νx is contained in U. △

PROPOSITION 5.10 *Let $p : Q \to \mathbb{R}_+$ and $U = \{x \in Q : p(x) \leq 1\}$. The following assertions are equivalent:*

1) p is positively homogeneous, nonnegative and continuous;

2) U is a radiative subset of Q and $p = \mu_U$.

Proof: First we show that 1) implies 2). Since p is positively homogeneous it follows that U is radiant. Let $x \in Q$ and $\nu(x) = \sup\{\nu > 0 : \nu x \in \mathrm{int}_Q U\}$. If $\nu(x) = 0$, then $\nu x \notin \mathrm{int}_Q U$ for all $\nu > 0$. Hence $R_x \cap \mathrm{bd}_Q U$ is empty. Indeed, if this intersection is not empty, then there exists $\nu > 0$ such that $\nu x \in \mathrm{bd}_Q U \subset U$. It follows from Lemma 5.1 that $\nu' x \in \mathrm{int}\, Q$ for $\nu' < \nu$ and we obtain a contradiction. If $\nu(x) = +\infty$, then $R_x \subset \mathrm{int}_Q U$ so $R_x \cap \mathrm{bd}_Q U = \emptyset$.

Consider now a point x such that $0 < \nu(x) < +\infty$. Assume without loss of generality that $\nu(x) = 1$. It is easy to check that the intersection $R_x \cap \mathrm{bd}_Q U$ contains the point x. Assume this intersection contains another point $x' = \nu' x$. If $\nu' < 1$, then (see Lemma 5.1) $\nu' x \in \mathrm{int}_Q U$ so $\nu' > 1$. Applying Lemma 5.1 again we conclude that $x = (1/\nu')x' \in$

$\text{int}_Q U$ which contradicts the definition of x'. Thus U is a radiative subset of Q. The equality $p = \mu_U$ is clear.

Next we show that 2) implies 1). Since U is radiant we can consider the Minkowski gauge $p = \mu_U$. We only need to check the continuity of p. Since U is closed, it follows that $p = \mu_U$ is lower semicontinuous. Thus we should check that the sets $U_c = \{x : p(x) < c\}$ are open for all $c > 0$. Since p is positively homogeneous, it is enough to consider the set U_1. If U_1 is not open, then there exist a point x and a sequence x_l such that $x_l \to x$, $p(x) < 1$, $p(x_l) \geq 1$ for all l. Since $0 \in \text{int}_Q U$, we conclude that $x \neq 0$. Let $p(x) < \lambda < 1$. Then $p(x/\lambda) < 1$, hence $x/\lambda \in U$. We also have

$$p\left(\frac{x_l}{\lambda}\right) > 1 \text{ and } \frac{x_l}{\lambda} \to \frac{x}{\lambda}.$$

Thus $x/\lambda \in \text{bd}_Q U$. Since $x \neq 0$ it follows that the intersection $R_x \cap U$ contains more than one point. △

We shall now study co-radiative subsets of a closed cone Q. We shall define such subsets in the following way.

DEFINITION 5.6 A closed nonempty set $V \subset Q$ is called a *co-radiative* subset of Q if

1) $0 \notin V$;

2) each ray R_x with $x \in Q$ does not intersect the boundary $\text{bd}_Q V$ more than once;

3) if the set $V \cap R_x$ is not empty, then this set contains at least two points.

REMARK 5.7 A co-radiative set is called a *star-shaped set with respect to infinity* in [160]. A convex co-radiant set is called a *shady* set in [122] and a *semiconical* set in [102, 103].

Let V be a co-radiative subset of Q. Since V is closed and $0 \notin V$, it follows that for each $x \in Q$, the set V does not contain the ray R_x. If $x \in Q$, then either $V \cap R_x = \emptyset$ or $(\text{bd}_Q V) \cap R_x$ contains a unique point.

PROPOSITION 5.11 *Let V be a co-radiative subset of Q, let $x \in V$ and let $\lambda > 1$. Then $\lambda x \in \text{int}_Q V$.*

Proof: Since $V \cap R_x$ is not empty it follows that $R_x \cap \text{bd}_Q V$ contains a unique point $\rho x = x'$ with $\rho > 0$. Note that $R_{x'} = R_x$. Let $\Lambda = \{\lambda > 1 : \lambda x' \notin V\}$ and assume that Λ is not empty. Since V is closed it follows that Λ is open, hence Λ is the union of disjoint open segments. Let (α, β) be one of such segments. If $\beta < +\infty$ then $\beta x' \in \mathrm{bd}_Q V$ and $\beta > \alpha \geq 1$. This contradicts the definition of a co-radiative subset of Q, hence $\beta = +\infty$ and Λ coincides with the segment $(\alpha, +\infty)$ with $\alpha \geq 1$. Since $\alpha x' \in \mathrm{bd}_Q V$ we conclude that $\alpha = 1$, so $\Lambda = (1, +\infty)$. Let $\bar{\lambda} = \inf\{\lambda > 0 : \lambda x' \in V\}$. Since $0 \notin V$ and V is closed it follows that $\bar{\lambda} > 0$. Clearly, $\bar{\lambda} x' \in \mathrm{bd}_Q V$ so $\bar{\lambda} = 1$. It follows from the definitions of Λ and $\bar{\lambda}$ that the intersection $R_x \cap V$ consists of the unique point $\{x'\}$. We have a contradiction since, by the definition of co-radiative subsets, this intersection contains at least two points. Thus Λ is empty, that is, $\lambda x' \in V$ for all $\lambda > 1$. Since $R_x \cap \mathrm{bd}_Q V = \{x'\}$, it follows that $\lambda x' \in \mathrm{int}_Q V$ for all $\lambda > 1$.

Recall that $x' = \rho x$. It is easy to show that $\rho \leq 1$ (compare with the proof of Proposition 5.9). So, if $\lambda > 1$, then $\lambda x = (\lambda/\rho)x' \in \mathrm{int}_Q V$. $\triangle$

COROLLARY 5.2 *A co-radiative subset V of a cone Q is a co-radiant subset of Q.*

Let X be a topological space. Recall the following definitions (for example see [169]).

DEFINITION 5.7 1) A closed set $Z \subset X$ is called *regular* if $Z = \mathrm{cl}\,\mathrm{int}\,Z$.

2) If Z is regular, then the regular set $C_r Z = \mathrm{cl}\,(CZ)$ is called the *regular complement* to the set Z.

Note (see [169]) that the class of all regular sets ordered by inclusion is a complete lattice. Moreover, this class is a Boolean algebra, where $C_r Z$ is the complement to an element Z. If Z is a regular set, then $\mathrm{bd}\,Z = \mathrm{bd}\,C_r Z$. A regular set Z is called *proper* if $Z \neq X$, $Z \neq \emptyset$.

It follows from Proposition 5.9 and Proposition 5.11 that both radiative subsets and co-radiative subsets of a closed conic set Q are regular subsets of this cone.

PROPOSITION 5.12 *A proper regular set $U \subset Q$ is a radiative subset of Q if and only if U is the regular complement (with respect to the topological space Q) to a co-radiative subset of Q.*

Proof: Note that the definition of an radiant subset of Q can be presented in the following form. A closed set $U \subset Q$ is a radiative subset of Q if and only if

1) $0 \in \mathrm{int}_Q U$;

2) for each $x \in Q$ either $R_x \subset \mathrm{int}_Q U$ or there exists $\lambda > 0$ such that $R_x \cap \mathrm{bd}_Q U = \{\lambda x\}$.

Let U be a radiative set and V be the regular complement to U, that is, $V = \mathrm{cl}\,(Q \setminus U) = Q \setminus \mathrm{int}_Q U$. Therefore, $0 \in \mathrm{int}_Q U$ if and only if $0 \notin V$. As was mentioned above, $\mathrm{bd}_Q U = \mathrm{bd}_Q V$. Thus the intersection $R_x \cap \mathrm{bd}_Q U$ coincides with the intersection $R_x \cap \mathrm{bd}_Q V$. Assume $R_x \cap \mathrm{bd}_Q U$ is nonempty. Since U is a radiative set it follows that this intersection contains a unique point x' and $\lambda x' \notin U$ for all $\lambda > 1$. Hence $R_x \cap \mathrm{bd}_Q V = \{x'\}$ and $R_x \cap V$ contains more than one point x'. If $R_x \cap \mathrm{bd}_Q U = \emptyset$, then $R_x \subset \mathrm{int}_Q U$ so $R_x \cap V = \emptyset$; that is, R_x does not intersect the boundary of V. Thus we have proved that V is co-radiative. Since $U = C_r(C_r U) = C_r V$, the necessity is proved.

Assume now that $U \subset Q$ is the regular complement to a co-radiative set V. The same argument as above demonstrates that $C_r V = U$ is a radiative set. △

COROLLARY 5.3 *Let U be a radiative subset of Q and $V = C_r U$. Then $\mu_U(x) = \nu_V(x)$ for all $x \in Q$.*

Proof: If $R_x \cap \mathrm{bd}_Q U = \emptyset$, then $\mu_U(x) = \nu_V(x) = 0$. If $R_x \cap \mathrm{bd}_Q U \neq \emptyset$, then there exists $\rho > 0$ such that $R_x \cap \mathrm{bd}_Q U = R_x \cap \mathrm{bd}_Q V = \{\rho x\}$. In such a case $R_x \cap U = \{\lambda x : \lambda \leq \rho\}$ and $R_x \cap V = \{\lambda x : \lambda \geq \rho\}$, hence

$$\mu_U(x) = \sup\{\lambda > 0 : \lambda x \in U\} = \rho; \quad \nu_V(x) = \inf\{\lambda > 0 : \lambda x \in V\} = \rho.$$

△

THEOREM 5.1 *1) A set $U \subset Q$ is a radiative subset of Q if and only if there exists a positively homogeneous, continuous, non-zero function $p : Q \to \mathbb{R}_+$ such that $U = \{x \in Q : p(x) \leq 1\}$.*

2) A set $V \subset Q$ is a co-radiative subset of Q if and only if there exists a positively homogeneous, continuous, non-zero function $p : Q \to \mathbb{R}_+$ such that $V = \{x \in Q : p(x) \geq 1\}$.

3) A closed subset U of Q is a radiative subset of Q if and only if its regular complement V (with respect to Q) is co-radiative. The equality $\mu_U = \nu_V$ holds.

Proof: The result follows from Proposition 5.10, Proposition 5.12 and Corollary 5.3. △

5.2.4. Radiant sets with Lipschitz continuous Minkowski gauges

We shall assume in this section that $Q = \mathbb{R}^n$. Let $p \neq +\infty$ be a nonnegative positively homogeneous function. Then the following holds:

- there exist a closed-along-rays radiant set U such that $p = \mu_U$ and a closed-along-rays co-radiant set V such that $p = \nu_V$ (see Proposition 5.1 and Proposition 5.6, respectively);

- if $p \neq +\infty$ is lower semicontinuous, then there exists a closed radiant set U such that $p = \mu_U$ (see Proposition 5.2);
- if $p \neq 0$ is upper semicontinuous, then there exists a closed co-radiant set V such that $p = \nu_V$ (see Proposition 5.8);
- if p is finite and continuous, then there exist a radiative set U and a co-radiative set V (the regular complement to U) such that $p = \mu_U$ and $p = \nu_V$ (see Theorem 5.1).

Here μ_U is the Minkowski gauge of a radiant set U and ν_V is the Minkowski co-gauge of a co-radiant set V.

We now describe radiant sets with the Lipschitz continuous Minkowski gauge. First, we consider an auxiliary result. We need the following definition.

DEFINITION 5.8 A positively homogeneous function $f : \mathbb{R}^n \to \mathbb{R}$ is called *min-sublinear* if for each $x \in \mathbb{R}^n$ there exists a finite sublinear function p_x such that $p_x(y) \geq f(y)$ for all $y \in \mathbb{R}^n$ and $p_x(x) = f(x)$.

PROPOSITION 5.13 *A positively homogeneous function $f : \mathbb{R}^n \to \mathbb{R}$ is min-sublinear if and only if for any $z \in \mathbb{R}^n$ there exists a number $k_z > 0$ such that*

$$f(x) - f(z) \leq k_z\|x - z\| \text{ for all } x \in \mathbb{R}^n. \tag{5.2.7}$$

Proof: Let f be a min-sublinear function, let $z \in \mathbb{R}^n$ and assume that p_z is a finite sublinear function with properties

$$p_z(x) \geq f(x) \text{ for all } x \in \mathbb{R}^n, \qquad p_z(z) = f(z).$$

We have

$$f(x) - f(z) \leq p_z(x) - p_z(z) \leq \|p_z\|\|x - z\|,$$

where $\|p_z\| = \max\{|p_z(x)| : \|x\| = 1\}$ is the norm of the sublinear function p_z.

Conversely, let f be a positively homogeneous function and assume that for each $z \in \mathbb{R}^n$ there exists k_z such that (5.2.7) holds. For each $z \in \mathbb{R}^n$ consider the function

$$l_z(x) = f(z) + k_z\|x - z\| \qquad (x \in \mathbb{R}^n). \tag{5.2.8}$$

It follows directly from the definition that l_z is convex and $l_z(x) \geq f(x)$ for each $x \in \mathbb{R}^n$. Moreover,

$$l_z(z) = f(z). \tag{5.2.9}$$

Let

$$p_z(x) = \inf_{s>0} s\, l_z\left(\frac{x}{s}\right) \quad \text{for all } x \in \mathbb{R}^n.$$

We now point out some properties of the function p_z.

1) p_z is positively homogeneous. Indeed, we have for $\lambda > 0$:

$$\begin{aligned} p_z(\lambda x) &= \inf\{s\, l_z\left(\frac{\lambda x}{s}\right) : s > 0\} \\ &= \inf\{s\, l_z\left(\frac{x}{s/\lambda}\right) : s > 0\} \\ &= \lambda \inf\left\{\frac{s}{\lambda}\, l_z\left(\frac{x}{s/\lambda}\right) : \frac{s}{\lambda} > 0\right\} = \lambda p_z(x). \end{aligned}$$

2) The following inequalities hold:

$$l_z(x) \geq p_z(x) \geq f(x) \text{ for all } x \in \mathbb{R}^n. \tag{5.2.10}$$

In fact

$$l_z(x) = 1 \cdot l_z\left(\frac{x}{1}\right) \geq \inf\{s\, l_z\left(\frac{x}{s}\right) : s > 0\} = p_z(x).$$

Since f is positively homogeneous and $f \leq l_z$, we conclude that

$$s\, l_z\left(\frac{x}{s}\right) \geq s\, f\left(\frac{x}{s}\right) = f(x).$$

Thus

$$p_z(x) = \inf_{s>0} s\, l_z\left(\frac{x}{s}\right) \geq f(x).$$

3) $p_z(z) = f(z)$. Indeed, this follows from (5.2.10) and (5.2.9).

4) p_z is finite. This follows, for example, from (5.2.10).

5) p_z is subadditive. Let $x, y \in \mathbb{R}^n$ and $\varepsilon > 0$. Take $t > 0$ and $s > 0$ such that

$$p_z(x) > t\, l_z\left(\frac{x}{t}\right) - \varepsilon, \quad p_z(y) > s\, l_z\left(\frac{y}{s}\right) - \varepsilon.$$

Since l_z is a convex function, we have

$$\begin{aligned} p_z(x) + p_z(y) &> t\, l_z\left(\frac{x}{t}\right) + s\, l_z\left(\frac{y}{s}\right) - 2\varepsilon \\ &= (s+t)\left\{\frac{t}{s+t}\, l_z\left(\frac{x}{t}\right) + \frac{s}{s+t}\, l_z\left(\frac{y}{s}\right)\right\} - 2\varepsilon \\ &\geq (s+t)\, l_z\left(\frac{x}{s+t} + \frac{y}{s+t}\right) - 2\varepsilon \\ &= (s+t)\, l_z\left(\frac{x+y}{s+t}\right) - 2\varepsilon \geq p_z(x+y) - 2\varepsilon. \end{aligned}$$

Since ε is an arbitrary positive number, we conclude that p_z is subadditive.

Thus for each z we have constructed a finite sublinear function p_z such that $p_z(x) \geq f(x)$ for all x and $p_z(z) = f(z)$. △

REMARK 5.8 Proposition 5.13 remains valid for normed spaces. (We accept the following definition for the infinite dimensional case. A function f defined on a normed space X is called *min-sublinear* if for each $x \in X$ there exists a *continuous* sublinear function p_x such that $p_x(y) \geq f(y)$ for all $y \in X$ and $p_x(x) = f(x)$.)

PROPOSITION 5.14 *Let f be a positively homogeneous Lipschitz function defined on $\mathbb{R}^n$ with a Lipschitz constant L (that is, $|f(x) - f(y)| \leq L\|x - y\|$ for all x, y). Let $S = \{x : \|x\| = 1\}$ be the unit sphere. Then there exists a family $(p_z)_{z\in S}$ of sublinear functions such that*

1) $f(x) = \min_{z\in S} p_z(x)$ for all $x \in \mathbb{R}^n$;

2) $p_z(x) \leq f(z) + L\|x - z\|$ for all $x \in \mathbb{R}^n$.

Proof: Let $k_z = L$ for all $z \in S$. It follows from Proposition 5.13 that for each $z \in S$, there exists a sublinear function p_z such that $p_z(z) = f(z)$ and $p_z(x) \geq f(x)$ for all $x \in \mathbb{R}^n$, in particular for every $x \in S$. Hence

$$f(x) = \min_{z\in S} p_z(x) \text{ for all } x \in S.$$

Since both the function f and the functions p_z are positively homogeneous, it follows that

$$f(x) = \min_{z\in S} p_z(x) \text{ for all } x \in \mathbb{R}^n. \tag{5.2.11}$$

It follows from the proof of Proposition 5.13 (see (5.2.10)) that we can choose the functions p_z in such a manner that

$$p_z(x) \leq l_z(x) = f(z) + L\|x - z\| \text{ for all } x \in \mathbb{R}^n.$$

△

We now show that the Minkowski gauge of a radiant subset of $\mathbb{R}^n$ is Lipschitz if and only if this set can be represented as the closure of the union of convex sets containing the same ball centred at the origin.

THEOREM 5.2 *Let U be a closed radiant subset of $\mathbb{R}^n$. The Minkowski gauge μ_U of the set U is Lipschitz if and only if there exists $\varepsilon > 0$, a* *set of indices T and a family $(U_t)_{t\in T}$ of convex sets containing the ball $B(0,\varepsilon) = \{x \in \mathbb{R}^n : \|x\| \le \varepsilon\}$ such that*

$$U = \text{cl} \bigcup_{t\in T} U_t. \tag{5.2.12}$$

Proof: Necessity. Assume that there exist $\varepsilon > 0$, a set of indices T and a family of convex sets U_t such that (5.2.12) holds and

$$B(0,\varepsilon) \subset U_t \quad (t \in T). \tag{5.2.13}$$

Let μ_t be the Minkowski gauge of the set U_t. The convexity of U_t implies the sublinearity of μ_t. Hence μ_t is a Lipschitz function with a Lipschitz constant $\|\mu_t\| = \max_{s\in S} \mu_t(s)$, where $S = \{x : \|x\| = 1\}$ is the unit sphere. Note that the Minkowski gauge of the ball $B(0,\varepsilon)$ coincides with $(1/\varepsilon)\|\cdot\|$. It follows from (5.2.13) that $\mu_t(x) \le (1/\varepsilon)\|x\|$ for all x, thus the Lipschitz constants $\|\mu_t\|$ of the functions μ_t are uniformly bounded by the constant $L = 1/\varepsilon$. Applying Proposition 5.4 and (5.2.12), we conclude that μ_U coincides with the lower regularization of the function $x \mapsto \inf_{t\in T} \mu_t(x)$. Since the Lipschitz constants of the functions μ_t are uniformly bounded, it follows that the function $x \mapsto \inf_{t\in T} \mu_t(x)$ is Lipschitz. Thus μ_U coincides with this function and therefore μ_U is Lipschitz.

Sufficiency. Let the Minkowski gauge μ_U of the set U be Lipschitz with a Lipschitz constant L. Applying Proposition 5.14, we can find a finite sublinear function p_z for each $z \in S = \{x : \|x\| = 1\}$, such that

$$\mu_U(x) = \min_{z\in S} p_z(x) \text{ for all } x \in \mathbb{R}^n \tag{5.2.14}$$

and

$$\mu_U(z) + L\|x - z\| \ge p_z(x) \text{ for all } x \in \mathbb{R}^n. \tag{5.2.15}$$

Consider the convex set $U_z = \{y : p_z(y) \le 1\}$. It follows from Proposition 5.2 that p_z is the Minkowski gauge of U_z. Combining (5.2.14) and Proposition 5.4 we conclude that $U = \text{cl} \bigcup_{z\in S} U_z$.

Applying (5.2.15) we deduce that for $z \in S$

$$\|p_z\| = \max_{y\in S} p_z(y) \le \max_{y\in S}(\mu_U(z) + L\|y - z\|) \le M,$$

where $M = \|\mu_U\| + 2L$ with $\|\mu_U\| = \max_{y\in S} \mu_U(y)$. Thus the Lipschitz constants $\|p_z\|$ of the functions p_z are uniformly bounded for all $z \in S$. Since $p_z(y) \le M\|y\|$ for all $y \in \mathbb{R}^n$ and the function $y \mapsto M\|y\|$ is the

Minkowski gauge of the ball $B(0,1/M)$ it follows that $U_z \supset B(0,1/M)$. $\triangle$

We shall need the following assertion.

PROPOSITION 5.15 *Let $(U_t)_{t\in T}$ be a family of convex sets containing the ball $B(0,\varepsilon)$ and $U = \bigcup_{t\in T} U_t$. Then the Minkowski gauge μ_U of the set U coincides with the Minkowski gauge $\mu_{\mathrm{cl}\,U}$ of the closure of this set, μ_U is a Lipschitz function and* $\mathrm{bd}\,U = \mathrm{bd}\,\mathrm{cl}\,U = \{x : \mu_U(x) = 1\}$.

Proof: Let μ_t be the Minkowski gauge of the set U_t, $(t \in T)$. Then (see Proposition 5.5) $\mu_U(x) = \inf_{t\in T} \mu_t(x)$. It has been proved (see the proof of Theorem 5.2) that μ_t are Lipschitz functions and their Lipschitz constants are uniformly bounded. Hence μ_U is a Lipschitz function as well. Let $U_0 = \{x : \mu_U(x) < 1\}$, $U_1 = \{x : \mu_U(x) \le 1\}$. Since μ_U is continuous it follows that U_0 is open and U_1 is closed. Since μ_U is positively homogeneous, $\mathrm{cl}\,U_0 = U_1$ and $\mathrm{int}\,U_1 = U_0$. Inclusions $U_0 \subset U \subset \mathrm{cl}\,U \subset U_1$ imply $\mu_U = \mu_{\mathrm{cl}\,U}$ and

$$\mathrm{bd}\,U = \mathrm{bd}\,(\mathrm{cl}\,U) = U_1 \setminus U_0 = \{x : \mu_U(x) = 1\}. \qquad \triangle$$

5.3. STAR-SHAPED SETS AND CO-STAR-SHAPED SETS

5.3.1. Overview

Shifts of radiant sets are called star-shaped sets. These sets and their kernels have been studied by many mathematicians from various points of view (see, for example [19, 20, 24, 25, 98, 183, 192]). Shifts of co-radiant sets are called co-star-shaped sets. We discuss in detail links between star-shaped and co-star-shaped sets and algebraic operations on these sets. We also examine similar questions for the so-called strongly star-shaped sets, which are shifts of radiative sets, and strongly co-star-shaped sets which are shifts of co-radiative sets. Main results of this section can be found in [145, 160].

5.3.2. Star-shaped sets and their kernels

Recall the following well-known definition (see, for example, [54, 183]).

DEFINITION 5.9 1) Let A be a nonempty subset of $\mathbb{R}^n$. The set $\mathrm{kern}\,A$ consisting of all $a \in A$ such that $(x \in A,\ 0 \le \lambda \le 1) \implies a+\lambda(x-a) \in A$ is called the *convex kernel* of A.
2) A nonempty set A is called *star-shaped* if $\mathrm{kern}\,A \ne \emptyset$.

Sometimes we will use the abbreviation *st-sh* for the word *star-shaped.* Clearly, $0 \in \operatorname{kern} A$ if and only if A is a radiant subset of $\mathbb{R}^n$ with $0 \in A$. If a set A is st-sh and $a \in \operatorname{kern} A$, then $a \in A$ and $A - a$ is a radiant set. It is known (see [24]) that the set $\operatorname{kern} A$ is convex for an arbitrary st-sh set A.

PROPOSITION 5.16 *Let A be a st-sh set. Then* $\operatorname{cl} A$ *is st-sh as well and* $\operatorname{cl}\operatorname{kern} A \subset \operatorname{kern}\operatorname{cl} A$.

Proof: Let $a \in \operatorname{cl}\operatorname{kern} A$. Then there exists a sequence $a_k \in \operatorname{kern} A$ such that $a_k \to a$. For each $y \in A$ and $\lambda \in (0,1)$ it follows that $\lambda y + (1-\lambda)a_k \in A$. Hence $\lambda y + (1-\lambda)a \in \operatorname{cl} A$. Let $z \in \operatorname{cl} A$. Then $z = \lim_k z_k$, where $z_k \in A$. Since $\lambda z_k + (1-\lambda)a \in \operatorname{cl} A$, we have $\lambda z + (1-\lambda)a \in \operatorname{cl} A$. Thus, $a \in \operatorname{kern}\operatorname{cl} A$. Since $\operatorname{kern}\operatorname{cl} A \neq \emptyset$ it follows that $\operatorname{cl} A$ is st-sh. △

COROLLARY 5.4 *If A is a closed st-sh set then* $\operatorname{kern} A$ *is closed.*

Recall (see Remark 5.6), that in contrast with radiant sets, the property of a set to be radiative or co-radiative depends on an outer cone Q, where this set is placed, so we speak about radiative (co-radiative) subsets of the cone Q. In this section we mainly assume that $Q = \mathbb{R}^n$, so we shall study radiative (co-radiative) subsets of $\mathbb{R}^n$. For the sake of simplicity, we shall use the term *radiative set* instead of *radiative subset of* $\mathbb{R}^n$ and the term *co-radiative set* instead of *co-radiative subset of* $\mathbb{R}^n$.

We now define strongly star-shaped sets.

DEFINITION 5.10 A closed proper set $A \subset \mathbb{R}^n$ is called *strongly star-shaped* if there exists a point $a \in A$ such that $A - a$ is a radiative set, that is, $a \in \operatorname{int} A$ and for each $x \in \mathbb{R}^n$, the ray $a + R_x$ does not intersect the boundary $\operatorname{bd} A$ more than once. The set of all points $a \in A$ which enjoy these properties is denoted by $\operatorname{kern}_* A$.

It follows from Definition 5.10 that $A \neq \mathbb{R}^n$ is a radiative subset of $\mathbb{R}^n$ if and only if $0 \in \operatorname{kern}_* A$. It is known (see [182] and references therein) that the set $\operatorname{kern}_* A$ is convex for an arbitrary strongly st-sh set A.

PROPOSITION 5.17 *Let U be a radiant subset of* $\mathbb{R}^n$ *and* $0 \in \operatorname{int}\operatorname{kern} U$. *Then there exist $\varepsilon > 0$ and a family of convex closed sets $(K_u)_{u \in U}$ such that*

1) $u \in K_u$ and $B(0,\varepsilon) \subset K_u$ for all $u \in U$;

2) $U = \bigcup_{u \in U} K_u$.

Proof: It follows from the definition of the convex kernel that $\lambda a + (1-\lambda)u \in U$ for all $a \in \operatorname{kern} U$, $u \in U$ and $\lambda \in (0,1)$. Since $B(0,\varepsilon) \subset \operatorname{kern} U$, the convex hull $K_u = \mathrm{co}\,(u \cup B(0,\varepsilon))$ of the set $u \cup B(0,\varepsilon)$ is contained in U for all $u \in U$. Hence

$$U = \bigcup_{u \in U} K_u. \tag{5.3.1}$$

$\triangle$

THEOREM 5.3 *Let $U \subset \mathbb{R}^n$ be a closed radiant set. Then the Minkowski gauge μ_U of the set U is Lipschitz if and only if $0 \in \mathrm{int\,kern}\, U$.*

Proof: Let U be a closed radiant subset of $\mathbb{R}^n$ and there exists $\varepsilon > 0$ such that $B(0,\varepsilon) \subset \mathrm{kern}\, U$. It follows from Proposition 5.17 that $U = \bigcup_{u \in U} K_u$, where K_u is convex and contains the ball $B(0,\varepsilon)$. It follows from Theorem 5.2 and (5.3.1) that μ_U is Lipschitz.

Assume now that μ_U is Lipschitz. Applying Theorem 5.2 we can find $\varepsilon > 0$ and a family $(U_t)_{t \in T}$ of convex sets such that $U = \mathrm{cl}\, U'$, where $U' = \bigcup_{t \in T} U_t$ and $U_t \supset B(0,\varepsilon)$. Let $u \in U'$. Then there exists $t \in T$ such that $u \in U_t$. Since U_t is convex, $\lambda u + (1-\lambda)y \in U_t \subset U$ for all $y \in B(0,\varepsilon)$ and $\lambda \in (0,1)$. Thus $B(0,\varepsilon) \subset \mathrm{kern}\, U' \subset \mathrm{cl\,kern}\, U'$. It follows from Proposition 5.16 that $\mathrm{cl\,kern}\, U' \subset \mathrm{kern\,cl}\, U' = \mathrm{kern}\, U$. Hence $B(0,\varepsilon) \subset \mathrm{kern}\, U$. $\triangle$

The following result holds.

PROPOSITION 5.18 *1) Let A be a strongly st-sh set. Then A is st-sh and* $\mathrm{kern}_* A \subset \mathrm{kern}\, A$.
2) Let $A \neq \mathbb{R}^n$ be a closed radiant subset of $\mathbb{R}^n$ and $\mathrm{int\,kern}\, A \neq \emptyset$. *Then A is a radiative set and* $\mathrm{int\,kern}\, A \subset \mathrm{kern}_* A$.

Proof: 1) Let $a \in \mathrm{kern}_* A$. Then the set $U = A - a$ is radiative, hence (see Corollary 5.1) this set is radiant. It follows from the definition of the convex kernel that $a \in \mathrm{kern}\, A$.
2) Let $a \in \mathrm{int\,kern}\, A$ and $U = A - a$. Then $0 \in \mathrm{int\,kern}\, U$, hence μ_U is a Lipschitz function. It follows from Theorem 5.1 that U is radiative, that is, $0 \in \mathrm{kern}_* U$. Hence $a \in \mathrm{kern}_* A$. $\triangle$

REMARK 5.9 Let $A \neq \mathbb{R}^n$ be a closed radiant set and $\mathrm{int\,kern}\, A \neq \emptyset$. Then

$$\mathrm{int\,kern}\, A \subset \mathrm{kern}_* A \subset \mathrm{kern}\, A. \tag{5.3.2}$$

Since both sets $\mathrm{kern}\, A$ and $\mathrm{kern}_* A$ are convex and (see Corollary 5.4) $\mathrm{kern}\, A$ is closed, it follows from (5.3.2) that

$$\mathrm{int\,kern}_* A = \mathrm{int\,kern}\, A, \qquad \mathrm{cl\,kern}_* A = \mathrm{kern}\, A.$$

Note that G. Beer [19] proved in 1955 that the Minkowski gauge is Lipschitz for a compact radiant set U such that $\operatorname{int} \operatorname{kern} U$ is nonempty.

We shall now study co-star-shaped and strongly co-star-shaped sets.

DEFINITION 5.11 1) Let B be a nonempty subset of $\mathbb{R}^n$. The set $\operatorname{kern}_\infty B$ of all $u \in \mathbb{R}^n$ such that $u \notin B$ and $(x \in B, \lambda \geq 1) \implies u + \lambda(x-u) \in B$ is called the ∞-*kernel* of the set B.

2) A nonempty set $B \in \mathbb{R}^n$ is called *co-star-shaped* if $\operatorname{kern}_\infty B \neq \emptyset$.

It follows from Definition 5.11 that $0 \in \operatorname{kern}_\infty B$ if and only if B is co-radiant with $0 \notin B$. If B is co-star-shaped and $b \in \operatorname{kern}_\infty B$, then $b \notin B$, and $B - b$ is co-radiant.

DEFINITION 5.12 1) The *upper kernel* $\operatorname{kern}^* B$ of a nonempty set $B \subset \mathbb{R}^n$ is the set of all $b \in \mathbb{R}^n$ such that

a) $b \notin B$;

b) each ray R starting from b does not intersect the boundary $\operatorname{bd} B$ more than once;

c) if the set $R \cap B$ is not empty then this set contains at least two points.

2) A closed nonempty set $B \subset \mathbb{R}^n$ is called *strongly co-star-shaped* if $\operatorname{kern}^* B \neq \emptyset$.

If B is strongly co-star-shaped and $b \in \operatorname{kern}^* B$, then the set $V = B - b$ is co-radiative. If V is co-radiative and $b \in \mathbb{R}^n$, then the set $B = b + V$ is strongly co-star-shaped and $b \in \operatorname{kern}^* B$. Thus a set is co-star-shaped if and only if this set is a shift of a co-radiative set. It follows directly from Corollary 5.2 that a strongly co-star-shaped set is co-star-shaped. If B is strongly co-star-shaped, then $\operatorname{kern}_0 B \subset \operatorname{kern}^* B$.

We present an important example of a strongly co-star-shaped set.

EXAMPLE 5.2 Let U be an open proper convex set. Then its complement $B = \mathbb{R}^n \setminus U$ is strongly co-star-shaped. This statement follows from Corollary 5.5 below.

Since both radiative and co-radiative sets are regular we conclude that strongly star-shaped and strongly co-star-shaped sets are regular as well. It follows from Proposition 5.12 that A is radiative if and only if its regular complement $C_r A$ is co-radiative. We now extend this result to include strongly star-shaped and strongly co-star-shaped sets.

LEMMA 5.2 *1) Let Z be a regular set and $z \in \mathbb{R}^n$. Then $C_r(z + Z) = z + C_r Z$.*

Proof: We have

$$\mathcal{C}_r(z+Z) = \mathrm{cl}\,(\mathcal{C}(z+Z)) = \mathrm{cl}\,(z+\mathcal{C}Z) = z + \mathrm{cl}\,(\mathcal{C}Z) = z + \mathcal{C}_r Z. \quad \triangle$$

COROLLARY 5.5 *1) A set is strongly star-shaped if and only if its regular complement is strongly co-star-shaped.*
2) A set is strongly co-star-shaped if and only if its regular complement is strongly star-shaped.

Proof: The result follows immediately from Lemma 5.2, and the following statements: 1) a set is strongly star-shaped (strongly co-star-shaped) if and only if this set is a shift of a radiative set (co-radiative set); 2) a set is radiative if and only if its regular complement is co-radiative. $\triangle$

We now compare the kernel $\mathrm{kern}_* A$ of a strongly star-shaped set A and the kernel $\mathrm{kern}^* B$ of its regular complement B (a strongly co-star-shaped set).

PROPOSITION 5.19 *1) If A is a strongly star-shaped set and $B = \mathcal{C}_r A$ then* $\mathrm{kern}_* A = \mathrm{kern}^* B$.
2) If B is a strongly co-star-shaped set and $A = \mathcal{C}_r B$ then $\mathrm{kern}^* B = \mathrm{kern}_* A$.

Proof: We only prove the second part of the proposition. Let $b \in \mathrm{kern}_* B$, then $B = b + V$, where V is a co-radiative set. Applying Lemma 5.2 we conclude that $\mathcal{C}_r B = b + \mathcal{C}_r V$. Since $\mathcal{C}_r V$ is a radiative set, $b \in \mathrm{kern}_* \mathcal{C}_r B$. Thus $\mathrm{kern}_* B \subset \mathrm{kern}_* \mathcal{C}_r B$. In the same manner, we can prove the reverse inclusion. $\triangle$

THEOREM 5.4 *Let V be a closed co-radiant subset of $\mathbb{R}^n$. Then its Minkowski co-gauge ν_V is Lipschitz if and only if*

$$0 \in \mathrm{int}\,\mathrm{kern}^* V. \tag{5.3.3}$$

Proof: Let V be a closed co-radiant set with the Lipschitz co-gauge ν_V. Then V is closed-along-rays, so according to Proposition 5.6 $V = \{x \in \mathbb{R}^n : \nu_V(x) \geq 1\}$. Since ν_V is continuous, we conclude that V is co-radiative, hence regular. Combining Theorem 5.1, Theorem 5.3 and Proposition 5.19 we deduce that (5.3.3) holds. The same argument demonstrates that (5.3.3) implies Lipschitz continuity of ν_V. $\triangle$

Let $X \subset \mathbb{R}^n$. Recall that the conic hull of the set X is the set

$$\mathrm{cone}\, X = \bigcup_{\lambda > 0} \lambda X.$$

PROPOSITION 5.20 *Let V be a closed co-radiant set and $0 \notin V$. Then $z \in \operatorname{cl} \operatorname{cone} V$ if and only if there exist sequences $v_k \in V$ and $\lambda_k \downarrow 0$ such that $\lambda_k v_k \to z$.*

Proof: Let $z \in \operatorname{cl} \operatorname{cone} V$. Then $z = \lim \lambda_k v_k$, where $v_k \in V$ and $\lambda_k > 0$ for all positive integers k . Clearly, $\|z\| = \lim \lambda_k \|v_k\|$. If the sequence v_k is unbounded, then we can assume without loss of generality that $\lambda_k \downarrow 0$. Let v_k be bounded. Assume that $v_k \to v \in V$. Note that $v \neq 0$. Clearly, $\lambda_k \to \lambda = \|z\| \|v\|^{-1}$, so $z = \lambda v$. If $\lambda = 0$, that is, $z = 0$, then $\lambda_k \downarrow 0$. Assume that $\lambda > 0$. Then $z = \lambda v = \lim_k (1/k)(k\lambda v)$. Since $k\lambda > 1$ for large k, it follows that $k\lambda v \in V$. Thus $z = \lim \lambda'_k v'_k$, where $\lambda'_k = 1/k$ and $v_k := k\lambda v \in V$. Hence the desired result follows. △

Recall the following well-known definition (see, for example [134], Section 8).

DEFINITION 5.13 Let C be a convex set. The *recession cone* $\operatorname{rc} C$ of the set C is the set of all vectors y such that $y + C \subset C$.

It is well known (see [134], Theorem 8.2) that for a closed convex set C

$$\operatorname{rc} C = \{z : \text{there exist } \lambda_k \downarrow 0,\ v_k \in C \text{ such that } z = \lim_k \lambda_k v_k\}. \quad (5.3.4)$$

Thus if V is a closed convex co-radiant set and $0 \notin V$, then $\operatorname{cl} \operatorname{cone} V$ coincides with $\operatorname{rc} V$.

We now define the recession cone for co-star-shaped sets.

DEFINITION 5.14 Let B be a co-star-shaped set. The cone

$$\operatorname{rc} B = \{x = \lim \lambda_i b_i : \lambda_i \downarrow 0, b_i \in B\}$$

is called the *recession cone* of the set B.

Clearly, $\operatorname{rc} B$ is a cone, $\operatorname{rc} B = \operatorname{rc}(\alpha B)$ for $\alpha > 0$ and $\operatorname{rc}(x+B) = \operatorname{rc} B$ for each x. If V is a closed co-radiant set and $0 \notin V$, then (see Proposition 5.20) $\operatorname{rc} V = \operatorname{cl} \operatorname{cone} V$. Thus the following assertion holds.

PROPOSITION 5.21 *If B is a closed co-star-shaped set and $b \in \operatorname{kern}_\infty B$ then $rc\, B = \operatorname{cl} \operatorname{cone}(B - b)$.*

5.3.3. Sum of star-shaped sets and sum of co-star-shaped sets

In this subsection we shall examine algebraic operations on star-shaped and co-star-shaped sets. We begin with star-shaped sets.

PROPOSITION 5.22 *Let A_1 and A_2 be star-shaped sets. Then the algebraic sum $A_1 + A_2$ is a star-shaped set and*

$$\text{kern}\,(A_1 + A_2) \supset \text{kern}\,A_1 + \text{kern}\,A_2. \qquad (5.3.5)$$

Proof: Let $a_1 \in \text{kern}\,A_1$, $a_2 \in \text{kern}\,A_2$. We check that $a := a_1 + a_2 \in \text{kern}\,(A_1 + A_2)$. Take $x \in A_1 + A_2$. There exist $x_1 \in A_1$ and $x_2 \in A_2$ such that $x = x_1 + x_2$. Let $\lambda \in (0, 1)$. Then (by the definition of the convex kernel) we have

$$a_1 + \lambda(x_1 - a_1) \in A_1, \qquad a_2 + \lambda(x_2 - a_2) \in A_2. \qquad (5.3.6)$$

It follows from (5.3.6) that $a + \lambda(x - a) \in A_1 + A_2$, hence $a \in \text{kern}\,(A_1 + A_2)$. Thus (5.3.5) holds. Since $\text{kern}\,(A_1 + A_2)$ is nonempty, it follows that $A_1 + A_2$ is star-shaped. △

The following simple assertion demonstrates that the sum of strongly star-shaped and star-shaped sets is strongly star-shaped if this sum is closed. (Recall that a strongly star-shaped set is closed by the definition.)

PROPOSITION 5.23 *Let A be a strongly star-shaped set and X be a star-shaped set. Assume the sum $A + X$ is not equal to $\mathbb{R}^n$ and is closed. Then $A + X$ is strongly star-shaped.*

Proof: Consider the sets $U = A - a$ and $Y = X - x$, where $a \in \text{kern}_* A$ and $x \in X$. Clearly, $0 \in \text{int}\,(U + Y)$. Assume there exists a point $z \neq 0$ such that the intersection $R_z \cap \text{bd}\,(U + Y)$ contains at least two points z and λz with $0 < \lambda < 1$. Since $z \in \text{bd}\,(U + Y)$, it follows that there exist $u \in \text{bd}\,U$ and $y \in \text{bd}\,Y$ such that $z = u + y$. Because U is strongly star-shaped, $\lambda u \in \text{int}\,U$; since Y is radiative, $\lambda y \in Y$. So $\lambda z = \lambda u + \lambda y \in \text{int}\,(U + Y)$. Thus we have a contradiction and $U + Y$ is radiative. Clearly $A + X$ is strongly star-shaped with respect to $a + x$. △

PROPOSITION 5.24 *Let A be a star-shaped set and $\lambda > 0$. Then the set λA is star-shaped and* $\text{kern}\,(\lambda A) = \lambda(\text{kern}\,A)$. *If A is strongly star-shaped then λA is also strongly star-shaped.*

Proof: The proof is a direct application of the definitions of star-shaped sets and strongly star-shaped sets. △

We now consider algebraic operations with co-star-shaped sets and strongly co-star-shaped sets. We begin with the multiplication by positive number.

PROPOSITION 5.25 *Let B be a co-star-shaped set and $\lambda > 0$. Then the set λB is co-star-shaped and* $\text{kern}_\infty(\lambda B) = \lambda(\text{kern}_\infty B)$. *If B is strongly star-shaped then λB is also strongly star-shaped.*

Proof: It follows directly from the definitions of co-star-shaped sets and strongly co-star-shaped sets. △

The situation with the summation of co-star-shaped sets and strongly co-star-shaped sets is much more complicated. It is possible that the sum of two co-radiative sets may contain zero, so we need some assumptions which allow us to exclude this difficulty. We begin with the following simple assertion.

PROPOSITION 5.26 *Let Z_1 and Z_2 be regular sets and $Z_1 \cap Z_2 \neq \emptyset$. Then either the intersection of these sets is equal to the intersection of their boundaries or* $\text{int } Z_1 \cap \text{int } Z_2$ *is not empty.*

Proof: It follows directly from the definition of regular sets. △

LEMMA 5.3 *Let V_1 be a co-radiative set and V_2 be a co-radiant set and let b be a vector such that the intersection of sets $b + V_1$ and V_2 is either empty or equal to the intersection of their boundaries. Then $(\lambda b + V_1) \cap V_2 = \emptyset$ for each $\lambda \in (0, 1)$.*

Proof: Assume that there exist $\lambda \in (0,1)$ and $v_1 \in V_1$ such that $c = \lambda b + v_1 \in V_2$. Since $(1/\lambda) > 1$, it follows that $(1/\lambda)v_1 \in \text{int } V_1$, so $(1/\lambda)c = b + (1/\lambda)v \in \text{int } (b + V_1)$. On the other hand $(1/\lambda)c \in \text{int } V_2$. Thus the intersection $\text{int } (b + V_1) \cap \text{int } V_2$ is not empty. △

PROPOSITION 5.27 *Let V_1 and V_2 be co-radiative sets, let $B_1 = b_1 + V_1$, $B_2 = b_2 + V_2$ and $B_1 \cap B_2 \neq \emptyset$. Then the following relations are equivalent*

1) $B_1 \cap B_2 = (\text{bd } B_1) \cap (\text{bd } B_2)$;

2) $b_1 - b_2 \in \text{bd } (V_2 - V_1)$.

Proof: 1) $\Longrightarrow$ 2). Since $B_1 \cap B_2 \neq \emptyset$ it easily follows that $b_1 - b_2 \in V_2 - V_1$. We also have that $V_2 \cap ((b_1 - b_2) + V_1) = \text{bd } V_2 \cap \text{bd } ((b_1 - b_2) + V_1)$. It follows from Lemma 5.3 that $(\lambda(b_1 - b_2) + V_1) \cap V_2 = \emptyset$ for all $\lambda \in (0, 1)$. Therefore, $\lambda(b_1 - b_2) \notin V_2 - V_1$ for all such λ.

2) $\Longrightarrow$ 1). If $B_1 \cap B_2 \neq \text{bd } B_1 \cap \text{bd } B_2$, then (see Proposition 5.26) there exists $c \in \text{int } B_1 \cap \text{int } B_2$. We have $c - b_1 \in \text{int } V_1$, $c - b_2 \in \text{int } V_2$, so $b_1 - b_2 \in \text{int } V_2 - \text{int } V_1 \subset \text{int } (V_2 - V_1)$ which contradicts 2). △

PROPOSITION 5.28 *Let V_1 be a co-radiative set and V_2 be a closed co-radiant set. If $0 \notin V_1 + V_2$ and $V_1 + V_2$ is closed, then $V_1 + V_2$ is co-radiative.*

Proof: Since V_1 is co-radiant it follows that $\operatorname{int} V_1$ is non-empty, hence $\operatorname{int}(V_1 + V_2)$ is also nonempty. Let $R = R_x$ be a ray such that $R \cap (V_1 + V_2) \neq \emptyset$. Since $V_1 + V_2$ is closed and $0 \notin V_1 + V_2$, we conclude that there exists $\lambda > 0$ such that $\lambda x \in \operatorname{bd}(V_1 + V_2)$. Then $\lambda x = x_1 + x_2$, where $x_1 \in \operatorname{bd} V_1$ and $x_2 \in \operatorname{bd} V_2$. Let $\lambda' > \lambda$. Since V_1 is co-radiative and V_2 is co-radiant, we conclude that $\lambda' x_1 \in \operatorname{int} V_1$ and $\lambda' x_2 \in V_2$. Hence $\lambda' x \in \operatorname{int} V_1 + V_2 \subset \operatorname{int}(V_1 + V_2)$. If $\lambda'' < \lambda$ then $\lambda'' x \notin V_1 + V_2$, otherwise $\lambda x \notin \operatorname{bd}(V_1 + V_2)$. Thus $V_1 + V_2$ is co-radiative. △

The further study of the sum of two strongly co-star-shaped sets is based on the properties of the conic hull of these sets.

PROPOSITION 5.29 *Let V_1 be co-radiative and let V_2 be co-radiant. The following assertions are equivalent:*

1) $0 \in V_1 + V_2$*;*

2) $V_1 + V_2 = \mathbb{R}^n$*;*

3) the intersection $\operatorname{cone} V_1 \cap (-\operatorname{cone} V_2)$ *is nonempty and not equal to* $\{0\}$.

Proof: 1) $\Longrightarrow$ 2). Let $0 \in V_1 + V_2$. Then there exists an element v such that $v \in V_1, -v \in (-V_2)$. Then $2v \in \operatorname{int} V_1, -2v \in V_2$, so $0 \in (\operatorname{int} V_1) + V_2 \subset \operatorname{int}(V_1 + V_2)$. Thus for each $z \in \mathbb{R}^n$ there exists $\varepsilon \in (0, 1)$ such that $\varepsilon z \in V_1 + V_2$. Hence $z = (1/\varepsilon)\varepsilon z \in V_1 + V_2$.

2) $\Longrightarrow$ 3). Since $0 \notin V_1$ it follows from 2) that there exists $v \neq 0$ such that $v \in V_1, -v \in V_2$. Clearly $v \in \operatorname{cone} V_1 \cap (-\operatorname{cone} V_2)$.

3) $\Longrightarrow$ 1). Let $v \neq 0, v \in \operatorname{cone} V_1$, $-v \in \operatorname{cone} V_2$. There exist $\lambda_i > 0$ such that $v \in \lambda_1 V_1$, $-v \in \lambda_2 V_2$. If $\lambda_1 \geq \lambda_2$, then $v \in \lambda_2 V_1$ and $-v \in \lambda_2 V_2$. Hence $0 \in V_1 + V_2$. △

REMARK 5.10 If both sets V_1 and V_2 are only co-radiant and at least one of them does not contain zero, then we can assert that 1) and 3) are equivalent.

PROPOSITION 5.30 *Let V_1, V_2 be closed co-radiant sets. Assume that at least one of these sets does not contain zero and*

$$\operatorname{cl}\operatorname{cone} V_1 \cap (-\operatorname{cl}\operatorname{cone} V_2) = \{0\}. \tag{5.3.7}$$

Then $V_1 + V_2$ is closed and $0 \notin V_1 + V_2$. If at least one of the sets V_1 and V_2 is co-radiative, then $V_1 + V_2$ is also co-radiative.

Proof: It follows from (5.3.7) that $\operatorname{cone} V_1 \cap \operatorname{cone} V_2$ is either empty or coincides with zero, so (see Remark 5.10) $0 \notin V_1 + V_2$. Let

$$v^k = v_1^k + v_2^k \to v \text{ with } v_i^k \in V_i, \ i = 1, 2. \tag{5.3.8}$$

Let one of these sequences, say v_1^k, be unbounded. Assume without loss of generality that $\|v_1^k\| \to +\infty$. It follows from (5.3.8) that

$$0 = \lim_k \frac{v}{v_1^k} = \lim_k \left(\frac{v_1^k}{\|v_1^k\|} + \frac{v_2^k}{\|v_1^k\|} \right). \tag{5.3.9}$$

Let $u_i^k = v_i^k / \|v_1^k\|$, $(i = 1, 2)$. Since $\|u_1^k\| = 1$ we assume without loss of generality that that there exists $\lim_k u_1^k = u_1 \neq 0$. It follows from (5.3.9) that $\lim_k u_2^k = -u_1$. Note that $u_i^k \in \operatorname{cone} V_i$ for all k and $i = 1, 2$, hence $u_1 \in \operatorname{cl cone} V_1, -u_1 \in \operatorname{cl cone} V_2$. We have proved that $\operatorname{cl cone} V_1 \cap (-\operatorname{cl cone} V_2) \neq \{0\}$, which contradicts (5.3.7). Thus the sequences v_1^k and v_2^k are bounded. We assume without loss of generality that there exists $\lim_k v_i^k \in V_i$, $i = 1, 2$. Thus the set $V_1 + V_2$ is closed. The second part of the proposition follows directly from Proposition 5.28. △

REMARK 5.11 Equality (5.3.7) is a sufficient but not necessary condition for $V_1 + V_2$ to be co-radiative. The following example confirms this statement. Consider subsets V_1 and V_2 of the space $\mathbb{R}^2$: $V_1 = (1, 1) + Q_1$, $V_2 = (1, -1) + Q_2$, where $Q_1 = \{(x, y) : x \geq 0\}$, $Q_2 = \{(x, y) : x \geq 0, y \leq 0\}$. Clearly, $V_1 + V_2 = (2, 0) + \{(x, y) : x \geq 0\}$ is co-radiative but (5.3.7) is not valid.

PROPOSITION 5.31 *If B_1 is a strongly co-star-shaped set, B_2 is a co-star-shaped set and*

$$\operatorname{rc} B_1 \cap \operatorname{rc}(-B_2) = \{0\} \tag{5.3.10}$$

then $B_1 + B_2$ is strongly star-shaped and $\operatorname{rc}(B_1 + B_2) = \operatorname{rc} B_1 + \operatorname{rc} B_2$.

Proof: Let $b_1 \in \operatorname{kern}^* B_1$, $b_2 \in \operatorname{kern}_\infty B_2$ and let $V_i = B_i - b_i$, $i = 1, 2$. Combining Proposition 5.21 and (5.3.10) we conclude that

$$\operatorname{rc} V_1 \cap (-\operatorname{rc} V_2) = \{0\}. \tag{5.3.11}$$

It follows from Proposition 5.20 and Proposition 5.30 that $V_1 + V_2$ is co-radiative, so $B_1 + B_2$ is strongly co-star-shaped.

We now demonstrate that

$$\operatorname{cl cone} V_1 + \operatorname{cl cone} V_2 = \operatorname{cl cone} (V_1 + V_2). \tag{5.3.12}$$

First, we check that

$$\operatorname{cone} V_1 + \operatorname{cone} V_2 = \operatorname{cone}(V_1 + V_2). \tag{5.3.13}$$

The inclusion $\operatorname{cone} V_1 + \operatorname{cone} V_2 \supset \operatorname{cone}(V_1 + V_2)$ is trivial. In order to verify the reverse inclusion, take points $\lambda_1 v_1 \in \operatorname{cone} V_1$, $\lambda_2 v_2 \in \operatorname{cone} V_2$, where $\lambda_1, \lambda_2 > 0$, $v_1 \in V_1$, $v_2 \in V_2$. Let $\lambda_2 \geq \lambda_1$. Then $(\lambda_2/\lambda_1)v_2 \in V_2$. Since

$$\lambda_1 v_1 + \lambda_2 v_2 = \lambda_1 \left(v_1 + \frac{\lambda_2}{\lambda_1} v_2 \right)$$

it follows that $\lambda_1 v_1 + \lambda_2 v_2 \in \operatorname{cone}(V_1 + V_2)$. Thus (5.3.13) has been proved. It follows from (5.3.13) that

$$\operatorname{cl}(\operatorname{cone} V_1 + \operatorname{cone} V_2) = \operatorname{cl}\operatorname{cone}(V_1 + V_2). \tag{5.3.14}$$

It follows directly from the definitions of the closure and the sum that

$$\operatorname{cl}(\operatorname{cone} V_1 + \operatorname{cone} V_2) \supset \operatorname{cl}\operatorname{cone} V_1 + \operatorname{cl}\operatorname{cone} V_2.$$

In order to check the reverse inclusion we can use the same argument as in the proof of Proposition 5.30, since, due to Proposition 5.20 formulae (5.3.11) and (5.3.7) are equivalent. Thus (5.3.14) implies (5.3.12). Applying Proposition 5.21, Proposition 5.20 and (5.3.12), we conclude that

$$\operatorname{rc}(B_1 + B_2) = \operatorname{rc}(V_1 + V_2) = \operatorname{rc} V_1 + \operatorname{rc} V_2 = \operatorname{rc} B_1 + \operatorname{rc} B_2. \qquad \triangle$$

5.4. SEPARATION

5.4.1. Overview

Many applications of convexity are based on the separation property. If the intersection of two convex sets is either empty or coincides with the intersection of their boundaries, then these sets can be separated by a linear function (geometrically, by a hyperplane) under fairly mild assumptions. In particular, the separation property plays a crucial role in the study of convex optimization problems. The separation property for two convex sets easily follows from a simpler fact. If a point does not belong to a closed convex set then this point can be separated from this set. Generalizations of this assertion are studied in the framework of abstract convexity. Note that in contrast with the classical case, the nonlinear separation of a point from a set does not imply the separation property for two sets.

A very interesting notion of separability of star-shaped sets by finite collections of linear functions has been introduced and studied in [182].

Actually this kind of separation can be studied in the framework of abstract convexity, by assuming that finite minima of linear functions are used as nonlinear separation functions. As it turned out ([182]) the separation of a point from a strongly star-shaped set by finite collections of linear functions can be extended to the separation of two strongly star-shaped sets, hence this kind of separation can be applied in the study of some nonconvex problems of global optimization (see [182], where possible applications are outlined). We show that the separation by a finite collection of linear functions can be studied in more general frameworks of cone-separation, that is, a separation of a point from a set by a convex solid cone. We also discuss separation properties of co-star-shaped sets.

We give a generalization of one of the main results from [182] related to the separability of two star-shaped sets. This result is based on the theorem from [182] describing conditions that guarantee that a boundary point of a star-shaped set can be separated from this set by a collection of n linearly independent linear functions. We present a new proof of this theorem.

Main results of this section can be found in [182, 145, 160].

5.4.2. Cone-separation and separation by a finite collection of linear functions

We begin with the following definition.

DEFINITION 5.15 Let $X \subset \mathbb{R}^n$ and $x \in \mathbb{R}^n$. Let Q be a closed solid convex cone.
1) The set X and the point x are said to be *separated* by the cone Q if $(x + \operatorname{int} Q) \cap X = \emptyset$.
2) The set X and the point x are said to be *cone-separated* if there exists a closed convex solid cone Q, which separates X and x.

It is convenient to present cone-separation in terms of separation by a finite collection of linear functions. Consider a collection of m elements $\ell = (l(1), \dots, l(m))$ of the space $\mathbb{R}^n$ and let

$$T_\ell = \{x : [l(i), x] < 0 \ (i \in J)\}, \qquad T^\ell = \{x : [l(i), x] > 0 \ (i \in J)\}, \tag{5.4.1}$$

where $J = \{1, \dots, m\}$. Clearly, both T_ℓ and T^ℓ are open convex cones. If $m \leq n$ and elements $l(i)$ are linearly independent, then both T_ℓ and T^ℓ are not empty.

PROPOSITION 5.32 *Let Q be a solid cone and $x \in \operatorname{int} Q$. Then there exists a collection $\ell = (l(1), \dots, l(n))$ of n linearly independent elements such that $T_\ell \subset \operatorname{int} Q$ and $[l(i), x] = -1$ for all $i = 1, \dots, n$.*

Proof: Denote by H the hyperplane $\{y : [x, y] = 0\}$. Since $x \in \operatorname{int} Q$, it follows that there exists a simplex $S \subset H$ such that zero is an interior point of S (with respect to H) and $x + S \subset \operatorname{int} Q$. Let $y^1, \dots, y^n$ be vertices of S, then points $x + y^1, \dots, x + y^n$ are vertices of $x + S$. We now verify that elements

$$x, x + y^1, \dots, x + y^{k-1}, x + y^{k+1}, \dots, x + y^n \tag{5.4.2}$$

are linearly independent for each $k = 1, \dots, n$. Assume that there exist numbers α_i, $(i = 1, \dots, n)$ such that

$$0 = \alpha_k x + \sum_{i \neq k} \alpha_i (x + y^i) = (\sum_{i=1}^{n} \alpha_i) x + \sum_{i \neq k} \alpha_i y^i. \tag{5.4.3}$$

Note that x is orthogonal to H and $y^i \in H$, so it follows from (5.4.3) that

$$\sum_{i=1}^{n} \alpha_i = 0. \tag{5.4.4}$$

Applying (5.4.3) again, we conclude that $\sum_{i \neq k} \alpha_i y^i = 0$. Since zero is an interior point of the simplex S (with respect to H), it follows that vertices $(y^i)_{i \neq k}$ are linearly independent, hence $\alpha_i = 0$ for all $i \neq k$. Combining this result with (5.4.4) we obtain $\alpha_k = 0$ as well. Thus vectors (5.4.2) are linearly independent. Then the system of linear equations

$$[l, x] = -1, \ [l, x + y^i] = 0 \ (i \neq k)$$

has the unique solution $l(k)$. We now check that the collection $\ell = (l(1), \dots, l(n))$ consists of linearly independent elements. Note that $[l(k), y^i] = 1$ for all $i \neq k$. It follows from this observation that for each k the restriction of the linear function $y \mapsto [l(k), y]$ to H is not equal to zero, hence the set $H_k = \{y \in H : [l(k), y] \geq 0\}$ does not coincide with H. Clearly that this set is a halfspace of the vector space H. First, we show that $[l(k), y^k] < 0$. These inequalities hold, since otherwise, all vertices of the simplex S are placed in the halfspace H_k, hence the simplex itself is placed in this halfspace as well. However, this is impossible since 0 is an interior point of S (with respect to H). Let β_k $(k = 1, \dots, n)$ be numbers such that $\sum_{k=1}^{n} \beta_k l(k) = 0$. Then

$$0 = \sum_{k=1}^{n} \beta_k [l(k), x] = \sum_{k=1}^{n} \beta_k (-1).$$

Therefore, we have for all $i = 1, \dots, n$

$$\sum_{k=1}^{n} \beta_k [l(k), y^i] \ = \ \sum_{k \neq i} \beta_k [l(k), y^i] + \beta_i [l(i), y^i] = \sum_{k \neq i} \beta_k + \beta_i [l(i), y^i]$$

$$= \sum_{k=1}^{n}(\beta_k - \beta_i) + \beta_i[l(i), y^i]$$
$$= -\beta_i + \beta_i[l(i), y^i] = \beta_i(-1 + [l(i), y^i]).$$

Since $[l(i), y^i] \neq 1$, it follows that $\beta_i = 0$.

Thus, the collection of n linearly independent vectors $l(1), \dots, l(n)$ enjoys the following property: for each k all vertices of the simplex, excluding the vertex y^k, are placed in the hyperplane $[l(k), y] = 1$ and y^k is located in the halfspace $[l(k), y] < 0$. This means that the simplex S can be represented in the following form: $S = \{y : [l(k), y] \leq 1, \ k = 1, \dots, n, \ [x, y] = 0\}$. Since $[l(k), x] = -1$, we also have that

$$\begin{aligned} x + S &= \{z : z = y + x; \ [l(k), y] \leq 1, \ k = 1, \dots, n, \ [x, y] = 0\} \\ &= \{z : [l(k), z] \leq 0, \ k = 1, \dots, n, \ [x, z] = \|x\|^2\}. \end{aligned}$$

This representation implies the equality

$$\text{cone}\,(x + S) \cup \{0\} = \{z : [l(k), z] \leq 0, k = 1, \dots, n\}.$$

Since $x + S \subset \text{int}\, Q$ it follows that $T_\ell = \{z : [l(1), z] < 0, \dots, [l(n), z] < 0\} \subset \text{int}\, Q$. △

REMARK 5.12 Let the conditions in Proposition 5.32 hold. Then there exists a collection $\ell' = \{l'(1), \dots, l'(n)\}$ of linearly independent vectors such that $[l'(i), x] = 1$ for all $i = 1, \dots, n$ and $T^{\ell'} \subset \text{int}\, Q$. Indeed, let ℓ be a collection as in Proposition 5.32. Then $\ell' = -\ell$ enjoys the required properties.

THEOREM 5.5 *Let $X \subset \mathbb{R}^n$ and $y \notin X$. Then the set X and the point y can be cone-separated by a cone Q such that $y \in \text{int}\, Q$ if and only if there exists a collection ℓ of linearly independent vectors $l(1), \dots, l(n)$ with the following properties:*

$$\min_{i=1,\dots,n} [l(i), y] > \sup_{x \in X} \min_{i=1,\dots,n} [l(i), x] \ \textit{and} \ \min_{i=1,\dots,n} [l(i), y] > 0. \qquad (5.4.5)$$

Proof: Assume that there exists a convex solid cone Q such that $(y + \text{int}\, Q) \cap X = \emptyset$ and let $y \in \text{int}\, Q$. We can find a positive $\varepsilon < 1$ such that $y' = (1 - \varepsilon)y \in \text{int}\, Q$ as well. It follows from Remark 5.12 that there exists a collection ℓ of linearly independent vectors $l(1), \dots, l(n)$ such that $[l(k), y'] = 1$ for all $k = 1, \dots, n$ and $T^\ell := \{x : [l(k), x] > 0, \ k = 1, \dots, n\} \subset \text{int}\, Q$. If $x \in y' + T^\ell$, then $[l(i), x] > 1$ for all i. Let $x \in X$.

Then $x \notin y' + \operatorname{int} Q$, therefore $x \notin y' + T^\ell$; this means that there exists an index i such that $[l(i), x] \le 1$. Thus we have

$$\min_{i=1,\dots,n} [l(i), x] \le 1 = \min_{i=1,\dots,n} [l(i), y'] = (1-\varepsilon) \min_{i=1,\dots,n} [l(i), y].$$

So $\min_{i=1,\dots,n}[l(i), y] > 0$ and

$$\sup_{x \in X} \min_{i=1,\dots,n} [l(i), x] \le (1-\varepsilon) \min_{i=1,\dots,n} [l(i), y] < \min_{i=1,\dots,n} [l(i), y].$$

Assume now that (5.4.5) holds for a collection $\ell = (l(1), \dots, l(n))$. Let $c > 0$ be the number such that

$$\sup_{x \in X} \min_{i=1,\dots,n} [l(i), x] < c < \min_{i=1,\dots,n} [l(i), y].$$

Consider the convex cone $T^\ell = \{z : [l(i), z] > 0,\ i = 1, \dots, n\}$. Since $[l(i), y] > 0$ for all i it follows that $y \in T^\ell$, hence T^ℓ is a nonempty open cone. Moreover, we have that

$$y + T^\ell \subset \{z : [l(i), z] > c,\ i = 1, \dots, n\} = \{z : \min_{i=1,\dots,n} [l(i), z] > c\}.$$

Hence $(y + T^\ell) \cap X = \emptyset$. △

It follows from Theorem 5.5 that the separation of a set and a point which does not belong to this set, can be accomplished by a collection of linear independent vectors. In the sequel we shall examine the separation of a set and its boundary point. We shall use the notion of Bouligand cone for this purpose (this is one of the main notions of nonsmooth analysis; see, for example, [8, 43]).

DEFINITION 5.16 Let $U \subset \mathbb{R}^n$ and $x \in \operatorname{cl} U$. We say that a vector u belongs to the Bouligand cone of the set U at the point x if for any positive numbers α_0 and ε there exist v and α such that $\|v - u\| \le \varepsilon$, $\alpha \in (0, \alpha_0]$ and $x + \alpha v \in U$. We denote the Bouligand cone of U at a point x by $\Gamma(U, x)$.

Subsequently, we will consider a point x such that x does not belong to the Bouligand cone of U at the same point x, that is, $x \notin \Gamma(U, x)$. It follows from the definition of the Bouligand cone that x possesses this property if and only if there exist $\alpha_0 > 0$ and $\varepsilon > 0$ such that

$$\left(x + \bigcup_{0 < \alpha \le \alpha_0} \alpha B(x, \varepsilon) \right) \cap U = \emptyset, \tag{5.4.6}$$

where $B(x, \varepsilon) = \{y : \|y - x\| < \varepsilon\}$.

Let $f : \mathbb{R}^n \to \mathbb{R}_{+\infty}$ and $x \in \operatorname{dom} f$. Consider a vector u such that $x + \alpha u \in \operatorname{dom} f$ for all small enough $\alpha > 0$. Recall (see, for example [43]) that the Dini lower derivative $f_D^{\downarrow}(x, u)$ and the Hadamard lower derivative $f_H^{\downarrow}(x, u)$ at the point x in the direction u are defined by

$$f_D^{\downarrow}(x,u) = \liminf_{\alpha \to +0} \frac{1}{\alpha}(f(x + \alpha u) - f(x)),$$

$$f_H^{\downarrow}(x,u) = \liminf_{u' \to u, \alpha \to +0} \frac{1}{\alpha}(f(x + \alpha u') - f(x)).$$

The Dini upper derivative and the Hadamard upper derivative are defined in a similar manner:

$$f_D^{\uparrow}(x,u) = \limsup_{\alpha \to +0} \frac{1}{\alpha}(f(x + \alpha u) - f(x)),$$

$$f_H^{\uparrow}(x,u) = \limsup_{u' \to u, \alpha \to +0} \frac{1}{\alpha}(f(x + \alpha u) - f(x)).$$

If f is a positively homogeneous function, then $f_D^{\downarrow}(x,x) = f_D^{\uparrow}(x,x) = f(x)$. If f is a locally Lipschitz function at the point x (that is, for some $\varepsilon > 0$ there exists C such that $|f(x'') - f(x')| \leq C\|x'' - x'\|$ whenever $\|x' - x\| < \varepsilon$ and $\|x'' - x\| < \varepsilon$), then $f_D^{\uparrow}(x,u) = f_H^{\uparrow}(x,u)$ and $f_D^{\downarrow}(x,u) = f_H^{\downarrow}(x,u)$. We need the following simple assertion.

PROPOSITION 5.33 *Let f be a positively homogeneous function and $\gamma := f(x_0) \in (0, +\infty)$. Let $S_\gamma(f) = \{x : f(x) \leq \gamma\}$. Then $x_0 \in \Gamma(S_\gamma, x_0)$ if and only if $f_H^{\downarrow}(x_0, x_0) \leq 0$.*

Proof: Assume without loss of generality that $\gamma = 1$. If x_0 belongs to the cone $\Gamma(S_1(f), x_0)$, then the inequality $f_H^{\downarrow}(x_0, x_0) \leq 0$ follows directly from the definitions of the Hadamard derivative and the Bouligand cone. Assume now that

$$f_H^{\downarrow}(x_0,x_0) = \lim \inf_{x' \to x_0, \alpha \to +0} \frac{1}{\alpha}(f(x_0 + \alpha x') - f(x_0)) \leq 0.$$

We only have to consider the case when this derivative is equal to zero. In this case there exist elements $x_k \to x_0$ and positive numbers $\alpha_k \to +0$, $k = 1, \ldots$, such that $(1/\alpha_k)(f(x_0 + \alpha_k x_k) - f(x_0)) = \varepsilon_k \to 0$. Let

$$x'_k = \frac{x_k - \varepsilon_k x_0}{1 + \alpha_k \varepsilon_k}.$$

Since $f(x_0 + \alpha_k x_k) = 1 + \alpha_k \varepsilon_k$, we have $f(x_0 + \alpha_k x'_k) = 1$, that is, $x_0 + \alpha_k x'_k \in S_1(f)$. Since $x'_k \to x_0$ and $\alpha_k \to +0$, it follows that x_0 belongs to the Bouligand cone of $S_1(f)$ at x_0. △

COROLLARY 5.6 *Let a positively homogeneous function f be locally Lipschitz at a point x_0 with $f(x_0) := \gamma > 0$. Let $S_\gamma(f) = \{u : f(u) \le \gamma\}$. Then $x_0 \notin \Gamma(S_\gamma(f), x_0)$.*

Proof: Applying positive homogeneity of f, we conclude that

$$\begin{aligned} f_D^{\downarrow}(x_0, x_0) &= \lim_{\alpha \to +0} \frac{1}{\alpha}(f(x_0 + \alpha x_0) - f(x_0)) \\ &= \lim_{\alpha \to +0} \frac{1}{\alpha}((1+\alpha) f(x_0) - f(x_0)) = \gamma > 0. \end{aligned}$$

Since f is locally Lipschitz, we have $f_H^{\downarrow}(x_0, x_0) = f_D^{\downarrow}(x_0, x_0) > 0$. △

PROPOSITION 5.34 *Let f be a positively homogeneous function and let $\gamma := f(x_0) \in (0, +\infty)$. Then $-x_0 \in \Gamma(V, x_0)$ if and only if $f_H^{\uparrow}(x_0, -x_0) \ge 0$.*

The proof is similar to that of Proposition 5.33 and so is omitted.

COROLLARY 5.7 *If a positively homogeneous function f is locally Lipschitz at a point x_0 with $f(x_0) \in (0, +\infty)$, then $-x_0 \notin \Gamma(V, x_0)$.*

5.4.3. Separation of star-shaped sets

In this subsection we shall study the separation of star-shaped sets by a finite collection of linearly independent vectors. We start with the following definitions, in which $[x, y]$ stands for the inner product of vectors x and y.

DEFINITION 5.17 Let A_1, A_2 be subsets of $\mathbb{R}^n$ and $l(1), \dots, l(m)$ be linearly independent vectors. The sets A_1, A_2 are said to be *weakly separated* by the vectors $l(1), \dots, l(m)$ if for each pair $a_1 \in A_1$, $a_2 \in A_2$ there exists an index $i \in \{1, \dots, m\}$ such that $[l(i), a_1] \le [l(i), a_2]$.

In other words, sets A_1 and A_2 are separated by linearly independent vectors $l(1), \dots, l(m)$ if $\min_{i=1,\dots,m}[l(i), a_1 - a_2] \le 0$ for each $a_1 \in A_1, a_2 \in A_2$.

Consider now a set A and a point x. Assuming that $A_1 = A$ and $A_2 = \{x\}$, we arrive at the following definition.

DEFINITION 5.18 A set A and a point x are said to be *weakly separated* by linearly independent vectors $l(1), \dots, l(m)$ if for each $a \in A$ there exists an index $i \in \{1, \dots, m\}$ such that $[l(i), x] \ge [l(i), a]$.

We now give some more definitions.

DEFINITION 5.19 A set A and a point x are said to be *separated* by vectors $l(1), \dots, l(m)$ if $\min_{i=1,\dots,m}[l(i), x] \geq \sup_{a \in A} \min_{i=1,\dots,m}[l(i), a]$.

Let a set A and a point x be separated by vectors $l(1), \dots, l(m)$. It follows from Definition 5.19 that for each point $a \in A$ there exists an index i such that $\min_j [l(j), x] \geq [l(i), a]$. In particular, $[l(i), x] \geq [l(i), a]$. Thus the separation by a collection $l(1), \dots, l(m)$ implies the weak separation by the same collection. Note that if two sets (in particular, a point and a set) are weakly separated by a collection of vectors, then the shifts of these sets are also separated by this collection. This is no longer true for the separation of a point from a set.

DEFINITION 5.20 Let A be a subset of $\mathbb{R}^n$ and $x \notin A$. We say that linearly independent vectors $l(1), \dots, l(m)$ *strictly separate* A and x if there exists $\varepsilon \in (0, 1)$ with the following property: for each $a \in A$ there exists i such that $[l(i), a] < 1 - \varepsilon$ and $[l(1), x] = \dots = [l(m), x] = 1$.

DEFINITION 5.21 Let U be a closed subset of $\mathbb{R}^n$ and $x \in \operatorname{bd} U$, $x \neq 0$. A collection of linearly independent vectors $l(1), \dots, l(m)$ is called a *support collection* to U at the point x if $[l(1), x] = \dots = [l(m), x] = 1$ and $\min_i [l(i), u] < 1$ for each $u \in U$, $u \neq x$.

If a closed set U has a support collection at a point $x \in U$ then x is a boundary point of U. The set U and the set $\{x\}$ can be separated by linearly independent vectors $l(1), \dots, l(m)$ with the following property: for each $u \in U$ there exists an index i such that the strict inequality $[l(i), u] < [l(i), x]$ holds.

PROPOSITION 5.35 *Suppose that a set U contains zero and possesses the following property: U and each point $x \notin U$ can be separated by a finite collection of linear functions. Then U is a closed radiant set.*

Proof: For each point $x \notin U$, there exist $c \in R$ and a collection $\ell = \{l(1), \dots, l(m)\}$ such that $U \subset U_{\ell,c}$, where

$$U_{\ell,c} = \{x' : \min_i [l(i), x'] \leq c\}. \tag{5.4.7}$$

Thus there exists a family $(U_{\ell,c})$ such that U coincides with the intersection of this family. Since $0 \in U$ it follows that the number c in (5.4.7) is nonnegative. The set $U_{\ell,c}$ in (5.4.7) is closed. Let us check that this set is radiant. Indeed, let $x' \in U_{\ell,c}$ and $0 < \lambda \leq 1$. If $\min_i [l(i), x'] \leq 0$, then $\min_i [l(i), \lambda x'] \leq 0 \leq c$. If $\min_i [l(i), x] > 0$, then $\min_i [l(i), \lambda x'] < c$, hence $\lambda x' \in U_{\ell,c}$ and $U_{\ell,c}$ is radiant. Since the intersection of an arbitrary family of closed radiant sets is also closed and radiant, it follows that U is closed and radiant. $\triangle$

We now demonstrate that the reverse statement to Proposition 5.35 is valid. Moreover, we prove that each point, which does not belong to a closed radiant set, can be ***strictly separated*** from this set.

THEOREM 5.6 ***Let $U \subset \mathbb{R}^n$ be a closed radiant set and let $x \notin U$. Then U and x are strictly separated by a collection of n linearly independent elements.***

Proof: First we prove that there exist a closed solid convex cone Q and a positive $\lambda < 1$ such that

$$x \in \operatorname{int} Q \text{ and } U \cap (\lambda x + Q) = \emptyset. \tag{5.4.8}$$

Since U is a closed radiant set, it follows that there exists a positive $\lambda < 1$ such that $x' = \lambda x \notin U$. We claim that U and x' can be cone-separated by a solid closed convex cone Q such that $x' \in \operatorname{int} Q$. If this assertion does not hold, then there exists a sequence Q_j of closed solid convex cones such that $Q_{j+1} \setminus \{0\} \subset \operatorname{int} Q_j$, $\bigcap_j Q_j = R_x$ and for each j there exists $y_j \in Q_j$ such that $x_j = x' + y_j \in U$. Consider two possible situations.

1) y_j is unbounded. Assume without loss of generality that $\|y_j\| \to +\infty$. Then for all $\rho > 0$ we have

$$\rho \frac{x_j}{\|y_j\|} = \rho \frac{x'}{\|y_j\|} + \rho \frac{y_j}{\|y_j\|}. \tag{5.4.9}$$

The first addend on the right-hand side of (5.4.9) tends to zero. The equality $\bigcap_j Q_j = R_x$ demonstrates that the second addend tends to $(\rho x)/\|x\|$. Since U is radiant and $x_j \in U$, we conclude that $(\rho x_j)/\|y_j\| \in U$ for large enough j. Since U is closed, it follows that $\rho x \in U$ for all $\rho > 0$, which is impossible.

2) y_j is bounded. Assume without loss of generality that $y_j \to y$. Then there exists $\gamma \geq 0$ such that $y = \gamma x$, hence $x_j \to x' + \gamma x = (\lambda + \gamma)x$. Since $x_j \in U$, we have $(\lambda + \gamma)x \in U$. On the other hand, $(\lambda + \gamma)x \notin U$ since $\lambda x \notin U$, $\gamma \geq 0$ and U is a radiant set. Thus we have a contradiction.

Consider a cone Q and a number λ such that (5.4.8) holds. Applying Proposition 5.32 we can find linearly independent elements $l(1), \dots, l(n)$ such that $T = \{y : [l(i), y] > 0, i = 1, \dots, n\} \subset \operatorname{int} Q$ and $[l(i), x] = 1$ for all i. It is easy to verify that $\lambda x + T = \{y : [l(i), y] > \lambda \text{ for all } i\}$. Since $U \cap (\lambda x + T) = \emptyset$, it follows that for each $u \in U$, there exists i such that $[l(i), u] \leq \lambda$. $\triangle$

We shall now study the separation of a radiant set and its boundary point.

THEOREM 5.7 *Let U be a closed radiant set and let $x \in U$ be a point, such that $\mu_U(x) = 1$. Assume that x does not belong to the Bouligand cone of U at the point x. Then there exists a support collection to the set U at the point x.*

Proof: First, we prove that there exists a solid convex cone Q such that

$$x \in \operatorname{int} Q \text{ and } U \cap (x + Q) = \{x\}. \tag{5.4.10}$$

Assume that such a cone does not exist. Then we can find a sequence of closed solid convex cones (Q_j) with the following properties:

1) $Q_{j+1} \setminus \{0\} \subset \operatorname{int} Q_j,\ j = 1, 2 \dots;$

2) $\bigcap_j Q_j = R_x;$

3) $x \in \operatorname{int} Q_j,\ \ j = 1, \dots;$

4) there exists $y_j \in Q_j$ such that $x_j = x + y_j \in U$ and $y_j \neq 0,\ j = 1, \dots$

If the sequence (y_j) is unbounded, then the same arguments as in the proof of Proposition 5.6 show that $\rho x \in U$ for all $\rho > 0$, which is impossible. Assume that the sequence (y_j) is bounded and there exists $\lim y_j = y$. Then $y = \gamma x$ with $\gamma \geq 0$. If $\gamma > 0$, then $x_j \to (1 + \gamma)x$ and hence $(1 + \gamma)x \in U$. However, this is impossible since $\mu_U(x) = 1$. Assume that $\gamma = 0$. Let $\beta_j = \|y_j\| / \|x\|$. We have $\beta_j > 0$ and $\beta_j \to 0$. Since $y_j \in Q_j$ and $\bigcap_j Q_j = R_x$, it follows that

$$\frac{y_j}{\beta_j} \to x. \tag{5.4.11}$$

Since $x + y_j \in U$ and x does not belong to the Bouligand cone of U at the point x, applying (5.4.6) we have

$$y_j \notin \bigcup_{0 < \alpha \leq \alpha_0} \alpha B(x, \varepsilon). \tag{5.4.12}$$

It follows from (5.4.12) that $\|(y_j/\beta_j) - x\| \geq \varepsilon$ for large enough j, which contradicts (5.4.11).

Let Q be a solid convex cone such that (5.4.10) holds. Applying Proposition 5.32 we find linearly independent vectors $l(1), \dots, l(n)$ such that $T = \{y : [l(i), y] > 0,\ i = 1, \dots, n\} \subset \operatorname{int} Q$ and $[l(1), x] = \dots = [l(n), x] = 1$. Since $x + T = \{y : [l(i), y] > 1,\ i = 1, \dots, n\}$, it follows that for each $u \in U$, $u \neq x$ there exists i such that $[l(i), u] < 1$. Since $\{z : [l(i), z] = 1, i = 1, \dots, n\} = \{x\}$, we have $\min_i [l(i), u] < 1$ for each $u \in U$, $u \neq x$. △

REMARK 5.13 If U is a radiative set, that each boundary point x of U such that $x \notin \Gamma(U,x)$ enjoys a support collection to the set U.

Thus the inclusion $x \notin \Gamma(U,x)$ implies the existence of a support collection. We now demonstrate that the reverse assertion holds without any additional assumptions.

PROPOSITION 5.36 *Let $X \subset \mathbb{R}^n$. If there exists a support collection to X at a point $x_0 \in \mathrm{cl}\, X$, then $x_0 \notin \Gamma(X,x_0)$.*

Proof: Assume that $x_0 \in \Gamma(X,x_0)$. Then there exist sequences $u_k \to x_0$ and $\lambda_k \downarrow 0$ such that $x_0 + \lambda_k u_k \in X$. Let $l(1),\dots,l(m)$ be a support collection to X at a point x_0. Then $[l(i),x_0] = 1$ for all i and $\min_i [l(i),x] \le 1$ for all $x \in X$. Since $x_0 + \lambda_k u_k \in X$, it follows that

$$1 \ge \min_i [l(i), x_0 + \lambda_k u_k] = \min_i (1 + [l(i), \lambda_k u_k]) = 1 + \lambda_k \min_i [l(i), u_k].$$

Thus $\min_i [l(i),u_k] \le 0$. Since $u_k \to x_0$, we have $\min_i [l(i),x_0] \le 0$, which is a contradiction. △

THEOREM 5.8 *Let A_1 and A_2 be star-shaped sets such that*
1) $\mathrm{int\,kern}\, A_1$ *is not empty;*
2) $(\mathrm{int}\, A_1) \cap A_2 = \emptyset$.
Then A_1 and A_2 are weakly separated by n linearly independent vectors.

The proof of Theorem 5.8 is based on the following assertion.

LEMMA 5.4 *Let A_1 be a st-sh set and* $\mathrm{int\,kern}\, A_1$ *be nonempty. Further, let A_2 be a radiant set,* $(\mathrm{int}\, A_1) \cap A_2 = \emptyset$ *and $A_1 \cap A_2 \ne \emptyset$. Then* $0 \in \mathrm{bd}\,(A_1 - A_2)$.

Proof of Lemma 5.4: First, we show that

$$(\lambda a + A_1) \cap A_2 = \emptyset \qquad \text{for } a \in \mathrm{int\,kern}\, A_1,\ \lambda > 0. \tag{5.4.13}$$

Assume on the contrary that there exist $a_1 \in \mathrm{int\,kern}\, A_1$, $z \in A_1$ and $\lambda > 0$ such that $a_2 := \lambda a_1 + z \in A_2$. Then

$$\frac{1}{1+\lambda} a_2 \in A_2, \quad \frac{1}{1+\lambda} a_2 = \frac{\lambda}{1+\lambda} a_1 + \frac{1}{1+\lambda} z \in \mathrm{int}\, A_1$$

and we have a contradiction. Clearly, (5.4.13) is equivalent to

$$\lambda a \notin A_2 - A_1 \qquad \text{for all } a \in \mathrm{int\,kern}\, A_1,\ \lambda > 0. \tag{5.4.14}$$

Since $A_1 \cap A_2 \ne \emptyset$, it follows that $0 \in A_2 - A_1$. Applying (5.4.14) we conclude that $0 \in \mathrm{bd}\,(A_2 - A_1)$. Clearly, $0 \in \mathrm{bd}\,(A_1 - A_2)$ as well. △

Proof of Theorem 5.8: Let $a_1 \in \operatorname{int}\operatorname{kern} A_1$ and $U_1 = A_1 - a_1$. Then U_1 is a radiant set and $0 \in \operatorname{int}\operatorname{kern} U_1$. Applying Proposition 5.17, we deduce that there exist $\varepsilon > 0$ and a family $(U^t)_{t\in T}$ of convex sets U^t such that $U_1 = \bigcup_{t\in T} U^t$ and $U^t \supset B(0,\varepsilon)$ for all $t \in T$. Let $U_2 := A_2 - a_2$ where $a_2 \in \operatorname{kern} U_2$. Then U_2 is a radiant set. We represent U_2 in the following form: $U_2 = \bigcup_{v\in U_2} D_v$ where $D_v = \{\alpha v : \alpha \in [0,1]\}$. Let $U = U_1 - U_2$. Then

$$U = U_1 - U_2 = \bigcup_{t\in T} U^t - \bigcup_{v\in U_2} D_v = \bigcup_{t\in T, v\in U_2} (U^t - D_v).$$

Sets $U^t - D_v$, $(t \in T, v \in U_2)$ are convex and each of them contains the ball $B(0,\varepsilon)$. It follows from Proposition 5.15 that the Minkowski gauge μ_U of the set U coincides with the Minkowski gauge $\mu_{\operatorname{cl} U}$ of the closure $\operatorname{cl} U$ of this set and that $\mu_{\operatorname{cl} U}$ is a Lipschitz function. Note that $\operatorname{cl} U$ is a radiative set.

Let $a = a_2 - a_1$. Consider the set $A_1 - A_2 = U_1 - U_2 - (a_2 - a_1) = U - a$. If $a \notin U$, then $\mu_U(a) \geq 1$, hence $\mu_{\operatorname{cl} U}(a) = \mu_U(a) \geq 1$. It follows from Proposition 5.15 that $\operatorname{cl} U = \{x : \mu_U(x) \leq 1\}$ and $\operatorname{bd}\operatorname{cl} U = \{x : \mu_U(x) = 1\}$. Thus either $a \notin \operatorname{cl} U$ or $a \in \operatorname{bd}\operatorname{cl} U$. Assume that $a \in U$. Then

$$0 \in U - a = (U_1 + a_1) - (U_2 + a_2) = A_1 - A_2,$$

hence $A_1 \cap A_2 \neq \emptyset$. Since $(\operatorname{int} A_1) \cap A_2 = \emptyset$ we conclude, by applying Lemma 5.4, that

$$0 \in \operatorname{bd}(A_1 - A_2) = \operatorname{bd}(U - a).$$

Hence a is a boundary point of U. It follows from Proposition 5.15 that a is a boundary point of the closed radiant set $\operatorname{cl} U$. We have verified that either $a \notin \operatorname{cl} U$ or $a \in \operatorname{bd}\operatorname{cl} U$. Assume that $a \notin \operatorname{cl} U$. Applying Theorem 5.6 we can strictly separate a and $\operatorname{cl} U$ by n linearly independent vectors. These vectors weakly separate a and $\operatorname{cl} U$ as well. It is easy to check that these vectors also weakly separate A_1 and A_2.

Assume now that a is a boundary point of the set $\operatorname{cl} U$. Since the Minkowski gauge $\mu_{\operatorname{cl} U}$ of $\operatorname{cl} U$ is Lipschitz, it follows (see Corollary 5.6) that the boundary point a of the set U does not belong to the Bouligand cone of $\operatorname{cl} U$ at the same point a. Using Theorem 5.7 we find a support collection $l(1), \dots, l(n)$ to $\operatorname{cl} U$ at a. Vectors $l(1), \dots, l(n)$ weakly separate $\operatorname{cl} U$ and a as well, hence they weakly separate the sets A_1 and A_2. $\triangle$

5.4.4. Separation of co-star-shaped sets

We can strictly separate a radiant set and a point, which does not belong to this set, without any additional assumptions. Unfortunately this is no longer true for co-star-shaped sets. Indeed, let $X = \{x : \|x\| \geq 1\}$ be the complement of the open unit ball and $x = 0$. Clearly, it is impossible to separate X and x by a cone. However, X is a co-radiant (and even co-radiative) set. Thus we need to impose some additional assumptions, relating to the location of a point with respect to a set.

Let $x \neq 0$. Denote the open ray $\{\lambda x : \lambda > 0\}$ by $\tilde{R}_x$. We start with the following assertion:

PROPOSITION 5.37 *Let X be a closed set and $x \notin X$. Suppose there exists $y \notin \operatorname{cl} \operatorname{cone} X$ such that $(x + \tilde{R}_y) \cap X = \emptyset$. Then X and x can be separated by a cone Q such that $y \in \operatorname{int} Q$.*

Proof: Since $x \notin X$, it follows that $(x + R_y) \cap X$ is empty. Let Q_i be a sequence of closed, solid, convex cones such that $\bigcap_i \operatorname{int} Q_i = R_y$ and $\operatorname{int} Q_i \supset Q_{i+1} \setminus \{0\}$. Assume that the intersection $(x + \operatorname{int} Q_i) \cap X$ is nonempty for all i. Then there exists a sequence $u_i \in X$ such that $u_i - x \in \operatorname{int} Q_i$, $i = 1, 2, \dots$. Since $\bigcap_i \operatorname{int} Q_i = R_y$, it follows that there exist a sequence $y_i \to y$ and a sequence of positive numbers λ_i such that $u_i = x + \lambda_i y_i$. If λ_i is bounded, we can assume without loss of generality that u_i tends to a point on the ray $x + R_y$. Since $u_i \in X$, it follows that $\lim_i u_i \in X$, which is a contradiction. If λ_i is unbounded we have

$$\frac{u_i}{\lambda_i} = \frac{x}{\lambda_i} + y_i \to y.$$

Since $u_i \in X$, it follows that $y \in \operatorname{cl} \operatorname{cone} X$, which is again a contradiction. Thus there exists i such that $(x + Q_i) \cap X = \emptyset$. It follows from the definition of the cone Q_i that $y \in \operatorname{int} Q_i$. △

The next example shows that the requirements "$(x + \tilde{R}_y) \cap X$ is nonempty" and "$y \in \operatorname{cl} \operatorname{cone} X$" are independent. There exists a closed set V, a point $x \notin V$ and a point y such that $(x + \tilde{R}_y) \cap V = \emptyset$ and $y \in \operatorname{cl} \operatorname{cone} V$. Moreover, there exists a co-radiative set V such that for some x, y this property is valid.

EXAMPLE 5.3 Let $n = 2$, $V = \{(x_1, x_2) \in \mathbb{R}^2 : x_2 \geq a - \exp(-|x_1|)\}$ where $a > 1$. It is easy to check that V is a co-radiative set and $0 \in \operatorname{int} \operatorname{kern}^* V$. If $x = (0, (a-1)/2)$ and $y = (1, 0)$ then $(x + R_y) \cap V$ is empty and $y \in \operatorname{cl} \operatorname{cone} V$.

COROLLARY 5.8 *Let X be a closed set and $x \notin \operatorname{cl} \operatorname{cone} X$. Then X and x can be cone-separated.*

Proof: We have $x + \tilde{R}_x = \{\lambda x : \lambda > 1\}$, so if $(x + \tilde{R}_x) \cap X \neq \emptyset$, then there exists $\lambda > 1$ such that $\lambda x \in X$. In such a case, $x \in \operatorname{cone} X \subset \operatorname{cl} \operatorname{cone} X$

and we have a contradiction. Thus $(x + \tilde{R}_x) \cap X = \emptyset$ and we can apply Proposition 5.37. $\triangle$

PROPOSITION 5.38 *Let V be a co-radiant closed set, $x \notin V$ and $-x \notin \operatorname{cl}\operatorname{cone} V$. Then V and x can be cone-separated by the cone Q such that $-x \in \operatorname{int} Q$.*

Proof: In order to apply Proposition 5.37 we need to show that $(x + \tilde{R}_{-x}) \cap V = \emptyset$. It is easy to check that $x + \tilde{R}_{-x} = \{\lambda x : \lambda < 1\}$. If $(x + \tilde{R}_{-x}) \cap V \neq \emptyset$, then there exists $\lambda < 1$ such that $\lambda x \in V$. Since V is a co-radiant set, it follows that $0 \notin V$. So either $0 < \lambda < 1$ or $\lambda < 0$. Assume $0 < \lambda < 1$ and let $x' = \lambda x$. Since $x' \in V$, $x = \lambda^{-1} x'$ and V is a co-radiant set it follows that $x \in V$, which is a contradiction. If $\lambda < 0$, then $-x \in |\lambda|^{-1} V \subset \operatorname{cl}\operatorname{cone} V$, which is again a contradiction. $\triangle$

The next assertion shows, in particular, that for a co-radiative set V, the conditions in Proposition 5.37 are not only sufficient but also necessary for the separability by a cone.

PROPOSITION 5.39 *Let V be a co-radiative set and $x \in \mathbb{R}^n$. If V and x can be cone-separated by a cone Q, then for each $y \in \operatorname{int} Q$ we have that $y \notin \operatorname{cl}\operatorname{cone} V$ and $(x + \tilde{R}_y) \cap V = \emptyset$.*

Proof: Let Q be a closed, solid, convex cone such that $(x + \operatorname{int} Q) \cap V$ is empty. Let $y \in \operatorname{int} Q$. Then $x + \tilde{R}_y \subset \operatorname{cl}(CV) = C_r V$. Theorem 5.1 demonstrates that $C_r V$ is a radiative set, so $\tilde{R}_y \subset C_r V$. Thus $\tilde{R}_y \cap \operatorname{int} V = \emptyset$. Since V is co-radiative, it follows that $\tilde{R}_y \cap V$ is empty. We have proved that $\operatorname{int} Q$ does not intersect V. Since $\operatorname{int} Q$ is an open cone, we have that $\operatorname{int} Q \cap \operatorname{cl}\operatorname{cone} V = \emptyset$. Hence, if $y \in \operatorname{int} Q$, then $y \notin \operatorname{cl}\operatorname{cone} V$ and $(x + \tilde{R}_y) \cap V = \emptyset$. $\triangle$

Let V be a co-radiative set and $x \notin V$. Corollary 5.8 and Proposition 5.38 show that V and x can be cone-separated if either x or $-x$ does not belong to $\operatorname{cl}\operatorname{cone} V$. Assume that $V = CU$, where U is an open, bounded, convex set with $0 \in U$. Then $\operatorname{cone} V = \mathbb{R}^n$ so $x, -x \in \operatorname{cl}\operatorname{cone} V$ for all $x \in \mathbb{R}^n$. Clearly V and each point $x \notin V$ cannot be cone-separated. We now give an example of the set V such that $\operatorname{cl}\operatorname{cone} V$ is a proper cone and there exists a point $x \notin V$ such that 1) $x \in \operatorname{cl}\operatorname{cone} V$, 2)$-x \in \operatorname{cl}\operatorname{cone} V$, 3) V and x cannot be separated.

EXAMPLE 5.4 Let φ be a positive, decreasing, convex function defined on $\mathbb{R}_+ = \{x \in \mathbb{R} : x \geq 0\}$ and such that $\varphi(0) = 2$, $\varphi(2) < \frac{1}{3}$, $\varphi(x) \to 0$ as $x \to +\infty$. Consider the following sets:

$$Y' = \{(x, y) \in \mathbb{R}^2 : x \geq 0,\ 0 \leq y \leq \varphi(x)\},$$

$$Y'' = \{(x,y) \in \mathbb{R}^2 : x \geq 0, \ -\frac{1}{3}x - \varphi(x) \leq y \leq 0\}.$$

Let $C_2 = \{(x,y) \in \mathbb{R}^2 : x^2 + y^2 \leq 4\}$. It is easy to show that the set $U = Y' \cup Y'' \cup C_2$ is radiative so $V = C_r U$ is co-radiative. It is also easy to check that $\operatorname{cl}\operatorname{cone} V = CQ$, where

$$Q = \{(x,y) \in \mathbb{R}^2 : x > 0, \ -\frac{1}{3}x < y < 0\}.$$

Consider the point $x = (0,1)$. We have $x \notin V$, $x \in \operatorname{cl}\operatorname{cone} V, -x \in \operatorname{cl}\operatorname{cone} V$. It is easily seen that $(x + \tilde{R}_y) \cap V \neq \emptyset$ for each $y \in Q$. It follows from Proposition 5.39 that x and V cannot be separated.

Let Q be a convex cone. Denote the conjugate cone to Q by Q^*. Recall that $Q^* = \{l \in \mathbb{R}^n : [l,x] \geq 0\}$ for all $x \in Q$.

PROPOSITION 5.40 *Let V be a co-radiant closed set such that* $\operatorname{cl}\operatorname{cone} V$ *is a proper, convex cone. Then V and each point $x \notin V$ can be cone-separated.*

Proof: It follows from Proposition 5.38 and Corollary 5.8, that we can consider only elements x such that $x \in \operatorname{cl}\operatorname{cone} V$, $-x \in \operatorname{cl}\operatorname{cone} V$. First assume that $x \neq 0$. Since $\operatorname{cl}\operatorname{cone} V$ is a proper cone, it easily follows that $x \in \operatorname{bd}\operatorname{cl}\operatorname{cone} V$. So there exists $l \in (\operatorname{cl}\operatorname{cone} V)^*$, $l \neq 0$ such that $[l,x] = 0$. We have that $-l \notin \operatorname{cl}\operatorname{cone} V$ and $[x - \lambda l, l] < 0$ for all $\lambda > 0$, that is, $(x + \tilde{R}_{-l}) \cap V = \emptyset$. Applying Proposition 5.37, we conclude that V and x can be cone-separated. Let $x = 0$. Consider a non-zero vector $l \in (\operatorname{cl}\operatorname{cone} V)^*$. Since $-l \notin \operatorname{cl}\operatorname{cone} V$, we have $(0 + \tilde{R}_{-l}) \cap V = \emptyset$, so 0 and V can be cone-separated by Proposition 5.37. △

We now study the cone-separation of a co-radiant closed set V and a point $x \in V$. Since it is impossible to separate by a cone a set and its interior point, we assume that x is a boundary point of V.

PROPOSITION 5.41 *Let V be a co-radiant closed set with the Minkowski co-gauge ν_V. Let $x \in V$ be a vector such that $\nu_V(x) = 1$ and $-x \notin$* $\operatorname{cl}\operatorname{cone} V$*. Then the following assertions are equivalent:*
1) V and x can be cone-separated by the cone Q such that $-x \in \operatorname{int} Q$;
2) $-x$ does not belong to the Bouligand cone $\Gamma(V,x)$ of the set V at the point x.

Proof: 1) $\Longrightarrow$ 2). Let Q be a closed solid convex cone such that

$$(x + \operatorname{int} Q) \cap V = \emptyset \tag{5.4.15}$$

and $-x \in \operatorname{int} Q$. If $-x \in \Gamma(V,x)$, then there exist sequences $u_k \to -x$ and $\lambda_k \to 0$, $\lambda_k > 0$ such that $x + \lambda_k u_k \in V$. It follows from (5.4.15) that $x+\lambda_k u_k \notin (x+\operatorname{int} Q)$, hence $u_k \notin \operatorname{int} Q$. Thus $\lim u_k = -x \notin \operatorname{int} Q$ and we have a contradiction.

2) $\Longrightarrow$ 1). Assume that $-x \notin \Gamma(V,x)$. Let Q_i be a sequence of closed, solid and convex cones such that $\operatorname{int} Q_i \supset Q_{i+1} \setminus \{0\}$ and $\bigcap_i \operatorname{int} Q_i = \tilde{R}_{-x} := \{\lambda(-x) : \lambda > 0\}$. Assume that $(x+\operatorname{int} Q_i) \cap V \neq \emptyset$ for all i. Then there exists a sequence $y_i \in V$ such that $y_i - x \in \operatorname{int} Q_i$, $i = 1, 2, \dots$. Since $\bigcap_i \operatorname{int} Q_i = \tilde{R}_{-x}$, it follows that there exist sequences $z_i \to -x$ and $\lambda_i > 0$ such that $y_i = x + \lambda_i z_i$. If λ_i is bounded, we can assume without loss of generality that there exists $\lambda \geq 0$ such that $y_i \to x + \lambda(-x) = (1-\lambda)x$. Since $-x$ is not a feasible direction at x, it follows that $\lambda > 0$. Assume that $0 < \lambda < 1$. Then $y = \lim y_i = (1-\lambda)x \in V$. We have $\nu_V(y) = (1-\lambda)\nu_V(x) < 1$, which is impossible since $V = \{x : \nu_V(x) \geq 1\}$. Let $\lambda \geq 1$. Since $-(\lambda-1)x \in V$, it follows that $-x \in \operatorname{cl} \operatorname{cone} V$, which is again a contradiction. Thus λ_i is unbounded. Assume without loss of generality that $\lambda_i \to +\infty$. Then

$$\frac{y_i}{\lambda_i} = \frac{x}{\lambda_i} + z_i \to -x.$$

Since $y_i \in V$ it follows that $-x \in \operatorname{cl} \operatorname{cone} V$. We have arrived at a contradiction, which shows that there exists i such that the intersection $(x+Q_i) \cap V$ is empty. △

COROLLARY 5.9 *Let V be a co-radiative set. Then the equality $\nu_V(x) = 1$ can be expressed in the form $x \in \operatorname{bd} V$.*

PROPOSITION 5.42 *Let V be a co-radiative set and let x be a boundary point of V. Assume that*

1) the Minkowski co-gauge ν_V of V is locally Lipschitz at the point x;

2) $-x \notin \operatorname{cl} \operatorname{cone} V$.

Then the set V and the point x can be cone-separated by a cone Q such that $-x \in \operatorname{int} Q$.

Proof: This follows from Proposition 5.41, Corollary 5.7 and Corollary 5.9. △

Recall (see Theorem 5.4) that the Minkowski co-gauge ν_V of a closed co-radiant set is Lipschitz if and only if $0 \in \operatorname{int} \operatorname{kern}_\infty V$. Thus if $0 \in \operatorname{int} \operatorname{kern}_\infty V$ and $-x \notin \operatorname{cl} \operatorname{cone} V$, then V and x can be cone-separated by a cone Q such that $-x \in \operatorname{int} Q$.

We now present one more result related to the cone-separability.

THEOREM 5.9 *Let V be a co-radiant set and $0 \in \operatorname{int} \operatorname{kern}_\infty V$, let $x \in \operatorname{bd} V$ and assume that there exists $y \notin \operatorname{cl} \operatorname{cone} V$ such that $(x+\tilde{R}_y) \cap V = \emptyset$. Then V and x can be cone-separated.*

Proof: Since $y \notin \operatorname{cl}\operatorname{cone} V$, it follows that there exists a closed solid convex cone $\tilde{Q}$ such that $y \in \operatorname{int}\tilde{Q}$ and the intersection $(\tilde{Q} \cap \operatorname{cl}\operatorname{cone} V$ is empty. Assume $(x + \tilde{R}_y) \cap \operatorname{int}\tilde{Q} = \emptyset$. Applying the separation theorem, we can find a vector $l \neq 0$ such that

$$\sup\{[l, z] : z \in \tilde{Q}\} \leq 0 \leq \inf\{[l, x + \lambda y] : \lambda > 0\}. \tag{5.4.16}$$

It follows from (5.4.16) that $l \in -\tilde{Q}^*$, so $[l, y] \leq 0$. If $[l, y] < 0$, then $\inf_{\lambda > 0}[l, x + \lambda y] = -\infty$, which contradicts (5.4.16). Therefore $[l, y] = 0$. However, relations $[l, y] = 0$ and $y \in \operatorname{int}\tilde{Q}$ imply $l = 0$. Thus we get a contradiction which shows that there exists $\lambda' > 0$ such that $x + \lambda' y \in \operatorname{int}\tilde{Q}$. On the other hand, the inclusion $y \in \operatorname{int}\tilde{Q}$ implies $x + \lambda' y \in x + \operatorname{int}\tilde{Q}$. Clearly there exists positive $\alpha < 1$ such that

$$z := \alpha(x + \lambda' y) \in (\operatorname{int}\tilde{Q}) \cap (x + \operatorname{int}\tilde{Q}). \tag{5.4.17}$$

Since $\operatorname{kern}_\infty V$ is convex and $0 \in \operatorname{int}\operatorname{kern}_\infty V$, there exists an open convex subset $\tilde{G}$ of $\operatorname{kern}_\infty V$ with $0 \in \tilde{G}$. Define

$$D = \bigcup_{\lambda > 0,\, 0 < \beta < 1} ((1 - \beta)(x + \lambda y) + \beta\tilde{G}).$$

Clearly D is open. Let $R = \{x + \lambda y : \lambda > 0\}$. Since $D = \bigcup_{0 < \beta < 1}((1 - \beta)R + \beta\tilde{G})$ and both sets R and $\tilde{G}$ are convex, it easily follows that D is a convex set. We now check that the intersection $D \cap V$ is empty. On the contrary, assume that there exists $v \in V$ such that $v = (1 - \beta)(x + \lambda y) + \beta g$ with $0 < \beta < 1$, $\lambda > 0$ and $g \in \tilde{G}$. The inclusion $g \in \operatorname{kern}_\infty V$ implies $g + \gamma(v - g) \in V$ for all $\gamma > 1$. In particular,

$$x + \lambda y = \frac{v - \beta g}{1 - \beta} = g + \frac{1}{1 - \beta}(v - g) \in V.$$

We arrive at a contradiction, since $(x + \tilde{R}_y) \cap V = \emptyset$. Hence $D \cap V = \emptyset$. It is clear that the point z defined by (5.4.17) belongs to D. It follows from (5.4.17) that

$$z \in \operatorname{int}\tilde{Q} \cap \operatorname{int}(x + \tilde{Q}) \cap D. \tag{5.4.18}$$

There exists positive $r < \|x - z\|$ such that the intersection in (5.4.18) contains the closed ball $B(z, r) = \{x' : \|x' - z\| \leq r\}$. Consider the closed solid convex cone

$$Q = \operatorname{cone}(B(z, r) - x) \cup \{0\} = \bigcup_{\lambda \geq 0} \lambda(B(z, r) - x). \tag{5.4.19}$$

An easy computation shows that

$$x + Q = \bigcup_{\lambda \geq 0} ((1-\lambda)x + \lambda B(z,r)).$$

We now check that $(x + \operatorname{int} Q) \cap V = \emptyset$. Let $x' = (1-\lambda)x + \lambda g \in x + \operatorname{int} Q$. Here $\lambda > 0$ and $g \in B(z,r)$. First assume that $\lambda > 1$. We have $x' = (1-\lambda)x + \lambda g = g + (\lambda - 1)(g - x)$. Since $g \in B(z,r) \subset x + \tilde{Q}$ and $\tilde{Q}$ is a convex cone, it follows that $(\lambda - 1)(g - x) \in \tilde{Q}$. Since $g \in \tilde{Q}$, we have $x' \in \tilde{Q}$, hence $x' \notin V$. Assume now that $0 < \lambda < 1$. An easy computation shows that $x' \in D$ in such a case, hence $x' \notin V$ again. △

Using the notion of the recession cone of a co-star-shaped set (see Definition 5.14) we can obtain a version of Theorem 5.9 for co-star-shaped sets.

PROPOSITION 5.43 *Let V be a closed co-star-shaped set with nonempty* $\operatorname{int} \operatorname{kern}_\infty V$ *and let $x \in \operatorname{bd} V$. Assume that there exists $y \notin \operatorname{rc} V$ such that $(x + \tilde{R}_y) \cap V = \emptyset$. Then V and x can be cone-separated.*

Proof: Let $a \in \operatorname{int} \operatorname{kern}_\infty V$. Then $-a + V$ is a closed co-radiant set, so $0 \notin -a + V$. Furthermore,

$$-a + x \in \operatorname{bd}(-a + V), \quad 0 \in \operatorname{int} \operatorname{kern}_\infty (V - a), \quad y \notin \operatorname{rc}(-a + V).$$

Proposition 5.21 demonstrates that $\operatorname{rc}(-a+V) = \operatorname{cl} \operatorname{cone}(-a+V)$. Hence applying Theorem 5.9 we can conclude that there exists a closed solid convex cone Q which separates $-a + x$ and $-a + V$. Clearly Q separates x and V as well. △

PROPOSITION 5.44 *Let V be a closed co-radiant set and* $\operatorname{int} \operatorname{kern}_\infty V \neq \emptyset$. *Let $x \in V$ with $\nu_V(x) = 1$ be a point such that $-x \notin \operatorname{cl} \operatorname{cone} V$. Then V and x can be cone-separated. If* $0 \in \operatorname{int} \operatorname{kern}_\infty V$ *then V and x can be separated by a cone Q such that* $-x \in \operatorname{int} Q$.

Proof: Let $y = -x$. First we show that $(x + \tilde{R}_y) \cap V = \emptyset$. We have $x + \tilde{R}_y = \{\lambda x : \lambda < 1\}$. Since V is co-radiant and $x \in \operatorname{bd} V$, it follows that $\lambda x \notin V$ for $\lambda \in [0,1)$. Clearly $V \cap \tilde{R}_y = \emptyset$, otherwise $-x \in \operatorname{cl} \operatorname{cone} V$. Proposition 5.43 shows that V and x can be cone-separated.

Let $0 \in \operatorname{int} \operatorname{kern}_\infty V$. It follows from the proof of Theorem 5.9 (see (5.4.19) and (5.4.17)) that V and x can be separated by the convex cone Q, where

$$Q = \bigcup_{\lambda > 0} \lambda(B(z,r) - x) \text{ with } z = \alpha(x + \lambda' y), \; 0 < \alpha < 1, \; \lambda' > 0.$$

Here $B(z,r) = \{x' : \|z - x'\| \leq r\}$. Since $y = -x$, it follows that $z - x = (\alpha - \alpha\lambda' - 1)x$. Since $\alpha - 1 < 0$ and $-\alpha\lambda' < 0$, it follows that $-x \in \operatorname{int} Q$. △

THEOREM 5.10 *Let V be a co-radiant set and* $\operatorname{int}\operatorname{kern}_\infty V \neq \emptyset$. *Let the cone* $\operatorname{cl}\operatorname{cone} V$ *be proper and convex. Then each point x with $\nu_V(x) = 1$ can be cone-separated from V.*

Proof: If $-x \notin \operatorname{cl}\operatorname{cone} V$ then we can apply Proposition 5.44. Assume $-x \in \operatorname{cl}\operatorname{cone} V$. Let $g \in (\operatorname{cl}\operatorname{cone} V)^*$, $g \neq 0$. Since both $x, -x \in \operatorname{cl}\operatorname{cone} V$, it follows that $[x, g] = 0$. Thus we have $[x - \lambda g, g] = -\lambda\|g\|^2 < 0$ for all $\lambda > 0$. On the other hand, $[g, v] \geq 0$ for all $v \in V$, so $(x + \tilde{R}_{-g}) \cap V = \emptyset$. Since $-g \in \operatorname{cl}\operatorname{cone} V = \operatorname{rc}\operatorname{cone} V$, we can apply Proposition 5.43. The proof is complete. △

Consider a collection $\ell = (l(i))_{i=1}^n$ of n linearly independent elements of the space $\mathbb{R}^n$. This collection determines the cones T_ℓ and T^ℓ defined by (5.4.1):

$$T_\ell = \{x \in \mathbb{R}^n : [l(i), x] < 0,\ i = 1, \dots, n\},$$
$$T^\ell = \{x \in \mathbb{R}^n : [l(i), x] > 0,\ i = 1, \dots, n\}.$$

Since vectors $l(1), \dots, l(n)$ are linearly independent, it follows that both T_ℓ and T^ℓ are nonempty. Clearly T_ℓ and T^ℓ are open convex cones.

PROPOSITION 5.45 *Let V be a closed set. Then the following assertions are equivalent:*

1) V is a co-radiant set, its complement $C(V)$ does not intersect the cone $(-\operatorname{cl}\operatorname{cone} V)$ *and $0 \notin V$.*

2) For each nonzero $x \notin V$ there exists a collection $\ell = (l(1), \dots, l(n))$ of linearly independent vectors such that

$$0 < [l(1), x] = \dots = [l(n), x] < 1 \leq \inf_{v \in V} \max_{i=1,\dots,n} [l(i), v]. \qquad (5.4.20)$$

Proof: 1) $\Longrightarrow$ 2). Let $x \notin V$. There exists $\varepsilon > 0$ such that $(1+\varepsilon)x \notin V$, $x \neq 0$, that is, $(1+\varepsilon)x \in C(V)$. Then $-(1+\varepsilon)x \notin \operatorname{cl}\operatorname{cone} V$. In the light of Proposition 5.38, there exists a closed solid cone Q such that $((1+\varepsilon)x + \operatorname{int} Q) \cap V = \emptyset$ and $-x \in \operatorname{int} Q$. It follows from Proposition 5.32 that there exists a collection $\tilde{\ell} = (\tilde{l}(i))_{i=1}^n$ such that $[\tilde{l}(i), -x] = -1$ for all i and $T_{\tilde{\ell}} \subset \operatorname{int} Q$. It is easy to check that

$$(1+\varepsilon)x + T_{\tilde{\ell}} = \{y : [\tilde{l}(i), y] < 1+\varepsilon,\ (i \in I)\} = \{y : \max_{i \in I}[\tilde{l}(i), y] < 1+\varepsilon\},$$

where $I = \{1, \ldots, n\}$. Since $(1+\varepsilon)x + T_{\tilde{\ell}} \subset (1+\varepsilon)x + \operatorname{int} Q$, it follows that $\max_{i=1,\ldots,n}[\tilde{l}(i), v] \geq 1+\varepsilon$ for all $v \in V$. Let $l(i) = (1/(1+\varepsilon))\tilde{l}(i)$. Then $0 < l(1)(x) = \ldots = l(n)(x) < 1$ and $\inf_{v \in V} \max[l(i), v] \geq 1$.

2) $\Longrightarrow$ 1). First we check that

$$(v \in V, v \neq 0, \lambda > 1) \implies \lambda v \in V. \tag{5.4.21}$$

If $\lambda v \notin V$ then we can find a collection $\ell = (l(i))_{i=1}^{n}$ such that for all k

$$\begin{aligned} \lambda \max_{i \in I}[l(i), v] &= \lambda[l(k), v] = [l(k), \lambda v] < 1 \\ &\leq \inf_{v' \in V} \max_{i \in I}[l(i), v'] \leq \max_{i \in I}[l(i), v]. \end{aligned}$$

We have a contradiction which shows that (5.4.21) is valid. If $0 \in V$, then

$$\inf_{v \in V} \max_{i \in I}[l(i), v] \leq 0$$

for each collection ℓ, which is impossible. Thus V is a co-radiant set and $v \neq 0$.

We now check that $\mathcal{C}(V) \cap (-\operatorname{cl} \operatorname{cone} V) = \{0\}$. Let $x \notin V$, $x \neq 0$. Then there exists a collection $\ell = (l(i))_{i=1}^{n}$ such that $\max_{i=1,\ldots,n}[l(i), v] \geq 1$ for each $v \in V$ and $[l(i), x] = \lambda > 0$ for all i. We have $\max_{i=1,\ldots,n}[l(i), y] \geq 0$ for each $y \in \operatorname{cl} \operatorname{cone} V$. On the other hand, $\max_{i=1,\ldots,n}[l(i), -x] = -\lambda < 0$. Thus $-x \notin \operatorname{cl} \operatorname{cone} V$. $\triangle$

THEOREM 5.11 *Let V be a closed co-radiant set and let*

$$\operatorname{cl} \operatorname{cone} V \cap (-\operatorname{cl} \operatorname{cone} V) = \emptyset.$$

Then for each non-zero $x \notin V$ there exists a collection $\ell = (l(i))_{i=1}^{n}$ of linearly independent vectors such that

$$[l(1), x] = \ldots [l(n), x] \leq \inf_{v \in V} \max_{i=1,\ldots,n} [l(i), v].$$

Proof: Let $x \in \operatorname{cl} \operatorname{cone} V$. Then $-x \notin \operatorname{cl} \operatorname{cone} V$. In such a case the proof is similar to the proof of 1) $\Longrightarrow$ 2) in Proposition 5.45. Assume now that $x \notin \operatorname{cl} \operatorname{cone} V$. Then we conclude, by applying Corollary 5.8, that V and x can be separated by a closed convex solid cone Q such that $x \in \operatorname{int} Q$. It follows from Proposition 5.32 that there exists a collection ℓ of linearly independent vectors $l(1), \ldots, l(n)$ such that $[l(1), x] = \ldots = [l(n), x] = -1$ and

$$T_\ell := \{x' : [l(i), x'] < 0, \ i = 1, \ldots, n\} \subset \operatorname{int} Q.$$

Clearly $x+T_\ell = \{y : [l(i), y] < -1,\ i = 1, \dots, n\}$. Since $(x+\operatorname{int} Q) \cap V = \emptyset$, it follows that $(x + T_\ell) \cap V = \emptyset$, hence for each $v \in V$ there exists i such that $[l(i), v] \geq -1$. Thus $\max_{i=1,\dots,n}[l(i), v] \geq -1$ for each $v \in V$. $\triangle$

We now discuss the weak separation of two co-star-shaped sets. Applying the results related to the separation of a co-star-shaped set and a point, we obtain different sufficient conditions for the weak separation of two co-star-shaped sets. We now present some of these conditions.

The following assertion is self-evident.

PROPOSITION 5.46 *If B_1 and B_2 are co-star-shaped sets and* $\operatorname{int} B_1 \cap \operatorname{int} B_2 \neq \emptyset$, *then these sets cannot be weakly separated.*

PROPOSITION 5.47 *Let B_1 and B_2 be closed co-star-shaped subsets of $\mathbb{R}^n$. Furthermore, let $B_1 \cap B_2 = \emptyset$ and*

$$(rc\, B_1)\cap(rc\, B_2) = \{0\}, \quad (rcB_1 - rcB_2)\cap(rcB_2 - rcB_1) = \{0\}. \quad (5.4.22)$$

Then B_1 and B_2 are weakly separated by a collection of n linearly independent vectors.

Proof: Let $b_i \in \operatorname{kern}^* B_i$ and $V_i = B_i - b_i$, $i = 1, 2$. Then V_1 and V_2 are co-radiative and $B_2 - B_1 = b_2 - b_1 + (V_2 - V_1)$. It follows from Proposition 5.21 and Proposition 5.30 that $V := V_2 - V_1$ is a co-radiative set, so V is closed and $0 \notin V$. Applying Proposition 5.21 and Proposition 5.31, we also have:

$$\operatorname{cl}\operatorname{cone} V = \operatorname{rc}(B_2 - B_1) = \operatorname{rc} B_2 - \operatorname{rc} B_1.$$

Since B_1 and B_2 are disjoint sets, it follows that $b_1 - b_2 \notin V$. Assume that $b_2 \neq b_1$. Applying Theorem 5.11 we conclude that there exists a collection $\ell = (l(i))_{i=1}^n$ of n linearly independent vectors such that $[l(1), b_1 - b_2] = \dots = [l(n), b_1 - b_2] \leq \inf_{v\in V} \max_{i=1,\dots,n}[l(i), v]$. It is easy to check that B_1 and B_2 are weakly separated by the collection ℓ. If $b_2 = b_1$, then we can obtain the required collection ℓ by applying Theorem 5.11 for arbitrary $x \notin V$, $x \neq 0$. $\triangle$

REMARK 5.14 The conditions in Proposition 5.47 are satisfied, for example, if there exists a convex closed cone Q such that $Q \cap (-Q) = \{0\}$ and $\operatorname{rc} B_1 = \operatorname{rc}(-B_2) = Q$.

PROPOSITION 5.48 *Let B_1 and B_2 be strongly co-star-shaped, let $B_1 \cap B_2 = (\operatorname{bd} B_1) \cap (\operatorname{bd} B_2)$ and let at least one of the sets* $\operatorname{int}\operatorname{kern}^* B_i$ $(i = 1, 2)$ *be nonempty. Further let $rc\, B_1 \cap rc\, B_2 = \{0\}$ and $rc\, B_2 - rc\, B_1$ be a proper convex cone. Then B_1 and B_2 are weakly separated.*

Proof: Assume that $\operatorname{int}\operatorname{kern}^* B_2$ is not empty. Let $b_1 \in \operatorname{kern}^* B_1, b_2 \in \operatorname{int}\operatorname{kern}^* B_2$ and $V_i = B_i - b_i$. It follows from Proposition 5.21 and Proposition 5.30 that $V := V_2 - V_1$ is co-radiative. In the light of Proposition 5.27, we have $b_1 - b_2 \in \operatorname{bd} V$. Clearly $0 \in (\operatorname{kern}^* V_1) \cap (\operatorname{int}\operatorname{kern}^* V_2)$, hence $0 \in \operatorname{int}\operatorname{kern}^* V_2 - \operatorname{kern}^* V_1$. Since $\operatorname{kern}^* V_2 - \operatorname{kern}^* V_1 \subset \operatorname{kern}^* (V_2 - V_1)$, we can conclude that $0 \in \operatorname{int}\operatorname{kern}^* V$. It follows from Proposition 5.31 and Proposition 5.21 that $\operatorname{cl}\operatorname{cone} V = \operatorname{rc} B_2 - \operatorname{rc} B_1$ is a proper convex cone. Due to Theorem 5.10, the point $b_1 - b_2$ can be cone-separated from the set $V_2 - V_1$. Proposition 5.32 demonstrates that there exists a collection $\ell = (l(i))_{i=1}^n$ of n linearly independent vectors with the following property: for each $v = v_2 - v_1 \in V$, there exists i such that $[l(i), b_1 - b_2] \leq [l(i), v_2 - v_1]$. Clearly, ℓ is a weakly separating collection for sets B_1 and B_2. △

Finally, we consider a separation of a point from a co-radiant set by a collection of $n+1$ vectors in the sense similar to that of Definition 5.19.

THEOREM 5.12 *Let V be a closed set. Then the following assertions are equivalent:*

a) V is co-radiant and $0 \notin V$;

b) for each point $x \notin V$ there exist $n+1$ vectors $l(1), \dots, l(n+1)$ such that

$$\min_{i=1,\dots,n+1} [l(i), x] \geq -1 \geq \sup_{y \in V} \min_{i=1,\dots,n+1} [l(i), y]. \tag{5.4.23}$$

Proof: a) $\Longrightarrow$ b). Let $x_0 \notin V$. Since V is co-radiant it follows that the compact segment $[0, x_0]$ has no common points with V. Since V is closed, we can find an open ball $B(0, \varepsilon)$ such that the intersection of the open set $G = [0, x_0] + B(0, \varepsilon)$ with V is empty. Noting that both 0 and x belong to G, we deduce that there exists an n-dimensional simplex $S \subset G$, which contains both 0 and x_0 at its interior. The inclusion $0 \in \operatorname{int} S$ provides the existence of a collection ℓ' of $n+1$ vectors $l'(1), \dots, l'(n), l'(n+1)$ such that $S = \{x \in \mathbb{R}^n : [l'(i), x] \leq 1, i = 1, \dots, n+1\}$. Let $\ell'(x) = \max_{i=1,\dots,n+1}[l'(i), x]$. The inclusion $x_0 \in \operatorname{int} S$ implies $\ell'(x_0) < 1$. Since $S \cap V$ is empty, we conclude that $\ell'(x) \geq 1$ for $x \in V$. Let $l(i) = -l'(i), i = 1, \dots, n+1$. Then $\min[l(i), x_0] > -1$ and $\min[l(i), x] \leq -1$. Thus the desired result follows.

b) $\Longrightarrow$ a). Let $\ell = (l(1), \dots, l(n+1))$ be a collection of vectors and $V_\ell = \{x \in \mathbb{R}^n : \min_{i=1,\dots,n+1}[l(i), x] \leq -1\}$. The set V_ℓ is closed and $0 \notin V_\ell$. If $x \in V_\ell$ and $\lambda > 1$, then

$$\min_{i=1,\dots,n+1} [\lambda l(i), x] = \lambda \min_{i=1,\dots,n+1} [l(i), x] < -1,$$

so $\lambda x \in V$. Hence V_ℓ is co-radiant. If b) holds, then V is the intersection of a family of sets (V_ℓ), hence V is a closed co-radiant set and $0 \notin V$. $\triangle$

REMARK 5.15 We have proved that for $x \notin V$ there exists a collection $l(1), \ldots, l(n+1)$ such that the strict inequalities hold in 5.4.23.

It is interesting to compare Theorem 5.12 with corresponding results for radiant sets (see Proposition 5.35 and Theorem 5.6). Recall, that according to these results, a set is closed and radiant if and only if each point, which does not belong to this set can be separated from it by a collection of n vectors.

5.5. ABSTRACT CONVEXITY WITH RESPECT TO GENERAL MIN-TYPE FUNCTIONS

5.5.1. Overview

In this section we shall study abstract convexity with respect to the classes $\mathcal{L}_k$ of all functions of the form

$$x \to \langle \ell, x \rangle_*, \tag{5.5.1}$$

where

$$\langle \ell, x \rangle_* = \min_{i=1,\ldots k} [l(i), x] \tag{5.5.2}$$

with $\ell = (l(1), \ldots, l(k)) \in (\mathbb{R}^n)^k$. Here $[l, x]$ stands for the inner product of vectors l and x. A function given by (5.5.1) will be called a *general min-type function.*We consider classes $\mathcal{L}_k$ with $k = n$ and $k = n+1$. We also study abstract convexity with respect to the class $\mathcal{H}_{n+1} := H_{\mathcal{L}_{n+1}}$ of abstract affine functions. It is easy to check that an abstract convex function with respect to $\mathcal{H}_{n+1}$ is lower semicontinuous and convex-along-rays. We shall give an example, which shows that the reverse statement is not true. Nevertheless lower semicontinuity and convexity-along-rays imply $\mathcal{H}_{n+1}$-convexity under very mild assumptions. In particular, any lower semicontinuous convex-along-rays function f with $f(0) < +\infty$ is $\mathcal{H}_{n+1}$-convex. If $f(0) = +\infty$ then we can guarantee $\mathcal{H}_{n+1}$-convexity for functions with a "regular" behaviour at infinity. The class of lower semicontinuous convex-along-rays functions is very broad and contains many functions that are important in applications.

One of the main problems of the theory of abstract convexity is to find conditions guaranteeing that the subdifferential of a function f at a given point is not empty. We describe a large class of convex-along-rays functions with this property at any given point.

We also consider abstract convex sets with respect to the classes $\mathcal{L}_n$ and $\mathcal{L}_{n+1}$. The results from the previous sections allow us to establish the following property: if a set U contains the origin, then this set is $\mathcal{L}_n$-convex if and only if it is closed and radiant. The main results from this section can be found in [164].

5.5.2. Positively homogeneous abstract convex functions

Let L_* be the set of all finite collections $\ell = (l(1), \dots, l(k))$, $k = 1, \dots,$ of vectors $l(i) \in \mathbb{R}^n$. We shall use the following notation

$$\ell(x) = \langle \ell, x \rangle_* = \min_{i=1\dots k} [l(i), x]. \tag{5.5.3}$$

Let us denote a general min-type function

$$x \to \langle \ell, x \rangle_* \tag{5.5.4}$$

by the same symbol as the collection of vectors $(l(1), \dots, l(k))$ generating this function, that is, by the letter ℓ. Denote by $\mathcal{L}$ the set of all functions l of the form (5.5.4) with $\ell \in L_*$. The set $\mathcal{L}$ possesses the property (A) from Section 1.4 (see (1.4.5)) because $\ell(0) = 0$ for any $\ell \in L_*$.

Note that a function ℓ defined by (5.5.3) is superlinear, that is, abstract concave with respect to the space $(\mathbb{R}^n)'$ of all linear functions defined on $\mathbb{R}^n$. As usual we identify $(\mathbb{R}^n)'$ with the space $\mathbb{R}^n$. The upper support set, $\mathrm{supp}_+(\ell, (\mathbb{R}^n)')$, of the function ℓ coincides with its superdifferential $\bar{\partial}\ell(0)$ at zero. Recall (see Definition 1.9) that

$$\mathrm{supp}^+(\ell, (\mathbb{R}^n)') = \{l \in \mathbb{R}^n : [l, x] \geq \ell(x) \text{ for all } x \in \mathbb{R}^n\}.$$

If $\ell(x) = \min_{i=1,\dots,k}[l(i), x]$, then $\mathrm{supp}^+(\ell, (\mathbb{R}^n)')$ coincides with the convex hull $\mathrm{co}\,(l(1), \dots, l(k))$ of points $l(1), \dots, l(k)$.

PROPOSITION 5.49 *The set $\mathcal{L}$ is a convex cone in the space C_* of all finite continuous positively homogeneous functions defined on $\mathbb{R}^n$.*

Proof: We shall only prove that the sum of two functions $\ell_1, \ell_2 \in \mathcal{L}$ belongs to $\mathcal{L}$. Let

$$\langle \ell_1, x \rangle_* = \min_i [l(i), x], \quad \langle \ell_2, x \rangle_* = \min_k [l'(k), x].$$

Then

$$\ell_1(x) + \ell_2(x) = \min_i [l(i), x] + \min_k [l'(k), x] = \min_{i,k} [l(i) + l'(k), x]. \quad \triangle$$

Let us denote by $\mathcal{L}_k$ the set of all functions $\ell \in \mathcal{L}$, which can be represented in the form $\langle \ell, x \rangle_* = \min_{i=1,\dots,j}[l(i), x]$ with $j \le k$. Clearly the set $\mathcal{L}_k$ is conic (if $\ell \in \mathcal{L}_k$ and $\lambda > 0$, then $\lambda \ell \in \mathcal{L}_k$). However this set is not a convex cone, since the sum of two functions belonging to $\mathcal{L}_k$ does not necessarily belong to this set.

Let $\ell^i \in \mathcal{L}$, $i = 1, 2, \dots$. We say that the sequence of functions (ℓ^i) converges if there exists a finite function q such that $\langle \ell^i, x \rangle_* \to q(x)$ for all $x \in \mathbb{R}^n$. In other words, we consider the pointwise convergence.

PROPOSITION 5.50 *The conic set $\mathcal{L}_k$ is pointwise closed.*

Proof: Let $\langle \ell^i, x \rangle_* \to q(x)$ for all $x \in \mathbb{R}^n$. Since the functions ℓ^i are superlinear it follows that q is superlinear as well. It follows from concavity of ℓ^i that ℓ^i converges to q uniformly with respect to the unit ball (see, for example [134], Theorem 10.8). Let U^i be the upper support set of the function ℓ^i, then U^i is a polyhedron with at most k vertices. Let $U = \text{supp}^+(q, (\mathbb{R}^n)')$. Since the uniform convergence of superlinear functions is equivalent to the convergence of their upper support sets (superdifferentials at zero) in the Hausdorff metric, we conclude that $U = \lim_i U^i$. Hence U is a polyhedron with vertices $l(1), \dots, l(j)$ and $j \le k$. It follows from the definition of the upper support set that $q(x) = \inf_{l \in U}[l, x] = \min_{i=1,\dots,j}[l(i), x]$ for all $x \in \mathbb{R}^n$. △

A set $V \subset \mathcal{L}_k$ is called *bounded* if $\sup_{\ell \in V} \|\ell\| < +\infty$, where $\|\ell\| = \max_{\|x\| \le 1} |\langle \ell, x \rangle_*|$. Let $\mathcal{U} = \{\text{supp}^+(q, (\mathbb{R}^n)') : q \in V\}$. It follows from the Minkowski duality (a one-to-one correspondence between superlinear functions and their upper support set) that V is bounded if and only if $\mathcal{U}$ is bounded, that is, if all sets $U \in \mathcal{U}$ are contained in the same ball.

A set $V \subset \mathcal{L}_k$ is called *compact* if any sequence of elements of V contains a subsequence converging to an element of V.

PROPOSITION 5.51 *A set $V \subset \mathcal{L}_k$ is compact if and only if it is closed and bounded.*

Proof: Let $\mathcal{U}$ be the totality of upper support sets of all functions from V. The set $\mathcal{U}$ consists of polyhedrons with at most k vertices. Clearly, a sequence $U_j \in \mathcal{U}$ contains a converging subsequence if and only if $\mathcal{U}$ is bounded. Thus the result follows. △

Let f be an $\mathcal{L}$-convex function. Since each function $\ell \in \mathcal{L}$ is positively homogeneous, the function f is also positively homogeneous. Since each $\ell \in \mathcal{L}$ is continuous, f is a lower semicontinuous function. In this subsection we shall show that a broad class of lower semicontinuous positively homogeneous functions is $\mathcal{L}_n$-convex and every lower semicontinuous positively homogeneous function is $\mathcal{L}_{n+1}$-convex.

Since $\ell(0) = 0$ for all $l \in \mathcal{L}$, we conclude that each $\mathcal{L}$-convex function f is proper that is $\operatorname{dom} f = \{x : f(x) < +\infty\}$ is not empty. It is easy to check that any lower semicontinuous positively homogeneous function $f : \mathbb{R}^n \to \mathbb{R}_{+\infty}$ is proper if and only if $f(0) = 0$. We shall consider only proper functions in this section. We begin with the following assertion.

THEOREM 5.13 *Each proper lower semicontinuous positively homogeneous nonnegative function is $\mathcal{L}_n$-convex.*

Proof: A lower semicontinuous positively homogeneous nonnegative function f coincides with the Minkowski gauge μ_U of the closed radiant set $U = \{x : f(x) \leq 1\}$ (see Proposition 5.2 for details). Consider a point $x^0 \in \mathbb{R}^n$. First assume that $0 < f(x_0) < +\infty$. Let ε be an arbitrary positive number and $\lambda = (1+\varepsilon)/f(x_0)$. Then $f(\lambda x_0) = 1+\varepsilon$, so $\lambda x_0 \notin U$. Due to Theorem 5.6, the point λx_0 can be strictly separated from the set U by a collection ℓ of n linearly independent vectors $l(1), \dots, l(n)$. By the definition, $[l(1), \lambda x_0] = \dots = [l(n), \lambda x_0] = 1$ and there exists $\delta \in (0,1)$ such that $\langle \ell, x \rangle_* \equiv \min_{i=1,\dots,n}[l(i), x] < 1 - \delta$ for all $x \in U$. We have

$$\langle \ell, x_0 \rangle_* = \frac{1}{\lambda} = f\left(\frac{x_0}{1+\varepsilon}\right) > f(x_0)(1-\varepsilon).$$

Let $x' \in \mathbb{R}^n$ be a point such that $0 < f(x') < +\infty$ and $x = x'/f(x')$. Then $x \in U$, hence $\langle \ell, x \rangle_* < 1 - \delta$. Keeping in mind the positive homogeneity of both functions f and $x \mapsto \langle \ell, x \rangle_*$ we conclude that $\langle \ell, x' \rangle_* < (1-\delta) f(x') < f(x')$. Thus $\langle \ell, x' \rangle_* < f(x')$ for all $x' \in \mathbb{R}^n$ with $f(x') > 0$. If $f(x') = 0$ then $\lambda x' \in U$ for all $\lambda > 0$, so $\langle \ell, x' \rangle_* = 0 = f(x')$. We have proved that $\ell \in \operatorname{supp}(f, \mathcal{L}_n)$. Thus for each $\varepsilon > 0$ there exists $\ell \in \operatorname{supp}(f, \mathcal{L}_n)$ such that $\langle l, x_0 \rangle_* > (1-\varepsilon) f(x_0)$, so

$$f(x_0) = \sup\{\langle \ell, x \rangle_* : \ell \in \operatorname{supp}(f, \mathcal{L}_n)\}. \tag{5.5.5}$$

Assume now that $f(x_0) = +\infty$. Then $\lambda x_0 \notin U$ for each $\lambda > 0$. The same argument as above demonstrates that for each $\lambda > 0$ there exists a collection $\ell \in \operatorname{supp}(f, \mathcal{L}_n)$ such that $\langle \ell, x_0 \rangle_* = (1/\lambda)$. Hence (5.5.5) holds for x_0 as well. It is much easier to cope with a point x_0 such that $f(x_0) = 0$. Indeed, we have $f(x_0) = \langle 0, x_0 \rangle_*$. Since f is nonnegative it follows that $0 \in \operatorname{supp}(f, \mathcal{L}_n)$, so (5.5.5) holds for this point x_0 as well. $\triangle$

COROLLARY 5.10 *Let f be a proper lower semicontinuous positively homogeneous function such that $f(x) \geq [v, x]$ for a vector v and all $x \in \mathbb{R}^n$. Then f is $\mathcal{L}_n$-convex.*

Indeed, we can apply Theorem 5.13 to the function $x \mapsto f(x) - [v, x]$, which is proper, lower semicontinuous, positively homogeneous and nonnegative.

Theorem 5.14 *A proper positively homogeneous function $f : \mathbb{R}^n \to \mathbb{R}_{+\infty}$ is $\mathcal{L}_{n+1}$-convex if and only if it is lower semicontinuous.*

Proof: We have to check that any proper lower semicontinuous positively homogeneous function f is $\mathcal{L}_{n+1}$-convex. First, note that $f(x)$ is well defined by its restriction to the unit sphere $S_* = \{x \in \mathbb{R}^n : \|x\| = 1\}$. This restriction is lower semicontinuous on the compact set S_* and hence bounded from below, that is, there exists a number A such that

$$f(x) \geq -A \text{ for all } x \in S_*. \tag{5.5.6}$$

Choose some $x_0 \in S_*$. We need to show that for each positive $\delta < f(x_0)$, there exists a function $\ell(x) = \min_{i=1,\dots,n+1}[l(i), x]$ with $l(i) \in \mathbb{R}^n$ such that

$$\ell(x_0) \geq f(x_0) - \delta \tag{5.5.7}$$

and

$$\ell(x) \leq f(x), \quad x \in \mathbb{R}^n. \tag{5.5.8}$$

Consider the linear function

$$v(x) = (f(x_0) - \delta)[x, x_0]. \tag{5.5.9}$$

We have $v(x_0) = (f(x_0) - \delta)\|x_0\| = f(x_0) - \delta < f(x_0)$. Since both f and v are positively homogeneous functions, we deduce that $v(x) \leq f(x)$ for all x from the ray $R_{x_0} = \{\alpha x_0 : \alpha \geq 0\}$.

Let X be the $(n-1)$-dimensional subspace of the space $\mathbb{R}^n$ orthogonal to the vector x_0. Since X is an $(n-1)$-dimensional space we can find n vectors $l'(1), \dots, l'(n) \in X$ such that 0 is an interior point (with respect to X) of their convex hull S, which is a simplex. The support function $x \mapsto \max_{l \in S}[l, x]$ of the simplex S (in X) coincides with the function $p(x) = \max_{i=1,\dots,n}[l'(i), x]$. Let S contain a ball $\{x \in X : \|x\| \leq C\}$. Then $p(x) \geq C\|x\|$ for all $x \in X$, hence

$$-p(x) = \min_{i=1,\dots,n}[-l'(i), x] \leq -C\|x\| \tag{5.5.10}$$

for all $x \in X$. Clearly, for any $C > 0$ we can choose vectors $l'(i)$ in such a way that (5.5.10) holds. For example we can find arbitrary vectors $\bar{l}(i)$ for which (5.5.10) holds with some C and then multiply these vectors by a sufficiently large number. Let $l(i) = (f(x_0) - \delta)x_0 - l'(i)$, $i = 1, \dots, n$ and $l(n+1) = dx_0$ with $d > \max(-f(-x_0), f(x_0) - \delta)$. Note that $[l(i), x] = v(x) - [l'(i), x]$ where v is defined by (5.5.9). Let

$$\ell_C(x) = \min_{i=1,\dots,n+1}[l(i), x] = \min(\min_{i=1,\dots,n}[-l'(i), x] + v(x), d[x, x_0]).$$

Note that $[l'(i), x_0] = 0$ and $v(x_0) = f(x_0) - \delta$. Hence $\ell_C(x_0)$ does not depend on C and

$$\ell_C(x_0) = \min(v(x_0), d\|x_0\|^2) = \min(v(x_0), d) = f(x_0) - \delta.$$

Thus we have proved (5.5.7). We now show that (5.5.8) holds. Consider the restriction of f to S_*. Recall that $v(x_0) < f(x_0)$. Due to the lower semicontinuity of f, there exists an open neighbourhood V of the point x_0 such that $v(x) < f(x)$ for all $x \in V$. Representing x in the form $x = \lambda x_0 + x'$ with $x' \in X$, $\lambda \in \mathbb{R}$ and keeping in mind that $[l'(i), x_0] = 0$ for $i \leq n$, we have due to (5.5.10):

$$\ell_C(x) \leq \min_{i=1,\dots,n+1}([(f(x_0) - \delta)x_0, x] - [l'(i), \lambda x_0 + x'])$$

$$= v(x) + \min_{i=1,\dots,n}[-l'(i), x'] \leq v(x) < f(x), \qquad (x \in V).$$

We also have

$$[l(n+1), -x_0] = -d\|x_0\|^2 = -d < f(-x_0).$$

Let V' be a neighbourhood of the point $-x_0$ such that $[l(n+1), x] < f(x)$ for $x \in V'$. We have

$$\ell_C(x) \leq [l(n+1), x] < f(x), \qquad (x \in V').$$

Let $Z = S_* \backslash (V \cup V')$ and let $\varepsilon = \min_{\alpha \in \mathbb{R}, z \in Z} \|\alpha x_0 - z\|$ be the Hausdorff distance between the set Z and the straight line $\{\alpha x_0 : \alpha \in \mathbb{R}\}$. Obviously, $\varepsilon > 0$. Let $\|x\| = 1$, $x = \lambda x_0 + x'$ with $x' \in X$. If $x \notin V \cup V'$ then $\|x'\| \geq \varepsilon$. Applying (5.5.10) we conclude that

$$\ell_C(x) \leq \min_i [l(i), x] = (f(x_0) - \delta)[x, x_0] + \min_i [-l'(i), x'] \leq f(x_0) - \delta - \varepsilon C.$$

Choosing $C > (A + f(x_0 - \delta))/\varepsilon$, where A is defined by (5.5.6), we get $\ell_C(x) < f(x)$ for all $x \in S_*$ and hence $\ell_C(x) \leq f(x)$ for all $x \in \mathbb{R}^n$. $\triangle$

REMARK 5.16 Set $f(x) = -|x|$, $x \in \mathbb{R}$. This example shows that the bound $n + 1$ in the hypothesis of Theorem 5.14 is strict.

We now turn to the study of $\mathcal{L}_k$-subdifferentials with $k = n$ and $k = n + 1$. Let L be a set of positively homogeneous functions defined on $\mathbb{R}^n$ and let f be a L-convex function. Recall (see Proposition 1.3) that L-subdifferential, $\partial_L f(x_0)$, of the function f at a point x_0 consists of all $l \in L$ such that $l(x) \leq f(x)$ for all $x \in \mathbb{R}^n$ and $l(x_0) = f(x_0)$.

PROPOSITION 5.52 *Let f be a lower semicontinuous nonnegative positively homogeneous function and x_0 be a point such that $0 < f(x_0) <$*

$+\infty$. Let $\gamma = f(x_0)$ and $S_\gamma(f) = \{x \in \mathbb{R}^n : f(x) \leq \gamma\}$ be a level set of the function f, corresponding to γ. The following assertions are equivalent:

1) $\partial_{\mathcal{L}_n} f(x_0)$ is nonempty;

2) there exists $\ell' = (l'(1), \dots, l'(n)) \in \mathcal{L}_n$, such that $[l'(1), x_0] = \dots = [l'(n), x_0] = 1$ and

$$\min_i [l'(i), x] < 1 \text{ for each } x \in S_\gamma(f),\ x \neq x_0. \tag{5.5.11}$$

(If the strict inequality for $x \neq x_0$ holds in (5.5.11) and $l'(1), \dots, l'(n)$ are linearly independent, then ℓ' is a support collection in the sense of Definition 5.21.)

Proof: 1) $\Longrightarrow$ 2). Assume without loss of generality that $\gamma = 1$. Let $\ell = (l(1), \dots, l(n)) \in \partial_{\mathcal{L}_n} f(x_0)$. Then $\langle \ell, x_0 \rangle_* = \min_{i=1,\dots,n}[l(i), x_0] = 1$. Let $\lambda_i = 1/[l(i), x_0]$ and $l'(i) = \lambda_i l(i)$. Then $[l'(i), x_0] = 1$ for all i. Note that $0 < \lambda_i \leq 1$ for all i. Consider a collection $\ell' = (l'(1), \dots, l'(n))$ and take $x \in S_\gamma(f)$. Assume that $\langle \ell, x \rangle_* \geq 0$. Then $[l(i), x] \geq 0$ for all i, hence

$$\langle \ell', x \rangle_* = \min_{1=\dots,n} \lambda_i [l(i), x] \leq \min_{i=1,\dots,n} [l(i), x] \leq f(x) \leq 1.$$

Next assume that $\langle \ell, x \rangle_* < 0$. Then there exists i_0 such that $[l(i_0), x] < 0$ and we have

$$\langle \ell', x \rangle_* \leq \lambda_{i_0} [l(i_0), x] < 1.$$

Thus $\langle \ell', x \rangle_* \leq 1$ for all $x \in S_\gamma(f)$, so 2) holds.

2) $\Longrightarrow$ 1). Let $\ell' = (l'(1), \dots, l'(n))$ be a collection such that 2) holds and $\ell = (\gamma l'(1), \dots, \gamma l'(n))$. Then $\langle l, x_0 \rangle_* = \gamma = f(x_0)$. Let $x \in \mathbb{R}^n$. If $f(x) > 0$ then $x' = x/f(x) \in S_\gamma(f)$, hence $\langle l, x \rangle_* \leq f(x)$. If $f(x) = 0$ then $\lambda x \in S_\gamma(f)$ for all $\lambda > 0$, hence $\langle l, x \rangle_* = 0 = f(x)$. Thus $l \in \partial_{\mathcal{L}_n} f(x_0)$. $\triangle$

Proposition 5.52, Proposition 5.33 and Theorem 5.7 allow us to give simple sufficient conditions so that the $\mathcal{L}_n$-subdifferential is nonempty.

COROLLARY 5.11 *Let $f \geq 0$ be a positively homogeneous function, and let x_0 be such that $f(x_0) \in (0, +\infty)$. Suppose that the Hadamard lower derivative $f_H^\downarrow(x_0, x_0)$ at the point x_0 is positive. Then $\partial_{\mathcal{L}_n} f(x_0) \neq \emptyset$.*

Proof: Let U coincide with the level set $S_\gamma(f) = \{x : f(x) \le \gamma\}$, where $\gamma = f(x_0)$. The set U is closed and

$$\begin{aligned}\mu_U(x_0) &= \inf\left\{\lambda > 0 : \frac{x_0}{\lambda} \in U\right\} = \inf\left\{\lambda > 0 : f\left(\frac{x_0}{\lambda}\right) \le \gamma\right\} \\ &= \inf\{\lambda > 0 : f(x_0) \le \lambda\gamma\} = 1.\end{aligned}$$

Proposition 5.33 and Theorem 5.7 imply the existence of a support collection, hence due to Proposition 5.52, $\partial_{\mathcal{L}_n} f(x_0) \ne \emptyset$. △

It follows from Corollary 5.6 that the subdifferential $\partial_{\mathcal{L}_n} f(x_0)$ is not empty for a nonnegative positively homogeneous function, which is locally Lipschitz at a point x_0 such that $0 < f(x_0) < +\infty$.

In order to formulate conditions for the nonemptiness of the $\mathcal{L}_{n+1}$-subdifferential for a positively homogeneous lower semicontinuous function, we need the notion of calmness.

Recall (see Definition 4.5) that a function $f : \mathbb{R}^n \to \mathbb{R}_{+\infty}$ is calm of degree one at a point $x_0 \in \operatorname{dom} f$ if

$$\liminf_{x \to x_0} \frac{f(x) - f(x_0)}{\|x - x_0\|} > -\infty.$$

In other words, f is calm of degree one if there exists $C > 0$ such that

$$f(x) - f(x_0) \ge -C\|x - x_0\|$$

for all x in some open neighbourhood of x_0.

THEOREM 5.15 *Let f be a lower semicontinuous and positively homogeneous function and $x_0 \ne 0$. Then the following statements are equivalent:*

a) $f^{\downarrow}_H(x_0, x_0) > -\infty$;

b) the subdifferential $\partial_{\mathcal{L}_{n+1}} f(x_0)$ is not empty ;

c) f is calm of degree one at x_0.

Proof: The implications *b)* $\Longrightarrow$ *c)* and *c)* $\Longrightarrow$ *a)* are obvious, so we shall only prove *a)* $\Longrightarrow$ *b)*. First, assume that $f(x_0) < 0$. Let $f^{\downarrow}_H(x_0, x_0) = c > -\infty$ and let $p(x) = \|x\|$. Clearly,

$$p^{\downarrow}_H(x_0, x_0) = p^{\downarrow}_D(x_0, x_0) = \lim_{\alpha \to +0} \frac{p(x_0 + \alpha x_0) - p(x_0)}{\alpha} = \|x_0\|.$$

The lower semicontinuous function f is bounded from below on the unit sphere, so there exists $C' > 0$ such that $g(x) := f(x) + Cp(x) > 0$ for all $x \in \mathbb{R}^n$ ad $C > C'$. Let $C > C'$ and $C > -c/\|x_0\|$. We have

$$g^{\downarrow}_H(x_0, x_0) \ge f^{\downarrow}_H(x_0, x_0) + Cp^{\downarrow}_H(x_0, x_0) \ge c + C\|x_0\| > 0.$$

Applying Proposition 5.52 we conclude that there exists a function $\ell' \in \mathcal{L}_n$ such that

$$\langle \ell', x_0 \rangle_* = f(x_0) + Cp(x_0); \quad \langle \ell', x \rangle_* \le f(x) + Cp(x) \text{ for all } x \in \mathbb{R}^n.$$

Let $r(x) = Cp(x) - \ell'(x)$ and $S = \{x : r(x) \le r(x_0)\}$. We have

$$r(x_0) = C\|x_0\| - \langle l', x_0 \rangle_* = C\|x_0\| - f(x_0) - C\|x_0\| = -f(x_0) > 0.$$

Since r is a sublinear function, it follows that S is a closed convex set. Since the sublinear function r is continuous, we conclude that S contains an open ball centred at zero. So we can find points $x_i \in S$, $(i = 1, \dots, n)$ such that the convex hull S_1 of the $n+1$ points $x_0, x_1, \dots, x_n$ is a simplex and such that zero is an interior point of this simplex. We can represent the simplex S_1 as the intersection of $n+1$ halfspaces, that is, there exist $l(0), \dots, l(n)$ such that $S_1 = \{x : [l(i), x] \le 1, \, i = 0, \dots, n\}$. Denote by r_1 the Minkowski gauge of S_1 . We have

$$\begin{aligned} r_1(x) &= \inf\left\{\lambda > 0 : \frac{x}{\lambda} \in S_1\right\} \\ &= \inf\left\{\lambda > 0 : \left[l(i), \frac{x}{\lambda}\right] \le 1, \, i = 1, \dots, n\right\} \\ &= \inf\{\lambda > 0 : [l(i), x] \le \lambda, i = 1, \dots, n\} = \max(0, \max_i [l(i), x]). \end{aligned}$$

The inclusion $0 \in \operatorname{int} S_1$ implies that $\max_i [l(i), x] > 0$. So $r_1(x) = \max_i [l(i), x]$. Since $S_1 \subset S$, it follows that $r_1(x) \ge r(x)$. Since x_0 is a boundary point of both S and S_1, we have $r_1(x_0) = r(x_0)$. Let $\hat{\ell}(x) = -r_1(x)$. Then $\hat{\ell} \in \mathcal{L}_{n+1}$ and

$$\langle \hat{\ell}, x_0 \rangle_* = -r(x_0) = f(x_0),$$

$$\langle \hat{\ell}, x \rangle_* \le -r(x) \le f(x) \text{ for all } x \in \mathbb{R}^n.$$

Thus $\hat{\ell} \in \partial_{\mathcal{L}_{n+1}} f(x_0)$ and the required result follows for x_0 such that $f(x_0) < 0$.

Assume now that $f(x_0) \ge 0$. Take a vector $l(0)$ such that $f(x_0) - [l(0), x_0] < 0$ and consider the function $f_0(x) = f(x) - [l(0), x]$. The function f_0 is lower semicontinuous and positively homogeneous. Clearly, $(f_0)^{\downarrow}_H(x_0, x_0) > -\infty$. So there exist vectors $l(1), \dots, l(n+1)$ such that

$$f(x_0) = [l(0), x_0] + \min_{i=1,\dots,n+1} [l(i), x_0]; \tag{5.5.12}$$

$$f(x) \ge [l(0), x] + \min_{i=1,\dots,n+1} [l(i), x] \text{ for all } x \in \mathbb{R}^n. \tag{5.5.13}$$

Let $\ell(x) = \min_{i=1,\dots,n+1}[l(0) + l(i), x]$. It follows from (5.5.12) and (5.5.13) that $\ell \in \partial_{\mathcal{L}_{n+1}} f(x_0)$. $\triangle$

REMARK 5.17 The property of f to be calm at a point x_0 depends only on the behaviour of f at a neighbourhood of x_0, so this is a local property. Theorem 5.15 demonstrates that this local property implies a global one, namely the existence of a $\mathcal{L}_{n+1}$-subgradient.

Finding an element of the $\mathcal{L}_{n+1}$- subdifferential is a difficult problem. Nevertheless, it is possible to solve this problem for some sufficiently broad classes of positively homogeneous functions. We shall point out one of them.

Let f be a finite nonpositive superlinear (that is, concave and positively homogeneous) function defined on the space $\mathbb{R}^n$, for example, $f(x) = -\|x\|$. We consider the sublinear function $p(x) = -f(x)$ and the set $S = \{x \in R^n : p(x) \leq 1\}$. Assume for the sake of simplicity that $x_0 = (0, \dots, 0, 1)$ and $p(x_0) = 1$. Let $x_i' = (x_{i,1}, \dots, x_{i,n-1})$, $i = 1, \dots, n$ be the vertices of a simplex $S_1 \in \mathbb{R}^{n-1}$ such that $0 \in \operatorname{int} S_1$; it is not very difficult to find these points. Set $x_i = (\varepsilon x_{i,1}, \dots, \varepsilon x_{i,n-1}, -\varepsilon_1)$, $i = 1, \dots, n$, where $\varepsilon, \varepsilon_1$ are chosen so that $x_i \in S$. Let us consider the convex hull S_2 of the points $x_0, x_1, \dots, x_n$. It is easy to check that S_2 is a simplex and $0 \in \operatorname{int} S_2 \subset S$. Suppose that a vector l_i is a solution of the linear system $[l_k, x] = 1$, $k \in \{0, 1, \dots, n\} \setminus i$, $i = 0, 1 \dots, n$. Clearly $S_2 = \{x : [l_i, x] \leq 1\}$. So $\max_{i=0,\dots,n}[l_i, x] \geq p(x)$ for all $x \in \mathbb{R}^n$. Also, $\max_{i=0,\dots,n}[l_i, x_0] = 1 = p(x_0)$. Let $l(x) = \min_{i=0,\dots,n}[-l_i, x]$. Clearly, $l \in \partial_{\mathcal{L}_{n+1}} f(x_0)$.

5.5.3. $\mathcal{H}$-convex functions

Here we consider abstract convexity with respect to the class $\mathcal{H} \equiv H_{\mathcal{L}}$ of all $\mathcal{L}$-affine functions and some of its subclasses. We shall denote by $\mathcal{H}_k$ the class of all $\mathcal{L}_k$-affine functions. By definition $\mathcal{H}_k$ consists of functions $h(x) = \ell(x) - c$, where $\ell \in \mathcal{L}_k$ and $c \in \mathbb{R}$. Recall (see Definition 3.3.1) that a function $f : \mathbb{R}^n \to \mathbb{R}_{+\infty}$ is called *convex-along-rays* (CAR) if, for all $x \in \mathbb{R}^n$, the function $\alpha \to f(\alpha x)$ is convex on the ray $[0, +\infty)$.

The next proposition follows directly from the definition of abstract convex functions.

PROPOSITION 5.53 *Any $\mathcal{H}$-convex function is lower semicontinuous and CAR.*

The example below demonstrates that the reverse statement is not true.

EXAMPLE 5.5 Consider a function g defined on $\mathbb{R}^2$ by

$$g(x_1, x_2) = \begin{cases} x_1^+/|x_2| - \sqrt{x_1^2 + x_2^2} & \text{if } x_2 \neq 0; \\ 0 & \text{if } x_2 = 0, \end{cases}$$

where $a^+ = \max(a, 0)$. It is easy to check that g is lower semicontinuous at each point $x \neq 0$. Let

$$f(x) = \begin{cases} +\infty & \text{if } \|x\| < 1; \\ g(x) & \text{otherwise.} \end{cases}$$

The function $f(x)$ is lower semicontinuous on $\mathbb{R}^2$. Its restriction $f_v(\alpha) = f(\alpha v)$ to any ray $R_v = \{\alpha v : \alpha \geq 0\}$, $\|v\| = 1$, has the form

$$f_v(\alpha) = \begin{cases} +\infty & \text{if } \alpha < 1; \\ a\alpha + b & \text{otherwise.} \end{cases}$$

Hence f is a CAR function. We now show that f is not $\mathcal{H}$-convex, in other words, there exists a point v such that $f(v_0) > \sup\{h(v_0) : h \in \text{supp}(f, \mathcal{H})\}$.

Let $h \in \text{supp}(f, \mathcal{H})$, that is, $h \in \mathcal{H}$, $h(x) = \langle \ell, x \rangle_* - c$ and $h(x) \leq f(x)$ for all $x \in \mathbb{R}^2$. Let $v_0 = (1, 0)$. Then for any $v \in S^1 = \{(x, y) : x^2 + y^2 = 1\}$, $v \neq v_0$ and $\alpha \geq 1$ we have

$$h(\alpha v) = \alpha \langle \ell, v \rangle_* - c \leq \frac{v_1^+}{v_2} - \alpha.$$

Thus $\langle \ell, v \rangle_* \leq -1$ and $\langle \ell, v_0 \rangle_* \leq -1$ as well. We also have $h(v_0) = \langle \ell, v_0 \rangle_* - c \leq f(v_0) = 0$ so $-c \leq -\langle \ell, v_0 \rangle_*$. Hence

$$h(2v_0) = 2\langle \ell, v_0 \rangle_* - c \leq \langle \ell, v_0 \rangle_* \leq -1 \tag{5.5.14}$$

and (5.5.14) holds for any $h \in \text{supp}(f, \mathcal{H})$. Thus $f(2v_0) = 0 > -1 \geq \sup\{h(2v_0) : h \in \text{supp}(f, \mathcal{H})\}$.

Let f be a lower semicontinuous convex-along-rays function. For $x \in \mathbb{R}^n$ set $f_x(\alpha) = f(\alpha x)$, $\alpha \geq 0$. Let U_x be the support set of f_x with respect to the set of affine functions: U_x consists of all functions

$$l_x(\alpha) = a + c\alpha, \qquad \alpha \geq 0 \tag{5.5.15}$$

such that $f_x(\alpha) \geq l_x(\alpha)$ for all $\alpha \geq 0$. We shall denote the function (5.5.15) by (a, c). Since f_x is lower semicontinuous and convex,

$$f_x(\alpha) = \sup\{a + c\alpha : (a, c) \in U_x\}, \qquad \alpha > 0.$$

In particular

$$f(0) = f_x(0) = \sup\{a : (a, c) \in U_x\}.$$

Let $a < f(0)$. Then there exists $c \in \mathbb{R}$ such that $(a, c) \in U_x$. Consider the quantity

$$b^a(x) = \inf_{\alpha>0} \frac{f_x(\alpha) - a}{\alpha} = \sup\{c : a + c\alpha \leq f_x(\alpha),\ \alpha \geq 0\}. \qquad (5.5.16)$$

We have $b^a(x) > -\infty$ and $b^a(x) = +\infty$ if and only if $f(\alpha x) = +\infty$ for all $\alpha > 0$. Suppose that there exists $\alpha > 0$ such that $f(\alpha x) < +\infty$. Then the set $\{c : a + c\alpha \leq f(\alpha x)\}$ is compact, so $b^a(x) < +\infty$ and

$$b^a(x) = \max\{c \in \mathbb{R} : a + c\alpha \leq f(\alpha x),\ \alpha > 0\}.$$

DEFINITION 5.22 A *lower affine approximation* of a lower semicontinuous CAR function f defined on $\mathbb{R}^n$, corresponding to a number $a < f(0)$, is a function g^a defined by

$$g^a(x) = a + b^a(x), \qquad x \in \mathbb{R}^n,$$

where b^a is defined by (5.5.16).

Note that

1) The lower affine approximation g^a is a minorant of the function f for any $a < f(0)$. Indeed, it follows from (5.5.16) that

$$g^a(x) = a + b^a(x) \leq f_x(1) = f(x).$$

2) The function $b^a(x)$ is positively homogeneous of degree one for any $a < f(0)$. For $\lambda > 0$ we have

$$b^a(\lambda x) = \inf_{\alpha>0} \frac{f(\alpha\lambda x) - a}{\alpha} = \lambda \inf_{\alpha\lambda>0} \frac{f(\alpha\lambda x) - a}{\alpha\lambda} = \lambda b^a(x).$$

REMARK 5.18 Assume

$$\inf_{\alpha>0} \frac{f(\alpha x) - a}{\alpha} < \lim_{\alpha\to+\infty} \frac{f(\alpha x) - a}{\alpha}. \qquad (5.5.17)$$

Then there exists $\alpha_0 \geq 0$ such that $f(\alpha_0 x) = g(\alpha_0 x)$. Indeed, it follows from (5.5.17) that for sufficiently large r

$$b^a(x) = \max\{c : a + c\alpha \leq f(\alpha x),\ 0 \leq \alpha \leq r\}. \qquad (5.5.18)$$

Suppose the lower semicontinuous function $x \to f_x(\alpha) - (a + b^a(x)\alpha)$ attains its minimum over the segment $[0, r]$ at a point α_0. It follows from (5.5.18) that $f(\alpha_0 x) = a + b^a(x)\alpha_0 = g^a(\alpha_0 x)$.

LEMMA 5.5 *A lower semicontinuous CAR function f is the supremum of its lower affine approximations, that is,*

$$f(y) = \sup_{a<f(0)} g^a(y). \tag{5.5.19}$$

Proof: Let $\|x\| = 1$. Set

$$U_x = \{(a, c) : \; a + c\alpha \leq f_x(\alpha) \text{ for all } \alpha \geq 0\}.$$

If $(a, c) \in U_x$, then $a \leq f(0)$. Since f_x is a lower semicontinuous convex function, it follows that

$$\begin{aligned} f_x(\alpha) &= \sup_{(a,c)\in U_x} (a + c\alpha) = \sup_{(a,c)\in U_x,\; a<f(0)} (a + c\alpha) \\ &= \sup_{a<f(0)} \sup_{c:\; a+c\alpha\leq f(\alpha x),\; \alpha\geq 0} (a + c\alpha) \\ &= \sup_{a<f(0)} (a + b^a(x)\alpha) = \sup_{a<f(0)} g^a(x). \end{aligned}$$

△

LEMMA 5.6 *Let f be a CAR function such that its lower affine approximations g^a are lower semicontinuous for all $a < f(0)$. Then $f \in \mathcal{H}_{n+1}$.*

Proof: It is sufficient to prove the lemma for proper functions f. Since g^a is a shift of the positively homogeneous lower semicontinuous function b^a, it follows from Theorem 5.14 that g^a is $\mathcal{H}_{n+1}$-convex. Applying Lemma 5.5 we deduce that f is also $\mathcal{H}_{n+1}$-convex. △

Now we are ready to prove the following important result.

THEOREM 5.16 *If $f : \mathbb{R}^n \to \mathbb{R}_{+\infty}$ is a lower semicontinuous CAR function and $f(0) < +\infty$, then f is $\mathcal{H}_{n+1}$-convex.*

Proof: Due to Lemma 5.6, it is enough to prove that any lower affine approximation $g^a(x)$ is lower semicontinuous on $\mathbb{R}^n$. Assume without loss of generality that $f(0) = 1$ and $a = 0$, so $g^a(x) = b^a(x)$. Assume that b^a is not lower semicontinuous, that is, for some $x_0 \in \mathbb{R}^n$ there exists a sequence (x_i) such that $x_i \to x_0$ as $i \to \infty$ and $\lim_i b^a(x_i) < b^a(x_0)$.

By the definition of b^a with $a = 0$ there exists a sequence $\alpha_i > 0$ and a number $c < b^a(x_0)$ such that

$$\frac{f(\alpha_i x_i)}{\alpha_i} \le c. \tag{5.5.20}$$

If the sequence α_i has a bounded subsequence, we can assume that $\alpha_i \to \alpha$ as $i \to +\infty$. Applying (5.5.20), we have

$$f(\alpha x_0) \le \liminf f(\alpha_i x_i) \le \liminf \alpha_i c = \alpha c.$$

If $\alpha = 0$, then $f(\alpha x_0) = 1 \le 0$, which is impossible. If $\alpha > 0$ then

$$b^a(x_0) \le \frac{f(\alpha x_0)}{\alpha} \le c < b^a(x_0)$$

which is again impossible. Therefore $\alpha_i \to +\infty$ as $i \to +\infty$. Let $t > 1/(b^a(x_0) - c)$ if $b^a(x_0) < +\infty$ and t be an arbitrary positive number if $b^a(x^0) = +\infty$. Then

$$b^a(x) > \frac{1+tc}{t}. \tag{5.5.21}$$

Due to the convexity-along-rays of the function f and (5.5.20), we have, for sufficiently large i:

$$\begin{aligned} f(tx_i) &\le \frac{\alpha_i - t}{\alpha_i} f(0) + \frac{t}{\alpha_i} f(\alpha_i x_i) \le \frac{\alpha_i - t}{\alpha_i} + \frac{t}{\alpha_i}(\alpha_i c) \\ &= 1 - \frac{t}{\alpha_i} + tc, \end{aligned}$$

which implies

$$b^a(x^0) = \frac{b^a(tx_0)}{t} \le \frac{1}{t} f(tx_0) \le \frac{1}{t} \liminf_i f(tx_i) \le \frac{1+tc}{t}.$$

However, these inequalities contradict (5.5.21). Thus b^a is a lower semicontinuous function. △

Let g be a finite convex function defined on $[a, +\infty)$. It follows from the convexity of g that there exists the right-hand side derivative $g'(\alpha)$ of the function g and that $g'(\alpha)$ is an increasing function. Therefore there exists the limit, $\lim_{\alpha \to +\infty} g'(\alpha)$, which is either a finite number or $+\infty$. We need the following well-known assertion.

LEMMA 5.7 *Let g be a finite convex function defined on $[a, +\infty)$. Let $\gamma = \lim_{\alpha \to +\infty} g'(\alpha)$. Then there exists* $\lim_{\alpha \to +\infty} g(\alpha)/\alpha = \gamma$.

Proof: Let $\alpha_0 \in (0, +\infty)$. Then $g(\alpha) \ge g(\alpha_0) + g'(\alpha_0)(\alpha - \alpha_0)$. Hence

$$\liminf_{\alpha \to +\infty} \frac{g(\alpha)}{\alpha} \ge \liminf_{\alpha \to +\infty} \frac{1}{\alpha}(g(\alpha_0) + g'(\alpha_0)(\alpha - \alpha_0)) = g'(\alpha_0).$$

Since g' is an increasing function, we have

$$\liminf_{\alpha\to+\infty}\frac{g(\alpha)}{\alpha}\geq\lim_{\alpha_0\to+0}g'(\alpha_0)=\gamma.$$

Let $\alpha>\alpha_0$. Then there exists $\theta\in(\alpha_0,\alpha)$ such that

$$g(a)-g(\alpha_0)=g'(\theta)(\alpha-\alpha_0)\leq\gamma(\alpha-\alpha_0).$$

Hence

$$\limsup_{\alpha\to+\infty}\frac{g(\alpha)}{\alpha}\leq\gamma. \tag{5.5.22}$$

Thus the result follows. △

Let $f:\mathbb{R}^n\to\mathbb{R}_{+\infty}$ be a lower semicontinuous CAR function. First, consider a point $x\in\mathbb{R}^n$ such that $f_x(\alpha)=f(\alpha x)<+\infty$ for all sufficiently large α. Lemma 5.7 shows that there exist limits

$$\gamma(x):=\lim_{\alpha\to+\infty}\frac{f(\alpha x)}{\alpha}=\lim_{\alpha\to+\infty}f'_x(\alpha). \tag{5.5.23}$$

Next, consider a point x such that $f_x(\alpha_k)=+\infty$ for a sequence $\alpha_k\to+\infty$. It follows from the convexity of f_x that $f_x(\alpha)=+\infty$ for all sufficiently large α. We set $\gamma(x)=+\infty$ for such x.

DEFINITION 5.23 A lower semicontinuous CAR function f is called *totally lower semicontinuous* if, for all x such that $\|x\|=1$, the following inequality holds:

$$\gamma(x)\leq\liminf_{x'\to x,\alpha\to+\infty}\frac{f(\alpha x')}{\alpha}. \tag{5.5.24}$$

PROPOSITION 5.54 *Let f be a totally lower semicontinuous CAR function. Then the function γ is lower semicontinuous*

Proof: It follows directly from the definition of lim inf that

$$\liminf_{x'\to x}\left(\lim_{\alpha\to+\infty}\frac{f(\alpha x')}{\alpha}\right)\geq\liminf_{x'\to x,\alpha\to+\infty}\frac{f(\alpha x)}{\alpha}.$$

Hence

$$\gamma(x)\leq\liminf_{x'\to x,\alpha\to+\infty}\frac{f(\alpha x')}{\alpha}\leq\liminf_{x'\to x}\gamma(x').$$ △

Let us describe the total lower semicontinuity in other terms. For any function $f(x)$ on $\mathbb{R}^n$, define the function

$$\tilde{f}(x)=(1-\|x\|)f\left(\frac{x}{1-\|x\|}\right) \tag{5.5.25}$$

on the open unit ball $\{x \in \mathbb{R}^n : \|x\| < 1\}$. It follows immediately from (5.5.23) that, for a CAR function f and any $x \neq 0$, there exists a limit $\lim_{\beta \to 1-0} \tilde{f}(\beta x) = \gamma(x)$. Hence we can consider the extension $\bar{f}$ of the function $\tilde{f}$ on the closed unit ball defined by the formula

$$\bar{f}(x) = \begin{cases} \tilde{f}(x) & \text{if } \|x\| < 1; \\ \gamma(x) & \text{if } \|x\| = 1. \end{cases}$$

PROPOSITION 5.55 *A CAR function f is totally lower semicontinuous if and only if the function $\bar{f}$ is lower semicontinuous on the unit ball.*

Proof: Let us check that the lower semicontinuity of $\bar{f}$ implies the total lower semicontinuity of f. First, we represent the function f by means of the function $\tilde{f}(x)$. Let $\|x\| < 1$ and $x' = x/(1 - \|x\|)$, hence $x = x'/(1 + \|x'\|)$. Applying (5.5.25), we get

$$f(x') = (1 + \|x'\|)\tilde{f}\left(\frac{x'}{1 + \|x'\|}\right) \tag{5.5.26}$$

for an arbitrary $x' \in \mathbb{R}^n$. Since $\tilde{f}$ is lower semicontinuous we can conclude by applying (5.5.26) that f is a lower semicontinuous function. Let us check inequality (5.5.24). For any y with $\|y\| = 1$, we have

$$\begin{aligned} \liminf_{y' \to y, \alpha \to +\infty} \frac{f(\alpha y')}{\alpha} &= \liminf_{y' \to y, \alpha \to +\infty} \frac{1 + \alpha\|y'\|}{\alpha} \tilde{f}\left(\frac{\alpha y'}{1 + \alpha\|y'\|}\right) \\ &= \liminf_{y' \to y, \alpha \to +\infty} \bar{f}\left(\frac{y'}{\frac{1}{\alpha} + \|y'\|}\right). \end{aligned}$$

Since $\bar{f}$ is a lower semicontinuous function and $\|y\| = 1$, we have

$$\liminf_{y' \to y, \alpha \to +\infty} \frac{f(\alpha y')}{\alpha} \geq \bar{f}(y) = \gamma(y).$$

Assume now that f is a totally lower semicontinuous function. Clearly, $\bar{f} = \tilde{f}$ is lower semicontinuous for all x such that $\|x\| < 1$. So, we have to check the lower semicontinuity of $\bar{f}$ only for x with $\|x\| = 1$. Let $\|x\| = 1$ and $x_k \to x$, $\|x_k\| \leq 1$. If $\|x_k\| = 1$ for all k, then $\bar{f}(x_k) = \gamma(x_k)$ and also $\bar{f}(x) = \gamma(x)$. Since γ is lower semicontinuous (see Proposition 5.54), we conclude that $\liminf_k \bar{f}(x_k) \geq \bar{f}(x)$. Assume without loss of generality that $\|x_k\| < 1$ for all k. Let $\alpha_k = 1/(1 - \|x_k\|)$. Then

$$\liminf_k \bar{f}(x_k) = \liminf_k \frac{f(\alpha_k x_k)}{\alpha_k} \geq \gamma(x) = \bar{f}(x).$$

Thus the required result follows. $\triangle$

COROLLARY 5.12 *A lower semicontinuous positively homogeneous function is totally lower semicontinuous.*

LEMMA 5.8 *Any lower affine approximation $g^a(x)$ of a totally lower semicontinuous CAR function f is a lower semicontinuous function.*

Proof: Assume without loss of generality that $f(0) > 0$ and $a = 0$. Then $g^a(y) = b^a(y)$ for all $y \in \mathbb{R}^n$. Let $S_* = \{x : \|x\| = 1\}$. We have

$$b^a(y) = \inf_{\alpha>0} \frac{f(\alpha y)}{\alpha} \leq \lim_{\alpha\to+\infty} \frac{f(\alpha y)}{\alpha} = \gamma(y) \quad \text{for all } y \in S.$$

If $b^a(x_i) = \gamma(x_i)$ for i large enough then the statement of Lemma follows from the lower semicontinuity of γ (see Proposition 5.54). Otherwise, choosing a subsequence if necessary, we can assume that $b^a(x_i) < \gamma(x_i)$, $i = 1, 2, \ldots$, and hence (see Remark 5.18) there exists a sequence $\{\alpha_i\}$, $\alpha_i \geq 0$, such that

$$\alpha_i b^a(x_i) = b^a(\alpha_i x_i) = f(\alpha_i x_i), \quad i = 1, \ldots \tag{5.5.27}$$

Suppose that there exists a number c such that

$$b^a(x) > c \geq b^a(x_i) \quad i = 1, \ldots \tag{5.5.28}$$

If the sequence (α_i) is bounded, then we can assume that it converges to some $\alpha^* \geq 0$. We have

$$f(\alpha^* x) \geq b^a(\alpha^* x) = \alpha^* b^a(x) > \alpha^* c$$

Applying (5.5.27) we conclude

$$\liminf_i f(\alpha_i x_i) = \liminf_i \alpha_i b^a(x_i) = \alpha_i \liminf_i b^a(x_i) \leq \alpha^* c.$$

Thus $\liminf_i f(\alpha_i x_i) \leq \alpha^* c < f(\alpha^* x)$, which contradicts the lower semicontinuity of f.

Otherwise, we can assume without loss of generality that $\alpha_i \to +\infty$ and hence

$$\liminf_i \frac{1}{\alpha_i} f(\alpha_i x_i) \geq \liminf_{x'\to x, \alpha\to+\infty} \frac{f(\alpha x')}{\alpha}.$$

Applying (5.5.27) and the total lower semicontinuity of f we get

$$c \geq \liminf_i b^a(x_i) = \liminf_i \frac{f(\alpha_i x_i)}{\alpha_i} \geq \gamma(x).$$

Also

$$\gamma(x) = \lim_{\alpha\to+\infty} \frac{f(\alpha x)}{\alpha} \geq \inf_\alpha \frac{f(\alpha x)}{\alpha} = b^a(x) > c$$

and we again have a contradiction. Thus (5.5.28) is not true, which implies the required lower semicontinuity. △

THEOREM 5.17 *Any totally lower semicontinuous CAR function f belongs to the class $\mathcal{H}_{n+1}$.*

Proof: The proof follows directly from Lemma 5.6 and Lemma 5.8. △

As the next example shows, functions in $\mathcal{H}_{n+1}$ are not necessarily totally lower semicontinuous.

EXAMPLE 5.6 The function g defined on $\mathbb{R}^2$ by the formula:

$$g(x,y) = \begin{cases} +\infty & \text{if } y > -1; \\ 0 & \text{otherwise} \end{cases}$$

is lower semicontinuous and convex, therefore this function belongs to $\mathcal{H}_2 \subset \mathcal{H}_{n+1}$. Since $\gamma(1,0) = +\infty$ and $\gamma(1,-\varepsilon) = 0$ for any $\varepsilon > 0$, it follows that g is not totally lower semicontinuous.

Consider now the classes T_1 of all totally locally semicontinuous CAR functions, T_2 of all $\mathcal{H}_{n+1}$-convex functions, and T_3 of all lower semicontinuous CAR functions. It has been shown that $T_1 \subset T_2 \subset T_3$ and $T_1 \neq T_2 \neq T_3$. Clearly, T_2 contains the class of all lower semicontinuous convex functions. Example 5.6 shows that T_1 does not contain this class. It has been shown (see Corollary 5.12) that each lower semicontinuous positively homogeneous function belongs to T_1. If a lower semicontinuous CAR function f is coercive in the sense that

$$\liminf_{x' \to x, \alpha \to +\infty} \frac{f(\alpha x')}{\alpha} = +\infty \text{ for all } x \text{ with } \|x\| = 1$$

then $f \in T_1$, which follows immediately from the definition of the total lower semicontinuity. Thus any lower semicontinuous positively homogeneous of degree $k > 1$ function is totally lower semicontinuous. Let us point out some properties of sets T_1, T_2 and T_3.

1) Each T_i $(i = 1,2,3)$ is a semilinear space (if $f_1, f_2 \in T_i$, then $f_1 + f_2 \in T_i$; if $f \in T_i$ and $\lambda > 0$ then $\lambda f \in T_i$).

2) T_2 and T_3 are complete upper semilattices. If A is an arbitrary set of indices and $(f_\alpha)_{\alpha \in A} \in T_i$ with $i = 2,3$, then the upper envelope f of the family $(f_\alpha)_{\alpha \in A}$: $f(x) = \sup_{\alpha \in A} f_\alpha(x)$ also belongs to the class T_i.

3) T_1 is an upper semilattice. If $f(x) = \max_{i=1,\dots,k} f_i(x)$ and $f_i \in T_1$ for all i, then $f \in T_1$.

The property 3) follows from Proposition 5.55. If $f = \max_{i=1\dots,k} f_i$, then $\gamma(f,x) = \max_i \gamma(f_i,x)$, where $\gamma(f,\cdot)$ is defined by (5.5.23). At the same time the maximum of a finite number of lower semicontinuous functions is again lower semicontinuous. If $f_\alpha \in T_1$ for an infinite family $\alpha \in A$ then it can happen that $\sup_\alpha f_\alpha \notin T_1$. In fact, we can represent a lower semicontinuous convex function which does not belong to T_1 as the upper envelope of affine functions belonging to this set.

5.5.4. Subdifferentials of $\mathcal{H}_{n+1}$-convex functions

In this section we present some conditions ensuring that the subdifferential $\partial_{\mathcal{L}_{n+1}} f(x_0)$ of a CAR function f at a point x_0 is not empty. Recall that

$$\partial_{\mathcal{L}_{n+1}} f(x_0) = \{\ell \in \mathcal{L}_{n+1} : \langle l, x\rangle_* - \langle l, x_0\rangle_* \leq f(x) - f(x_0),\ x \in \mathbb{R}^n\}.$$

First, we point out some properties of the subdifferential.

PROPOSITION 5.56 *Let f be a bounded function on the ball $B_\varepsilon = \{x : \|x - x_0\| \leq \varepsilon\}$. Assume that the subdifferential $\partial_{\mathcal{L}_{n+1}} f(x_0)$ of the function f at the point x_0 is not empty. Then $\partial_{\mathcal{L}_{n+1}} f(x_0)$ is a closed bounded subset of the space $\mathcal{L}_{n+1}$ and the following inequality holds for $\ell \in \mathcal{L}_{n+1}$:*

$$\|\ell\| \leq \frac{1}{\varepsilon} \sup_{u \in B_\varepsilon} (f(x_0 + u) - f(x_0)). \tag{5.5.29}$$

Proof: It is clear that the subdifferential is a closed set. Let us check that it is a bounded set. Let $\ell \in \partial\mathcal{L}_{n+1} f(x_0)$ and $\|u\| \leq \varepsilon$. Note that the function $x \mapsto \langle \ell, x\rangle_*$ is superlinear, so

$$\begin{aligned} \langle \ell, u\rangle_* &= \langle \ell, x_0\rangle_* + \langle \ell, u\rangle_* - \langle \ell, x_0\rangle_* \\ &\leq \langle \ell, x_0 + u\rangle_* - \langle \ell, x_0\rangle_* \leq f(x_0 + u) - f(x_0). \end{aligned}$$

Hence

$$\|l\| = \sup_{\|u\| \leq 1} \langle \ell, u\rangle_* \leq \frac{1}{\varepsilon} \sup_{\|u\| \leq \varepsilon} (f(x_0 + u) - f(x_0)).$$

Thus the inequality (5.5.29) holds. △

We now consider a function f and a point $x_0 \in \operatorname{dom} f = \{x : f(x) < +\infty\}$ such that f is locally Lipschitz on the ray $R_{x_0} = \{\alpha x_0 : \alpha > 0\}$ in the following sense: there exists a number $\delta > 0$ such that

1) $\{\alpha x : x \in \mathbb{R}^n, \|x - x_0\| < \delta, 0 \leq \alpha \leq 1\} \subset \operatorname{dom} f$;

2) for each $r > 0$, there exists a number L_r satisfying

$$|f(\alpha x) - f(\alpha x_0)| \le L_r \|\alpha x - \alpha x_0\| \tag{5.5.30}$$

whenever $\|x - x_0\| < \delta$, $0 < \alpha < r$.

Let f be a lower semicontinuous CAR function and $x_0 \in \operatorname{int}\operatorname{dom} f$. Then (see Remark 3.10) there exist the directional derivatives $f'(x_0, x_0)$ and $f'(x_0, -x_0)$ at the point x_0 in the directions x_0 and $-x_0$ respectively and $f'(x_0, x_0) = (f_{x_0})'_+(1)$, $f'(x_0, -x_0) = -(f_{x_0})'_-(1))$. We have

$$\partial f_{x_0}(1) = [f'(x_0, -x_0), f'(x_0, x_0)].$$

LEMMA 5.9 *Let f be a lower semicontinuous CAR function and $x_0 \in \operatorname{int}\operatorname{dom} f$. Suppose that f is locally Lipschitz on the ray R_{x_0} and $f'(x, x)$ is bounded in a neighbourhood of the point x_0. Let also*

$$\lim_{x \to x_0, \alpha \to +\infty} \frac{f(\alpha x)}{\alpha} = +\infty \tag{5.5.31}$$

and suppose the function f is not affine on the segment $[0, x]$. Then $\partial_{\mathcal{L}_{n+1}} f(x_0)$ is not empty.

Proof: Since $x_0 \in \operatorname{int}\operatorname{dom} f$ it follows that the subdifferential $\partial f_{x_0}(1)$ is nonempty. Take $v \in \partial f_{x_0}(1)$. Then

$$\alpha v \le f_{x_0}(\alpha) - (f_{x_0}(1) - v) \text{ for all } \alpha \ge 0. \tag{5.5.32}$$

In particular $a := f_{x_0}(1) - v \le f_{x_0}(0) = f(0)$. Since f is not affine on $[0, x_0]$, we conclude that the function f_{x_0} is not affine on $[0, 1]$. In this case $a < f(0)$. It follows from (5.5.32) that

$$f(x_0) - a = f_{x_0}(1) - a = v \le \frac{f_{x_0}(\alpha) - a}{\alpha} \text{ for all } \alpha > 0. \tag{5.5.33}$$

Thus

$$b^a(x_0) = \inf_{\alpha > 0} \frac{f(\alpha x) - a}{\alpha} \ge f(x_0) - a. \tag{5.5.34}$$

Let $\gamma > f(x_0) - a$. It follows from 5.5.30 that f is continuous in a neighbourhood of the point x_0. Hence for all x sufficiently close to x_0 we have $f(x) - a < \gamma$. This inequality implies

$$b^a(x) = \inf_{\alpha > 0} \frac{f(\alpha x) - a}{\alpha} < \gamma.$$

We now check that if x is sufficiently close to x_0, then

$$|b^a(x) - b^a(x_0)| \le L_r \|x - x_0\|. \tag{5.5.35}$$

It follows from (5.5.31) that there exist numbers $\delta > 0$ and $r > 0$ such that

$$\frac{f(\alpha x) - a}{\alpha} > \gamma > b^a(x)$$

for all $\alpha > r$ and for all x such that $\|x_0 - x\| < \delta$. Thus, if $\|x - x_0\| < \delta$, then

$$b^a(x) = \inf_{0<\alpha<r} \frac{f(\alpha x) - a}{\alpha}.$$

Assume also that δ is so small that (5.5.30) is valid if $\|x - x_0\| < \delta$ and $\alpha < r$. For these x_0 and α, we have:

$$\left|\frac{f(\alpha x) - a}{\alpha} - \frac{f(\alpha x_0) - a}{\alpha}\right| = \left|\frac{f(\alpha x) - f(\alpha x_0)}{\alpha}\right| \leq L_r \|x - x_0\|.$$

Thus (5.5.35) holds if $\|x - x_0\| < \delta$. Let $g^a(\alpha x) = a + b^a(x)\alpha$ be a lower affine approximation of the function f and $s(\alpha x) = g^a(\alpha x) - a$ be a shift of the function g^a. It follows from (5.5.31) that f is totally lower semicontinuous, hence (see Lemma 5.8) s is lower semicontinuous. Applying Theorem 5.15 and (5.5.35), we conclude that the subdifferential $\partial_{\mathcal{L}_{n+1}} s(x_0)$ is not empty. Let $\ell \in \partial_{\mathcal{L}_{n+1}} s(x_0)$. Then

$$g^a(x) - g^a(x_0) = s(x) - s(x_0) \geq \langle \ell, x\rangle_* - \langle \ell, x_0\rangle_*.$$

Since g^a is a minorant of the function f, (5.5.34) implies that

$$f(x) - f(x_0) \geq g^a(x) - f(x_0) \geq g^a(x) - (a + b^a(x_0))$$

$$= g^a(x) - g^a(x_0) \geq \langle \ell, x\rangle_* - \langle \ell, x_0\rangle_*$$

for any $x \in \mathbb{R}^n$. Thus $\ell \in \partial_{\mathcal{L}_{n+1}} f(x_0)$. △

THEOREM 5.18 *Let f be a lower semicontinuous CAR function. Suppose $x_0 \in \operatorname{dom} f$ and f is locally Lipschitz on the ray R_{x_0}. Then the subdifferential $\partial_{\mathcal{L}_{n+1}} f(x_0)$ is not empty.*

Proof: For each positive integer k we shall consider the function $f^k(x) = f(x) + \frac{1}{k}\|x\|^2$. Clearly the hypothesis of Lemma 5.9 holds for f^k. Applying Lemma 5.9, we find a function $\ell^k \in \partial_{\mathcal{L}_{n+1}} f^k(x_0)$ satisfying the inequality

$$\langle \ell^k, x\rangle_* - \langle \ell^k, x_0\rangle_* \leq f(x) - f(x_0) + \frac{1}{k}(\|x\|^2 - \|x_0\|^2). \qquad (5.5.36)$$

Proposition 5.56 shows that the sequence (ℓ_k) is bounded in the space $\mathcal{L}_{n+1}$. Due to Proposition 5.51 we can assume without loss of generality that there exists the limit ℓ of this sequence. Passing to the limit in (5.5.36), we conclude that $\ell \in \partial_{\mathcal{L}_{n+1}} f(x_0)$. $\triangle$

We now describe local conditions guaranteeing that the subdifferential $\partial_{\mathcal{L}_{n+1}} f(x_0)$ is nonempty in terms of directional derivatives.

LEMMA 5.10 *Let f be a lower semicontinuous CAR function and $x_0 \in \operatorname{int} \operatorname{dom} f$. Suppose that the function $x \to f'(x,x)$ is calm of degree one at the point x_0 (see Definition 4.5), the equality (5.5.31) holds and f is not affine on the segment $[0, x_0]$. Then $\partial_{\mathcal{L}_{n+1}} f(x_0)$ is not empty.*

Proof: Let $v = f'(x_0, x_0)$ and $a = f_{x_0}(1) - v$. Then $v \in \partial f_{x_0}(1)$ and $a < f(0)$. It follows from (5.5.33) and (5.5.34) that $b^a(x_0) = v = f'(x_0, x_0)$. So the function $b^a(x)$ is calm of degree one at the point x_0. The rest of the proof runs as in the proof of Lemma 5.9. $\triangle$

LEMMA 5.11 *Let f be a lower semicontinuous CAR function and $x_0 \in \operatorname{int} \operatorname{dom} f$. Suppose that the function $x \to -f'(x, -x)$ is calm of degree one at the point x_0, the equality (5.5.31) holds and f is not affine on the segment $[0, x_0]$. Then $\partial_{\mathcal{L}_{n+1}} f(x_0)$ is not empty.*

Proof: Take $v = -f'(x_0, -x_0)$ and repeat the proof of Lemma 5.10. $\triangle$

THEOREM 5.19 *Let f be a lower semicontinuous CAR function and $x_0 \in \operatorname{int} \operatorname{dom} f$. Suppose that either the function $x \mapsto f'(x,x)$ or the function $x \mapsto -f'(x, -x)$ is calm of degree one at the point x_0. Then the subdifferential $\partial_{\mathcal{L}_{n+1}} f(x_0)$ is not empty.*

Proof: The result follows from the proof of Theorem 5.18, where either Lemma 5.10 or Lemma 5.11 is used instead of Lemma 5.9. $\triangle$

EXAMPLE 5.7 Let f be a function considered in Example 5.5. We have $\operatorname{dom} f = \{x : \|x\| \geq 1\}$ where $\|\cdot\|$ is the Euclidean norm. Let $X = \{x \in \operatorname{dom} f : x \neq (\lambda, 0),\ \lambda \geq 1\}$. It is easy to check that $f'(x,x) = -\|x\|$ for all $x \in X$ and

$$f'(x, -x) = \begin{cases} \|x\| & \text{if } x \in X,\ \|x\| > 1; \\ +\infty & \text{if } x \in X,\ \|x\| = 1. \end{cases}$$

It follows from Theorem 5.19 that the subdifferential $\partial_{\mathcal{L}_{n+1}} f(x)$ is not empty for all $x \in X$. Obviously, the subdifferential $\partial_{\mathcal{L}_{n+1}} f((\lambda, 0))$ is empty for all $\lambda \geq 1$. Recall (see Example 5.5) that the function f is not $\mathcal{H}$-convex.

5.5.5. Abstract convex sets

In this subsection we shall examine abstract convex sets with respect to either $\mathcal{L}_n$ or $\mathcal{L}_{n+1}$. By the definition, the set $U \subset \mathbb{R}^n$ is abstract convex with respect to $\mathcal{L}_n$ ($\mathcal{L}_{n+1}$) if for each $x_0 \notin U$ there exists a collection $\ell \in \mathcal{L}_n$ ($\ell \in \mathcal{L}_{n+1}$), such that $\ell(x_0) > \sup_{x \in U} \ell(x)$.

PROPOSITION 5.57 *Let $0 \in U \subset \mathbb{R}^n$. Then U is $\mathcal{L}_n$-convex if and only if U is a closed radiant set.*

Proof: The proof follows directly from Proposition 5.35 and Theorem 5.6. △

In order to study $\mathcal{L}_{n+1}$-convex sets we shall use the following statement, which is a special case of Theorem 7.2.

PROPOSITION 5.58 *A set $U \subset \mathbb{R}^n$ is abstract convex with respect to $\mathcal{L}_{n+1}$ if and only if there exists an $\mathcal{H}_{n+1}$-convex function f and $c \in \mathbb{R}$ such that U coincides with the level set $S_c(f) = \{x \in \mathbb{R}^n : f(x) \leq c\}$.*

THEOREM 5.20 *1) Assume that a closed set W can be represented as the intersection $W = U \cap V$, where U is a closed radiant set and V is a co-radiant set. (It is possible that either U or V coincides with $\mathbb{R}^n$.) Then W is an $\mathcal{L}_{n+1}$-convex subset of $\mathbb{R}^n$.*

2) Let W be a bounded $\mathcal{L}_{n+1}$-convex subset of $\mathbb{R}^n$. Then there exists a closed radiant set U and a co-radiant set V such that $W = U \cap V$.

Proof: 1) Consider a closed set $W = U \cap V$, where U is a closed radiant set and V is a co-radiant set. Let $x^0 \notin W$ and let $[0, x_0) = \{\lambda x_0 : 0 \leq \lambda < 1\}$. We consider two cases.

Case 1. Assume that $[0, x_0) \cap W = \emptyset$. Let $G = \mathbb{R}^n \setminus W$. The set G is open and $[0, x_0] \subset G$. Then the set G contains a simplex S with $0 \in \operatorname{int} S, x_0 \in \operatorname{int} S$. Since $0 \in \operatorname{int} S$, there exist vectors $l(1), \dots, l(n+1)$ such that $S = \{x : [l(i), x] \leq 1\} = \{x : [-l(i), x] \geq -1\}$. The collection $-l(1), \dots, -l(n+1)$ separates the point x_0 from the set W.

Case 2. Assume that $[0, x_0) \cap W \neq \emptyset$. Then there exists $\lambda \in (0, 1)$ such that $\lambda x_0 \in W$. Since $\lambda x_0 \in V$ and V is co-radiant, we conclude that $x_0 \in V$. Since $x_0 \notin W = U \cap V$, it follows that $x_0 \notin U$. Since U is a closed radiant set, we can separate the point x_0 from U by a collection of n linearly independent vectors. Clearly, this collection also separates x_0 from the set $W \subset U$.

2) Let W be a bounded $\mathcal{L}_{n+1}$-convex subset of $\mathbb{R}^n$. It follows from Proposition 5.58 that $W = \{x \in \mathbb{R}^n : f(x) \leq c\}$, where f is an $\mathcal{H}_{n+1}$-convex function and $c \in \mathbb{R}$. For $x \in \mathbb{R}^n, x \neq 0$ consider the intersection $O_x = R_x \cap W$, where R_x is the closed ray starting from zero and passing through x. We have

$$O_x = \{x' : f(x') \leq c, x' = \lambda x, \lambda \geq 0\} = \{\lambda x : f_x(\lambda) \leq c, \lambda \geq 0\},$$

where $f_x(\lambda) = f(\lambda x)$. The convexity of the function f_x implies the convexity of the set O_x. Thus if O_x is nonempty, then O_x is a segment. In other words, there exist $\rho(x), \sigma(x) \in \bar{\mathbb{R}}_+$ such that $\rho(x) \leq \sigma(x)$ and

$$O_x = [\rho(x), \sigma(x)]x := \{y : y = \alpha x : \rho(x) \leq \alpha \leq \sigma(x)\}.$$

(If $\sigma(x) = +\infty$, then $[\rho(x), \sigma(x)]x = \{\alpha x : \rho \leq \alpha < \sigma(x)\}$.) Note that the functions $\rho(x), \sigma(x)$ are defined on the conic hull $Q = \operatorname{cone} W$ of the set W.

Let $\lambda > 0$. For $x \in Q$ we have

$$\begin{aligned} &[\rho(\lambda x), \sigma(\lambda x)](\lambda x) \\ &\qquad = O_{\lambda x} = W \cap R_{\lambda x} = W \cap \{\beta(\lambda x) : \beta > 0\} \\ &\qquad = W \cap \{(\beta\lambda)x : \beta\lambda > 0\} = [\rho(x), \sigma(x)]x. \end{aligned}$$

Thus

$$\rho(\lambda x) = \frac{1}{\lambda}\rho(x), \qquad \sigma(\lambda x) = \frac{1}{\lambda}\sigma(x).$$

In particular, if $x \neq 0$ and $x' = x/\|x\|$, then

$$\sigma(x')\|x'\| = \sigma\left(\frac{x}{\|x\|}\right)x' = \sigma(x)\|x\|x' = \sigma(x)x. \tag{5.5.37}$$

The same argument shows that $\rho(x')x' = \rho(x)x$. Recall that the set W is bounded, that is, there exists a constant $c > 0$ such that $\|\sigma(x)x\| \leq c$ for all $x \in Q$. It follows from (5.5.37) that the function σ is bounded on the set $\{x' \in Q : \|x'\| = 1\}$. Let

$$U = \bigcup_{\|x'\|=1, x' \in Q} [0, \sigma(x')]x', \qquad V = \bigcup_{\|x'\|=1, x' \in Q} [\rho(x'), +\infty)x.$$

Let us indicate some properties of the sets U and V.

1) U is a radiant set. Indeed, let $x \in U$ and $0 < \lambda \leq 1$. Let $\|x'\| = x/\|x\|$. Then $\sigma(x')x' = \sigma(x)x \in U$, $\lambda x \in [0, \sigma(x)]x \in U$.

2) V is a co-radiant set. The proof is similar.

3) $W = U \cap V$. Let $x \in W$ and $x' = x/\|x\|$. Then $x \in [\rho(x'), \sigma(x')]x'$, hence $x \in [\rho(x'), +\infty)x' \subset V$ and $x \in [0, \sigma(x')]x' \subset U$. Thus $W \subset U \cap V$. The same argument shows that the reverse inclusion holds.

4) The set U is closed. In fact, let $x_k \in U$ and $x_k \to x$. Let $x'_k = x_k/\|x_k\|$. Then there exists $\lambda_k \in [0, 1]$ such that $x_k = \lambda_k \sigma(x'_k)x'_k$. We have $x'_k \to x' := x/\|x\|$. Since the sequence $\sigma(x'_k)$ is bounded we can assume that there exists $\lim \sigma(x'_k) := \sigma$. We also assume

that there exists $\lim \lambda_k = \lambda$. Thus $x = \lambda \sigma x'$. Since $\sigma(x'_k)x'_k \in W$ and W is closed, it follows that $\sigma x' \in W$, hence $\sigma \leq \sigma(x')$. Thus $x \in [0, \sigma(x')]x' \in U$.

Thus we have proved the existence of sets U and V with the required properties. △

5.5.6. Other classes of abstract convex functions

Let $z \in \mathbb{R}^n$. Consider the class $\mathcal{L}_k^z$ of all functions of the form $x \mapsto \min_{i=1,\dots,k}[l(i), x - z]$ with $k = n, n+1$. All results related to $\mathcal{L}_k$-convex functions and sets can naturally be transformed to $\mathcal{L}_k^z$-convex functions and sets by shifts on the vector z. For example, it follows from Proposition 5.57 that the following assertion holds. Let $z \in U \subset \mathbb{R}^n$. Then the set U is abstract convex with respect to $\mathcal{L}_n^z$ if and only if U is closed and star-shaped and $z \in \operatorname{kern} U$.

Consider now the set $\mathcal{L}_k^*$ of all functions of the form

$$\ell(x) = \max_{i=1,\dots,k} [l(i), x - z],$$

where $((l(1), \dots, l(k)) \in (\mathbb{R}^n)^k$ and $z \in \mathbb{R}^n$. In other words, $\mathcal{L}_k^* = \bigcup_{z \in \mathbb{R}^n} \mathcal{L}_k^z$. Clearly, $\mathcal{L}_k^* \subset H_k$, where H_k is the class of all functions h given by

$$h(x) = \min_{i=1,\dots,j} ([l(i), x] - c_i) \text{ with } l(i) \in \mathbb{R}^n, \ \ c_i \in \mathbb{R}, \ \ j \leq k.$$

The classes of abstract convex functions with respect to H_k with $k = n+1$ and $k = n+2$ are very broad.

THEOREM 5.21 *A function f is H_{n+2}-convex if and only if f is lower semicontinuous.*

Proof: Clearly each H_{n+2}-convex function is lower semicontinuous. To prove the reverse assertion, consider a proper lower semicontinuous function $f : \mathbb{R}^n \to \mathbb{R}_{+\infty}$. Let $\hat{f}$ be a positively homogeneous extension of the function f. Then (see Definition 7.10)

$$\hat{f}(x, \lambda) = \lambda f\left(\frac{x}{\lambda}\right) \quad \text{for all } x \in X \text{ and } \lambda > 0.$$

Denote by $\bar{f}$ the lower regularization of $\hat{f}$ (see Definition 2.2). Let $x \in \mathbb{R}^n$. Since f is lower semicontinuous at the point x it follows that the function $\hat{f}$ is lower semicontinuous at the point $(x, 1)$. Hence

$$\bar{f}(x, 1) = \hat{f}(x, 1) = f(x).$$

Denote a vector $(x_1, \dots, x_n, x_{n+1}) \in \mathbb{R}^{n+1}$ by (x, x_{n+1}), where $x = (x_1, \dots, x_n) \in \mathbb{R}^n$. Applying Theorem 5.14, we conclude that a lower semicontinuous positively homogeneous function $\bar{f}$ defined on $\mathbb{R}^{n+1}$, can be represented as the upper envelope of a set D consisting of collections $\ell = (l'(1), \dots, l'(n+2))$ of vectors $l'(i) \in \mathbb{R}^{n+1}$. Representing $l'(i)$ as $(l(i), l_{n+1}(i))$, we have for $(x, x_{n+1}) \in \mathbb{R}^{n+1}$:

$$\bar{f}(x, x_{n+1}) = \sup_{l \in D} \min_{i=1,\dots,n+2} ([l(i), x] + l_{n+1}(i) x_{n+1}), \quad (x, x_{n+1}) \in \mathbb{R}^{n+1}.$$

In particular,

$$f(x) = \bar{f}(x, 1) = \sup_{l \in D} (\min_i [l(i), x] + l_{n+1}(i)) \text{ for all } x \in \mathbb{R}^n.$$

$\triangle$

Applying Theorem 5.13 and using the same argument we infer that the following result holds.

THEOREM 5.22 *A bounded from below function f is H_{n+1}-convex if and only if f is lower semicontinuous.*

Proof: Let $f(x) \geq c > -\infty$ for all $x \in \mathbb{R}^n$. Then the function $f_c(x) = f(x) - c$ is nonnegative, hence its positively homogeneous extension $\hat{f}$ is also nonnegative and the closure $\bar{f}_c$ of this extension is also nonnegative. It follows from Theorem 5.13 that $\bar{f}_c$ can be represented as the upper envelope of a set D consisting of collections of $n+1$ vectors of $n+1$ dimension. It follows from this representation that f_c is H_{n+1}-convex, hence f is also H_{n+1}-convex. $\triangle$

Chapter 6

SUPREMAL GENERATORS AND THEIR APPLICATIONS

6.1. INTRODUCTION

A supremal generator of a set X of functions is a subset H of X such that each function from X is abstract convex with respect to H. In other words, H is a supremal generator of X if each function from X can be represented as the upper envelope of a subset of H. As it turns out there exist very large sets with very small supremal generators. For example, the space of all lower semicontinuous functions defined on a segment of the real line has supremal generators, which are cones spanned by three functions only. If H is a supremal generator of a set of functions, then some properties of H allow one to study some properties of functions belonging to X. Properties of small supremal generators very often can be verified "manually", that is by direct calculation. Thus the existence of small supremal generators is very helpful. In this chapter we first present a description of supremal generators for some sets of lower semicontinuous functions and then describe some small generators for these sets. We shall show that the description of supremal generators for the set of *all* lower semicontinuous functions is much easier than for proper subsets of this set.

The results, discussed in this chapter, can be applied in the study of various problems. In particular we shall apply them in the study of 'thin' supremal generators of the spaces of positively homogeneous functions (see Subsection 6.3.3 and Subsection 6.3.4). Positively homogeneous functions arise in many areas of mathematics. In particular they are an important tool of modern nonsmooth analysis. Various kinds of generalized derivatives of the first order (Dini derivatives, Hadamard derivatives, Clarke derivatives etc.) are used in nonsmooth analysis (see, for example, [43]). All these kinds of generalized derivatives are positively homogeneous functions of degree one. Symmetric positively homogeneous functions of degree two arise naturally in the study of different kinds of approximation of the second order for nonsmooth functions (see, for example, [49] for details).

The representation of a lower semicontinuous positively homogeneous function as the upper envelope of a set of sufficiently simple functions allows us to simplify many constructions of nonsmooth analysis. This problem has been studied in [43] for positively homogeneous of degree one functions defined on the Euclidean space in connection with the so-called exhaustive families of upper convex approximations. It was shown in [43] (see the proof of Theorem 9.2) that each continuous positively homogeneous of degree one function f defined on the Euclidean space can be represented as the supremum of a subset of the set $H = \{h : h(x) = -a\|x\| + [u, x];\ a \geq 0, u \in \mathbb{R}^n\}$. Here $\|\cdot\|$ is the Euclidean norm and $[u, x]$ is the inner product of vectors u and x.

The representation of a symmetric (even) positively homogeneous of degree two function as the upper envelope of a set of symmetric quadratic forms has been studied in [49, 50]. A finite-dimensional version of this representation is discussed in [49]. The special approach to the study of this problem for functions defined on a Banach space has been developed in [50]

Supremal generators for the space of continuous functions defined on a compact set arise naturally in the study of some problems of the approximation theory (convergence of the sequences of positive operators) and topology. We shall present some examples of such applications in Section 6.4.

An interesting area of applications of supremal generators for some classes of functions is Hadamard type inequalities (see, for example, [45] and references therein). In the classical convex case the Hadamard inequality closely related to the so-called Choquet theory (see, for example, [128]). Its proof is based on the following fundamental fact of the theory of convex functions. The set of continuous affine functions, defined on a convex compact set is a supremal generator of the set of lower semicontinuous convex functions defined on this set. In order to obtain Hadamard type inequalities for various classes of functions we should describe small supremal generators of these classes. This kind of application is discussed in detail in Section 6.5.

6.2. CONTINUOUS AND LOWER SEMICONTINUOUS FUNCTIONS

6.2.1. Overview

In this section we give a simple criterion for a set H to be a supremal generator of the set of all lower semicontinuous functions bounded by H. This criterion is based on the notion of the *support to Urysohn peak* which was introduced in [99, 100]. The first version of this criterion was presented in [99, 100] for a special case. (In [99, 100] the authors considered only convex cones contained in the space $C(Z)$ of all continuous functions defined on a compact space Z.) The proof of this criterion in general situation, presented in this section, is a simple modification of the proof given in [100]. Various generalizations of results from [99, 100] are given in many papers (see, for example [15, 44]). We shall provide some interesting examples of very thin supremal generators. These examples can be found in [99, 100, 143].

6.2.2. Lower semicontinuous functions

We begin with the following definition.

DEFINITION 6.1 Let X be a set of functions $Z \to \bar{\mathbb{R}}$. A set $H \subset X$ is called a *supremal generator* of X if each function $f \in X$ is H-convex.

First we give a classical example of a supremal generator. Recall the well-known Hanh-Banach theorem. We shall need only a version of this theorem for Banach spaces.

THEOREM 6.1 (Hahn -Banach) *Assume that we have*

1) a Banach space X with the conjugate space X^*;

2) a bounded sublinear function p defined on X;

3) a linear subspace Y of the space X;

4) a linear function l defined on Y and bounded by p, that is, $l(y) \le p(y)$ for each $y \in Y$.

Then there exists a linear function $l' \in X^*$ such that $l'(y) = l(y)$ for $y \in Y$ and $l'(x) \le p(x)$ for all $x \in X$.

Recall that a sublinear function $p : X \to \mathbb{R}$ is called bounded if

$$\sup_{\|x\| \le 1} |p(x)| = 1.$$

It is well known (and easy to check) that the function p is bounded if and only if it is continuous.

The Hahn-Banach theorem is equivalent to the following statement (see, for example, [40, 39]).

THEOREM 6.2 *Let X be a Banach space and $p : X \to \mathbb{R}_{+\infty}$ be a lower semicontinuous sublinear function. Then there exists a w^*-closed convex set $U \subset X^*$ such that $p(x) = \sup_{l \in U} l(x)$ for all $x \in X$.*

Recall that a convex set $U \subset X^*$ is w^*-closed if for each $l \in X^* \setminus U$ there exists $x \in X$ and $\varepsilon > 0$ such that $U \cap \{l' : |(l - l')(x)| < \varepsilon\} = \emptyset$.

EXAMPLE 6.1 Let X be a Banach space. Then the conjugate space $L = X^*$ is a supremal generator of the set of all lower semicontinuous sublinear functions $p : X \to \mathbb{R}_{+\infty}$. Indeed this assertion is a reformulation of Theorem 6.2.

Example 6.1 demonstrates that the problem of describing a fairly small supremal generators is not simple. It turns out that it is much easier to describe supremal generators of the set of *all* lower semicontinuous functions bounded from below (in a certain sense) than supremal generators of proper subsets of this set. We shall give a special criterion which often allows us to easily find supremal generators of this set. We begin with the following definition.

DEFINITION 6.2 Let H be a set of function defined on a set Z and let f be a function defined on Z. We say that the function f is *minored* by the set H if there exists $\tilde{h} \in H$ such that

$$f(x) > \tilde{h}(x) \text{ for all } x \in Z. \tag{6.2.1}$$

A function f is *strictly minored* by H if it is possible to find a function $\tilde{h} \in H$ such that (6.2.1) holds and, in addition,

$$h + \tilde{h} \in H \text{ for all } h \in H. \tag{6.2.2}$$

REMARK 6.1 If H is a convex cone then f is strictly minored by H if and only if f is minored by H.

Let $z \in Z$, where Z is a metric space. Denote by $\mathcal{N}(z)$ the family of all open neighbourhoods of the point z .

We need the following simple assertion.

PROPOSITION 6.1 *Let $z \in Z$ and $U \in \mathcal{N}(z)$. Then there exists a continuous function f defined on Z such that*

$$f(z) = 1; \quad f(x) = 0 \text{ for all } x \notin U; \quad 0 \le f(x) \le 1 \text{ for all } x \in Z. \tag{6.2.3}$$

Proof: Let V be the complement to the open set U and $\rho_V(x) = \inf_{v \in V} \rho(x, v)$ be the distance between x and V (here ρ is the metric in the space Z). Set

$$f(x) = \min \left(1, \frac{\rho_V(x)}{\rho_V(z)}\right). \tag{6.2.4}$$

The triangle inequality implies that for each $v \in V$ the function $x \mapsto \rho(x, v)$ is Lipschitz with a Lipschitz constant $L = 1$. Therefore, the function ρ_V is Lipschitz, hence continuous. It follows from continuity of this function that the function f defined by (6.2.4) is also continuous. An easy calculation shows that (6.2.3) holds for the function f. △

REMARK 6.2 It is well known (see for example [168], Theorem 6.4.8), that if Z is a normal topological space then there exists a function with properties (6.2.3). Note that the results presented in this section hold for normal topological spaces as well. However, we restrict ourselves only to metric spaces.

A continuous function f with properties (6.2.3) is referred to as an *Urysohn peak* at the point z.

Let $C(Z)$ ($LSC(Z)$) be the space of all continuous (lower semicontinuous) functions defined on the metric space Z. First, we show that the following result holds.

PROPOSITION 6.2 *Let Z be a metric space, $H \subset C(Z)$ and the set $P(H, Z)$ of all H-convex functions contain the set of all nonnegative bounded continuous functions defined on Z. Then for any triplet (ε, z, U) with $\varepsilon > 0, z \in Z$ and $U \in \mathcal{N}(z)$, there exists a function $h \in H$ such that*

$$h(z) > 1 - \varepsilon; \quad h(x) \le 1 \text{ for all } x \in U; \quad h(x) \le 0 \text{ for all } x \notin U. \tag{6.2.5}$$

Proof: Let $\varepsilon > 0, z \in Z$ and $U \in \mathcal{N}(z)$. Consider an Urysohn peak, which is generated by the triplet (ε, z, U), that is, a function $f \in C(Z)$ such that

$$f(z) = 1; \quad f(x) = 0 \text{ for all } x \notin U; \quad 0 \le f(x) \le 1 \text{ for all } x \in Z.$$

The function f is H-convex therefore for the given $\varepsilon > 0$ there exists a function $h \in H$ such that $h(x) \leq f(x)$ for all $x \in Z$ and $h(z) > f(z) - \varepsilon$. Clearly, (6.2.5) is valid for the function h. Thus the result follows. △

DEFINITION 6.3 A function h with properties (6.2.5) is called a *support to an Urysohn peak* corresponding to the triplet (ε, z, U).

Let $H \subset C(Z)$. We denote by LSC_H the set of all lower semicontinuous functions defined on Z that are strictly minored by H.

Assume that H contains a strictly negative function. Then LSC_H contains the set of all bounded nonnegative continuous functions. If $P(H, Z) \supset LSC_H$, then, due to Proposition 6.2, we can conclude that for each triplet (ε, z, U) with $\varepsilon > 0, z \in Z$ and $U \in \mathcal{N}(z)$ there exists $h \in H$, which is the support to an Urysohn peak corresponding to this triplet. We now prove the following statement.

LEMMA 6.1 *Let H be a set of continuous functions defined on a metric space Z with the following properties :*

(i) H is a conic set (if $h \in H$, then $\lambda h \in H$ for all $\lambda > 0$);

(ii) for each triplet (ε, z, U), where $\varepsilon > 0$, $z \in Z$ and $U \in \mathcal{N}(z)$ there exists a function $h \in H$, which is the support to an Urysohn peak corresponding to this triplet that is:

$$h(z) > 1 - \varepsilon; \quad h(x) \leq 1 \text{ for all } x \in U; \quad h(x) \leq 0 \text{ for all } x \notin U. \tag{6.2.6}$$

Then the set $P(H, Z)$ of all H-convex functions contains the set LSC_H.

Proof: Let $f \in LSC_H$ and $z \in Z$. Since f is strictly minored by H we can find a function $\tilde{h} \in H$ such that $h + \tilde{h} \in H$ for all $h \in H$ and $f(x) > \tilde{h}(x)$ for all $x \in Z$. Let $g = f - \tilde{h}$. The function g is positive. Since f is lower semicontinuous and $\tilde{h}$ is continuous, g is lower semicontinuous. Assume that the equality

$$g(z) = \sup\{h(z) : \; h \in \mathrm{supp}(g, H)\} \tag{6.2.7}$$

has been already proved. Then we have

$$\begin{aligned} f(z) &= g(z) + \tilde{h}(z) = \sup\{h(z) : h \in \mathrm{supp}(g, H)\} + \tilde{h}(z) \\ &= \sup\{(h + \tilde{h})(z) : \; h \in \mathrm{supp}(g, H)\} \\ &= \sup\{h'(z) : \; h' = h + \tilde{h} \text{ where } h \in \mathrm{supp}(g, H)\}. \end{aligned}$$

Since $h + \tilde{h} \in H$ for all $h \in H$ we have that

$$f(z) = \sup\{h'(z) : h' \in \text{supp}(f, H)\}.$$

So we only need to prove the equality (6.2.7) for a lower semicontinuous positive function g and all $z \in Z$. Let $z \in Z$ and

$$\bar{g}(x) = \frac{g(x)}{g(z)}. \tag{6.2.8}$$

Let $\varepsilon > 0$. Since $\bar{g}$ is lower semicontinuous we can find a neighbourhood $U \in \mathcal{N}(z)$ such that $\bar{g}(x) > 1 - \frac{\varepsilon}{2}$ for all $x \in U$. Applying property (*ii*) we can find a function $\bar{h} \in H$ such that

$$\bar{h}(z) > 1 - \frac{\varepsilon}{2}; \quad \bar{h}(x) \le 1 \text{ for all } x \in U; \quad \bar{h}(x) \le 0 \text{ for all } x \notin U.$$

Let $h = (1 - \frac{\varepsilon}{2})\bar{h}$. We have

$$h(z) = \left(1 - \frac{\varepsilon}{2}\right)\bar{h}(z) > \left(1 - \frac{\varepsilon}{2}\right)^2 > 1 - \varepsilon. \tag{6.2.9}$$

On the other hand,

$$\bar{g}(x) > 1 - \frac{\varepsilon}{2} \ge h(x) \quad \text{for all } x \in U$$

and

$$\bar{g}(x) \ge 0 \ge h(x) \quad \text{for all } x \notin U.$$

Applying (6.2.8) we obtain the inequality

$$h(x) \le \frac{g(x)}{g(z)}, \qquad x \in Z. \tag{6.2.10}$$

Since H is a conic set it follows that the function

$$h'(x) = g(z)h(x) = g(z)\left(1 - \frac{\varepsilon}{2}\right)\bar{h}(x), \qquad x \in Z$$

belongs to the set H. The inequality (6.2.10) shows that $h'(x) \le g(x)$ for all $x \in Z$ and (6.2.9) implies that $h'(z) > g(z)(1 - \varepsilon)$. Thus the result follows. △

COROLLARY 6.1 *Let Z be a metric compact space. Let $H \subset C(Z)$ be a convex cone with a strongly negative function $\bar{h}$. Assume that for each triplet (ε, z, U) there exists $h \in H$, which is the support to an*

Urysohn peak, corresponding to this triplet. Then a function f is H-convex on Z if and only if f is lower semicontinuous. In other words, $P(H,Z) = LSC(Z)$.

Proof: A lower semicontinuous function is bounded from below on the compact space Z. Let $\bar{h} \in H$ be a strictly negative function. Then for each lower semicontinuous function f there exists a number $\lambda > 0$ such that $f(x) > \lambda\bar{h} := \tilde{h}$, hence (6.2.1) holds. Since H is a convex cone, (6.2.2) also holds. Thus every lower semicontinuous function on Z is minored by H. So the result follows from Lemma 6.1. △

LEMMA 6.2 *Let Z be a compact metric space and H be a set of continuous functions defined on Z with the following properties:*

(i) H is a conic set;

(ii) $h + \lambda\mathbf{1} \in H$ for each $h \in H$ and $\lambda < 0$ (here $\mathbf{1}(x) = 1$ for all $x \in Z$);

(iii) for each $z \in Z$ there exists a function $h \in H$ such that

$$h(z) = 0, \ h(x) < 0 \text{ for } x \neq z, \ h+\delta\mathbf{1} \in H \text{ for all small enough } \delta > 0. \tag{6.2.11}$$

Then a function f defined on Z is H-convex if and only if f is lower semicontinuous.

Proof: We shall apply Lemma 6.1. Since H is a conic set, then only condition (*ii*) in Lemma 6.1 should be checked. Let $z \in Z$ and $h \in H$ be a function such that $h(z) = 0$ and $h(x) < 0$ for $x \neq z$. For $\delta > 0$, let $h_\delta = h + \delta\mathbf{1}$ and $V_\delta = \{x : h_\delta(x) \geq 0\} = \{x : h(x) \geq -\delta\}$. The family $(V_\delta)_{\delta>0}$ consists of compact sets and $V_{\delta'} \subset V_{\delta''}$ if $\delta' < \delta''$. We also have $\bigcap_{\delta>0} V_\delta = \{x : h(x) \geq 0\} = \{z\}$. Applying well-known Blaschke Theorem (see for example [71]) we conclude that $V_\delta \to \{z\}$ in the Hausdorff metric as $\delta \to 0$. Therefore for an arbitrary neighbourhood U of the point z there exists $\delta > 0$ such that $V_\delta \subset U$. By the definition of V_δ we have $h_\delta(x) < 0$ for $x \notin U$. Let $\tilde{h} = (1/h_\delta(z))h_\delta$. Then

$$\tilde{h}(z) = 1, \quad \tilde{h}(x) \leq 1 \text{ for all } x \in Z, \quad \tilde{h}(x) < 0 \text{ for all } x \notin U.$$

Thus condition (*ii*) of Lemma 6.1 holds. Applying Lemma 6.1 we conclude that each lower semicontinuous function strictly minored by H is abstract convex with respect to H. Note that a lower semicontinuous function is bounded from below on the compact set Z. Since $\lambda\mathbf{1}+h \in H$ for all $h \in H$ and $\lambda < 0$, each lower semicontinuous function is strictly minored by H. Thus the desired result follows. △

6.2.3. Examples

Now let us give some examples.

EXAMPLE 6.2 Let Z be a subset of a Hilbert space X and let H be the set of all quadratic functions h of the form:

$$h(x) = a\|x\|^2 + [\ell, x] + c \qquad (x \in Z), \tag{6.2.12}$$

where $a \le 0$, $\ell \in X$, ($[\cdot,\cdot]$ denotes the inner product on $X \times X$), $c \in \mathbb{R}$. Clearly, H is a convex cone.

PROPOSITION 6.3 *Let H be as above. Then $P(H, X)$ coincides with the set of all lower semicontinuous functions p defined on Z, which are minored by H.*

Proof: Clearly H is a convex cone, therefore condition (i) of Lemma 6.1 holds. We now check that condition (ii) of this lemma holds as well. Take $z \in Z$ and a neighbourhood $U \in \mathcal{N}(z)$. Without loss of generality, assume that $U = \{x \in Z : \ \|x - z\| < \eta\}$. Consider the function h', defined by

$$h'(x) = -\alpha\|x - z\|^2 + \delta$$

with $\alpha, \delta > 0$ and $\delta < \alpha\eta^2$. It is easy to check that $h' \in H$. We have

$$h'(z) = \delta; \ h'(x) \le \delta \text{ for all } x \in U; \ h'(x) \le -\alpha\eta^2 + \delta < 0 \text{ for all } x \notin U. \tag{6.2.13}$$

Let $h = (1/\delta)h'$ and let ε be an arbitrary positive number. It follows from (6.2.13) that

$$h(z) = 1 > 1 - \varepsilon; \quad h(x) \le 1 \text{ for all } x \in U; \quad h(x) \le 0 \text{ for all } x \notin U.$$

Thus condition (ii) of Lemma 6.1 holds. Since H is a convex cone it follows (see Remark 6.1) that the function f is strictly minored by H if and only if f is minored by H. Thus the result follows from Lemma 6.1. $\triangle$

COROLLARY 6.2 *If Z is a compact subset of the space X, then $P(H, Z)$ coincides with the set $LSC(Z)$ of all lower semicontinuous functions defined on Z.*

EXAMPLE 6.3 The following proposition describes very thin supremal generators of the space of continuous functions $C(Z)$ defined on the n-dimensional compact set Z. Recall the following definition. Let X be a vector space. A cone $H \subset X$ is said to be spanned by elements $x_i \in X$, $i = 1, \dots n$ if $H = \{h : h = \sum_{i=0}^{n+1} \alpha_i y_i, \ \alpha_i \ge 0, i = 1, \dots, n\}$.

PROPOSITION 6.4 *Let Z be a compact subset of the space $X = \mathbb{R}^n$. Then there exists a supremal generator of the space $C(Z)$, which is a convex cone spanned by $(n+2)$ functions.*

Proof: Without loss of generality assume that $Z \subset \mathbb{R}^n_{++} = \operatorname{int} \mathbb{R}^n_+$. Let H be a convex cone spanned by the following functions:

$$h_0 = -\mathbf{1}, \quad h_i(x) = x_i, \ i = 1, \ldots, n, \quad h_{n+1} = -\| \cdot \|^2, \qquad (6.2.14)$$

where $\|x\|^2 = \sum_{k=1}^n x_k^2$. We shall verify that conditions in Lemma 6.2 hold for the cone H. Since $h_0 = -\mathbf{1} \in H$ and H is a convex cone we should only check that for each $z \in Z$ there exists a function $h \in H$ such that (6.2.11) holds. Let $z \in Z$ and

$$h(x) = -\|x - z\|^2 = -\|x\|^2 + \sum_{k=1}^n 2x_k z_k - \|z\|^2 = \sum_{i=0}^{n+1} \alpha_i h_i(x),$$

where $\alpha_0 = \|z\|^2$, $\alpha_1 = 2z_1, \ldots, \alpha_n = 2z_n, \alpha_{n+1} = 1$. Since $z \in \operatorname{int} \mathbb{R}^n_+$ it follows that $\alpha_k > 0$ for all $k = 0, 1, \ldots, n+1$, so $h \in H$. We also have

$$h(z) = 0, \quad h(x) < 0 \text{ for all } x \neq z, \quad h + \delta\mathbf{1} \in H \text{ if } \delta < \|z\|^2.$$

Thus (6.2.11) holds for the function h and desired result follows from Lemma 6.2. △

EXAMPLE 6.4 We now describe some supremal generators of the space $C([a, b])$, where $[a, b]$ is a segment, $-\infty < a < b < +\infty$. Let f and g be strictly concave differentiable functions defined on the segment $[a', b']$, where $a' < a, b' > b$. Moreover, let f be a decreasing function and g be an increasing function and let both f and g be positive on $[a, b]$. Then the convex cone H spanned by functions f, g and $-\mathbf{1}$ is a supremal generator of the space $C([a, b])$. We shall apply Lemma 6.2. Let $z \in [a, b]$. Note that $f'(z) < 0$, $g'(z) > 0$, so there exist numbers $\alpha > 0$ and $\beta > 0$ such that $\alpha f'(z) + \beta g'(z) = 0$. Since $f(z) > 0, g(z) > 0$ it follows that $\gamma := \alpha f(z) + \beta g(z) > 0$. Let $h = \alpha f + \beta g + \gamma(-\mathbf{1})$. We have

$$h(z) = 0; \qquad h'(z) = \alpha f'(z) + \beta g'(z) = 0.$$

Since f and g are strictly concave functions it follows that h is strictly concave as well. Hence z is the unique maximizer of h over $[a, b]$. Thus $h(x) < 0$ for all $x \neq z$. Let $\delta \in (0, \gamma)$. Then $h + \delta\mathbf{1} = \alpha f + \beta g + (\gamma - \delta)(-\mathbf{1}) \in H$. Hence all conditions in Lemma 6.2 hold and the desired result follows directly from this lemma.

EXAMPLE 6.5 We now demonstrate that a convex cone $H \subset \mathbb{R}^2$ is a supremal generator of $\mathbb{R}^2$ if and only if there exists $\delta > 0$ such that

$(-1, \delta) \in H$ and $(\delta, -1) \in H$. Indeed, we identify $\mathbb{R}^2$ with the space $C(Z)$, where Z is a compact metric space consisting of two points. Note that each supremal generator of $\mathbb{R}^2$ contains a vector with negative coordinates (otherwise the support set of an arbitrary vector with negative coordinates is empty). Applying Corollary 6.2 and Proposition 6.2 we easily conclude that the result holds.

EXAMPLE 6.6 Let Z be a metric space with the metric ρ and H be the set of all functions h of the form

$$h(x) = a\rho(x, z) - c \qquad (x \in Z)$$

with $z \in Z$, $a \leq 0$ and $c \in \mathbb{R}$. Then each lower semicontinuous function $f : Z \to \mathbb{R}_{+\infty}$ such that $\inf_{x \in Z} f(x) > -\infty$ is abstract convex with respect to the set H.

Note that H is a conic set and for each $h \in H$ the function $h + \lambda \mathbf{1}$ belongs to H for all $\lambda \in \mathbb{R}$. It follows from this assertion that f is strictly minored by H. Let $z \in Z$ and $U \in \mathcal{N}(z)$. Assume without loss of generality that $U = \{x : \rho(z, x) < \eta\}$. Let $h'(x) = -\alpha\rho(x, z) + \delta$ with $\delta < \alpha\eta$. Clearly, $h' \in H$. We have

$$h'(z) = \delta; \quad (\forall x \neq z)\, h'(x) < \delta; \quad (\forall x \notin U)\, h'(x) \leq -\alpha\eta + \delta < 0.$$

Let $h = (1/\delta)h'$. Then $h(z) = 1$, $h(x) \leq 1$ for $x \in U$ and $h(x) \leq 0$ for $x \notin U$. Thus all conditions in Lemma 6.1 hold and the desired result follows directly from this Lemma. $\triangle$

6.2.4. ICAR extensions of functions defined on the unit simplex

We now apply the results obtained in Subsection 6.2.2 to study ICAR functions (for definitions and properties of these functions see Section 3.3). The following statement holds:

PROPOSITION 6.5 *Let f be a lower semicontinuous function defined on the unit simplex $S = \{x \in \mathbb{R}^n_+ : \sum_i x_i = 1\}$. Then there exists an ICAR extension of f, that is an ICAR function $\tilde{f} : \mathbb{R}^n_+ \to \mathbb{R}_{+\infty}$ such that $\tilde{f}(x) = f(x)$ for all $x \in S$.*

Proof: Let H_S be the set of all functions h_S defined on the simplex S by $h_S(x) = \langle l, x \rangle - c$ with $l \in \mathbb{R}^n_+$, $c \in \mathbb{R}$ where $\langle l, x \rangle = \min_{i \in I_+(l)} l_i x_i$ and $I_+(l) = \{i : l_i > 0\}$. Clearly, conditions (i) and (ii) from Lemma 6.2 hold for the set H_S. Let us check that condition (iii) holds as well. Let $z \in S$. Consider the vector $l = 1/z$, where

$$\left(\frac{1}{z}\right)_i = \begin{cases} \frac{1}{z_i} & \text{if } z_i > 0; \\ 0 & \text{if } z_i = 0. \end{cases}$$

It is clear that $\langle l, z\rangle = 1$. Since

$$\sum_{i=1}^{n} x_i = \sum_{i=1}^{n} z_i$$

for $x \in S$, it follows that for $x \neq z$ there exists an index j such that $x_j < z_j$. Clearly, $j \in I_+(z) = I_+(l)$. Therefore,

$$\langle l, x\rangle = \min_{i \in I_+(z)} \frac{x_i}{z_i} < 1.$$

Consider the function h defined on S by $h(x) = \langle l, x\rangle - 1$. We have $h(z) = 0$ and $h(x) < 0$ for $x \neq z$. We also have $h + \delta\mathbf{1} \in H_S$ for all δ. Thus the condition (iii) from Lemma 6.2 holds. Let $f : S \to \mathbb{R}_{+\infty}$ be a lower semicontinuous function. It follows from Lemma 6.2 that there exists a set $U \subset H_S$ such that $f(x) = \sup_{h \in U} h(x)$ for all $x \in S$. Consider now the function $\tilde{f}$ defined on $\mathbb{R}^n_+$ by

$$\tilde{f}(x) = \sup\{h(x) : h \in U\}.$$

It follows from Proposition 3.5 that $\tilde{f}$ is an ICAR function. We also have $\tilde{f}(x) = f(x)$ for $x \in S$. △

Obviously there exists the greatest ICAR extension $\bar{f}$ of a lower semicontinuous functions function f defined on S. This extension is given by

$$\bar{f}(x) = \sup\{h(x) : h \in \operatorname{supp}(f, H_S)\}.$$

6.3. SUPREMAL GENERATORS FOR SPACES OF HOMOGENEOUS FUNCTIONS

6.3.1. Overview

In this section we apply the results obtained in Section 6.2 to study supremal generators for the space of positively homogeneous functions of degree one and for the space of symmetric positively homogeneous of degree two functions. We use the results, based on an application of the *supports to Urysohn peak* functions. This technique allows us to find very thin sets that serve as supremal generators for spaces of positively homogeneous functions and leads to simple and transparent proofs.

In particular, for a Banach space X with a locally uniformly convex norm $\|\cdot\|$, we show that the set

$$H = \{h : (\forall x \in X)\ h(x) = -a\|x\| + l(x),\ a \geq 0, l \in X^*\}$$

is a supremal generator of the set $PH_1(X)$ of all lower semicontinuous positively homogeneous of degree one functions bounded from below on the unit ball. Therefore, the set

$$H_* = \{h : (\forall x \in X)\ h(x) = -a\|x\|^2 + \|x\| \cdot l(x),\ a \geq 0, l \in X^*\}$$

is a supremal generator of the set $PH_2(X)$ of all lower semicontinuous positively homogeneous of degree two functions bounded from below on the unit ball. At the same time the set

$$H_s = \{h : (\forall x \in X)\ h(x) = -a\|x\|^2 + l(x)^2,\ a \geq 0, l \in X^*\} \quad (6.3.1)$$

is a supremal generator of the set $PHS_2(X)$ of all lower semicontinuous symmetric positively homogeneous of degree two functions bounded from below on the unit ball. Main results of this subsection can be found in [142].

6.3.2. Preliminaries

Let X be a Banach space and let X^* denote the conjugate space. We need the following definitions (see, for example, [40]).

DEFINITION 6.4 Let C be a closed convex subset of a Banach space X. A point $z \in C$ is called *strongly exposed* if there exists $l \in X^*$ such that $\sup_{x \in C} l(x) = l(z) = 1$ and $\sup_{x \in C, \|x-z\| \geq \varepsilon} l(x) < 1$ for each $\varepsilon > 0$.

DEFINITION 6.5 Let X be a Banach space equipped with a norm $\|\cdot\|$. The norm $\|\cdot\|$ is called *locally uniformly convex* if for each z with $\|z\| = 1$ and each $\varepsilon > 0$ we have

$$\sup_{\|x\|=1, \|x-z\| \geq \varepsilon} \frac{\|x+z\|}{2} < 1.$$

If the norm $\|\cdot\|$ is locally uniformly convex, then each element z of the unit sphere $Z = \{x : \|x\| = 1\}$ is a strongly exposed point of the unit ball B. Indeed, let $\varepsilon > 0$ and

$$\sup_{\|x\|=1, \|x-z\| \geq \varepsilon} \frac{\|x+z\|}{2} = \eta < 1.$$

Let $l \in \partial\|\cdot\|(x)$, that is, $\|l\| = l(x) = 1$. Then for each $x \in Z$ with $\|x - z\| \geq \varepsilon$ we have $l(x+z) \leq \|l\|\|x+z\| \leq 2\eta$. Therefore,

$$\sup_{x \in B, \|x-z\| \geq \varepsilon} l(x) \leq 2\eta - l(z) = 2\eta - 1 < 2 - 1 = 1.$$

It is well known (see, for example, [40]) that each reflexive Banach space can be supplied with an equivalent locally uniformly convex norm.

Let $k \in \mathbb{R}$. A function $f : X \to \mathbb{R}_{+\infty}$ is called *positively homogeneous of degree k* if $f(\lambda x) = \lambda^k f(x)$ for all $x \in X$. A function f is called *symmetric (even)* if $f(x) = f(-x)$ for all $x \in X$. We shall use the following notation:

$PH_k(X)$ is the set of all lower semicontinuous positively homogeneous of degree k functions $f : X \to \mathbb{R}_{+\infty}$ such that $\inf_{\|x\|=1} f(x) > -\infty$.

$PHS_2(X)$ is the subset of $PH_2(X)$ consisting of all symmetric functions.

6.3.3. Homogeneous functions of degree one

We shall apply results from Section 6.2 to study supremal generators of the space $PH_1(X)$ of all lower semicontinuous positively homogeneous of degree one functions defined on a Banach space X and bounded from below on the unit ball $B = \{x \in X : \|x\| \leq 1\}$.

THEOREM 6.3 *Let a Banach space X contain a closed convex bounded set C with $0 \in \operatorname{int} C$ such that each boundary point of C is also a strictly exposed point of C. Let μ_C be the Minkowski gauge of the set C, that is, $\mu_C(x) = \inf\{\lambda > 0 : x \in \lambda C\}$. Then the cone*

$$H = \{h : h = -a\mu_C + l, \quad a > 0, l \in X^*\} \tag{6.3.2}$$

is a supremal generator of the set $PH_1(X)$.

Proof: Let Z be the boundary of C and $\bar{H}$ be the set of all functions $\bar{h}$ defined on Z by the formula $\bar{h}(x) = -a + l(x)$ with $a > 0$ and $l \in X^*$. Clearly, $\bar{H}$ is a convex cone which contains all functions $\tilde{h}(x) = h(x) - \lambda\mathbf{1}$ with $h \in \bar{H}$ and $\lambda > 0$. Note that a positively homogeneous of degree one function p is bounded from below on the unit ball if and only if p is bounded from below on the set C and if and only if p is bounded from below on the set Z. If $f : Z \to \mathbb{R}_{+\infty}$ is a function bounded from below, then f is strictly minored by H. Let us check that condition (ii) in Lemma 6.1 holds for $\bar{H}$, that is, for each triplet (ε, z, U) there exists a support to an Urysohn peak corresponding to this triplet. Let $z \in Z$, $\varepsilon > 0$ and $U = \{x : \|x - z\| < \varepsilon\}$. Since z is a strongly exposed point it follows that there exist $l \in X^*$ and $\delta > 0$ such that $l(z) = 1$, $l(x) \leq 1$ for $x \in Z$ and $l(x) < 1 - \delta$ for $x \in Z \setminus U$. Let $\tilde{h}(x) = -(1-\delta) + l(x)$. Then $\tilde{h} \in \bar{H}$ and

1) $\tilde{h}(z) = \delta > 0$;

2) $\tilde{h}(x) - \tilde{h}(z) = l(x) - l(z) \leq 0$ for all $x \in Z$;

3) $\tilde{h}(x) < 0$ for $x \in Z \setminus U$.

Let $\bar{h} = (1/\delta)\tilde{h}$. Then $\bar{h}(z) = 1$, $\bar{h}(x) \leq 1$ for all $x \in U$ and $\bar{h}(x) \leq 0$ for all $x \notin U$. Thus condition (ii) in Lemma 6.1 holds. Applying Lemma 6.1 we conclude that $\bar{H}$ is a supremal generator of the space of all lower semicontinuous functions $f : Z \to \mathbb{R}_{+\infty}$ bounded from below. Thus for an arbitrary $f \in PH_1(X)$ and $z \in Z$ we have

$$f(z) = \sup_{\bar{h} \in \bar{H}, \bar{h} \leq f} \bar{h}(z). \tag{6.3.3}$$

Let $h \in H$. Since the restriction $\bar{h}$ of h to Z belongs to $\bar{H}$ and the inequality $h(x) \leq f(x)$ for all $x \in X$ is equivalent to the inequality $\bar{h}(z) \leq f(z)$ for all $z \in Z$ it follows that

$$f(z) = \sup_{h \in H, h \leq f} h(z) \text{ for all } z \in Z. \tag{6.3.4}$$

Since the functions f and $h \in H$ are positively homogeneous of degree one, it follows that (6.3.4) holds also for all $x \in X$. △

REMARK 6.3 We can consider the conjugate space X^* as a subspace of $PH_1(X)$. This subspace is a supremal generator of the set $P \subset PH_1(X)$ consisting of all lower semicontinuous sublinear functions $p : X \to \mathbb{R}_{+\infty}$ (see Example 6.1). Theorem 6.3 shows that if we add only one ray $\{-a\mu_C : a > 0\}$ to X^*, then we get a supremal generator of the set $PH_1(X)$ of all lower semicontinuous positively homogeneous functions bounded from below on the unit ball.

REMARK 6.4 Let L be a w^*-closed half-space of the conjugate space X^*. Then the set

$$H^L = \{h : h = -a\mu_C + l,\ a \geq 0,\ l \in L\}$$

is not a supremal generator of the space $PH_1(X)$.

Indeed, there exist $x_0 \in X$ and $c \in \mathbb{R}$ such that $L = \{l \in X^* : l(x_0) \leq c\}$. We have that $h(x_0) = -a\mu_C(x_0) + l(x_0) \leq c$ for all $h \in H^L$ and hence $\sup_{h \in H^L} h(x_0) \leq c$. Thus the cone H defined by (6.3.2) is a minimal supremal generator in the following sense: for each half-space L of X^*, the subset H^L of H is not a supremal generator.

REMARK 6.5 Let $(X, \|\cdot\|)$ be a reflexive Banach space and let $\|\cdot\|_0$ be a locally uniformly convex norm equivalent to the norm $\|\cdot\|$. Then the set

$$H = \{h : h(x) = -a\|x\|_0 + l(x);\ a \geq 0,\ l \in X^*\}$$

is a supremal generator of the space $PH_1(X)$.

REMARK 6.6 Let a space X, a set C and a cone H be as in Theorem 6.3. Let k be a real number. Then the cone

$$H_*^k = \{h_* : h_*(x) = \|x\|^{k-1}h(x),\ h \in H\}$$

is a supremal generator of the space $PH_k(X)$. Indeed, $f_* \in PH_k(X)$ if and only if $f_*(x) = \|x\|^{k-1}f(x)$ with $f \in PH_1(X)$.

Consider now the space $PHS_2(X)$ of all lower semicontinuous positively homogeneous of degree two symmetric functions. Let $f \in PHS_2(X)$ and $h_*(x) = -a\|x\|^2 + l(x)\|x\| \leq f(x)$ for all $x \in X$. Then $h_*(-x) = -a\|x\|^2 - l(x)\|x\| \leq f(-x) = f(x)$ and

$$f(x) \geq -a\|x\|^2 + |l(x)| \cdot \|x\| \text{ for all } x \in X.$$

Let H^2_{sm} be the set of all functions h of the form $h(x) = -a\|x\|^2 + |l(x)| \cdot \|x\|$ with $a \geq 0$ and $l \in X^*$. Since

$$-a\|x\|^2 + |l(x)| \cdot \|x\| \geq -a\|x\|^2 + l(x)\|x\| \text{ for all } x \in X$$

and

$$f(x) = \sup\{h_*(x) : h_* \in H^2_*,\ h_* \leq f\}$$

it follows that H^2_{sm} is a supremal generator of the set $PHS_2(X)$.

We will describe more interesting supremal generators of $PHS_2(X)$ in the next subsection.

6.3.4. Symmetric positively homogeneous functions of degree two

In order to describe some supremal generators of $PHS_2(X)$ we will present a symmetric version of Lemma 6.1.

LEMMA 6.3 *Let Z be a symmetric subset of a Banach space. Let H be a conic set of continuous functions defined on Z with the following property: for an arbitrary triplet (ε, z, U), where $\varepsilon > 0$, $z \in Z$ and U is a symmetric neighbourhood of zero, there exists a function $h \in H$ such that*

$$h(z) = h(-z) > 1 - \varepsilon \tag{6.3.5}$$

and

$$h(x) \leq 1 \text{ for all } x \in Z; \qquad h(x) \leq 0 \text{ for all } x \notin (z + U) \cup (-z + U)). \tag{6.3.6}$$

Then a symmetric lower semicontinuous function, strictly minored by H is abstract convex with respect to H.

Proof: Let f be a symmetric lower semicontinuous function minored by H and $z \in Z$. Since f is minored by H it follows that there exists a function $\tilde{h} \in H$ such that $f(x) > \tilde{h}(x)$ for all $x \in Z$. Let $g = f - \tilde{h}$. The function g is positive on Z. Since f is lower semicontinuous and $\tilde{h}$ is continuous, g is lower semicontinuous. Assume that the equality

$$g(z) = \sup\{h(z): \ h \in H, h \leq f\} \tag{6.3.7}$$

has already been proved. Then we have

$$\begin{aligned} f(z) &= g(z) + \tilde{h}(z) = \sup\{(h+\tilde{h})(z): \ h \in H, h \leq f\} \\ &= \sup\{h'(z): \ h' = h + \tilde{h}, \ h \in H, \ h \leq g\}. \end{aligned}$$

Since $h + \tilde{h} \in H$ for all $h \in H$ (by the definition of a strict minoration) we have

$$f(x) = \sup\{h'(z) : h' \in H, h' \leq f\}.$$

So we only need to prove equality (6.3.7) for a lower semicontinuous positive function g. We now consider the function

$$\bar{g}(x) = \frac{g(x)}{g(z)}. \tag{6.3.8}$$

Let $\eta > 0$. Since $\bar{g}$ is lower semicontinuous and symmetric and $g(z) = g(-z) = 1$ we can find a symmetric neighbourhood U of zero such that $\bar{g}(x) > 1 - \frac{\eta}{2}$ for all $x \in (z+U) \cup (-z+U)$. It follows from (6.3.5) and (6.3.6) that there exists a function $\tilde{h} \in H$ such that

$$\tilde{h}(z) = \tilde{h}(-z) > 1 - \frac{\eta}{2}$$

and

$$\tilde{h}(x) \leq 1 \text{ for all } x \in Z; \qquad \tilde{h}(x) \leq 0 \text{ for all } x \notin (z+U) \cup (-z+U).$$

Let $h = (1 - \frac{\eta}{2})\tilde{h}$. We have that

$$h(z) = h(-z) = (1 - \frac{\eta}{2})\tilde{h}(z) > \left(1 - \frac{\eta}{2}\right)^2 > 1 - \eta. \tag{6.3.9}$$

Also

$$\bar{g}(x) > 1 - \frac{\eta}{2} \geq h(x) \text{ for all } x \in (z+U) \cup (-z+U)$$

and

$$\bar{g}(x) \geq 0 \geq h(x) \text{ for all } x \notin (z+U) \cup (-z+U).$$

Applying (6.3.8) we conclude that

$$h(x) \leq \frac{g(x)}{g(z)} \text{ for all } x \in Z. \tag{6.3.10}$$

Since H is a conic set it follows that $h' \in H$, where

$$h'(x) = g(z)h(x) = g(z)\left(1 - \frac{\eta}{2}\right)\tilde{h}(x), \qquad x \in Z.$$

It follows from (6.3.10) and (6.3.9) respectively that $h'(x) \leq g(x)$ for all $x \in Z$ and $h'(z) > g(z)(1-\eta)$. Thus the desired result follows. △

Let X be a Banach space and let H_s be the set of all functions h defined on X by

$$h(x) = -a\|x\|^2 + l(x)^2, \tag{6.3.11}$$

with $l \in X^*$ and $a \leq 0$. If X is a Hilbert space, then a function h defined by (6.3.11) is the quadratic form of the symmetric operator $A : X \to X$ given by the formula $A(x) = -ax + [l, x]l$, where $[l, x]$ is the inner product of vectors l and x. Clearly, such operators form a very thin subset of the set $\mathcal{L}(X)$ of all symmetric bounded operators.

THEOREM 6.4 *Let X be a Banach space such that each point x with $\|x\| = 1$ is a strongly exposed point of the unit ball. Then the set H_s of all functions (6.3.11) with $l \in X^*$ and $a \leq 0$, is a supremal generator of the set $PHS_2(X)$.*

Proof: Let $\bar{H}_s$ be the set of all functions $\bar{h}$ defined on the unit sphere Z by

$$\bar{h}(x) = -a + l(x)^2$$

with $l \in X^*, a \leq 0$. Clearly, $\bar{h} \in \bar{H}_s$ if and only if there exists a function $h \in H_s$ such that $\bar{h}$ is the restriction of h to Z. We now check that conditions in Lemma 6.3 hold for the set $\bar{H}_s$. It follows immediately from the definition that $\bar{H}_s$ is a conic set. Let us check that for an arbitrary $z \in Z$ and $\varepsilon > 0$ there exists a function $\bar{h} \in \bar{H}_s$ such that $\bar{h}(z) = 1$ and (6.3.6) holds for this function with $U = \{x : \|x\| < \varepsilon\}$. Since z is a strongly exposed point we can find $l \in X^*$ and $\delta > 0$ such that $l(z) = 1$, $l(x) \leq 1$ for all $x \in Z$ and $l(x) < 1-\delta$ for $x \in Z \setminus (z+U)$. Let $x \notin -z+U$. Then $-x \notin z+U$ so $l(-x) < 1-\delta$. Thus if $x \notin (z+U) \cap (z-U)$ then

$$|l(x)| < 1 - \delta < 1 - \frac{\delta}{2}. \tag{6.3.12}$$

Let s be a function defined on X by

$$s(x) = -\left(1 - \frac{\delta}{2}\right)^2 \|x\|^2 + l(x)^2.$$

Clearly, $s \in PHS_2(X)$ and $s(x) = -(1-\frac{\delta}{2})^2 + l(x)^2$ for $x \in Z$.

We have for sufficiently small $\delta > 0$:

1) $s(z) = s(-z) = -1 + \delta - \frac{\delta^2}{4} + 1 > 0$;

2) $s(x) \leq s(z)$ for all $x \in Z$;

3) If $x \in Z$ and the inequalities

$$\|z - x\| \geq \varepsilon \quad \text{and} \quad \|-z - x\| \geq \varepsilon \tag{6.3.13}$$

hold then $s(x) < 0$.

The latter follows immediately from (6.3.12) and the definition of s. Set $h(x) = s(x)/s(z)$. Since $s(z) > 0$, we conclude that $h \in H_s$, so the restriction of h to Z, $\bar{h}$, belongs to $\bar{H}_s$. It follows from 1) and 2) that $\bar{h}(x) \leq \bar{h}(z) = 1$ for all $x \in Z$. Applying 3) we can deduce that $\bar{h}(x) \leq 0$ if $\|z - x\| \geq \varepsilon$ and $\|-z-x\| \geq \varepsilon$.

Thus we can apply Lemma 6.3 to the set $\bar{H}_s$. So if f is a symmetric lower semicontinuous function defined on Z and minored by $\bar{H}_s$, then f is an abstract convex with respect to $\bar{H}_s$. Let $F : X \to \mathbb{R}_{+\infty}$ be a symmetric and positively homogeneous function of degree two, which is lower semicontinuous and bounded from below on Z. Let $\bar{F}$ be the restriction of F to the sphere Z and $\inf_{x \in S} \bar{F}(x) = m$. Clearly $\bar{F}$ is a lower semicontinuous symmetric function. The constant function $\tilde{h}(x) = m - 1$ defined on Z enjoys the following properties:

1) $\bar{F}(x) > \tilde{h}(x)$ for all $x \in Z$;

2) $\tilde{h} \in \bar{H}_s$ and $h + \tilde{h} \in \bar{H}_s$ for all $h \in H_s$.

It follows from Definition 6.2 that $\bar{F}$ is strictly minored by $\bar{H}_s$, so $\bar{F}(x) = \sup\{\bar{h}(x) : \bar{h} \in \bar{H}_s, \bar{h} \leq \bar{F}\}$ for all $x \in Z$. Since F as well as functions $h \in H_s$ are positively homogeneous of degree two, we have $F(x) = \sup\{h(x) : h \in H_s, h \leq F\}$ for all $x \in X$. $\triangle$

6.4. SOME APPLICATIONS OF SUPREMAL GENERATORS

6.4.1. Overview

Supremal generators can be applied in the study of various problems of mathematical analysis and topology. In particular, they are a convenient tool for some parts of the theory of approximation. Recall the notion of the Korovkin property. Let Z be a compact space. We say that a subspace $H \subset C(Z)$ possesses the *Korovkin property* if for each sequence of linear positive operators $A_n : C(Z) \to C(Z)$ such that $A_n h \to h$ for $h \in H$, we have $A_n f \to f$ for all $f \in C(Z)$. This property was explicitly introduced by P.P. Korovkin [96]. A typical example of a sequence of linear positive operators $A_n : C([a,b] \to C([a,b])$, such that $A_n f \to f$ for all $f \in C([a,b])$, is a sequence of polynomial operators $A_n(f)$, where $A_n(f)$ is a polynomial of degree n, which approximates the function f. The well-known example of such a sequence is represented by the so-called Bernstein polynomials. It was shown in [99, 100] that the Korovkin property is closely related to the supremal generation. In Subsection 6.4.2 we present some results from [99], however we shall only consider the pointwise convergence, which is equivalent to the convergence of sequences of functionals. The uniform convergence (that is the convergence in the Banach space $C(Z)$) can be obtained from the pointwise convergence (see [99, 100] for details).

Another application of the supremal generators is considered in Subsection 6.4.3. We show that some topological properties of a finite-dimensional compact space Z can be described in terms of supremal generators of the space $C(Z)$. This result can be found in [99, 100] as well.

6.4.2. Convergence of sequences of positive functionals

Consider the space $C(Z)$ of continuous functions defined on a metric space Z. We assume that $C(Z)$ is equipped with the natural order relation: $g_1 \geq g_2$ if and only if $g_1(x) \geq g_2(x)$ for all $x \in Z$.

Recall that a functional g defined on a vector space Y is called *odd* if $f(-y) = -f(y)$ for all $y \in Y$.

PROPOSITION 6.6 *Let a set H be a supremal generator of the space $C(Z)$. Let $x \in Z$ and T_n be a sequence of odd increasing finite functionals defined on $C(Z)$, such that* $\liminf T_n(h) \geq h(x)$ *for all $h \in H$. Then there exists* $\lim T_n(g) = g(x)$ *for all $g \in C(Z)$.*

Proof: Let h belong to the support set, supp(g, H), of a function $g \in C(Z)$. Since the functionals T_n are increasing and $h \leq g$, we have $T_n(h) \leq T_n(g)$ and so

$$h(x) \leq \liminf_{n \to +\infty} T_n(h) \leq \liminf_{n \to +\infty} T_n(g).$$

Since H is a supremal generator we conclude that

$$g(x) = \sup\{h(x) : h \in \operatorname{supp}(g, H)\} \leq \liminf_{n\to+\infty} T_n(g). \tag{6.4.1}$$

Consider now the function $-g$. The same argument shows that $-g(x) \leq \liminf_{n\to+\infty} T_n(-g)$. Since the functionals T_n are odd, we have

$$g(x) \geq -\liminf_{n\to+\infty}(-T_n(g)) = \limsup_{n\to+\infty} T_n(g). \tag{6.4.2}$$

The result follows directly from (6.4.1) and (6.4.2). △

A linear functional l defined on $C(Z)$ is called *positive* if $l(g) \geq 0$ for all $g \geq 0$. Clearly, a linear positive functional is increasing.

COROLLARY 6.3 *Let l_n be a sequence of linear positive functionals defined on $C(Z)$. Let H be a supremal generator of $C(Z)$ and $x \in Z$. Then the following assertions are equivalent:*

1) $\liminf_n l_n(h) \geq h(x)$ *for all* $h \in H$;

2) $\lim_n l_n(g) = g(x)$ *for all* $g \in C(Z)$.

Proof: Implication 2) $\implies$ 1) holds trivially. Implication 1) $\implies$ 2) follows from Proposition 6.6, since a positive linear functional is odd and increasing. △

COROLLARY 6.4 *Let l be a linear positive functional defined on $C(Z)$ and let H be a supremal generator of $C(Z)$ and $x \in Z$. Then the following assertions are equivalent:*

1′) $l(h) \geq h(x)$ *for all* $h \in H$;

2′) $l(g) = g(x)$ *for all* $g \in C(Z)$.

Corollary 6.4 follows immediately from Corollary 6.3 with $l_n = l$, $(n = 1, \ldots)$. Let H be a supremal generator of the space $C(Z)$. Then each function $g \in G$ is minored by H. In this case H is a *minorant* set, that is, for each $g \in C(Z)$ there exists a function $h \in H$ such that $h \leq g$.

In the rest of this subsection we assume that Z is a metric compact space. Then each function $f \in C(Z)$ is bounded. It is easy to check that a conic set H is minorant if and only if H contains a strictly negative function.

Let H be a convex minorant cone. We now demonstrate that if assertions 1′) and 2′) from Corollary 6.4 are equivalent for all $x \in Z$ and all positive linear functions l, then H is a supremal generator. Thus the supremal generation is a necessary and sufficient condition for the effect described in Corollary 6.4.

We shall need the following simple assertion. If p is a sublinear functional defined on a vector space Z, then

$$p(-x) \geq -p(x) \text{ for all } x \in Z. \tag{6.4.3}$$

Indeed, it follows from sublinearity of p that $0 = p(0) = p(x + (-x)) \leq p(x) + p(-x)$.

Theorem 6.5 *Let Z be a metric compact set and let $H \subset C(Z)$ be a convex cone such that assertions 1′) and 2′) from Corollary 6.4 are equivalent for all $x \in Z$ and all linear positive functionals l. Then H is a supremal generator of the space $C(Z)$.*

Proof: Assume on the contrary that H is not a supremal generator. Then there exists a point $x' \in Z$ and a function $g' \in C(Z)$ such that

$$\sup\{h(x') : h \in \text{supp}(g', H)\} < g'(x'). \tag{6.4.4}$$

Consider a functional q defined by

$$q(g) = \sup\{h(x') : h \in \text{supp}(g, H)\}, \qquad g \in C(Z).$$

Since H is a minorant set it follows that $q(g) > -\infty$ for all $g \in C(Z)$. We also have

$$q(x') \leq g(x') \text{ for all } g \in C(Z). \tag{6.4.5}$$

Therefore, q is a finite functional. Since H is a convex cone we have

$$\text{supp}(\lambda g, H) = \lambda \text{supp}(g, H) \qquad (\lambda > 0,\ g \in C(Z))$$

$$\text{supp}(g_1 + g_2, H) \supset \text{supp}(g_1, H) + \text{supp}(g_2, H) \qquad (g_1, g_2 \in C(Z)).$$

Hence

$$q(\lambda g) = \lambda q(g), \qquad \lambda > 0,\ g \in C(Z),$$

$$q(g_1 + g_2) \geq q(g_1) + q(g_2), \qquad g_1, g_2 \in C(Z).$$

Thus q is a finite superlinear functional. Let $p(g) = -q(-g)$. It follows from the superlinearity of q that p is sublinear. Consider the one-dimensional subspace $Y = \{\alpha g' : \alpha \in \mathbb{R}\}$ of the space $C(Z)$ and let $\tilde{l}$ be a linear functional defined on Y by $\tilde{l}(\alpha g') = \alpha p(g')$. We now verify that $p(g) \leq \tilde{l}(g)$ for all $g \in Y$. If $\alpha \geq 0$, then $\tilde{l}(\alpha g') = \alpha \tilde{l}(g') = \alpha p(g') = p(\alpha g')$. If $\alpha < 0$, then we have, by applying (6.4.3),

$$\tilde{l}(\alpha g') = \alpha p(g') = |\alpha|(-p)(g') \leq |\alpha| p(-g') = p(\alpha g').$$

Applying Hanh-Banach Theorem (see Theorem 6.1) we find a linear functional l defined on $C(Z)$ such that $l(g) \leq p(g)$ for all $g \in C(Z)$. We also have $l(g) = -l(-g) \geq -p(-g) = q(g)$. Let us verify that l is positive. In fact, if $g \geq 0$, then the support set $\operatorname{supp}(g, H)$ contains zero, so $l(g) \geq q(g) = \sup\{h(x) : h \in \operatorname{supp}(g, H)\} \geq 0$. Note that

$$q(h) = \sup\{\tilde{h}(x') : \tilde{h} \leq h, \tilde{h} \in H\} = h(x'), \qquad h \in H.$$

Hence $l(h) \geq h(x')$ for all $h \in H$. It follows from the definitions of q, g' and x' that

$$l(g') = \tilde{l}(g') = q(g') = \sup\{h(x') : h \in \operatorname{supp}(g', H)\} < g'(x'). \quad (6.4.6)$$

Consider the linear positive functional $\hat{\ell}$ defined by $\hat{\ell}(g) = g(x')$. It follows from (6.4.6) that $l(g') \neq \tilde{l}(g')$. Thus there exist two different linear positive functionals which are greater than or equal to $h(x')$ for all $h \in H$, which is impossible. △

6.4.3. The supremal rank of a compact set

In this subsection we shall demonstrate that the dimension of convex cones, which are supremal generators of the space $C(Z)$ of continuous functions defined on the compact Z, allows one to estimate a dimension of Z.

Let Z be a finite-dimensional compact space and $C(Z)$ be the space of continuous functions defined on Z. Then $C(Z)$ contains supremal generators, which are convex cones spanned by a finite number of functions. A finite family $(f_0, f_1, \dots, f_m)$ with $f_0(x) < 0$ for all $x \in Z$ is called the *supremal basis* of the space $C(Z)$ if the cone $\{\sum_{i=1}^m \alpha_i f_i : \alpha_i \geq 0\}$ is a supremal generator of $C(Z)$. The least number m such that there exists a supremal basis consisting of m functions is called the *supremal rank* of Z. We shall denote the supremal rank of Z by $\operatorname{sim}(Z)$.

PROPOSITION 6.7 *Let Z be a compact space homeomorphic to a subset of $\mathbb{R}^n$. Then $sim(Z) \leq n + 2$.*

Proof: If $Z \subset \mathbb{R}^n$ then (see Proposition 6.4) there exists a supremal generator spanned by $n + 2$ functions $f_0, \dots, f_{n+1}$ such that $f_0 = -1$. Hence $\operatorname{sim}(Z) \leq n+2$. Assume that a compact space Z is homeomorphic to a compact subset Z' of $\mathbb{R}^n$ and let $\psi : Z \to Z'$ be a corresponding homeomorphism. Consider an operator $A : C(Z') \to C(Z)$ defined by $(Af)(z) = f(\psi(z))$. Then A is a linear ordered isomorphism between spaces $C(Z')$ and $C(Z)$ and $A(\mathbf{1}) = \mathbf{1}$. It follows directly from the definition that a cone $H' \subset C(Z')$ is a supremal generator of $C(Z')$ if and only if $H = \{Ah' : h' \in H'\}$ is a supremal generator of $C(Z)$. Thus $\mathrm{sim}(Z)$ coincides with $\mathrm{sim}(Z')$, therefore $\mathrm{sim}(Z) \leq n+2$. △

We shall show in this subsection that the reverse assertion holds as well. If Z is a compact space and $\mathrm{sim}(Z) \leq n+2$, then Z is homeomorphic to a compact subset $Z' \subset \mathbb{R}^n$.

Let $H \subset C(Z)$, where Z is a compact space. Let $z_1, z_2 \in Z$ and

$$H_{z_1,z_2} = \{(h(z_1), h(z_2)) \in \mathbb{R}^2 : h \in H\}. \tag{6.4.7}$$

A set $H \subset C(Z)$ is said to be *two-separating* if, for each $z_1, z_2 \in Z$, the set H_{z_1,z_2} defined by (6.4.7) is a supremal generator of the space $\mathbb{R}^2$.

PROPOSITION 6.8 *A supremal generator H of $C(Z)$ is a two-separating set.*

Proof: Let $z_1, z_2 \in Z$. We need to check that the set H_{z_1,z_2} defined by (6.4.7) is a supremal generator of $\mathbb{R}^2$. Let $x = (x_1, x_2) \in \mathbb{R}^2$. There exists a continuous function f defined on Z such that $f(z_1) = x_1, f(z_2) = x_2$. Since H is a supremal generator it follows that

$$x_1 = f(z_1) = \sup\{h(z_1) : h \in \mathrm{supp}(f, H)\}$$

and

$$x_2 = f(z_2) = \sup\{h(z_2) : h \in \mathrm{supp}(f, H)\}.$$

Let $U = \{(h(z_1), h(z_2)) : h \in \mathrm{supp}(f, H)\}$. Then

$$x = (x_1, x_2) = \sup\{h(z_1), h(z_2) : h(z_1, h(z_2)) \in U\}.$$

Thus the result follows. △

Let $H \subset C(Z)$ be a convex cone. Then the set H_{z_1,z_2} is a convex cone as well. It follows from Example 6.5 that H_{z_1,z_2} is a supremal generator of $\mathbb{R}^2$ if and only if this cone contains points $(-1, \delta)$ and $(\delta, -1)$ with some positive $\delta > 0$. Thus a convex cone $H \subset C(Z)$ is a two-separating set if and only if for each pair $(z_1, z_2) \in Z$ there exists $\delta > 0$ such that $(-1, \delta) \in H_{z_1,z_2}$ and $(\delta, -1) \in H_{z_1,z_2}$.

PROPOSITION 6.9 *Let Z be a compact space such that the space $C(Z)$ contains a convex two-separating cone H, spanned by $n+2$ functions $f_0, f_1, \dots, f_{n+1}$ with $f_0(x) < 0$ for all $x \in Z$. Then Z is homeomorphic to a subset Z' of the space $\mathbb{R}^n$.*

Proof: Consider a mapping $\psi : Z \to \mathbb{R}^{n+1}$, where

$$\psi(z) = (f_0(z), f_1(z), \dots, f_n(z)).$$

Let $Z' = \psi(Z)$. Since H is a two-separating cone it follows that $\psi : Z \to Z'$ is a one-to-one correspondence. Indeed, assume that there exists two points $z_1, z_2 \in Z$ such that $\psi(z_1) = \psi(z_2)$. Then the cone H_{z_1,z_2} is a ray, which is impossible. Since $f_i \in C(Z)$, $(i = 0, 1, \ldots, n)$ it follows that ψ is a continuous mapping, hence a homeomorphism. Note that $0 \notin Z'$, since $f_0(z) < 0$ for all $z \in Z$.

Claim: *Each ray $R \subset \mathbb{R}^{n+1}$ starting from zero does not contain more than one point of the set Z'.*

Assume on the contrary that there exist points $z_1, z_2 \in Z$ and $\alpha > 0$ such that $\psi(z_1) = \alpha\psi(z_2)$, that is, $f_i(z_1) = \alpha f_i(z_2)$ for all $i = 0, 1 \ldots, n$. Then $(n + 1)$ points $(f_i(z_1), f_i(z_2)) \in \mathbb{R}^2$ are located on the straight line Q passing through the origin and the strictly negative element $(f_0(z_1), f_0(z_2))$. Consider the cone H_{z_1,z_2}. This cone is contained in the convex hull of the line Q and the ray R starting from zero and passing through the point $(f_{n+1}(z_1), f_{n+1}(z_2))$. Since Q cuts the interior of the negative orthant, the convex hull of Q and R does not contain this orthant, hence this convex hull does not contain a cone spanned by points $(-1, \delta)$ and $(\delta, -1)$ for an arbitrary δ. Thus the cone H_{z_1,z_2} is not a supremal generator of $\mathbb{R}^2$, which contradicts the definition of a two-separating set. The claim is proved.

Let $S_* = \{x \in \mathbb{R}^{n+1} : \|x\| = 1\}$ be the unit sphere (here $\|\cdot\|$ is the Euclidean norm). Consider the continuous mapping $\phi : Z \to S_*$ defined by

$$\phi(z) = \frac{\psi(z)}{\|\psi(z)\|}, \qquad z \in Z.$$

Let $Z'' = \phi(Z)$. Since each ray starting from zero does not contain more than one point $\psi(z)\,(z \in Z)$, it follows that $\phi : Z \to Z''$ is a one-to-one correspondence, hence a homeomorphism. Since $f_0(z) < 0$ for all $z \in Z$, Z'' is located in the hemisphere $\{x \in S_* : x_0 \leq 0\}$, so Z'' is homeomorphic to a subset of $\mathbb{R}^n$. $\triangle$

THEOREM 6.6 *Let Z be a compact space. Then* $\mathrm{sim}(Z) = n + 2$ *if and only if Z is homeomorphic to a subset of* $\mathbb{R}^n$ *and is not homeomorphic to a subset of* $\mathbb{R}^{n-1}$.

Proof: This follows directly from Proposition 6.9 and Proposition 6.7. $\triangle$

6.5. APPLICATION TO HADAMARD-TYPE INEQUALITIES

6.5.1. Overview

We begin with the *Principle of Preservation of Inequalities* ([99]).

PROPOSITION 6.10 *Let Y be a set of functions defined on a set Z and equipped with the natural order relation. Let H be a supremal generator of Y. Furthermore, let ψ be an increasing functional defined on Y and let $u \in Z$. Then*

$$(h(u) \leq \psi(h) \text{ for all } h \in H) \iff (f(u) \leq \psi(f) \text{ for all } f \in Y).$$

Proof: We have

$$\begin{aligned} f(u) &= \sup\{h(u) : h \in \operatorname{supp}(f, H)\} \leq \sup\{\psi(h) : h \in \operatorname{supp}(f, H)\} \\ &\leq \psi(\sup\{h : h \in \operatorname{supp}(f, H)\}) = \psi(f). \end{aligned}$$

△

Thus in order to establish that a certain inequality holds for a fairly large class of functions Y we only need to check this inequality for functions belonging to a supremal generator of Y. There exist very small supremal generators of very broad sets and these generators often consist of simple functions. For small sets of simple functions h, an inequality $h(u) \leq a(h)$ can be often checked 'manually'.

Let f be a convex function defined on the segment $[a, b]$ of the real line. Then the following inequality holds:

$$f\left(\frac{a+b}{2}\right) \leq 2\int_a^b f(x)dx \leq 2(f(a) + f(b)). \tag{6.5.1}$$

This inequality is known as the *Hadamard inequality* (see [45, 46] and references therein.) Its generalizations for some classes of nonconvex functions can be found in [45, 46] as well. We only examine here a generalization of the inequality from the left-hand side of (6.5.1). Such a generalization for convex functions defined on a convex compact subset of a vector topological space can be found in [128].

We shall study in this section various generalizations of the Hadamard inequality, using the principle of preservation of inequalities. In particular we consider Hadamard type inequalities for the following functions:

- convex functions (Subsection 6.5.2)

- ICAR functions (Subsection 6.5.5),
- quasiconvex functions and P-functions (Subsection 6.5.4).

The class of P-functions was introduced in [45]. A nonnegative function p defined on a segment S of the real line $\mathbb{R}$ is said to be a *P-function* if

$$p(\lambda x + (1-\lambda)y) \leq p(x) + p(y), \quad x, y \in S, \ 0 \leq \lambda \leq 1.$$

The class $\mathcal{P}(S)$ of P-functions defined on a segment S is extremely broad. For example, as it will be shown, if h is a bounded function defined on S and c is a sufficiently large number, then the function $f(x) = h(x) + c$ belongs to $\mathcal{P}(S)$. We shall study properties of P-functions and links between P-functions and nonnegative quasiconvex functions. It is well known that the sum of quasiconvex functions is not necessarily quasiconvex. The convex hull of the set of all quasiconvex functions defined on a segment S is a very large set containing, for example, all functions of bounded variation. The convex hull of the set $Q_+(S)$ of all *nonnegative* quasiconvex functions defined on S is also very broad, but one can find nonnegative functions of bounded variation which do not belong to this set. The pointwise supremum of a family of elements of $Q_+(S)$ is again an element of $Q_+(S)$. One of important problems of the theory of quasiconvex functions is to describe the least convex cone containing $Q_+(S)$ which is closed in the topology of pointwise convergence and contains pointwise suprema of all families of its elements. It will be proved that this cone coincides with the set $\mathcal{P}(S)$ of all P-functions defined on S.

Our approach is based on the description of small supremal generators of the sets under consideration. These supremal generators allow us to obtain Hadamard type inequalities for quasiconvex functions and P-functions. Our results demonstrate that from the validity of a Hadamard-type inequality with respect to an atomless measure for the class of quasiconvex functions, the validity of this inequality for a much more general class of P-functions follows. Main results of Subsection 6.5.3 and Subsection 6.5.4 can be found in [121].

6.5.2. Hadamard-type inequalities for convex functions

Let I be a finite set of indices and $Z \subset \mathbb{R}^I$ be a convex compact set. Let μ be a Borel nonnegative normed measure defined on Z, that is, μ is a σ-additive function defined on the σ-algebra of all Borel subsets of Z, such that $\mu(e) \geq 0$ for each Borel set $e \subset Z$ and $\mu(Z) = 1$.

Consider the point

$$\bar{z} = \int_Z x d\mu. \tag{6.5.2}$$

Note that $\int_Z x d\mu$ is a limit of the sums of the form $\sigma_k = \sum_k x_k \mu(e_k)$, where e_k is a family of disjoint subsets of Z such that $\bigcup_k e_k = Z$. Since Z is a convex compact set, $\mu(e_k) \geq 0$ and $\sum_k \mu(e_k) = 1$ it follows that $\sigma_k \in Z$, hence $\bar{z} \in Z$. The equality (6.5.2) is equivalent to $\bar{z}_i = \int_Z x_i d\mu$ for all $i \in I$. Let $l \in \mathbb{R}^I$. Then

$$[l, \bar{z}] = [l, \int_Z x d\mu] = \sum_{i \in i} l_i \int_Z x_i d\mu = \int_Z \sum_{i \in I} l_i x_i d\mu = \int_Z [l, x] d\mu.$$

Consider now an affine function h defined on Z. Then there exist $l \in \mathbb{R}^I$ and $c \in \mathbb{R}$ such that $h(x) = [l, x] - c$ for all $x \in Z$, hence

$$h(\bar{z}) = [l, \bar{z}] - c = \int_Z [l, x] d\mu - c \int_Z d\mu = \int_Z h(x) d\mu. \tag{6.5.3}$$

PROPOSITION 6.11 *Let $Z \subset \mathbb{R}^I$ be a convex compact set and let $\bar{z}$ be the point defined by (6.5.2). Then $f(\bar{z}) \leq \int_Z f(x) d\mu$ for each lower semicontinuous convex function f defined on Z.*

Proof: Note that the set H of all affine functions defined on Z is a supremal generator of the set of all lower semicontinuous convex functions. Thus the result follows from (6.5.3) and Proposition 6.10 (the principle of preservation of inequalities). △

6.5.3. Quasiconvex functions and P-functions

First, we recall the following definition.

DEFINITION 6.6 Let $S \subset \mathbb{R}$ be a segment A function $f : S \to \mathbb{R}_{+\infty}$ is called *quasiconvex* if

$$f(\lambda x + (1 - \lambda) y) \leq \max(f(x), f(y)) \text{ for all } x, y \in S \text{ and } 0 \leq \lambda \leq 1.$$

Note that a function f is quasiconvex if and only if its lower level sets $S_c(f) = \{x : f(x) \leq c\}$ are segments for all $c \in \mathbb{R}$. It follows directly from the definition that convex functions, increasing functions and decreasing functions are quasiconvex.

Quasiconvex functions can be defined in a much more general environment. Some results related to quasiconvex functions defined on Banach spaces will be discussed in Subsection 7.3.5. Here we shall consider only nonnegative quasiconvex functions defined on a segment S. Denote by $Q_+(S)$ the set of all such functions. We shall omit the symbol S, if $S = \mathbb{R}$, so the set of all nonnegative quasiconvex functions will be denoted by Q_+. We now indicate some properties of $Q_+(S)$ (see Subsection 7.3.5 for details.) It is easy to check that $Q_+(S)$ is a conic set, that is $(q \in Q_+(S), \lambda > 0) \implies \lambda q \in Q_+(S)$, however $Q_+(S)$ is not a convex set. The set $Q_+(S)$ is an upper complete semilattice: if $g_\alpha \in Q_+(S)$, $(\alpha \in A)$, where A is an arbitrary index set and $q(x) := \sup_{\alpha \in A} q_\alpha(x)$ then $q \in Q_+(S)$.

We now describe a supremal generator for the set Q_+. The following proposition is a particular case of well-known results from quasiconvex analysis (see, for example, [151] for a version of this result). Versions of these results in a more general environment can be found in Subsections 7.3.5, see Theorem 7.8 and Theorem 7.9.

PROPOSITION 6.12 Let H_1 be a set of two-step functions h of the form

$$h(x) = \begin{cases} c & \text{if } vx \geq d; \\ 0 & \text{if } vx < d \end{cases}$$

with $v \in \{1, -1\}$, $c \geq 0$, $d \in \mathbb{R}$. Then H_1 is a supremal generator of the set Q_+ of all nonnegative quasiconvex functions.

Proof: Clearly $H_1 \subset Q_+$. Since the pointwise supremum of a family of quasiconvex functions is again a quasiconvex function, each H_1-convex function belongs to Q_+. Consider now a function $q \in Q_+$ and the family of level sets $S_c = \{x : q(x) \leq c\}$ with $c \geq 0$. Since q is quasiconvex it follows that S_c is a segment for each $c \geq 0$. Let $x_0 \in \mathbb{R}$. Assume for the sake of definiteness that $q(x_0) > 0$ and let $0 < c < q(x_0)$. Since $x_0 \notin S_c$ it follows that there exists $v \in \{-1, 1\}$ such that $vx_0 > vx$ for all $x \in S_c$. Let $d = vx_0$. Then $\bar{c} := \inf_{vx \geq d} q(x) \leq q(x_0)$. Since $S_c \subset \{x : vx < d\}$ it follows that the inequality $vx \geq d$ implies $q(x) > c$. Hence $c \leq \bar{c}$. Let

$$h(x) = \begin{cases} \bar{c} & \text{if } vx \geq d; \\ 0 & \text{if } vx < d. \end{cases}$$

Clearly $h \leq q$ and $h(x_0) \geq \bar{c} \geq c$. Since c is an arbitrary number such that $0 < c < q(x_0)$ it follows that $q(x) = \sup\{h(x) : h \in \text{supp}(f, H_1)\}$ for all $x \in \mathbb{R}$. △

REMARK 6.7 Let H_2 be the set of all two-step functions h given by

$$h(x) = \begin{cases} c & \text{if } vx \geq d; \\ c' & \text{if } vx < d. \end{cases} \tag{6.5.4}$$

with $c \geq c' \geq 0$, $v \in \{1, -1\}$ and $d \in \mathbb{R}$. Since $H_1 \subset H_2 \subset Q_+$ it follows that H_2 is a supremal generator of Q_+ as well.

Definition 6.7 Let $S \subset \mathbb{R}$ be a segment. A nonnegative function p defined on S is called a *function of P-type* or simply a *P-function* if for all $x, y \in S$ and $0 \leq \lambda \leq 1$ we have

$$p(\lambda x + (1-\lambda)y) \leq p(x) + p(y).$$

Denote by $\mathcal{P}(S)$ the set of all P-functions $f : S \to \bar{\mathbb{R}}_+$. If $S = \mathbb{R}$ then we omit the symbol S, so the set of all P-functions defined on $\mathbb{R}$ will be denoted by $\mathcal{P}$. It follows immediately from Definition 6.6 and Definition 6.7 that $Q_+(S) \subset \mathcal{P}(S)$.

Let us point out some properties of a function $f \in \mathcal{P}(S)$.

1) If $\lambda_i > 0$ $(i = 1, \dots, m)$ and $\sum_{i=1}^m \lambda_i = 1$, then $f(\sum_{i=1}^m \lambda_i x_i) \leq \sum_i f(x_i)$. This can be proved by induction.

2) The set $\operatorname{dom} f = \{x \in S : f(x) < +\infty\}$ is a segment. Indeed, let $x_- = \inf \operatorname{dom} f$, $x_+ = \sup \operatorname{dom} f$. Suppose $x_- < x < x_+$. Then there exist points $x_1, x_2 \in \operatorname{dom} f$ such that $x \in (x_1, x_2)$. It follows from the definition of $\mathcal{P}$ that $f(x) < +\infty$, that is, $x \in \operatorname{dom} f$.

3) If the set $\{x \in S : f(x) = 0\}$ is nonempty, then it is a segment.

Let $S \subset \mathbb{R}$ be a segment, $S \neq \mathbb{R}$. For each function f defined on S, consider its extension $f_{+\infty}$ defined by

$$f_{+\infty}(x) = \begin{cases} f(x) & \text{if } x \in S; \\ +\infty & \text{if } x \notin S. \end{cases} \tag{6.5.5}$$

Clearly, $f \in \mathcal{P}(S)$ if and only if $f_{+\infty} \in \mathcal{P}$.

Let S be a segment. It is easy to check that the class $\mathcal{P}(S)$ enjoys the following properties.

1) $\mathcal{P}(S)$ is a convex cone: if $f_1, f_2 \in \mathcal{P}(S)$, then $f_1 + f_2 \in \mathcal{P}(S)$; if $\lambda > 0, f \in \mathcal{P}(S)$, then $\lambda f \in \mathcal{P}(S)$;

2) $\mathcal{P}(S)$ is a complete upper semilattice: if A is an arbitrary index set, $f_\alpha \in \mathcal{P}(S)$ for each $\alpha \in A$ and $f(x) = \sup_{\alpha \in A} f_\alpha(x)$ for all $x \in S$, then $f \in \mathcal{P}(S)$.

3) $\mathcal{P}(S)$ is closed under pointwise convergence.

The class $\mathcal{P}(S)$ is extremely broad. The following proposition confirms this statement.

PROPOSITION 6.13 *Let h be a bounded function defined on S. Then there exists a number $c > 0$ such that the function $h + c\mathbf{1} \in \mathcal{P}(S)$. (Here $\mathbf{1}(x) = 1$ for all $x \in S$.)*

Proof: Let

$$c = \sup_{x,y \in S,\ z \in [x,y]} (h(z) - h(x) - h(y)).$$

We have for each $x, y \in S$ and $\alpha, \beta \in [0,1]$ with $\alpha + \beta = 1$:

$$h(\alpha x + \beta y) \leq h(x) + h(y) + c.$$

Let $f(x) = h(x) + c$. Then

$$f(\alpha x + \beta y) = h(\alpha x + \beta y) + c \leq (h(x) + c) + (h(y) + c) = f(x) + f(y),$$

that is, $f \in \mathcal{P}(S)$. △

We now describe a small supremal generator of the set $\mathcal{P}$. Applying this generator and the extension defined by (6.5.5), we can easily describe a supremal generator of the class $\mathcal{P}(S)$ for a segment $S \subset \mathbb{R}$.

Let T be the set of all collections $t = \{u; c_1, c_2\}$ with $u \in \mathbb{R}$ and nonnegative c_1, c_2. For $t = \{u; c_1, c_2\} \in T$, consider the function h_t defined on $\mathbb{R}$ by

$$h_t(x) = \begin{cases} c_1 & \text{if } x < u; \\ c_1 + c_2 & \text{if } x = u; \\ c_2 & \text{if } x > u. \end{cases} \tag{6.5.6}$$

It follows directly from (6.5.6) that $h_t \in \mathcal{P}$ for all $t \in T$. Let H be the set of all functions of the form h_t with $t \in T$. Clearly H is a conic set. The following statement describes a certain extremal property of elements $h \in H$.

PROPOSITION 6.14 *Let $h \in H$, $h = h_t$ with $t = \{u; c_1, c_2\}$. If $f \in \mathcal{P}$, $f \leq h$ and $f(u) = h(u)$, then $f = h$.*

Proof: Let $f \in \mathcal{P}$, $f \leq h$ and $f(u) = h(u)$. For each point $y < u$ there exist a point $z > u$ and a number $\lambda \in (0,1)$ such that $u = \lambda y + (1 - \lambda) z$. Since $f \in \mathcal{P}$ it follows that

$$c_1 + c_2 = h(u) = f(u) \leq f(y) + f(z) \leq h(y) + h(z) = c_1 + c_2.$$

Hence $f(y) + f(z) = c_1 + c_2$. Since $f(y) \leq c_1$ and $f(z) \leq c_2$ it follows that $f(y) = c_1$. In the same manner we can show that $f(v) = c_2$ for an arbitrary point $v > u$. △

Let $\bar{H}$ be the set of all pointed functions, that is, functions h of the form

$$h(x) = \begin{cases} c & \text{if } x = u; \\ 0 & \text{if } x \neq u \end{cases} \tag{6.5.7}$$

with $u \in \mathbb{R}$, $c > 0$. It is clear that $\bar{H}$ a supremal generator of the set F_+ of all nonnegative functions defined on $\mathbb{R}$.

PROPOSITION 6.15 *Let $H = \{h_t : t \in T\}$. Then $H = H_2 + \bar{H}$, where H_2 is a supremal generator of the set Q_+ of nonnegative quasiconvex functions, consisting of all two-step functions (6.5.4) and $\bar{H}$ is a supremal generator of the set F_+ of nonnegative functions, consisting of all pointed functions (6.5.7).*

Proof: Let $h \in H$, $h = h_t$ with $t = \{u; c_1, c_2\}$. Assume for the sake of definitness that $c_2 > c_1$. Then $h = h' + h''$, where

$$h'(t) = \begin{cases} c_2 & \text{if } x \geq u; \\ c_1 & \text{if } x \leq u, \end{cases} \qquad h''(t) = \begin{cases} c_1 & \text{if } x = u; \\ 0 & \text{if } x \neq u. \end{cases}$$

Obviously, $h' \in H_2, h'' \in \bar{H}$, therefore the result follows. △

Thus H is the sum of supremal generators for sets Q_+ and F_+. We now show that H is a supremal generator of $\mathcal{P}$.

PROPOSITION 6.16 *H is a supremal generator of $\mathcal{P}$.*

Proof: Let $f \in \mathcal{P}$ and $u \in \mathbb{R}$. First assume that $u \in \operatorname{dom} f$. Let $\varepsilon > 0$ and $c_1' = \inf_{x<u} f(x)$, $c_2' = \inf_{x>u} f(x)$. We now check that $f(u) - 2\varepsilon \leq c_1' + c_2'$. Let points $x_1 < u$ and $x_2 > u$ be such that $f(x_1) \leq c_1' + \varepsilon$ and $f(x_2) \leq c_2' + \varepsilon$, respectively. Then $f(u) - 2\varepsilon \leq f(x_1) + f(x_2) - 2\varepsilon \leq c_1' + c_2'$. Take nonnegative numbers c_1 and c_2 such that $c_1 \leq c_1', c_2 \leq c_2'$ and $c_1 + c_2 = f(u) - 2\varepsilon$. Consider the function h_t with $t = \{u; c_1, c_2\} \in T$. It follows from the definition of t that $h_t \leq f$ and $h_t(u) = f(u) - 2\varepsilon$. Thus $f(u) = \sup\{h(u) : h \in H, h \leq f\}$ for all $u \in \operatorname{dom} f$. Consider now a point $u \notin \operatorname{dom} f$. Assume for the sake of definiteness that $u \leq \inf \operatorname{dom} f$. Let $t = \{u, c_1, c_2\} \in T$, where $c_2 = \inf_{x \in \mathbb{R}} f(x)$ and c_1 is an arbitrary positive number. Then $h_t \leq f$. So $f(u) = +\infty = \sup_{h \in H, h \leq f} h(u)$. △

We now show that the set $\mathcal{P}$ is the least set P, which enjoys the following properties: P is a convex cone, P is a complete upper semilattice, P is closed under pointwise convergence and $P \supset Q_+$.

Let $\operatorname{co} Q_+$ be the convex hull of the set Q_+ of all nonnegative quasiconvex functions. Since Q_+ is a conic set, $\operatorname{co} Q_+$ is the set of all functions f of the form $f = q_1 + q_2$, where $q_1, q_2 \in Q_+$.

PROPOSITION 6.17 $H \subset \operatorname{co} Q_+$.

Proof: Let $t = \{u; c_1, c_2\} \in T$. Consider the functions h_t^1 and h_t^2, where

$$h_t^1(x) = \begin{cases} 0 & \text{if } x < u; \\ c_2 & \text{if } x \geq u, \end{cases} \qquad h_t^2(x) = \begin{cases} c_1 & \text{if } x \leq u; \\ 0 & \text{if } x > u. \end{cases}$$

Then $h_t = h_t^1 + h_t^2$. Since h_t^1 is an increasing function and h_t^2 is a decreasing function it follows that $h_t^1 \in Q_+$, $h_t^2 \in Q_+$. △

Let $\tilde{H}$ be the upper semilattice generated by H, that is, the set of all functions $\tilde{h}$ of the form

$$\tilde{h}(x) = \max_{i=1,\dots,m} h_i(x), \qquad h_i \in H,\ i = 1,\dots,m; \qquad m = 1,2,\dots$$

and $\bar{H}$ be the upper semilattice generated by $\operatorname{co} Q_+$, that is, the set of all functions $\bar{h}$ of the form

$$\bar{h}(x) = \max_{i=1,\dots,m} \tilde{q}_i(x) \qquad \tilde{q}_i \in \operatorname{co} Q_+,\ i = 1,\dots,m; \qquad m = 1,2,\dots.$$

PROPOSITION 6.18 *$\mathcal{P} = \operatorname{cl} \tilde{H} = \operatorname{cl} \bar{H}$, where $\operatorname{cl} A$ is the closure of a set A in the topology of pointwise convergence.*

Proof: Let $f \in \mathcal{P}$ and $\operatorname{supp}(f, \tilde{H}) = \{\tilde{h} \in \tilde{H} : \tilde{h} \leq f\}$ be the support set of f with respect to $\tilde{H}$. We can consider $\operatorname{supp}(f, \tilde{H})$ as a directed set with respect to the natural order relation: $\tilde{h}_1 \geq \tilde{h}_2$ if $\tilde{h}_1(x) \geq \tilde{h}_2(x)$ for all $x \in \mathbb{R}$. Since for all $x \in R$

$$f(x) = \sup\{h(x) : h \in \operatorname{supp}(f, H)\} = \sup\{\tilde{h}(x) : \tilde{h} \in \operatorname{supp}(f, \tilde{H})\}$$

and the net (generalized sequence) $\{\tilde{h} : \tilde{h} \in \operatorname{supp}(f, \tilde{H})\}$ is increasing, it follows that $f(x)$ is the pointwise limit of this net. Thus

$$\mathcal{P} \subset \operatorname{cl} \tilde{H}. \tag{6.5.8}$$

Since $H \subset \operatorname{co} Q_+$ it follows that $\tilde{H} \subset \bar{H}$. As $\mathcal{P}$ is a convex cone and an upper semilattice, and the set Q_+ of quasiconvex nonnegative functions is contained in $\mathcal{P}$, it follows that $\bar{H} \subset \mathcal{P}$. Since $\mathcal{P}$ is closed in the topology of pointwise convergence, $\operatorname{cl} \tilde{H} \subset \operatorname{cl} \bar{H} \subset \mathcal{P}$. The desired result follows from this inclusion and (6.5.8). △

REMARK 6.8 We have proved that a function f belongs to $\mathcal{P}$ if and only if f can be represented as the pointwise limit of a net (f_α), where each f_α is a finite maximum of the functions represented as the sum of two nonnegative quasiconvex functions. It also follows from Proposition 6.16 and Proposition 6.17 that each $f \in \mathcal{P}$ can be represented as the supremum of a family of functions belonging to $\operatorname{co} Q_+$. Since $\mathcal{P}$ is a convex cone and a complete upper semilattice and $Q_+ \subset \mathcal{P}$, it follows that $\mathcal{P}$ coincides with the set of all functions which can be represented in such a form.

6.5.4. Inequalities of Hadamard type for P-functions and quasiconvex functions

We now establish some inequalities of Hadamard type for P-functions by applying the principle of preservation of inequalities.

Consider the σ-algebra Σ of Borel subsets of the segment $(0,1)$ and a nonnegative normed measure μ defined on Σ. (A measure μ is called normed if $\mu((0,1)) = 1$).

Let $\mathcal{P}'$ be the set of all summable (with respect to μ) P-functions defined on the segment $(0,1)$. Note that each function h_t defined by (6.5.6) is Borel-measurable on $\mathbb{R}$. We denote by h'_t the restriction of h_t to $(0,1)$. Let $T' = \{t = (u; c_1, c_2) : 0 < u < 1\}$ and $H' = \{h'_t : t \in T'\}$. Thus H' consists of Borel-measurable functions on $(0,1)$. It follows from Proposition 6.16 that H' is a supremal generator of $\mathcal{P}(S)$, where $S = (0,1)$. Hence H' is a supremal generator of $\mathcal{P}'$.

Consider the functional $J : \mathcal{P}' \to R$ given by

$$J(f) = \int_0^1 f d\mu.$$

where $\int_0^1$ stands for $\int_{(0,1)}$. For $y \in (0,1)$ consider the functions

$$e_y^1(x) = \begin{cases} 1 & \text{if } x \leq y; \\ 0 & \text{if } x > y, \end{cases} \qquad e_y^2(x) = \begin{cases} 0 & \text{if } x \leq y; \\ 1 & \text{if } x > y. \end{cases}$$

Let

$$g_1(y) = \int_0^1 e_y^1 d\mu = \mu((0,y]), \quad g_2(y) = \int_0^1 e_y^2 d\mu = \mu((y,1)), \quad y \in (0,1),$$

Clearly, g_1 is increasing, g_2 is decreasing and $g_1(y)+g_2(y) = \mu((0,1)) = 1$ for all $y \in (0,1)$. Recall that a measure μ defined on σ-algebra Σ is called *atomless* if $\mu(\{x\}) = 0$ for each $x \in (0,1)$. Let us calculate $J(h)$ for $h \in H'$ and an atomless measure μ.

LEMMA 6.4 *Let μ be an atomless Borel normed measure and let $h'_t \in H'$ be the function corresponding to a collection $t = \{y; c_1, c_2\}$. Then $J(h'_t) = c_1 g_1(y) + c_2 g_2(y)$.*

Proof: Consider the function $\tilde{h}_t$ given by

$$\tilde{h}_t(x) = \begin{cases} c_1 & \text{if } x \leq y; \\ c_2 & \text{if } x > y \end{cases} = \begin{cases} c_1 e_y^1(x) & \text{if } x \leq y; \\ c_2 e_y^2(x) & \text{if } x > y. \end{cases}$$

Since μ is atomless, it follows that $J(h'_t) = J(\tilde{h}_t)$. We have

$$J(h'_t) = \int_0^1 \tilde{h}_t d\mu = \int_0^y c_1 e_y^1 d\mu + \int_y^1 c_2 e_y^2 d\mu = c_1 g_1(y) + c_2 g_2(y).$$

Thus the result follows. △

For $u \in (0,1)$, consider the number

$$\gamma_u = \min_{c_1 \geq 0, c_2 \geq 0} \frac{c_1 g_1(u) + c_2 g_2(u)}{c_1 + c_2}. \tag{6.5.9}$$

It easy to check that

$$\gamma_u = \min(g_1(u), g_2(u)). \tag{6.5.10}$$

In fact, if $g_1(u) \geq g_2(u)$ then $\gamma_u = g_2(u)$ and if $g_1(u) \leq g_2(u)$ then $\gamma_u = g_1(u)$. Thus (6.5.10) holds.

THEOREM 6.7 *Let μ be an atomless measure and $u \in (0,1)$. Then*

$$f(u) \leq \frac{1}{\min(g_1(u), g_2(u))} \int_0^1 f d\mu \tag{6.5.11}$$

for all $f \in \mathcal{P}'$.

Proof: Clearly, J is an increasing function defined on the set $\mathcal{P}'$. First we check that (6.5.11) holds for all $h' \in H'$. We consider separately functions h'_t which are defined by collections $t = \{u; c_1, c_2\}$ and by collections $t = \{y; c_1, c_2\}$ with $y \neq u$.

Let $t = \{u; c_1, c_2\}$. It follows directly from (6.5.9), (6.5.10), Lemma 6.4 and the equality $h'_t(u) = c_1 + c_2$ that

$$\begin{aligned} h'_t(u) &= c_1 + c_2 \leq \frac{1}{\gamma_u}(c_1 g_1(u) + c_2 g_2(u)) \\ &= \frac{1}{\gamma_u} \int_0^1 h'_t d\mu = \frac{1}{\min(g_1(u), g_2(u))} \int_0^1 h'_t d\mu. \end{aligned} \tag{6.5.12}$$

Assume now that $t = \{y; c_1, c_2\}$ with $y \neq u$. It follows from Lemma 6.4 that

$$J(h'_t) = \int_0^1 h'_t(x) dx = c_1 g_1(y) + c_2 g_2(y).$$

Let us calculate $h'_t(u)$. Since $y \neq u$, then either $h'_t(u) = \min(c_1, c_2)$ or $h'_t(u) = \max(c_1, c_2)$. In the first case we have, taking into account that $g_1(y) \geq 0$, $g_2(y) \geq 0$ and $g_1(y) + g_2(y) = 1$, that

$$h'_t(u) = \min(c_1, c_2) \leq c_1 g_1(y) + c_2 g_2(y) = J(h'_t). \tag{6.5.13}$$

Since $g_1(u) \leq 1$, $g_2(u) \leq 1$, it follows that

$$h'_t(u) \leq J(h'_t) \leq \frac{1}{\min(g_1(u), g_2(u))} J(h'_t).$$

Assume now that $h'_t(u) = \max(c_1, c_2)$. If $c_1 \geq c_2$, then $h'_t(u) = c_1$ and $y > u$, the latter following directly from the definition of the function h'_t. Since g_1 is an increasing function, we have for $y > u$

$$g_1(u)h'_t(u) \leq g_1(y)c_1 \leq g_1(y)c_1 + g_2(y)c_2 = J(h'_t). \tag{6.5.14}$$

Thus

$$h'_t(u) \leq \frac{1}{g_1(u)} J(h'_t) \leq \frac{1}{\min(g_1(u), g_2(u))} J(h'_t).$$

If $c_2 \geq c_1$, then $h'_t(u) = c_2$ and $y < u$. In the same manner we have

$$h'_t(u) \leq \frac{1}{g_2(u)} J(h'_t) \leq \frac{1}{\min(g_1(u), g_2(u))} J(h'_t). \tag{6.5.15}$$

Thus we have verified that the desired inequality (6.5.11) holds for all $h' \in H'$. Since H' is a supremal generator of $\mathcal{P}'$ and J is an increasing function, we conclude, by applying the principle of preservation of inequalities, that (6.5.11) holds for all $f \in \mathcal{P}'$. △

REMARK 6.9 Let $t = \{u; c_1, c_2\}$, where $\gamma_u(c_1 + c_2) = c_1 g_1(u) + c_2 g_2(u)$. It follows from (6.5.12) that

$$h'_t(u) = \frac{1}{\min(g_1(u), g_2(u))}.$$

Thus the inequality (6.5.11) cannot be improved for all P-functions.

Let us give an example.

EXAMPLE 6.7 Let μ be the Lebesgue measure, i.e., $J(f) = \int_0^1 f(x)dx$. Then $g_1(y) = y$, $g_2(y) = 1 - y$. It follows from Theorem 6.7 that

$$f(u) \leq \frac{1}{\min(u, 1-u)} \int_0^1 f(x)dx$$

for all $f \in \mathcal{P}'$. In particular, we have

$$f\left(\frac{1}{2}\right) \leq 2\int_0^1 f(x)dx \qquad (f \in \mathcal{P}'). \tag{6.5.16}$$

This result was established in [45].

REMARK 6.10 In a similar manner we can prove that (6.5.11) holds for each nonnegative quasiconvex function and for each (not necessarily atomless) nonnegative measure μ such that $\mu((0,1)) = 1$. We can use the supremal generator H_1 described in Proposition 6.12 for this purpose. Indeed, the atomlessness of the measure μ has been used only for eliminating the special value of the function h'_t with $t = \{u; c_1, c_2\}$ at the point u. We do not need to eliminate this value in the case under consideration.

6.5.5. Inequality of Hadamard type for ICAR functions

Let H be a set of all functions h of the form $h(x) = \langle l, x\rangle$ defined on $\mathbb{R}^n_{++}$, where $\langle l, x\rangle = \min_{i\in I} l_i x_i$, with $l \in \mathbb{R}^n_{++}$. As it was shown in Section 3.3 (see Corollary 3.2), the set of H-convex finite functions coincides with the set of all ICAR (increasing convex-along-rays) functions. This fact allows us to examine inequalities of Hadamard type for ICAR functions defined on some subsets of $\mathbb{R}^n_{++}$.

In this subsection we establish the inequality of Hadamard type for ICAR functions defined on a special domain $\mathcal{D} \in \mathbb{R}^2_{++}$. We shall consider the usual Lebesgue measure. Then the inequality of Hadamard type for the functions defined on $\mathcal{D}$ has the following form: for a certain point $u \in \mathcal{D}$ and all ICAR functions f

$$f(u) \le \frac{1}{A_{\mathcal{D}}} \int_{\mathcal{D}} f(x, y)dxdy \tag{6.5.17}$$

where $A_{\mathcal{D}}$ is the area of D.

We consider the triangle $\mathcal{D}$ with vertices $(0,0)$, $(a,0)$ and (a, va) with $a, v > 0$ (see Fig. 1). It is convenient for us to assume that $\mathcal{D}$ is an open set. Hence

$$\mathcal{D} = \left\{(x,y) : 0 < x < a,\ 0 < \frac{y}{x} < v\right\}.$$

Let

$$z^0 = \left(\frac{a}{3}, \frac{va}{3}\right), \qquad z_1 = \left(\frac{2a}{3}, \frac{va}{3}\right). \tag{6.5.18}$$

Fugure 1.

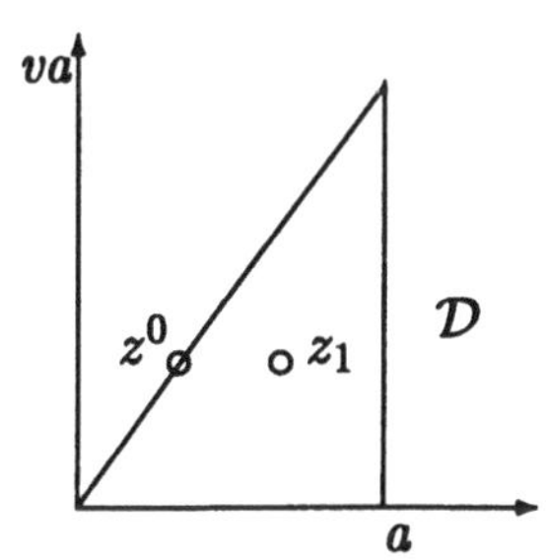

We begin with convex functions.

PROPOSITION 6.19 *Let f be a finite convex function defined on $\mathcal{D}$. Then*

$$f(z_1) \le \frac{1}{A_{\mathcal{D}}} \int_{\mathcal{D}} f(x, y)\, dxdy, \tag{6.5.19}$$

Proof: We have

$$\begin{aligned}\int_{\mathcal{D}} xdxdy &= \int_0^a xdx \int_0^{vx} dy = v\int_0^a x^2 dx = \frac{va^3}{3};\\ \int_{\mathcal{D}} ydxdy &= \int_0^a dx \int_0^{vx} ydy = \int_0^a \frac{v^2x^2}{2}dx = \frac{v^2a^3}{6}.\end{aligned}$$

Note that $A_{\mathcal{D}} = (1/2)a^2v$. Hence

$$\frac{1}{A_{\mathcal{D}}}\int_{\mathcal{D}} xdxdy = \frac{2a}{3} \equiv z_{11}, \qquad \frac{1}{A_{\mathcal{D}}}\int_{\mathcal{D}} ydxdy = \frac{va}{3} \equiv z_{12}.$$

The result follows directly from Proposition 6.11. △

Recall that a function f defined on $\mathbb{R}^2_{++} = \{(x,y) : x > 0, y > 0\}$ is said to be ICAR (increasing convex-along-rays) if

1) f is increasing: $x_1 \geq x_2,\ y_1 \geq y_2 \Rightarrow f(x_1, y_1) \geq f(x_2, y_2)$;

2) for each $z = (x,y)$ the function $f_z(t) = f(tz)$ is convex on $(0, +\infty)$.

The following assertion holds.

Theorem 6.8 *Let f be an ICAR function. For each $(\bar{x}, \bar{y}) \in \mathbb{R}^2_{++}$ there exist numbers $b \geq 0$ and c, such that*

1) $\min\left(\frac{b}{\bar{x}}x, \frac{b}{\bar{y}}y\right) - c \leq f(x,y)$ *for all* $(x,y) \in \mathbb{R}^2_{++}$;

2) $b - c \equiv \min\left(\frac{b}{\bar{x}}\bar{x}, \frac{b}{\bar{y}}\bar{y}\right) - c = f(\bar{x}, \bar{y})$.

Proof: This statement is a particular case of Theorem 3.4. △

We shall prove in this subsection the following statement.

Theorem 6.9 *1) Let f be the restriction of an ICAR function to the triangle $\mathcal{D}$. Then*

$$f\left(z^0\right) \leq \frac{1}{A_{\mathcal{D}}}\int_{\mathcal{D}} f(x,y)\,dxdy. \qquad (6.5.20)$$

2) There exists a non constant ICAR function l such that the equality holds in (6.5.20).

We assume later that $\frac{0}{0} = 0$. Consider the set L_v of all min-type functions defined on $\mathbb{R}^2_{++}$ by $l(x,y) = \min(l_1x, l_2y)$ with $0 \leq \frac{l_1}{l_2} < v$

(here v is the number from the definition of $\mathcal{D}$). Let H_v be the set of all shifts of functions $l \in L_v$. In other words, $h \in H_v$ if and only if there exist $l \in L_v$ and $c \in \mathbb{R}$ such that $h(x,y) = l(x,y) - c$.

Let F be the set of all restrictions of ICAR functions to $\mathcal{D}$. In other words, $f \in F$ if f is defined on $\mathcal{D}$ and there exists an ICAR function $\tilde{f}$ defined on $\mathbb{R}^2_{++}$ such that $f(x,y) = \tilde{f}(x,y)$ for $(x,y) \in \mathcal{D}$.

PROPOSITION 6.20 *H_v is a supremal generator of F.*

Proof: Let $f \in F$ and $(\bar{x},\bar{y}) \in \mathcal{D}$. It follows from Theorem 6.8 that there exist numbers $b \geq 0$ and c such that

$$\begin{aligned} l(x,y) - c &\leq f(x,y) \quad \forall (x,y) \in \mathcal{D} \\ l(\bar{x},\bar{y}) - c &= f(\bar{x},\bar{y}), \end{aligned}$$

where $l(x,y) = \min(l_1 x, l_2 y)$ with $l_1 = \frac{b}{\bar{x}}$, $l_2 = \frac{b}{\bar{y}}$. If $b = 0$, then $l_1 = l_2 = 0$, hence $\frac{l_1}{l_2} = 0 < v$ and $l \in L_v$. If $b > 0$, then $\dfrac{l_1}{l_2} = \dfrac{\bar{y}}{\bar{x}} < v$ (since $(\bar{x},\bar{y}) \in \mathcal{D}$), hence $l \in L_v$. Thus the function $h(x,y) = l(x,y) - c$ belongs to H_v. We have

$$h(x,y) \leq f(x,y) \quad ((x,y) \in \mathcal{D}); \quad h(\bar{x},\bar{y}) = f(\bar{x},\bar{y}). \tag{6.5.21}$$

It follows from (6.5.21) that

$$f(\bar{x},\bar{y}) = \max_{h \in H_v, h \leq f} h(\bar{x},\bar{y}) \quad \text{for all} \quad (\bar{x},\bar{y}) \in \mathcal{D}. \qquad \triangle$$

Note that

$$A(f) = \int_{\mathcal{D}} f(x,y)\, dx dy$$

is an increasing functional on the set of ICAR functions. Thus we can prove Theorem 6.9 by using the principle of preservation of inequalities. According to this principle, the inequality (6.5.20) will be established for all $f \in F$, if we prove it only for $h \in H_v$. Since $1/A_{\mathcal{D}} \int_{\mathcal{D}} c\,dxdy = c$ for all $c \in \mathbb{R}$, it is sufficient to prove (6.5.20) only for $l \in L_v$. So we begin with the following assertion.

LEMMA 6.5 *The inequality (6.5.20) holds for $l \in L_v$.*

Proof: Let $l \in L_v$, that is,

$$l(x,y) = \min(l_1 x, l_2 y)$$

with $l_1, l_2 > 0$, $(l_1/l_2) < v$. If either l_1 or l_2 is equal to zero, the result is trivial. Therefore, we assume that $l_1 > 0$, $l_2 > 0$.

Consider the line $l_1 x = l_2 y$. Since $\frac{l_1}{l_2} < v$, it follows that this line passes through $\mathcal{D}$, so $\mathcal{D}$ is the union of the two triangles $\mathcal{D}_1 \cup \mathcal{D}_2$ (see Fig. 2),

Figure 2.

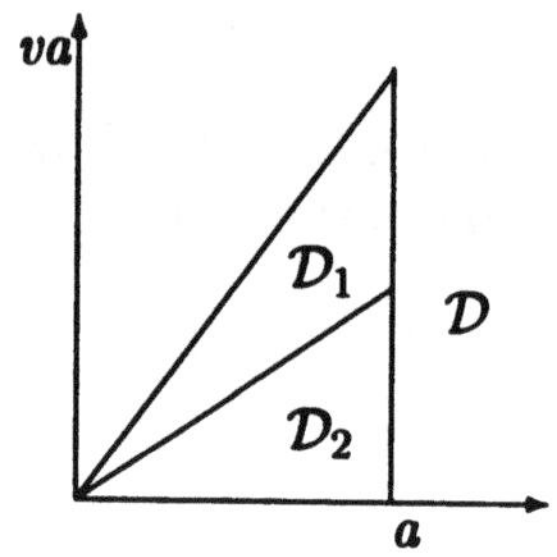

It is easy to check that

$$l(x,y) = \begin{cases} l_1 x & \text{if } (x,y) \in \mathcal{D}_1; \\ l_2 y & \text{if } (x,y) \in \mathcal{D}_2. \end{cases}$$

so

$$\int_{\mathcal{D}} l(x,y)\,dxdy = \int_{\mathcal{D}_1} l_1 x dxdy + \int_{\mathcal{D}_2} l_2 y dxdy.$$

An easy calculation shows that

$$\int_{\mathcal{D}_1} l_1 x dxdy = l_1 \left(v - \frac{l_1}{l_2} \right) \frac{a^3}{3},$$

and

$$\int_{\mathcal{D}_2} l_2 y dxdy = \frac{l_1^2}{6 l_2} a^3.$$

Thus

$$\int_{\mathcal{D}} l(x,y)\,dxdy = \left(\frac{l_1}{3} \left(v - \frac{l_1}{l_2} \right) + \frac{l_1^2}{6 l_2} \right) a^3 = \frac{a^3}{3} \left(l_1 v - \frac{1}{2} \cdot \frac{l_1^2}{l_2} \right)$$

and

$$\frac{1}{A_{\mathcal{D}}} \int_{\mathcal{D}} l(x,y)\,dxdy = \frac{2}{3} l_1 a \left(1 - \frac{1}{2} \frac{l_1}{l_2} \frac{1}{v} \right). \tag{6.5.22}$$

Since $l \in L_v$, it follows that $\frac{l_1}{l_2} \leq v$. Hence

$$1 - \frac{1}{2} \frac{l_1}{l_2} \frac{1}{v} \geq 1 - \frac{1}{2} = \frac{1}{2}$$

and we have

$$\frac{1}{A_{\mathcal{D}}} \int_{\mathcal{D}} l\,(x,y)\,dxdy \geq \frac{1}{3} l_1 a. \tag{6.5.23}$$

Consider now the point $z^0 = \left(\frac{a}{3}, \frac{va}{3}\right)$. We have

$$l(z^0) = \min\left(l_1 \frac{a}{3}, l_2 \frac{va}{3}\right).$$

Since $\frac{l_1}{l_2} < v$,

$$l_1 \frac{a}{3} < \frac{l_2 va}{3}$$

so

$$l(z^0) = l_1 \frac{a}{3}. \tag{6.5.24}$$

The result follows directly from (6.5.24) and (6.5.23). △

Proof of Theorem 6.9: 1) First we prove that the inequality (6.5.20) holds for all $f \in F$. It follows from Lemma 6.5 that

$$l(z^0) \leq \frac{1}{A_{\mathcal{D}}} \int\int_{\mathcal{D}} l\,(x,y)\,dxdy$$

for all $l \in L_v$. We also have

$$c = \frac{1}{A_{\mathcal{D}}} \int_{\mathcal{D}} c dxdy \text{ for all } c \in \mathbb{R}.$$

Therefore,

$$l\,(z_0) - c \leq \frac{1}{A_{\mathcal{D}}} \int\int_{\mathcal{D}} (l\,(x,y) - c)\,dxdy$$

for all $l \in L_v$, $c \in \mathbb{R}$. Hence

$$h(z^0) \leq \frac{1}{A_{\mathcal{D}}} \int_{\mathcal{D}} h\,(x,y)\,dxdy$$

for all $h \in H_v$. Since H_v is a supremal generator of F, we conclude, by applying the principle of preservation of inequalities, that

$$f(z^0) \leq \frac{1}{A_{\mathcal{D}}} \int_{\mathcal{D}} f\,(x,y)\,dxdy$$

for all $f \in F$.

2) We now show that there exists $f \in F$ such that

$$f(z^0) = \frac{1}{A_{\mathcal{D}}} \int_{\mathcal{D}} f\,(x,y)\,dxdy.$$

We take as f the min-type function $l(x,y) = \min\left(\frac{3}{a}x, \frac{3}{va}y\right)$. Then (see (6.5.22))

$$\frac{1}{A_{\mathcal{D}}}\int_{\mathcal{D}} l(x,y)\,dxdy = \frac{2}{3}l_1 a\left(1 - \frac{1}{2}\frac{l_1}{l_2}\frac{1}{v}\right)2\left(1 - \frac{1}{2}v \cdot \frac{1}{v}\right) = 1$$

and

$$l\left(\frac{a}{3}, \frac{va}{3}\right) = \min\left(\frac{3}{a}\cdot\frac{a}{3}, \frac{3}{va}\cdot\frac{va}{3}\right) = 1.$$

Hence

$$\frac{1}{A_{\mathcal{D}}}\int_{\mathcal{D}} l(x,y)\,dxdy = l\left(\frac{a}{3}, \frac{va}{3}\right).$$

The proof is complete. △

REMARK 6.11 Let $f \in F$. Since f is increasing, and $(0,0) < (x,y) \le (a, va)$ for all $(x,y) \in \mathcal{D}$, it follows that

$$f(0,0) \le f(x,y) \le f(a,va),$$

so

$$f(0,0) \le \frac{1}{A_{\mathcal{D}}}\int_{\mathcal{D}} f(x,y)\,dxdy \le f(a,va).$$

However, the equality holds here if and only if f is constant.

REMARK 6.12 It is interesting to compare points z^0 and z^1. The point z^1 is the centre of gravity of the triangle $\mathcal{D}$. At the same time z^0 is a boundary point of this triangle. Note that $z_1 \ge z_0$, so $f(z^1) \ge f(z^0)$ for increasing functions.

Chapter 7

FURTHER ABSTRACT CONVEXITY

7.1. INTRODUCTION

Some definitions related to abstract convexity have been introduced in Chapter 1. In Chapters 2 – 6 we concentrated mainly upon examples of abstract convexity and its applications. In this chapter we continue the examination of abstract convexity in a general situation. For some applications it is convenient to consider abstract convex functions defined only on a subset of the domain of elementary functions. We introduce the notion of abstract convex functions, abstract convex sets and corresponding hulls for this situation and provide many examples of abstract convexity in such a setting. We examine in detail the Fenchel-Moreau conjugacy, subdifferentials and approximate subdifferentials (known also as ε-subdifferentials) and present some links between these crucial notions of abstract convexity. We also examine the Minkowski duality, which is a one-to-one correspondence between abstract convex functions and corresponding support sets. This kind of duality has found many applications, some of which are studied in Chapter 8.

Section 7.2 contains some results related to abstract convexity with respect to a subset Z of the domain X. In particular subdifferentials and approximate subdifferentials are defined and studied in this case. The Fenchel-Moreau conjugacy is also considered, however we examine it only for $Z = X$. We provide some applications of abstract convexity to global optimization and min-max theory.

Special attention is paid to abstract convexity with respect to sets of positively homogeneous elementary functions. The description of subdifferentials and approximate subdifferentials can be substantially simplified in this case. We also consider a special type of polarity for such kinds of functions and sets. In particular we discuss dual relations between the Minkowski gauge of a radiant set and its support function. Note that the Minkowski gauge does not depend on the outer dual object, while the support function does depend upon it. We show that many results obtained for positively homogeneous functions can be extended to much more general situations with the help of the so-called positively homogeneous extension.

Section 7.3 contains many examples of abstract convex functions and sets for various classes of elementary functions and various subsets of the domain of these functions. In particular, we shall consider some classes of sublinear, convex and quasiconvex functions. We shall provide an example, which demonstrates that the subdifferential of an abstract convex function can be empty at all points of the domain, with the sole exception of global minimizers. (Note that the subdifferential at a global minimizer contains zero, hence is not empty, for all functions.)

The Minkowski duality, which is examined in Section 7.4, demonstrates that complete lattices is a natural environment in the study of totalities of all abstract convex functions and all abstract convex sets with a given set of elementary functions and a subset of the domain of these functions. In particular, the classical construction of the closure in the sense of Moore is a very useful tool in this situation. If a set of elementary functions is equipped with pointwise multiplication by a positive number and the pointwise summation, then the corresponding algebraic operation can be introduced in the sets of abstract convex functions and abstract convex sets and the Minkowski duality becomes an algebraic isomorphism. The natural frameworks in the study of such kind of isomorphism are the so-called c^2-lattices (conic complete lattices) and semilinear lattices. We provide examples of such kind of complete lattices.

7.2. ABSTRACT CONVEXITY WITH RESPECT ON A SUBSET

7.2.1. Overview

In this section we introduce a class of abstract convex functions on a subset of the domain of finite elementary functions and corresponding abstract convex sets. These notions can be considered as a special case of the abstract convexity, presented in Chapter 1. Abstract convexity on a subset leads to some modifications of the abstract convex hull for an abstract convex function and for an abstract convex set and also to a modification of the notion of subdifferential.

We examine some properties of the Fenchel-Moreau conjugacy operator and its relations with the subdifferential and the approximate subdifferential. We also present some simple applications of the discussed notions to the study of problems of global optimization and the min-max theory.

Another topic discussed in this section is abstract convexity with respect to sets of positively homogeneous elementary functions. We examine subdifferentials and approximate subdifferentials and also a special kind of polarity in this case. The notion of positively homogeneous extension will also be discussed. The positively homogeneous extension allows one to apply some results obtained for positively homogeneous functions in a general situation.

Main results from this section related to abstract convexity on a subset can be found in [155]. The Fenchel-Moreau conjugacy is well-known (see [175] and references therein), we follow presentation in [100]. The notion of the positively homogeneous extension is well-known, and may almost be considered as a part of mathematical folklore. Polarity for positively homogeneous functions on conic sets in a finite dimensional space has been studied in [145].

7.2.2. Basic definitions and properties

We begin with some essential definitions and an introduction to concepts used throughout this chapter.

DEFINITION 7.1 Let Z be a nonempty subset of a set X and let H be a set of finite functions defined on X. A function $f : Z \to \mathbb{R}_{+\infty}$, is called ***abstract convex*** or ***H-convex*** on the set Z if there exists a set $U \subset H$ such that

$$f(z) = \sup\{h(z) : h \in U\} \text{ for all } z \in Z.$$

We assume that the function $-\infty$, where $-\infty(z) = -\infty$ for all $z \in Z$, is also abstract convex.

The set H in Definition 7.1 is called the set of elementary functions.

DEFINITION 7.2 Let $f : Z \to \mathbb{R}_{+\infty}$. The set of H-minorants of f on Z is called the ***support set*** of f and is denoted by $\text{supp}(f, H, Z)$. Thus

$$\text{supp}(f, H, Z) = \{h \in H : h(z) \leq f(z) \text{ for all } z \in Z\}.$$

The function $\text{co}_H^Z f : X \to \mathbb{R}_{+\infty}$ defined by

$$(\text{co}_H^Z f)(z) := \sup\{h(z) : h \in \text{supp}(f, H, Z)\} \text{ for all } z \in Z$$

is called the *H-convex hull* of the function f.

We shall omit the symbol Z if $Z = X$, in order to be consistent with the notation adopted in Chapter 1. Thus we shall write $\mathrm{supp}(f, H)$ instead of $\mathrm{supp}(f, H, X)$ and $\mathrm{co}_H f$ instead of $\mathrm{co}_H^X f$.

Clearly, $\mathrm{supp}(f, H, Z) = \mathrm{supp}(\mathrm{co}_H^Z f, H, Z)$ for all $Z \subset X$. It follows immediately from Definition 7.2 that $\mathrm{co}_H^Z f$ is an H-convex function. Moreover, $\mathrm{co}_H^Z f$ is the greatest H-convex function minorized by f. So $f = \mathrm{co}_H^Z f$ if and only if f is an H-convex function.

REMARK 7.1 The definitions 7.2 and 7.1 are given in the situation, which differs from the situation considered in Chapter 1. If $Z = X$, we obtain the definitions related to "standard" abstract convexity, which was discussed in that chapter. In contrast with Chapter 1 we consider here only sets of finite elementary functions. Actually, the situation from the current chapter can be considered as a particular case of the situation from Chapter 1. Indeed, let $Z \subset X$ and H be a set of functions $h : X \to \mathbb{R}$. For each $h \in H$ consider a function $\bar{h}$ defined on X by

$$\bar{h}(x) = \begin{cases} h(x) & \text{if } x \in Z; \\ +\infty & \text{otherwise}. \end{cases}$$

Let $\bar{H} = \{\bar{h} : h \in H\}$. Then a function $p : Z \to \mathbb{R}_{+\infty}$ is H-convex on Z if and only if the function $\bar{p}$ defined on X by

$$\bar{p}(x) = \begin{cases} p(x) & \text{if } x \in Z; \\ +\infty & \text{if } x \in X \setminus Z \end{cases}$$

is $\bar{H}$-convex on X. We also have $h \in \mathrm{supp}(p, H, Z)$ if and only if $\bar{h} \in \mathrm{supp}(\bar{p}, \bar{H}, X)$. Thus abstract convexity on Z with respect to a set of finite elementary functions can be studied by means of abstract convexity on X with respect to a set of elementary functions taking the value $+\infty$. However, it is more convenient at times to consider only finite elementary functions. This indicates the reason for a special study of abstract convexity on a subset Z of the set X.

REMARK 7.2 In the current chapter we shall mainly consider examples of abstract convexity having in mind functions defined on a Banach space or on subsets of a Banach space. Many results, related to these examples, can easily be generalized for a locally convex Hausdorff topological space. We shall leave corresponding generalizations to the reader.

To illustrate the nature of support sets and their dependence on Z we include the following illustrative examples.

EXAMPLE 7.1 Let l be a continuous linear function defined on a Banach space X (so $l \in X^*$, where X^* denotes the *conjugate space* to X). Then we have the following:

1) If $H = X^*$, $Z = X$, then $\text{supp}(l, X^*, X) = \{l\}$.

2) If $H = X^*$, $Z = Q$ (a closed convex cone in X), then $\text{supp}(l, X^*, Q) = l - Q^*$. Here Q^* denotes the *conjugate cone* to Q, that is, $Q^* = \{l \in X^* : l(x) \geq 0 \text{ for all } x \in Q\}$.

3) Let H be the set of continuous affine functions defined on X (that is $h \in H$ if $h(x) = l(x) - c$ with $l \in X^*$ and $c \in \mathbb{R}$) and let $Z = X$. Then $\text{supp}(l, H, X) = \{l + c\mathbf{1} : c \leq 0\}$, where $\mathbf{1}$ denotes the function with the constant value 1 on X.

4) Let H be the set of continuous affine functions defined on X and let $Z = B(0,1) := \{x \in X : \|x\| \leq 1\}$. Then

$$\text{supp}(l, H, B(0,1)) = \{h \in H : h(x) = l'(x) - c;\ l' \in X^*,\ c \geq \|l' - l\|\}.$$

Thus, $\text{supp}(l, H, B(0,1))$ is the epigraph of the function $f(l') = \|l' - l\|$.

We shall only prove assertion 4). For $h(x) = l'(x) - c$ we have

$$\begin{aligned} h \in \text{supp}(l, H, B(0,1)) &\iff l'(x) - c \leq l(x) \quad \text{for all } x \in B(0,1) \\ &\iff (l - l')(x) \leq c \quad \text{for all } x \in B(0,1) \\ &\iff \|l - l'\| \leq c. \end{aligned}$$

It is easy to see that $\text{co}_H^Z\, l = \{l\}$ in cases 1) to 4) above.

EXAMPLE 7.2 We now consider the function $\mathbf{1}$, where $\mathbf{1}(x) = 1$ for all $x \in X$.

1) Let $H = X^*$ and let $Z = Q$ be a closed convex cone. It is easy to see that $\text{supp}(\mathbf{1}, X^*, Q) = -Q^*$ and $\text{co}_H^Z(\mathbf{1}) = \{0\}$.

2) Let $H = X^*$ and let $Z = B(0,1)$. Then

$$\text{supp}(\mathbf{1}, X^*, B(0,1)) = \{l : l(x) \leq 1 \text{ for all } x \in B(0,1)\} = B^*(0,1)$$

(here $B^*(0,1)$ is the unit ball in X^*). We also have $\text{co}_H^Z(\mathbf{1})(x) = \|x\|$.

3) Let $H = X^*$ and let Z be an arbitrary set. Then $\text{supp}(\mathbf{1}, X^*, Z) = Z^\circ$, where $Z^\circ = \{l \in X^* : l(x) \leq 1 \text{ for all } z \in Z\}$ is the polar set of Z. The function $\text{co}_H^Z(\mathbf{1})$ coincides with the support function p_{Z° of the convex set Z°.

We shall only check that assertion 3) holds. (Clearly 2) follows from 3) with $Z = B(0,1)$). We have

$$l \in \text{supp}(1, X^*, Z) \iff l(x) \le 1 \text{ for all } x \in Z \iff l \in Z^\circ$$

and

$$\begin{aligned} \text{co}_H^Z(1)(x) &= \sup\{l(x) : x \in \text{supp}(1, X^*, Z)\} \\ &= \sup\{l(x) : l \in Z^\circ\} = p_{Z^\circ}(x). \end{aligned}$$

EXAMPLE 7.3 We now consider a lower semicontinuous sublinear function $p : X \to \mathbb{R}_{+\infty}$ where X is a Banach space. Let $H = X^*$ and $Z = X$. Then

$$\text{supp}(p, X^*, X) = \partial p(0) \tag{7.2.1}$$

is the *subdifferential* of p at zero, i.e., $\partial p(0) = \{l \in X^* : p(x) \ge l(x) \text{ for all } x \in X\}$.

If Z coincides with a closed convex cone $Q \subset X$, then

$$\text{supp}(p, X^*, Q) = \text{cl}\,(\partial p(0) - Q^*), \tag{7.2.2}$$

where cl stands for the closure in the weak* topology of X^*. For the proof of (7.2.2) see Subsection 7.3.2.

Let h be an affine function defined on a Banach space X. Then there exists a unique pair (l, c) with $l \in X^*$, $c \in \mathbb{R}$ such that

$$h(x) = l(x) - c \text{ for all } x \in X. \tag{7.2.3}$$

Denote the function h by (l, c). We shall identify also the set of all affine functions defined on X with $X^* \times \mathbb{R}$. This identification is accomplished by (7.2.3).

EXAMPLE 7.4 Let X be a Banach space and $H = X^* \times \mathbb{R}$ be the set of all affine functions defined on X. If $f : X \to \mathbb{R}_{+\infty}$ is a lower semicontinuous convex function then

$$\text{supp}(f, H, X) = \text{epi}\, f^*.$$

Here f^* denotes *the Fenchel conjugate* of f. For the proof (in a more general situation) see Proposition 7.8.

We now define abstract convex sets in the situation under consideration.

DEFINITION 7.3 A set $U \subset H$ is called *X-convex on Z* or *(H, X)-convex on Z* or *(H, X, Z)-convex* if there exists a function $f : X \to \bar{\mathbb{R}}$ such that $U = \text{supp}(f, H, Z)$.

Note that the empty subset of H is (H, X, Z)-convex; namely $\emptyset = \text{supp}(-\infty, H, Z)$. A nonempty subset U of H is (H, X, Z)-convex if and only if $U = \text{supp}(p_U, H, Z)$, where,

$$p_U(z) = \sup\{h(z) : h \in U\} \qquad (z \in Z). \tag{7.2.4}$$

DEFINITION 7.4 Let $U \subset H, U \neq \emptyset$ and let p_U be the function defined by (7.2.4). The set $\text{supp}(p_U, H, Z)$ is called the *abstract convex hull on* Z or the (H, X, Z)-*convex hull* of the set U

We shall denoted the (H, X, Z)-convex hull of a set U by $\text{co}^Z_{H,X} U$. For the sake of simplicity we also shall use the notation $\text{co}^Z_H U$ (compare with Definition 1.5 and Remark 1.2). If $Z = X$ we shall omit the symbol Z and use the symbol $\text{co}_H U$.

The following lemma illustrates the separation property for abstract convex sets (compare with Lemma 1.1).

LEMMA 7.1 *Let U be a nonempty subset of H. The set U is (H, X)-convex on Z if and only if for each $h' \in H$, $h' \notin U$ there exists $x \in Z$ such that $h'(x) > \sup_{h \in U} h(x)$.*

Proof: Let $f(x) = \sup_{h \in U} h(x)$. Assume that U is (H, X, Z)-convex. Then

$$U = \{h \in H : \ h(x) \leq f(x) \text{ for all } x \in Z\},$$

hence for $h' \in H \setminus U$, there exists a point $x \in Z$ such that $h'(x) > f(x)$. Assume that for any $h' \in H$, $h' \notin U$ there exists a point $x \in Z$ such that $h'(x) > f(x)$. It means that $(h' \notin U) \implies (h' \notin \text{supp}(f, H, Z))$. Thus $U \supset \text{supp}(f, H, Z)$. The inclusion $U \subset \text{supp}(f, H, Z)$ is obvious. Since U coincides with the support set $\text{supp}(f, H, Z)$, it follows that U is (H, X)-convex on Z. △

We now introduce the notation to be used throughout this chapter.

We shall denote the totality of all H-convex functions defined on $Z \subset X$ by $P(H, X, Z)$ and the totality of all (H, X, Z)-convex sets by $S(H, X, Z)$. We shall omit the symbol Z if $Z = X$. We also denote by F_Z the union of the set of all functions $f : Z \to \mathbb{R}_{+\infty}$ and the function $-\infty$.

Clearly, $P(H, X, Z) \subset F_Z$.

We can define, in a symmetric manner, H-concave functions and sets. A function $q : Z \to \mathbb{R} \cup \{-\infty\}$ is said to be *abstract concave* or *H-concave* on a set Z if there exists a set $U \subset H$ such that,

$$q(z) = \inf\{h(z) : h \in U\} \text{ for each } z \in Z.$$

A set $U \subset H$ is called *(H, X, Z)-concave* if

$$U = \{h \in H : h(z) \geq \inf_{h' \in U} h'(z) \text{ for all } z \in Z\}.$$

The set of H-concave functions defined on a set $Z \subset X$ will be denoted by $\underline{P}(H, X, Z)$ and the set of all (H, X, Z)-concave sets by $\underline{S}(H, X, Z)$.

We shall be particularly interested in the class of functions expressible as the pointwise infimum of a family of H-convex on Z functions, in the notation above we are referring to the class $\underline{P}(P(H, X, Z), X, Z)$. Thus we require the following definition.

DEFINITION 7.5 Let X be an arbitrary set and let Z be a nonempty subset of X. Let H be a set of functions defined on X and mapping into $\mathbb{R}$. A function $q : Z \to \mathbb{R}_{+\infty}$ is called *inf-H-convex* on Z if there exists a set $U \subset P(H, X, Z)$ such that, for all $z \in Z$,

$$q(z) = \inf \{h(z) : h \in U\}.$$

If X is a Banach space and H is the set of continuous affine functions defined on X then we shall call inf-H-convex functions on Z as merely inf-convex functions on Z.

7.2.3. Subdifferentiability

Let L be a set of elementary functions defined on a set X. We shall consider L as a set of abstract linear functions. (See Section 1.4 for discussions related to abstract linear and abstract affine functions.) Recall that a function h of the form

$$h(x) = l(x) - c, \; x \in X \tag{7.2.5}$$

with $l \in L$, $c \in \mathbb{R}$ is called an *L-affine function*. The set of all L-affine functions is denoted by H_L. We shall often denote the function (7.2.5) by (l, c). As a rule we shall consider sets L of abstract linear functions which possesses the following property (A) (see (1.4.5):

$$(A) \qquad l - c\mathbf{1} \notin L \text{ for all } l \in L \text{ and for all } c \neq 0. \tag{7.2.6}$$

Here $\mathbf{1}(x) = 1$ for all $x \in X$. If L enjoys the property (A) then it is possible to represent an L-affine function h in the form (7.2.5) in a unique way. We can identify the function h with the pair (l, c) via (7.2.5). Thus we shall identify the set H_L with $L \times \mathbb{R}$.

Let H be an arbitrary set of functions defined on a set X and $Z \subset X$. Let $f : Z \to \mathbb{R}_{+\infty}$ be an H-convex function on Z. Then

$$f(x) = \sup\{h(x) : h \in \text{supp}(f, H, Z)\} \text{ for all } x \in Z.$$

Consider a point $x_0 \in \text{dom}\, f \equiv \{z \in Z : f(z) < +\infty\}$. We are interested in the set

$$\partial^*_{H,Z} f(x_0) = \{h \in \text{supp}(f, H, Z) : h(x_0) = f(x_0)\}. \qquad (7.2.7)$$

Note that $\partial^*_{H,Z} f(x_0) \neq \emptyset$ if and only if the supremum $\sup\{h(x_0) : h \in \text{supp}(f, H, Z)\}$ is attained. If $Z = X$ then we omit the symbol Z, so we shall write $\partial^*_H f(x_0)$ instead of $\partial^*_{H,X} f(x_0)$.

In this section we assume that H coincides with the set H_L of abstract affine functions, where L is a set of abstract linear functions. There is a close connection between the set $\partial^*_{H,Z} f(x_0)$ and the subdifferential of a function at the point x_0. We now give the definition of the subdifferential for the case under consideration (compare with Definition 1.7).

DEFINITION 7.6 Let L be a set of functions defined on X and let f be an H_L-convex function on a set $Z \subset X$. Let $x_0 \in Z$. Then the set

$$\partial_{L,Z} f(x_0) = \{l \in L : l(x) - l(x_0) \leq f(x) - f(x_0) \text{ for all } x \in Z\}$$

is called the *L-subdifferential of the function f at the point x_0 on the set Z*.

We shall denote by $\partial_L f(x_0)$ the subdifferential $\partial_{L,X} f(x_0)$ on the set X.

We now indicate some simple properties of the subdifferential.

1) If $Z_1 \subset Z_2$ then $\partial_{L,Z_1} f(x_0) \supset \partial_{L,Z_2} f(x_0)$.

In particular,

2) $\partial_L f(x_0) \subset \partial_{L,Z} f(x_0)$ for all $Z \subset X$.

3) If $L_1 \subset L_2$, then $\partial_{L_1,Z} f(x_0) \subset \partial_{L_2,Z} f(x_0)$.

The following point indicates links between the subdifferential $\partial_L f(x_0)$ and the level set $S_c(f) = \{x \in Z : f(x) \leq c\}$ with $c = f(x_0)$:

4) If $l \in \partial f(x_0)$ then $l(x_0) = \max\{l(x) : x \in S_c(f)\}$ with $c = f(x_0)$.

The following simple proposition describes one of the main properties of the subdifferential.

PROPOSITION 7.1 *Let $f \in P(H_L, X, Z)$, $x_0 \in Z$ and $l \in L$. Let $c = l(x_0) - f(x_0)$, $h = (l, c)$ (i.e., $h(x) = l(x) - c$ for all $x \in Z$). Then the following inclusions are equivalent:*

(i) $l \in \partial_{L,Z} f(x_0)$;

(ii) $h \in \partial^*_{H_L,Z} f(x_0)$.

Proof: (i) $\Longrightarrow$ (ii). Since $l \in \partial_{L,Z} f(x_0)$ we have, for all $x \in Z$

$$h(x) = l(x) - c = l(x) - (l(x_0) - f(x_0)) \leq f(x).$$

Therefore, $h \in \operatorname{supp}(f, H, Z)$ and

$$h(x_0) = l(x_0) - c = f(x_0).$$

(ii) $\Longrightarrow$ (i). We have, for all $x \in Z$,

$$l(x) - c \leq f(x); \quad c = l(x_0) - f(x_0).$$

Therefore $l(x) - l(x_0) \leq f(x) - f(x_0)$ as required. △

Next we consider some examples.

EXAMPLE 7.5 Consider the segment $Z = [-1, 1]$ on the line $X = \mathbb{R}$. Let H be the set of all functions h given by

$$h(x) = ax^2 + bx + c$$

with $a, b, c \in \mathbb{R}$, $a \leq 0$. We can consider H as the set H_L of all L-affine functions with respect to the set $L = \{l_{a,b} \in H : a \leq 0,\ b \in \mathbb{R}\}$, where $l_{a,b}(x) = ax^2 + bx$ for all $x \in \mathbb{R}$. It follows from Proposition 6.3 that the set of all abstract convex functions $P(H, X, Z)$ coincides with the set of all lower semicontinuous functions on $[-1, 1]$. For a lower semicontinuous function f and $x_0 \in [-1, 1]$, we have

$$\partial_{L,Z} f(x_0) = \{l_{a,b} : a(x^2 - x_0^2) + b(x - x_0) \leq f(x) - f(x_0), \quad -1 \leq x \leq 1\}.$$

Consider now the set

$$\partial^*_{H,Z} f(x_0) = \{h \in \operatorname{supp}(f, H, Z) : h(x_0) = f(x_0)\}.$$

Proposition 7.1 demonstrates that we can apply this set in the study of the subdifferential. There exist functions f such that $\partial^*_{H,Z} f(x_0) = \emptyset$. For instance, $\partial^*_{H,Z} f(x_0) = \emptyset$ if $f(x) = -|x|$, $x_0 = 0$. It is easy to check (compare with Lemma 7.3 below) that $\partial^*_{H,Z} f(x_0) \neq \emptyset$ if f is twice continuously differentiable at $x_0 \in (-1, 1)$. Applying Proposition 7.1 we conclude that the L-subdifferential $\partial_{L,Z} f(x_0)$ with respect to $Z = [-1, 1]$ is not empty for all twice continuously differentiable functions f if x_0 is an interior point of the segment Z.

EXAMPLE 7.6 Let X be a Banach space and L be the set of all functions l given by

$$l(x) = a\|x - x_0\|, \qquad x \in X,$$

where $a \leq 0$, $x_0 \in X$. The set L enjoys the property (A) and $H_L = L \times \mathbb{R}$ is the set of all functions $h = (l, c) : x \mapsto a\|x - x_0\| - c$ with $a \leq 0$, $x_0 \in X$, $c \in \mathbb{R}$. Let Z be a subset of X. For a function $f : Z \to \mathbb{R}$ and a point $x_0 \in Z$ consider the quantity

$$\beta(f, x_0) = \inf_{x \in Z, x \neq x_0} \frac{f(x) - f(x_0)}{\|x - x_0\|}. \tag{7.2.8}$$

Let f be a Lipschitz function on Z with the Lipschitz constant $\beta(f)$:

$$\beta(f) = \sup_{x \neq y, x, y \in Z} \frac{|f(x) - f(y)|}{\|x - y\|}.$$

Then $\beta(f, x_0) \geq -\beta(f) > -\infty$ for all $x_0 \in Z$.

PROPOSITION 7.2 *Let f be a Lipschitz function defined on Z. Then the set $\partial_{L,Z} f(x_0)$ is nonempty for all $x_0 \in Z$ and*

$$\partial_{L,Z} f(x_0) = \{l \in L : l(x) = a\|x - x_0\|, a \leq \beta(f, x_0)\},$$

where $\beta(f, x_0)$ is defined by (7.2.8).

Proof: Let $h(x) = \beta(f, x_0)\|x - x_0\| + f(x_0)$. Then $h(x_0) \leq f(x_0)$ for all $x \in Z$, and $h(x_0) = f(x_0)$. Therefore (see Proposition 7.1) for $l(x) = \beta(f, x_0)\|x - x_0\|$ we have $l \in \partial_{L,Z} f(x_0)$. So $\partial_{L,Z} f(x_0) \neq \emptyset$. Let $h(x) = a\|x - x_0\| - c$. An easy calculation shows that $h \in \text{supp}(f, H, Z)$ and $h(x_0) = f(x_0)$ if and only if $-c = f(x_0)$ and $a \leq \beta(f, x_0)$. △

Let Z be a subset of a set X and L be a set of functions defined on X. The L-subdifferential $\partial_{L,Z} f(x_0)$ was defined only for H_L-convex functions on Z. However, the definition of the subdifferential (see Definition 7.6) can be formally applied for an arbitrary function $g : Z \to \mathbb{R}_{+\infty}$. Thus the subdifferential $\partial_{L,Z} g(x_0)$ at a point $x_0 \in Z$ is defined by

$$\partial_{L,Z} g(x_0) = \{l \in L : l(x) - l(x_0) \leq g(x) - g(x_0) \text{ for all } z \in X\}.$$

We now show that the subdifferential for an arbitrary function can be expressed in terms of the subdifferential of (H_L, X, Z)-convex hull $\text{co}^Z_{H_L}$ of this function. Hence the study of the subdifferential for an arbitrary function can be reduced to the case of an abstract convex function.

PROPOSITION 7.3 *Let X, Z and L be as above, let g be a function defined on Z and mapping into $\mathbb{R}_{+\infty}$ and let $f = \text{co}^Z_{H_L} g$. Let $x_0 \in Z$ be a point such that $\partial_{L,Z} f(x_0) \neq \emptyset$. Then the following assertions are equivalent:*

(a) $g(x_0) = f(x_0)$;

(b) $\partial_{L,Z}g(x_0) = \partial_{L,Z}f(x_0)$;

(c) $\partial_{L,Z}g(x_0) \neq \emptyset$.

Proof: We will prove the implications:

$$(a) \implies (b) \implies (c) \implies (a).$$

(a) $\implies$ (b). Let $l \in \partial_{L,Z}f(x_0)$. Since $f \leq g$ and $f(x_0) = g(x_0)$, we have for all $x \in Z$

$$l(x) - l(x_0) \leq f(x) - f(x_0) \leq g(x) - g(x_0).$$

Thus $l \in \partial_{L,Z}g(x_0)$. Conversely, for $l' \in \partial_{L,Z}g(x_0)$ we have

$$h'(x) := l'(x) - l'(x_0) + g(x_0) \leq g(x) \text{ for all } x \in Z. \tag{7.2.9}$$

The function h' is L-affine, hence $h' \in \text{supp}(g, H, Z)$. Thus we have for all $x \in Z$

$$\begin{aligned} f(x) &= \text{co}^Z_{H_L}\, g(x) = \sup\{h(x) : h \in \text{supp}(g, H, Z)\} \geq h'(x) \\ &= l'(x) - l'(x_0) + g(x_0) = l'(x) - l'(x_0) + f(x_0). \end{aligned}$$

(b) $\implies$ (c). The proof is straightforward.

(c) $\implies$ (a). Let $l \in \partial_{L,Z}g(x_0)$. Since the function $h(x) = l(x) - l(x_0) + g(x_0)$ is L-affine and $h \leq g$, we have for all $x \in Z$:

$$h(x) = l(x) - l(x_0) + g(x_0) \leq \sup\{h(x) : h \in \text{supp}(g, H, Z)\} = f(x).$$

Since $h(x_0) = g(x_0)$, it follows that $g(x_0) \leq f(x_0)$. On the other hand, the inequality $f \leq g$ implies $f(x_0) \leq g(x_0)$. Thus $f(x_0) = g(x_0)$. △

7.2.4. Conjugation and approximate subdifferentials

In this subsection we shall discuss some properties of the Fenchel-Moreau conjugate functions. We shall also study approximate subdifferentials (known also as ε-subdifferentials) and their connections with conjugate functions.

Consider a pair (X, L) of sets with a coupling function $\langle\cdot|\cdot\rangle : L \times X \to \mathbb{R}$. As usual (see Remark 1.3), we shall consider an element $l \in L$ as a function defined on X and, similarly, an element $x \in X$ as a function defined on L. Sometimes we shall write either $l(x)$ or $x(l)$

interchangeably with $\langle l|x\rangle$. Thus we can consider L as a set of elementary functions defined on X and X as a set of elementary functions defined on L. Let H_L (H_X) be the set of all L-affine functions defined on X (the set of all X-affine functions defined on L). Recall that an L-affine function h where $h(x) = \langle l|x\rangle - c$ is denoted by $h = (l, c)$. In the same manner an X-affine function y, where $y(l) = \langle l|x\rangle - c$, will be denoted by $y = (x, c)$.

Recall that the union of the set of all functions $f: X \to \mathbb{R}_{+\infty}$ and the function $-\infty$ is denoted by F_X.

DEFINITION 7.7 Let $f \in F_X$. The function f^* defined by

$$f^*(l) = \sup_{x \in X}(l(x) - f(x)) \text{ for all } l \in L$$

is called *L-conjugate* to f or *Fenchel-Moreau conjugate function to f with respect to L.*

Clearly, if $f = -\infty$, then $f^* = +\infty$, if $f = +\infty$, then $f^* = -\infty$, and if f maps into $\mathbb{R}_{+\infty}$ and $f \not\equiv +\infty$ then f^* maps into $\mathbb{R}_{+\infty}$. Thus $f^* \in F_L$ for all $f \in F_X$.

We now point out some properties of conjugate functions, which follow directly from the definition.

a) Young inequality: if $f \in F_X$, then

$$f(x) + f^*(l) \geq \langle l|x\rangle \text{ for all } l \in L \text{ and } x \in X; \tag{7.2.10}$$

b) for $f_1, f_2 \in F_L$ we have

$$f_1 \geq f_2 \implies f_1^* \leq f_2^*; \tag{7.2.11}$$

c) let $(f_\alpha)_{\alpha \in A}$ be an arbitrary bounded from below family of functions and $f(x) = \inf_{\alpha \in A} f_\alpha(x)$. Then $f^*(l) = \sup f^*_{\alpha \in A}(l)$ for all $l \in L$;

d) $(f + c\mathbf{1})^* = f^* - c\mathbf{1}$.

PROPOSITION 7.4 *Let $f \in F_X$. Then the function f^* is abstract convex with respect to H_X.*

Proof: We consider only functions $f \in F_X$ such that $\operatorname{dom} f$ is nonempty. Let $x \in \operatorname{dom} f$ and let the function $y_x : L \to \mathbb{R}$ be defined by

$$y_x(l) = \langle l|x\rangle - f(x) \equiv x(l) - f(x).$$

Clearly, $y_x \in H_X$. Let $U = \{y_x : x \in \operatorname{dom} f\}$. For $l \in L$ we have

$$f^*(l) = \sup_{x \in X}(\langle l|x\rangle - f(x)) = \sup_{x \in \operatorname{dom} f}(x(l) - f(x)) = \sup_{x \in \operatorname{dom} f} y_x(l).$$

$\triangle$

Let $f \in F_X$. Since $f^* \in F_L$, we can consider the X-conjugate $(f^*)^*$ of the function f^*. The function $f^{**} := (f^*)^*$ is called the *second conjugate (or biconjugate)* of f. By definition,

$$f^{**}(x) = \sup_{l \in L}(\langle l|x\rangle - f^*(l)), \qquad (x \in X).$$

PROPOSITION 7.5 *If $f \in F_X$, then $f \geq f^{**}$.*

Proof: Applying (7.2.10), we can conclude that for all $l \in L$

$$f(x) \geq \langle l, x\rangle - f^*(l).$$

So

$$f(x) \geq \sup_{l \in L}(\langle l, x\rangle - f^*(l)) = f^{**}(x).$$ $\triangle$

THEOREM 7.1 (Fenchel-Moreau) Let $f \in F_X$. Then $f = f^{**}$ if and only if f is an H_L-convex function.

Proof: If $f = f^{**}$, then (see Proposition 7.4) f is H_L-convex. Assume now that f is H_L-convex. Then there exists a subset U of the set H_L such that $f(x) = \sup_{(l,c) \in U}(\langle l|x\rangle - c)$ for all $x \in X$. For $h = (l, c) \in U$, we have $f(x) \geq \langle l, x\rangle - c$ for all $x \in X$, so

$$f^*(l) = \sup_{x \in X}(\langle l, x\rangle - f(x)) \leq c.$$

Therefore,

$$f^{**}(x) \geq \langle l, x\rangle - f^*(l) \geq l(x) - c = h(x).$$

Since $f^{**}(x) \geq h(x)$ for all $h \in U$ it follows that $f^{**}(x) \geq \sup_{h \in U} h(x) = f(x)$. The reverse equality follows from Proposition 7.5. $\triangle$

Let $f \in F_X$ and let f_L be the H_L-convex hull of f on the set $Z = X$. By the definition of the abstract convex hull

$$f_L(x) = \sup\{\langle l|x\rangle - c : h = (l, c) \in \operatorname{supp}(f, H_L)\}, \qquad x \in X.$$

Recall that f_L is an H_L-convex function. If g is H_L-convex and $g \leq f$, then $g \leq f_L$ and $f = f_L$ if and only if f is an H_L-convex function.

PROPOSITION 7.6 *Let $f \in F_X$. Then f^{**} coincides with the H_L-convex hull f_L of the function f.*

Proof: It follows from Proposition 7.4 that f^{**} is an H_L-convex function. Proposition 7.5 shows that $f^{**} \leq f$. Therefore $f^{**} \leq f_L$. On the other hand, $f_L \leq f$. Applying (7.2.11) twice and Fenchel-Moreau Theorem, we have, that $f_L = f_L^{**} \leq f^{**}$. △

Now we point out some links between the conjugate with respect to L function and the subdifferentials $\partial_L f(x_0) \equiv \partial_{L,X} f(x_0)$ for $x_0 \in X$.

PROPOSITION 7.7 *Let $l_0 \in L$ and $x_0 \in X$. The following assertions are equivalent:*

(i) $\langle l_0 | x_0 \rangle = f^*(l_0) + f(x_0)$ *(that is, the equality holds in the Young inequality);*

(ii) $l_0 \in \partial_L f(x_0)$.

Proof: (*i*) $\Longrightarrow$ (*ii*). Since $f^*(l_0) \geq \langle l_0, x \rangle - f(x)$ for all $x \in X$, applying (*i*) we have

$$\langle l_0, x_0 \rangle - f(x_0) \geq \langle l_0, x \rangle - f(x) \text{ for all } \ x \in X,$$

that is, $l_0 \in \partial_L f(x_0)$
(*ii*) $\Longrightarrow$ (*i*). By the definition of the subdifferential we have

$$\langle l_0 | x_0 \rangle - f(x_0) \geq \langle l_0 | x \rangle - f(x) \text{ for all } x \in X,$$

therefore $\langle l_0, x_0 \rangle - f(x_0) \geq f^*(l)$. Applying (7.2.10) we get (*i*). △

Now we will give a description of the support set of a function $f \in F_X$ in terms of the epigraph of the conjugate function. Recall that the set

$$\text{epi}\, f = \{(x, \alpha) : x \in \text{dom}\, f, \ \alpha \geq f(x)\}$$

is called the *epigraph* of a function $f : X \to \mathbb{R}_{+\infty}$. If $f = -\infty$, then by definition, $\text{epi}\, f = X \times \mathbb{R}$.

PROPOSITION 7.8 *Let $f \in F_X$. Then*

$$\text{epi}\, f^* = \text{supp}(f, H_L),$$

where $\text{supp}(f, H_L) \equiv \text{supp}(f, H_L, X)$ *is the support set of the function f with respect to H_L.*

Proof: We have

$$(h, \lambda) \in \text{epi}\, f^* \iff \lambda \geq f^*(h) \iff \lambda \geq h(x) - f(x) \text{ for all } x \in X$$

$$\iff f(x) \geq h(x) - \lambda \text{ for all } x \in X \iff (h, \lambda) \in \text{supp}(f, H_L). \quad \triangle$$

We shall show that the set epi f^* (in other words, the support set of a function f) can easily be expressed in terms of so-called ε- subdifferentials. First, we shall define ε-subgradients and ε- subdifferentials.

DEFINITION 7.8 Let Z be a subset of a set X and let f be a function defined on Z. Furthermore, let ε be a nonnegative number. A function $l \in L$ is called an *ε-subgradient* of the function f at the point x_0 if, for all $x \in Z$,

$$l(x) - l(x_0) \leq f(x) - f(x_0) + \varepsilon.$$

The set $\partial_{L,Z,\varepsilon} f(x_0)$ of all ε-subgradients of f at x_0 is called the *ε-subdifferential* of f at x_0 with respect to L.

Sometimes ε-subdifferentials with $\varepsilon > 0$ are called *approximate* subdifferentials. The ε-subdifferential with $\varepsilon = 0$ coincides with the subdifferential $\partial_{L,Z} f(x_0)$.

If $Z = X$ we omit the symbol Z in this notation, so we denote $\partial_{L,X,\varepsilon} f(x_0)$ by $\partial_{L,\varepsilon} f(x_0)$.

It follows directly from Definition 7.8, that

$$\partial_{L,Z,\varepsilon'} f(x_0) \subset \partial_{L,Z,\varepsilon} f(x_0)$$

for $0 \leq \varepsilon' < \varepsilon$.

We shall examine links between support sets and approximate subdifferentials. Let $h \in H_L$. Then there exist $l \in L$ and $\lambda \in \mathbb{R}$ such that $h(x) = l(x) - \lambda$. First, we consider H_L-convex functions. If f is H_L-convex then for each $\varepsilon > 0$ there exists a function $l \in L$ and a number $\lambda \in \mathbb{R}$ such that

$$l(x) - \lambda \leq f(x) \text{ for all } x \in X \text{ and } l(x_0) - \lambda > f(x_0) - \varepsilon.$$

Thus $l(x) - l(x_0) \leq f(x) - f(x_0) + \varepsilon$. It follows from this inequality that $\partial_{L,Z,\varepsilon} f(x_0) \neq \emptyset$ for all $\varepsilon > 0$.

PROPOSITION 7.9 *Let $f : Z \to \mathbb{R}_{+\infty}$ be a function with the nonempty support set* supp(f, H_L, Z). *Let $x_0 \in$* dom $f = \{x \in Z : f(x) < +\infty\}$. *Then*

$$\text{supp}(f, H_L, Z) = \bigcup_{\varepsilon \geq 0} \{(l, \lambda) : l \in \partial_{L,Z,\varepsilon} f(x_0),\ \lambda = -(f(x_0) - l(x_0)) + \varepsilon\}. \tag{7.2.12}$$

Proof: Denote by V the set on the right-hand side of (7.2.12). If $(l, \lambda) \in$ supp(f, H_L, Z), then $f(x) \geq l(x) - \lambda$ for all $x \in Z$. In particular $f(x_0) \geq$

$l(x_0) - \lambda$. Let $\varepsilon = (f(x_0) - l(x_0)) + \lambda$. Then $\varepsilon \geq 0$ and $f(x) - f(x_0) \geq l(x) - l(x_0) - \varepsilon$. Thus $l \in \partial_{L,Z,\varepsilon} f(x_0)$ and $(l, \lambda) \in V$.

Conversely, if $(l, \lambda) \in V$ then there exists $\varepsilon \geq 0$ such that $l \in \partial_{L,Z,\varepsilon} f(x_0)$ and $\lambda = -(f(x_0) - l(x_0)) + \varepsilon$. Hence

$$f(x) \geq f(x_0) + l(x) - l(x_0) - \varepsilon \quad \text{and} \quad -\lambda = [f(x_0) - l(x_0)] - \varepsilon \text{ for all } x \in X.$$

Thus $(l, \lambda) \in \text{supp}(f, H_L, Z)$ as required. △

PROPOSITION 7.10 *Let f be an H_L-convex function and $\varepsilon \geq 0$. Let $x_0 \in X$. Then*

$$l \in \partial_{L,\varepsilon} f(x_0) \iff f^*(l) + f(x_0) - l(x_0) \leq \varepsilon.$$

Proof: The proof follows from the definitions of the L-conjugate function and the ε-subgradient. △

Let $U \subset L$. The indicator function $\delta_U : H \to \mathbb{R}_{+\infty}$ of the set U is defined by

$$\delta_U(l) = \begin{cases} 0 & \text{if } l \in U; \\ +\infty & \text{if } l \notin U. \end{cases} \tag{7.2.13}$$

The support function $p_U : X \to \mathbb{R}_{+\infty}$ of the set U is defined by

$$p_U(x) = \sup\{\langle l|x\rangle : l \in U\} \tag{7.2.14}$$

PROPOSITION 7.11 *Let $U \subset H$. Then $\delta_U^* = p_U$.*

Proof: For $x \in X$ we have

$$\begin{aligned} \delta_U^*(x) &- \sup\{\langle l|x\rangle - \delta_U(x) : l \in L\} \\ &= \sup\{\langle l|x\rangle - \delta_U(x) : l \in U\} \\ &= \sup\{\langle l|x\rangle : l \in U\} = p_U(x). \end{aligned}$$

Thus the result follows. △

It is convenient to present Lemma 7.1, which expresses the separation property of abstract convex sets, in the following form:

PROPOSITION 7.12 *A set $U \subset L$ is (L, X)-convex if and only if $\delta_U^{**}(l) > 0$ for each $l \in L \setminus U$.*

Proof: It follows from Proposition 7.11 that

$$\delta_U^{**}(l) = \sup_{x \in X}(\langle l|x\rangle - \delta_U^*(x)) = \sup_{x \in X}(\langle l|x\rangle - \sup_{l' \in U}\langle l'|x\rangle).$$

Let U be (L, X)-convex and $l \notin U$. It follows from Lemma 7.1 that there exists $x \in X$ such that $\langle l|x\rangle > \sup_{l' \in U}\langle l'|x\rangle$, hence $\delta_U^{**}(l) > 0$. The similar argument shows that the reverse assertion holds as well. △

The following result (see [100]) is useful in the study of L-convex sets.

THEOREM 7.2 *A set $U \subset L$ is (L, X)-convex if and only if there exist an H_X-convex function f and a number $c \in \mathbb{R}$ such that U coincides with the level set $S_c(f) = \{l \in L : f(l) \leq c\}$.*

Proof: Assume that U is an (L, X)-convex set. If $l \in U$, then $\delta_U^{**}(l) \leq \delta_U(l) = 0$. If $l \notin U$ then, by applying Proposition 7.12, we conclude that $\delta_U^{**}(l) > 0$. Thus $U = \{l : \delta_U^{**}(l) \leq 0\}$. It follows from the Fenchel-Moreau theorem, that the function δ_U^{**} is H_X-convex.

Assume now that $U = \{l \in L : f(l) \leq c\}$, where f is an H_X-convex function and $c \in \mathbb{R}$. We have

$$f(l) = \sup\{\langle l|x\rangle - c' : y = (x, c') \in \operatorname{supp}(f, H_X)\}.$$

If $l' \notin U$ then $f(l') > c$. Hence there exists $y = (x, c') \in \operatorname{supp}(f, L)$ such that $\langle l'|x\rangle - c' > c$. Thus we have $\langle l'|x\rangle > c + c'$. On the other hand, we have for all $l \in U$

$$\langle l|x\rangle - c' \leq f(l) \leq c,$$

so $\sup_{l \in U}\langle l|x\rangle \leq c + c' < \langle l', x\rangle$. The result follows from Lemma 7.1. △

REMARK 7.3 Since the set $U = \{l \in L : f(l) \leq -\infty\} = \emptyset$ is (L, X)-convex and the set $U = \{l \in L : f(l) \leq +\infty\} = L$ is also (L, X)-convex, we can put $c \in \bar{\mathbb{R}}$ instead of $c \in \mathbb{R}$ in the statement of Theorem 7.2.

7.2.5. Abstract convexity and global minimization

The notions of subdifferential and ε-subdifferential are useful in the study of some unconstrained problems of global minimization. Let L be a set of abstract linear functions defined on a set X and containing the function 0. The following assertion follows directly from the definition of the subdifferential.

PROPOSITION 7.13 *Let f be an H_L-convex function defined on X and $Z \subset X$. A point $x_0 \in Z$ is a global minimizer of the function f over the set Z if and only if $0 \in \partial_{L,Z} f(x_0)$.*

REMARK 7.4 If x_0 is a global minimizer of a function f over a set Z then for an arbitrary L with $0 \in L$, the subdifferential $\partial_{L,Z} f(0)$ is not empty.

PROPOSITION 7.14 *1) Let f be a bounded from below function, defined on a set $Z \subset X$. Then*

$$\inf_{x \in Z} f(x) = \inf_{x \in Z} \operatorname{co}^Z_{H_L} f(x), \tag{7.2.15}$$

where $\operatorname{co}^Z_{H_L} f$ is the abstract convex hull of the function f with respect to the set H_L of all L-affine functions.

2) Each global minimizer of f is also a global minimizer of the function $\operatorname{co}^Z_{H_L} f$.

Proof: 1) We have

$$\inf\{f(x) : x \in Z\} = \sup\{-c : (0, c) \in \operatorname{supp}(f, H_L, Z)\}.$$

Since $\operatorname{supp}(f, H_L, Z) = \operatorname{supp}(\operatorname{co}^Z_{H_L} f, H_L, Z)$ it follows, that the equality (7.2.15) holds.

2) Let x_0 be a global minimizer of the function f. Applying the equality (7.2.15), we have

$$\operatorname{co}^Z_{H_L} f(x_0) \le f(x_0) = \min\{f(x) : x \in Z\} = \inf\{\operatorname{co}^Z_{H_L}(x) : x \in Z\}.$$

Thus the desired result follows. △

In some instances we can take advantage of the special structure of a function in order to obtain necessary and sufficient conditions for global optimality. The result of this kind has been obtained by Hiriart-Urruty [74] for the difference $q - p$ of convex functions p and q in terms of ε-subdifferentials. This result was generalized by Tuy and Oettli [191] for the difference $q - p$ where only p is convex. We now present a result of this kind, where p is a function with the nonempty support set.

THEOREM 7.3 *Let X be an arbitrary set and $Z \subset X$. Let L be a set of functions defined on X. Let $f = q - p$, where p is a function such that the support set* $\operatorname{supp}(p, H_L, Z)$ *is nonempty.*

Then a point $x_0 \in Z$ is a global minimizer of the function f over the set Z if and only if, for all $\varepsilon > 0$,

$$\partial_{L,Z,\varepsilon} p(x_0) \subset \partial_{L,Z,\varepsilon} q(x_0). \tag{7.2.16}$$

Proof: Assume without loss of generality that $\min_{x \in Z} f(x) = 0$. Then a point x_0 is a global minimizer of the function f over Z if and only if $p(x_0) = q(x_0)$ and $q(x) \ge p(x)$ for all $x \in Z$. We have

$$q \ge p \iff \operatorname{supp}(q, H_L, Z) \supset \operatorname{supp}(p, H_L, Z) \ne \emptyset.$$

Applying Proposition 7.9, we can represent the last inclusion in the following form:

$$\bigcup_{\varepsilon\geq 0}\{(\ell,\lambda):\ \ell\in\partial_{L,Z,\varepsilon}q(x_0),\ \lambda=-(q(x_0)-\ell(x_0)+\varepsilon)\}\supset$$

$$\bigcup_{\varepsilon\geq 0}\{(\ell,\lambda):\ \ell\in\partial_{L,Z,\varepsilon}p(x_0),\ \lambda=-(p(x_0)-\ell(x_0)+\varepsilon)\}. \qquad (7.2.17)$$

Since $q(x_0)=p(x_0)$, the equality (7.2.17) holds if and only if (7.2.16) is valid. △

7.2.6. Minimax result for abstract convex functions

Consider another application of abstract convexity. Let H be a set of elementary functions defined on X such that the following properties hold:

1) $0\in H$.

2) H is closed under vertical shifts: if $h\in H$ and $\lambda\in\mathbb{R}$, then $h+\lambda\mathbf{1}\in H$.

THEOREM 7.4 *Let Z be an arbitrary subset of X and let U be an H-convex set with respect to Z. If*

$$\inf_{z\in Z}\sup_{h\in U}h(z)<+\infty,$$

then

$$\inf_{z\in Z}\sup_{h\in U}h(z)=\sup_{h\in U}\inf_{z\in Z}h(z).$$

Proof: The following inequality is always valid:

$$\sup_{h\in U}\inf_{z\in Z}h(z)\leq\inf_{z\in Z}\sup_{h\in U}h(z). \qquad (7.2.18)$$

Let $\psi(z)=\sup_{h\in U}h(z)$ and let

$$v=\inf_{z\in Z}\sup_{h\in U}h(z)=\inf_{z\in Z}\psi(z).$$

Consider the following two cases:

Case 1. $v=0$. We have $\inf_{z\in Z}\psi(z)=0$. Therefore, $\psi(z)\geq 0$ for all $z\in Z$ and $0\in\operatorname{supp}(\psi,H,Z)$. Since U is H-convex, we have $\operatorname{supp}(\psi,H,Z)=U$. Hence $0\in U$ and $\sup_{h\in U}\inf_{z\in Z}h(z)\geq 0$.

Case 2. Now assume $v \neq 0$. By (7.2.18) the result holds trivially if $v = -\infty$. We suppose that $v > -\infty$. By assumption, $v < +\infty$. Thus $v \in \mathbb{R}$ and we have $\inf_{z\in Z} \sup_{h\in U}(h(z) - v) = 0$. Let $h'(z) = h(z) - v = (h - v\mathbf{1})(z)$ and $U' = \{h' : h' = h - v\mathbf{1},\ h \in U\}$. Let $\psi'(z) = \psi(z) - v$.

We now verify that $U' = \operatorname{supp}(\psi', H, Z)$. This follows since

$$h \in U \implies h \leq \psi, h \in H \implies h - v \leq \psi', h \in H$$

and

$$\begin{aligned}\tilde{h} \in \operatorname{supp}(\psi', H, Z) \quad &\implies \quad \tilde{h}(z) \leq \psi'(z) \text{ for all } z \in Z,\\ &\implies \quad \tilde{h} + v\mathbf{1} \leq \psi \implies \tilde{h} + v\mathbf{1} \in U\\ &\implies \quad \tilde{h} = h - v\mathbf{1} \in U'.\end{aligned}$$

Therefore U' is an H-convex set. We have

$$\inf_{z\in Z} \sup_{h'\in U'} h'(z) = 0.$$

Therefore $\sup_{h'\in U'} \inf_{z\in Z} h'(z) = 0$ and so $\sup_{h\in U} \inf_{z\in Z} h(z) = v$ as required. △

7.2.7. Positively homogeneous functions

Notions and results of this section are simplified for positively homogeneous functions defined on a conic set. Recall (see Definition 1.8) that a subset X of a vector space is called conic if $(x \in X, \lambda > 0) \implies (\lambda x \in X)$. Now we define conic sets in a more general situation.

DEFINITION 7.9 A set X is called conic if the multiplication by a positive scalar is defined on X such that

$$1u = u; \quad \lambda(\mu u) = (\lambda\mu)u \text{ for all } u \in U,\ \lambda, \mu > 0.$$

Let X be a conic set. A function $f : X \to \mathbb{R}_{+\infty}$ is called *proper* if $\operatorname{dom} f = \{x \in X : f(x) < +\infty\}$ is nonempty; f is called positively homogeneous of degree one if $f(\lambda x) = \lambda f(x)$ for all $\lambda > 0$. Denote by $PH(X)$ the set of all proper positively homogeneous of degree one functions $f : X \to \mathbb{R}_{+\infty}$.

Assume that a conic set X contains an element v such that $f(v) = 0$ for all $f \in PH(X)$ and for each $x \in X, x \neq v$ there exists $f \in PH(X)$ such that $f(x) \neq 0$. In such a case the element v is called zero and is denoted by 0. We accept the following definition: $0 \cdot x = 0$ for all $x \in X$.

Let X be a conic set containing zero and let $L \subset PH(X)$ be a set of finite functions. We consider L as a set of elementary functions. It follows from positive homogeneity that L enjoys the property (A) (see (7.2.6)). Let Z be a nonempty conic subset of the set X. We now show that the set $P(L, X, Z)$ of all L-convex functions on Z coincides with the set $P(H_L, X, Z)$ of all H_L-convex functions, where H_L is the set of all L-affine functions.

PROPOSITION 7.15 *Let $f \in PH(Z)$. Then f is H_L-convex if and only if f is L-convex.*

Proof: It is obvious that $P(L, X, Z) \subset P(H_L, X, Z)$, so we only need to check the reverse inclusion. For a function $f \in PH(Z)$, a function $h = (l, c) \in \text{supp}(f, H_L, Z)$, a point $x_0 \in Z$ and a number $\lambda > 0$, we have

$$\begin{aligned} \lambda f(x_0) &= f(\lambda x_0) = \sup\{l(\lambda x_0) - c : (l, c) \in \text{supp}(f, H_L, Z)\} \\ &= \lambda \left(\sup \left\{ l(x_0) - \frac{c}{\lambda} : (l, c) \in \text{supp}(f, H_L, Z) \right\} \right). \end{aligned}$$

Dividing by λ and turning λ to $+\infty$ we conclude that

$$f(x_0) = \sup\{l(x_0) : (l, c) \in \text{supp}(f, H_L, Z)\} = \sup\{l(x_0) : l \in U\},$$

where U is the set of all l such that $(l, c) \in \text{supp}(f, H_L, Z)$ for some number c. Thus f is L-convex. △

It is easy to check that Proposition 1.3 is valid for an arbitrary conic set with zero, that is

$$\text{supp}(f, L, Z) = \{l \in L : l(z) \le f(z) \quad \text{for all } z \in Z\},$$

hence

$$\text{supp}(f, L, Z) = \partial_{L,Z} f(0). \tag{7.2.19}$$

We now describe the support set, $\text{supp}(f, H_L, Z)$.

PROPOSITION 7.16 *Let $f \in PH(Z)$. Then* $\text{supp}(f, H_L, Z) = \partial_{L,Z} f(0) \times \mathbb{R}_+$.

Proof: We have:

$$\begin{aligned} (l, c) \in \text{supp}(f, H_L, Z) & \\ &\Longrightarrow l(x) - c \le f(x) \text{ for all } x \in Z \\ &\Longrightarrow l(\lambda x) - c \le f(\lambda x) \text{ for all } x \in Z, \lambda > 0 \\ &\Longrightarrow l(x) - \frac{c}{\lambda} \le f(x) \text{ for all } x \in Z, \lambda > 0 \\ &\Longrightarrow l(x) \le f(x) \text{ for all } x \in Z \\ &\Longrightarrow l \in \partial_{L,Z} f(0). \end{aligned}$$

Since l and f are positively homogeneous, it follows that $l(0) = f(0) = 0$ so the inequality $l(0) - c \leq f(0)$ implies $c \geq 0$. Thus $\operatorname{supp}(f, H_L, Z) \subset \partial_{L,Z} f(0) \times \mathbb{R}_+$. The reverse inclusion is trivial. △

PROPOSITION 7.17 *Let $f \in PH(Z)$ and let $x_0 \in Z$. Then*

$$\partial_{L,Z} f(x_0) = \{l \in \partial_{L,Z} f(0) : l(x_0) = f(x_0)\}.$$

Proof: Clearly, relations $l \in \partial_{L,Z} f(0)$ and $l(x_0) = f(x_0)$ imply $l \in \partial_{L,Z} f(x_0)$. Conversely, assume $l \in \partial_{L,Z} f(x_0)$. For $x \in Z$ and $\lambda > 0$ we have

$$l(\lambda x) - l(x_0) \leq f(\lambda x) - f(x_0). \tag{7.2.20}$$

Representing (7.2.20) in the form

$$l(x) - \frac{1}{\lambda} l(x_0) \leq f(x) - \frac{1}{\lambda} f(x_0)$$

and turning λ to $+\infty$ we conclude that $l(x) \leq f(x)$. So, $l \in \partial_{L,Z} f(0)$. In particular, the inequality $l(x_0) \leq f(x_0)$ holds. On the other hand, it follows from (7.2.20), with $\lambda = 0$, that $l(x_0) \geq f(x_0)$. Thus $l(x_0) = f(x_0)$. △

Since $\partial_{L,Z} f(0) = \operatorname{supp}(f, L, Z)$ (see (7.2.19)), we can present Proposition 7.17 in the following form.

PROPOSITION 7.18 *Let f be an L-convex function. The following assertions are equivalent:*

(i) $l \in \partial_{L,Z} f(x_0)$;

(ii) $l \in \operatorname{supp}(f, L, Z)$ *and* $l(x_0) = f(x_0)$.

We now give a description of approximate subdifferentials for positively homogeneous functions.

PROPOSITION 7.19 *Let $Z \subset X$ be a conic set, $f : Z \to \mathbb{R}_{+\infty}$, $\varepsilon > 0$ and $x_0 \in Z$. Then*

$$\partial_{L,\ Z,\ \varepsilon} f(x_0) = \{l \in \operatorname{supp}(f, L, Z) : l(x_0) \geq f(x_0) - \varepsilon\}.$$

Proof: Let l be an ε-subgradient of the function f at the point x_0, that is,

$$l(x) - l(x_0) \leq f(x) - f(x_0) + \varepsilon \text{ for all } x \in Z.$$

Since Z is a conic set, we have for an arbitrary $\lambda > 0$ and for all $x \in Z$

$$l(\lambda x) - f(\lambda x) \leq l(x_0) - f(x_0) + \varepsilon. \qquad (7.2.21)$$

So

$$l(x) - f(x) \leq \frac{1}{\lambda}\left(l(x_0) - f(x_0) + \varepsilon\right) \longrightarrow_{\lambda \to +\infty} 0.$$

Thus $l(x) \leq f(x)$ for all $x \in Z$, that is, $l \in \mathrm{supp}(f, L, Z)$. It follows from (7.2.21) with $x = 0$ that $f(x_0) - \varepsilon \leq f(x_0)$.

Assume now that $l \in \mathrm{supp}(f, L, Z)$ and $l(x_0) \geq f(x_0) - \varepsilon$. Then

$$f(x_0) - l(x_0) - \varepsilon \leq 0 \leq f(x) - l(x) \text{ for all } x \in Z,$$

hence $l \in \partial_{L,\ Z,\varepsilon} f(x_0)$. △

Consider now a pair (X, L) of conic sets with a coupling function $\langle \cdot | \cdot \rangle : L \times X \to \mathbb{R}$. Assume that $\langle l | \lambda x \rangle = \lambda \langle l | x \rangle$ for all $\lambda > 0$.

PROPOSITION 7.20 *Let $f \in PH(X)$ be a proper function. Then the Fenchel-Moreau conjugate function f^* coincides with the indicator function $\delta_{\mathrm{supp}(f,L)}$ of the support set* $\mathrm{supp}(f, L)$.

Proof: Since $X = \{\lambda x : x \in X, \lambda > 0\}$, we have

$$\begin{aligned} f^*(l) &= \sup_{x \in X} (\langle l | x \rangle - f(x)) = \sup_{x \in X} \sup_{\lambda > 0} (\langle l | \lambda x \rangle - f(\lambda x) \\ &= \sup_{x \in X, \lambda > 0} \lambda (\langle l | x \rangle - f(x)). \end{aligned}$$

Hence

$$f^*(l) = \begin{cases} 0 & \text{if } \langle l | x \rangle \leq p(x) \text{ for all } x \in X; \\ +\infty & \text{otherwise}. \end{cases}$$

that is

$$f^* = \delta_{\mathrm{supp}(p,L)}.$$

△

7.2.8. Positively homogeneous extension

In many instances we can apply properties of positively homogeneous functions in the study of more complicated functions. The construction of a *positively homogeneous extension* can be used for this purpose. (This construction has already been applied in Subsection 3.2.3.)

Let X be a conic set. Consider the set

$$\hat{X} = \{(x, \lambda) : x \in X, \lambda > 0\} = X \times \mathbb{R}_{++}$$

with the algebraic operation of multiplication by a positive number defined by:

$$\mu(x, \lambda) = (\mu x, \mu \lambda), \qquad (x, \lambda) \in \hat{X},\ \mu > 0.$$

Obviously $\hat{X}$ is a conic set.

DEFINITION 7.10 Let X be a conic set and $f : X \to \mathbb{R}_{+\infty}$. The function $\hat{f} : \hat{X} \to \mathbb{R}_{+\infty}$:

$$\hat{f}(x, \lambda) = \lambda f\left(\frac{x}{\lambda}\right)$$

is called the *positively homogeneous extension* of the function f.

Below there are some properties of the function $\hat{f}$:

1) $\hat{f}(\mu(x,\lambda)) = \hat{f}(\mu x, \mu\lambda) = \mu\lambda f\left(\frac{\mu x}{\mu\lambda}\right) = \mu \hat{f}(x,\lambda)$. Thus $\hat{f}$ is a positively homogeneous function.
2) $\hat{f}(x, 1) = f(x)$.
3) If f is positively homogeneous, then $\hat{f}(x,\lambda) = f(x)$ for all $\lambda > 0$.

Let L be a set of finite positively homogeneous functions defined on a conic set X and let H_L be the set of L-affine functions. If $h = (l, c) \in H_L$, then

$$\hat{h}(x,\lambda) = \lambda h\left(\frac{x}{\lambda}\right) = \lambda\left(l\left(\frac{x}{\lambda}\right) - c\right) = l(x) - c\lambda. \qquad (7.2.22)$$

The notion of the conic (cone) hull of a subset of a vector space is well known. Only the conic structure is used under the definition of the conic hull. So we can also give this definition for subsets of conic sets.

DEFINITION 7.11 If U is a subset of a conic set, then the set $\operatorname{cone} U = \{\lambda x : x \in U, \lambda > 0\}$ is called the *conic hull* or *cone hull* of the set U.

The notion of conic hull allows us to explain the geometrical sense of the positively homogeneous extension.

Let f be a function defined on a conic set X and mapping into $\mathbb{R}_{+\infty}$. Consider the epigraph $\operatorname{epi} \hat{f}$ of the function $\hat{f}$. By the definition of the epigraph,

$$\begin{aligned} \operatorname{epi} \hat{f} &= \{((x,\lambda),\mu) \in X \times \mathbb{R}_{++} \times \mathbb{R} : \mu \geq \hat{f}(x,\lambda)\} \\ &= \left\{((x,\lambda),\mu) \in X \times \mathbb{R}_{++} \times \mathbb{R} : \frac{\mu}{\lambda} \geq f\left(\frac{x}{\lambda}\right)\right\}. \qquad (7.2.23) \end{aligned}$$

Consider now the epigraph, $\operatorname{epi} f$, of the function f and the conic hull of the subset $\operatorname{epi} f \times \{1\}$ of the conic set $\hat{X}$. Assume that $(x, \mu, \lambda) \in \operatorname{cone} \operatorname{epi} f \times \{1\}$. Then

$$\left(\frac{x}{\lambda}, \frac{\mu}{\lambda}\right) \in \operatorname{epi} f,$$

hence

$$\frac{\mu}{\lambda} \geq f\left(\frac{x}{\lambda}\right). \tag{7.2.24}$$

It follows from (7.2.23) and (7.2.24) that we can identify epi $\hat{f}$ with the conic hull of the set (epi f) $\times \{1\}$.

Consider the set $\hat{H}_L$ of positively homogeneous extensions of functions from H_L. This set consists of positively homogeneous functions defined on $\hat{X}$. It follows from (7.2.22) that

$$\hat{H}_L = \{\hat{h} : (\forall x \in X, \lambda > 0)\ \hat{h}(x,\lambda) = l(x) - \lambda c,\ l \in L, c \in \mathbb{R}\}.$$

We shall identify the function $\hat{h}(x,\lambda) = l(x) - \lambda c$ with the pair (l, c).

PROPOSITION 7.21 *Let Z be a conic subset of X and $f : Z \to \mathbb{R}_{+\infty}$ be a function defined on Z. Then f is an H_L-convex function if and only if $\hat{f}$ is an $\hat{H}_L$-convex function and the following holds:*

a)

$$\operatorname{supp}(\hat{f}, \hat{H}_L, Z) = \operatorname{supp}(f, H_L, Z). \tag{7.2.25}$$

b)

$$\begin{aligned} \partial_{\hat{H}_L,\hat{Z}} \hat{f}(x_0, 1) &= \{(\ell, c) \in \operatorname{supp}(f, H_L, Z) : \ell \in \partial_{L,Z} f(x_0), \\ &\qquad c = \ell(x_0) - f(x_0)\}. \end{aligned} \tag{7.2.26}$$

Proof: Let f be H_L-convex. Then for all $x \in Z$ we have

$$\begin{aligned} \hat{f}(x,1) &= f(x) = \sup\{h(x) : h \in \operatorname{supp}(f, H_L, Z)\} \\ &= \sup\{\hat{h}(x,1) : h \in \operatorname{supp}(f, H_L, Z)\}. \end{aligned}$$

Since $\hat{f}$ is positively homogeneous, it follows that

$$\hat{f}(x,\lambda) = \sup\{\hat{h}(x,\lambda) : h \in \operatorname{supp}(f, H_L, Z)\}$$

for all $x \in Z$ and $\lambda > 0$. Hence $\hat{f}$ is abstract convex with respect to $\hat{H}_L$ on $\hat{Z}$. The same argument shows that the abstract convexity of $\hat{f}$ implies the abstract convexity of f. We now prove (7.2.25).

We have:

$$\begin{aligned} \operatorname{supp}(f, H_L, Z) &= \{(l,c) : l(x) - c \leq f(x) \text{ for all } x \in Z\} \\ &= \left\{(l,c) : l\left(\frac{x}{\lambda}\right) - c \leq f\left(\frac{x}{\lambda}\right) \text{ for all } x \in Z,\ \lambda > 0\right\} \\ &= \{(l,c) : l(x) - c\lambda \leq \hat{f}(x,\lambda) \text{ for all } x \in Z,\ \lambda > 0\} \\ &= \operatorname{supp}(\hat{f}, \hat{H}_L, Z). \end{aligned}$$

Thus (7.2.25) holds; (7.2.26) follows directly from (7.2.25), Proposition 7.18 and Proposition 7.1. △

7.2.9. Polarity for functions and sets, which are abstract convex with respect to a conic set of positively homogeneous functions

Many results from classical convex polarity can be extended for the abstract convex setting, assuming that a set of elementary functions is a conic set consisting of positively homogeneous finite functions. (Compare with Section 2.2.) We shall use the theory of radiant sets for this purpose. See Chapter 5 for the presentation of this theory in a finite-dimensional space. In particular we shall use the Minkowski gauge μ_U of a radiant set U.

Note that only a conic structure of the space $\mathbb{R}^n$ was exploited under radiant set and its Minkowski gauge definitions; consequently, we are able to extend these definitions to include an arbitrary conic set.

DEFINITION 7.12 Let X be a conic set. A subset U of the set X is called *radiant* if $(x \in U, \lambda \in (0,1]) \implies \lambda x \in U$.

Let U be a radiant subset of X. The function $\mu_U : X \to \bar{\mathbb{R}}_+$ defined by

$$\mu_U(x) = \inf\{\lambda > 0 : x \in \lambda U\}$$

is called the *Minkowski gauge* of the set U (here $\lambda U = \{\lambda x : x \in U\}$.)

It follows directly from Definition 7.12 that

$$\{x : \mu_U(x) < 1\} \subset U \subset \{x : \mu_U(x) \le 1\}.$$

The subset U of the conic set Y is called *closed-along-rays* if $(\lambda_n > 0, \lambda_n \to \lambda, \lambda_n x \in U) \implies (\lambda x \in U)$. Let U be a radiant subset of $\mathbb{R}^n$. It has been shown (see Proposition 5.1) that $U = \{x \in \mathbb{R}^n : \mu_U(x) \le 1\}$ if and only if U is closed-along-rays. The same argument as in the proof of Proposition 7.12 demonstrates that the following assertion holds:

PROPOSITION 7.22 *Let U be a radiant subset of a conic set Y. Then $U = \{x \in X : \mu_U(x) \le 1\}$ if and only if U is closed-along-rays.*

Let us consider a pair (X, L) of conic sets with the coupling function $\langle l|x\rangle$ defined on $L \times X$ and mapping into the set $\mathbb{R}_+$ of nonnegative numbers. Assume that both sets X and L contain zero. Occasionally we denote the zero element of X (L) by 0_X (0_L). The coupling function allows us to consider an element $l \in L$ as a function defined on X and, similarly, an element $x \in X$ as a function defined on L. We shall write either $l(x)$ or $x(l)$ interchangeably with $\langle l|x\rangle$ in the sequel.

We assume that the coupling function $\langle\cdot|\cdot\rangle$ is positively homogeneous in both variables:

$$\langle l|\lambda x\rangle = \langle \lambda l|x\rangle = \lambda\langle l|x\rangle \text{ for all } l \in L, x \in X, \lambda > 0.$$

Thus $L \subset PH(X)$ and $X \subset PH(L)$. It follows from the definition of zero that

$$0_L(x) = 0_X(l) = 0$$

for all $x \in X$ and all $l \in L$. Since the coupling function maps into the set of nonnegative numbers, it follows that $l(x) \geq 0$ for all functions $l \in L$ and all points $x \in X$ and $x(l) \geq 0$ for all functions $x \in X$ and points $l \in L$.

Note that the definition of the Minkowski gauge of a radiant set $U \subset X$ does not require an outer (dual) conic set L. Having this set and a coupling function, we can define the support function of an arbitrary set $U \subset X$.

DEFINITION 7.13 Let $U \subset X$. The function

$$p_U(l) = \sup_{x\in U}\langle l|x\rangle \equiv \sup_{x\in U} x(l), \qquad (l \in L) \tag{7.2.27}$$

is called the *support function* of the set U (with respect to $(X, L, \langle\cdot|\cdot\rangle)$).

Since the functions $l(x) = \langle l|x\rangle$ are nonnegative and positively homogeneous, it follows that the support function p_U possesses the same properties. Note that $p_U(0_L) = 0$. It follows directly from the definition that the support function p_U is X-convex for each $U \subset X$. The following simple statement indicates links between radiant subsets of X and abstract convexity.

PROPOSITION 7.23 *Any (X, L)–convex set U is radiant and closed-along-rays.*

Proof: It follows immediately from the definition of abstract convexity that $U = \{l : l(x) \leq p_U(l) \text{ for all } l \in L\}$. The closeness-along-rays follows from this equality. Let $x \in U$, $\alpha \in [0,1]$ and $l \in L$. Since the function l is nonnegative and positively homogeneous, we have $l(\alpha x) = \alpha l(x) \leq l(x) \leq p_U(l)$. Thus $l \in U$. △

We now give the definitions of polar sets with respect to the coupling function $\langle\cdot|\cdot\rangle$ (see, for example [175]).

DEFINITION 7.14 Let $U \subset X$. The set

$$U^\circ = \{l \in L : \langle l|x\rangle \leq 1 \text{ for all } x \in U\}$$

is called the *polar set* to U (with respect to the coupling function $\langle\cdot|\cdot\rangle$).

It follows from this definition that $U^\circ = \{l : p_U(l) \le 1\}$. We can define polar sets for subsets of a conic set L in the same manner. We now present a version of the Fenchel-Moreau theorem for positively homogeneous functions.

PROPOSITION 7.24 *Let $p : X \to \mathbb{R}_{+\infty}$ be a positively homogeneous function. Then $p^{**} = p$ if and only if p is an L-convex function. Here p^{**} stands for the second Fenchel-Moreau conjugate to the function p.*

Proof: The Fenchel-Mourea theorem demonstrates that $p^{**} = p$ if and only if p is H_L-convex. It follows from Proposition 7.15 that p is L-convex if and only if p is H_L-convex. Combining these two statements, we obtain the desired result. △

Let $U \subset X$. Consider the indicator function of the set U, defined by (7.2.13):

$$\delta_U(x) = \begin{cases} 0 & \text{if } x \in U; \\ +\infty & \text{if } x \notin U \end{cases}$$

It follows from Proposition 7.11 that $\delta_U^* = p_U$.

PROPOSITION 7.25 *For any $U \subset X$, the polar set U° is (L, X)-convex.*

Proof: Let $l \in L$ and $l \notin U^\circ$. Then $p_U(l) > 1$. Since the support function p_U is X-convex, it follows that there exists an element $x \in Q$ such that $x(l) = \langle l|x\rangle > 1$ and $x(l') \le p_U(l')$ for all $l' \in L$. In particular, $\langle l'|x\rangle \le p_U(l') \le 1$ for $l' \in U^\circ$. Thus

$$\langle l|x\rangle > 1 \ge \sup_{l' \in U^\circ} \langle l'|x\rangle.$$

Hence the set U° is (L, X)-convex. △

PROPOSITION 7.26 *Let $U \subset X$. Then $U = U^{\circ\circ}$ if and only if U is (X, L)-convex.*

Proof: Proposition 7.25 demonstrates that the set $U^{\circ\circ} = (U^\circ)^\circ$ is (X, L)-convex. Now consider an (X, L)-convex set U. The inclusion $U \subset U^{\circ\circ}$ follows immediately from the definition of the polar set. Hence, if $U \ne U^{\circ\circ}$, then there exists $x \in U^{\circ\circ}$ which does not belong to U. It follows from the (X, L)-convexity of U that there exist $l \in L$ and a number s such that

$$\langle l|x\rangle > s > \sup\{\langle l|y\rangle : y \in U\} \ge 0. \tag{7.2.28}$$

Let $l' = l/s$. Then (7.2.28) implies that $(l', x) > 1$ and $l' \in U^\circ$, which is a contradiction. △

Recall (see Definition 1.5) that the (X, L)-convex hull $\mathrm{co}_X U$ of a set $U \subset X$ coincides with the intersection of all (X, L)-convex sets $V \supset U$.

COROLLARY 7.1 *Let $U \subset X$. Then $U^{\circ\circ}$ coincides with $\mathrm{co}_X U$.*

Proof: Let $V \supset U$ and let V be (X, L)-convex. Then $V = V^{\circ\circ} \supset U^{\circ\circ}$. Since $U^{\circ\circ}$ is an (X, L)-convex set, it follows that $U^{\circ\circ} = \mathrm{co}_X U$. △

The following statement indicates links between the Minkowski gauges and the indicator functions of an (X, L)-convex set U and its polar U°.

PROPOSITION 7.27 *Let $U \subset X$ be a radiant set. Then*

(i) $\delta_{U^\circ} = \mu_U^*$;

(ii) $\mu_{U^\circ} = \delta_U^*$.

*where * stands for the Fenchel-Moreau conjugacy operator with respect to the pair (X, L).*

Proof: (i) For $l \in L$, we have

$$\begin{aligned} \mu_U^*(l) &= \sup_{x \in X}(\langle l|x\rangle - \mu_U(x)) = \sup_{x \in X}(\langle l|x\rangle - \inf_{\lambda>0, x \in \lambda U} \lambda) \\ &= \sup_{(x,\lambda):\lambda>0, x\in\lambda U} (\langle l|x\rangle - \lambda) = \sup_{\lambda>0} \sup_{x \in \lambda U} (\langle l|x\rangle - \lambda) \\ &= \sup_{\lambda>0} \lambda \sup_{\frac{x}{\lambda} \in U} \left(\left\langle l, \frac{x}{\lambda}\right\rangle - 1\right) = \begin{cases} 0 & x' \in U^\circ; \\ +\infty & x' \notin U^\circ \end{cases} \\ &= \delta_{U^\circ}(x'). \end{aligned}$$

(ii) It has been shown (see Proposition 7.11) that $\delta_U^* = p_U$. Since $U^\circ = \{l \in L : p_U(l) \le 1\}$, we have for $l \in L$:

$$\begin{aligned} \mu_{U^\circ}(l) &= \inf\left\{\lambda > 0 : \frac{l}{\lambda} \in U^\circ\right\} = \inf\left\{\lambda > 0 : p_U\left(\frac{l}{\lambda}\right) \le 1\right\} \\ &= \inf\{\lambda > 0 : p_U(l) \le \lambda\} = p_U(l) = \delta_U^*(l). \end{aligned}$$

REMARK 7.5 A similar result in a more general situation was proved in [100].

COROLLARY 7.2 *Let $U \subset X$ be a radiant set. Then the indicator function, δ_U, of the set U is L-convex if and only if the set U is (X, L)-convex.*

Proof: It follows from Proposition 7.27 that

$$(\delta_U)^{**} = (\mu_{U^\circ})^* = \delta_{U^{\circ\circ}}.$$

The Fenchel-Moreau theorem (Proposition 7.24) implies that δ_U is L-convex if and only if $\delta_U = \delta_{U^{\circ\circ}}$. The corollary follows directly from Proposition 7.26. $\triangle$

We need the following definition.

DEFINITION 7.15 Let $p : X \to \bar{\mathbb{R}}_+$ be a positively homogeneous function. The function $p^\circ : L \to \bar{\mathbb{R}}_+$ defined by

$$p^\circ(l) = \inf\{\lambda > 0 : \langle l|x\rangle \leq \lambda p(x) \text{ for all } x \in X\}$$

is called the *polar function* of p.

Polar functions are well known in convex analysis (see, for example [134], Section 15).

Subsequently, we assume that $0/0 = 0$. We also need to define the *closure-along-rays*, $\mathrm{cl}_r U$, of a set $U \subset X$. By definition, $\mathrm{cl}_r U$ is the intersection of all closed-along-rays sets $U' \supset U$. If U is a radiant set and $\mu_U(x) = 1$, then $x \in \mathrm{cl}_r U$.

PROPOSITION 7.28 *Let μ_U be the Minkowski gauge of a nonempty radiant set $U \subset X$ such that there exists $x \in X$ with the property $0 < \mu_U(x) < +\infty$. Then $U^\circ = \{l \in L : \mu^\circ(l) \leq 1\}$.*

Proof: Let $(\mu_U)^\circ(l) \leq 1$. It follows from the definition of the polar function that for each $\varepsilon > 0$ and each $x \in X$ the inequality $\langle l|x\rangle \leq (1+\varepsilon)\mu_U(x)$ holds. Since $\mu_U(x) \leq 1$ for $x \in U$ it follows that $l \in U^\circ$. Conversely, let $l \in U^\circ$. Then $\mu_U(x) = 0$ implies $\langle l|x\rangle = 0$. Indeed, if $\mu_U(x) = 0$, then $\lambda x \in U$ for all $\lambda > 0$. Therefore $\langle l|\lambda x\rangle \leq 1$ for all $\lambda > 0$. Hence, we have

$$\begin{aligned}
\mu_{U^\circ}(l) &= \inf\{\lambda > 0 : \langle l|x\rangle \leq \lambda\mu_U(x) \text{ for all } x \in X\} \\
&= \inf\{\lambda > 0 : \langle l, x\rangle \leq \lambda\mu_U(x) \text{ for all } x \in X \text{ with } \mu_U(x) > 0\} \\
&= \sup_{x\in X, \mu_U(x)>0} \frac{\langle l|x\rangle}{\mu_U(x)} \\
&= \sup\left(\sup_{x\in X, +\infty>\mu_U(x)>0} \frac{\langle l, x\rangle}{\mu_U(x)}, \sup_{x\in X, \mu_U(x)=+\infty} \frac{\langle l|x\rangle}{\mu_U(x)}\right) \\
&= \sup_{x\in X, 0<\mu_U(x)<+\infty} \frac{\langle l, x\rangle}{\mu_U(x)}.
\end{aligned}$$

The set $\{x \in X : 0 < \mu_U(x) < +\infty\}$ is not empty, so

$$\mu_{U^\circ}(l) = \sup_{x\in X, 0<\mu_U(x)<+\infty} \frac{\langle l|x\rangle}{\mu_U(x)} = \sup_{x\in X, 0<\mu(x)<+\infty} \left\langle l \middle| \frac{x}{\mu_U(x)}\right\rangle.$$

Since $\mu_U\left(\frac{x}{\mu_U(x)}\right) = 1$ it follows that $\frac{x}{\mu_U(x)}$ belongs to the closure-along-rays $\mathrm{cl}_r U$ of the set U. Hence

$$\mu_U^\circ(l) \leq \sup_{x \in \mathrm{cl}_r U} \langle l|x\rangle = \sup_{x \in U} \langle l|x\rangle.$$

The inclusion $l \in U^\circ$ implies $\langle l, x\rangle \leq 1$ for $x \in U$. Hence $\mu_U^\circ(l) \leq 1$. △

COROLLARY 7.3 *Let $p : X \to \bar{\mathbb{R}}_+$ be a positively homogeneous function such that the set $\{x \in X : 0 < p(x) < +\infty\}$ is nonempty. Then*

$$\{x \in X : p(x) \leq 1\}^\circ = \{l \in L : p^\circ(l) \leq 1\}.$$

The result follows from Proposition 7.28, since $p = \mu_U$ with $U = \{x \in X : p(x) \leq 1\}$. We now present two examples.

EXAMPLE 7.7 Consider a pair (X, L) where X is a Banach space, and $L = X^*$ is the conjugate to X space. Define the coupling function $\langle l|x\rangle$ by

$$\langle l|x\rangle = \max(l(x), 0).$$

It is easy to check that a set $U \subset X$ is (X, L)-convex if and only if U is convex, closed and $0 \in U$. A function $p : X \to \mathbb{R}_{+\infty}$ is L-convex if and only if p is a lower semicontinuous sublinear nonnegative function.

EXAMPLE 7.8 Let $X = L = \mathbb{R}^n_{++} \cup \{0\}$ with the coupling function $\langle l|x\rangle = \langle l, x\rangle = \min_{i=1,\dots,n} l_i x_i$. (This function has been studied in Chapter 2.) Assume that X is equipped with the coordinate-wise order relation. Then a set $U \subset X$ is (X, L)-convex if and only if this set is closed (in the topological space X) and normal. A positively homogeneous function $p : X \to \bar{\mathbb{R}}$ is L-convex if and only if p is IPH. These assertions easily follow from results of Chapter 2, where the cone $\mathbb{R}^n_{++} = X \setminus \{0\}$ has been considered.

Finally we consider H_L−convex and H_X− convex functions, where H_L (H_X) is the set of all L-affine functions defined on X (the set of all X-affine functions defined on L). Let X and L be a pair of conic sets with a coupling function $\langle l|x\rangle : L \times X \to \mathbb{R}$, such that

$$\langle \lambda l|x\rangle = \langle l|\lambda x\rangle = \lambda\langle l|x\rangle.$$

Note that in contrast with the previous discussion, we do not assume that $\langle l|x\rangle \geq 0$ for all $l \in L$ and $x \in X$. We also do not assume that both X and L contain zero. The coupling function $\langle l|x\rangle$ allows us to define the order relations $\succeq$ in both sets X and L. Let $l_1, l_2 \in L$ and $x_1, x_2 \in X$. By definition

$$l_1 \succeq l_2 \iff \langle l_1|x\rangle \geq \langle l_2|x\rangle \text{ for all } x \in X, \tag{7.2.29}$$

$$x_1 \succeq x_2 \iff \langle l|x_1\rangle \geq \langle l|x_2\rangle \text{ for all } l \in L. \tag{7.2.30}$$

Having the conic structure of the set X and the order relation (7.2.30) we can consider increasing convex-along-rays (ICAR) functions defined on X. A function $f : X \to \mathbb{R}_{+\infty}$ is called ICAR if this function is increasing (with respect to $\succeq$) and for each $x \in X$ the function $t \to f(tx)$ is convex on $(0 + \infty)$. In the same manner we define an ICAR function on the set L.

PROPOSITION 7.29 *The Fenchel-Moreau conjugate of an arbitrary function $f : X \to \mathbb{R}_{+\infty}$ is an ICAR function.*

Proof: The isotonicity of f^* follows immediately from the definition of the Fenchel-Moreau conjugacy. Let $l \in L$ and $t > 0$. Then

$$f^*(tl) = \sup_{x \in X}(\langle tl|x\rangle - f(x)) = \sup_{x \in X}(t\langle l|x\rangle - f(x)).$$

The function $f_l^*(t) = f^*(tl)$ is the supremum of a family of affine functions $t \to \langle l|x\rangle t - f(x)$, hence f_l^* is convex. $\triangle$

COROLLARY 7.4 *Let a function $f : X \to \mathbb{R}_{+\infty}$ be H_L-convex. Then f is ICAR.*

Proof: The proof follows from the Fenchel-Moreau theorem and Proposition 7.29.

Note that the order relation $\succeq$ essentially depends on a coupling function.

EXAMPLE 7.9 Let X be a Banach space and let $L = X^*$ be a conjugate space. Consider a coupling function $\langle l|x\rangle = l(x)$. Let $x_1, x_2 \in X$. Then $x_1 \succeq x_2$ if and only if $x_1 = x_2$. Hence an arbitrary function $f : X \to \mathbb{R}_{+\infty}$ is increasing with respect to $\succeq$.

EXAMPLE 7.10 Let $X = L = \mathbb{R}^n_{++} \cup \{0\}$ and

$$\langle l|x\rangle = \langle l, x\rangle = \begin{cases} \min_{i=1,\dots,n} l_i x_i & \text{if } l \in \mathbb{R}^n_{++}; \\ 0 & \text{if } l = 0. \end{cases}$$

Then $\succeq$ coincides with the coordinate-wise order relation $\geq$, hence a function f is increasing with respect to $\succeq$ if and only if f is increasing with respect to $\geq$.

7.3. SOME CLASSES OF ABSTRACT CONVEX FUNCTIONS

7.3.1. Overview

In this section we shall study some important examples of abstract convex functions and sets, abstract convex hulls and subdifferentials. Note that as a rule, it is more difficult to describe abstract convex sets than functions. A description of abstract convex hull in verifiable terms (i.e., involving standard notions of convex hull or cone hull or normal hull) is, in general, a very difficult problem. Examples related to convex sets, sublinear, convex and quasiconvex functions are based on convex analysis, so it is assumed that the reader is familiar with main notions and results of this discipline, in particular, with subdifferential calculus.

Examples related to sublinear and convex functions and also to infima of families of abstract convex functions can be found in [155]. Results from Subsection 7.3.5, related to the description of abstract convex functions can be found in [159, 153]. Note that some of these results can easily be obtained by using some known approaches developed in quasiconvex analysis (see for example, [107, 175, 125, 193]). The description of abstract convex sets presented in Subsection 7.3.6 can be found in [153]. Results from Subsection 7.3.7 related to the description of the subdifferential have not been published before. Note that there exist various schemes for the definition of subdifferentials for quasiconvex functions; the survey of these schemes can be found in [126]. Our example is only an illustration of the approach based on abstract convexity and we do not compare this approach with others.

7.3.2. Linear functions - generating sublinearity

Let X be a Banach space and let $L = X^*$ be the set of continuous linear functions defined on X. The following well-known result ([80]) is a corollary from Hahn-Banach theorem. Note that Theorem 6.2 is a particular case of this result.

THEOREM 7.5 *A function $f : X \to \mathbb{R}_{+\infty}$ is L-convex if and only if f is a sublinear lower semicontinuous function. A set $U \subset L$ is (L, X)-convex if and only if U is convex and w^*-closed.*

We now describe the set $P(L, X, K)$ of all L-convex functions on K, where $K \subset X$ is a closed convex cone.

PROPOSITION 7.30 *A function $p : K \to \mathbb{R}_{+\infty}$ is L-convex if and only if p is lower semicontinuous and sublinear.*

Proof: It is easy to check that an L-convex function is lower semicontinuous and sublinear. So we need only to verify the inverse assertion. Let a function p be lower semicontinuous and sublinear on K. Define an extension $\tilde{p}$ of the function p on the entire space X given by

$$\tilde{p}(x) = \begin{cases} p(x) & \text{if } x \in K; \\ +\infty & \text{if } x \notin K. \end{cases} \tag{7.3.1}$$

Since K is a closed convex cone, it follows that $\tilde{p}$ is a lower semicontinuous sublinear function on X. Applying Theorem 6.2, we deduce that there exists a subset U of the set $L = X^*$ such that $\tilde{p}(x) = \sup_{l \in U} l(x)$ for all $x \in X$. In particular,

$$p(x) = \tilde{p}(x) = \sup_{l \in U} l(x) \text{ for all } x \in K.$$

So p is a L-convex function on K. $\triangle$

We now describe the totality $S(L, X, K)$ of all (L, X, K)-convex subsets of L with respect to a closed convex cone K. We need the following definition.

DEFINITION 7.16 Let M be a convex cone in a vector space Y with $0 \in M$. A set $U \subset Y$ is called *M-normal* if $U = (U - M) \cap M$. A set U is called *M-conormal* if $U + M = U$.

Since $0 \in M$ it follows that $(U - M) \cap M \supset U$ and $U + M \supset U$ always. Therefore U is M-normal if and only if $(U - M) \cap M \subset U$ and U is M-conormal if and only if $U + M \subset U$.

PROPOSITION 7.31 *Let $L = X^*$. Then $U \in S(L, X, K)$ if and only if U is w^*-closed convex and $-K^*$-conormal, where $K^* = \{l \in L : l(x) \geq 0$ for all $x \in K\}$ is the cone conjugate to K.*

Proof: First we assume that $U \in S(L, X, K)$. Then

$$U = \text{supp}(p, L, K) = \{l \in L : l(x) \leq p(x) \text{ for all } x \in K\},$$

where $p(x) = \sup\{l(x) : l \in U\}$ for all $x \in K$. The set U is w^*-closed and convex. Let us consider the extension $\tilde{p}$ of the function p to the entire space X defined on X by (7.3.1). Clearly,

$$\partial\tilde{p} = \text{supp}(\tilde{p}, L, X) = \text{supp}(p, L, K) = U.$$

Let δ_K be the indicator function of the set K (see (7.2.13) for the definition of the indicator function). Note that $\partial\delta_K(0) = -K^*$. We have $\tilde{p} = \tilde{p} + \delta_K$, so, by applying the subdifferential calculus we conclude that

$$U = \partial\tilde{p}(0) = \partial(\tilde{p} + \delta_K)(0) = \text{cl}\,(\partial\tilde{p}(0) + \partial\delta_K(0)) = \text{cl}\,(U - K^*) \supset U - K^*.$$

Thus U is a w^*-closed convex and $-K^*$-conormal set.

Next we suppose that $U \subset L$ a w^*-closed convex and $-K^*$-conormal set. Since U is closed and convex, it follows from Theorem 6.2 that $U = \operatorname{supp}(p, X, X^*)$, where $p(x) = \sup\{l(x) : l \in U\}$ for all $x \in X$. If $x \notin K$, then applying the separation theorem we find $l \in X^*$ such that $l(x) > \sup\{l(u) : u \in K\} = 0$. So $l \in -K^*$. Since U is $-K^*$-conormal, we have $l' + \lambda l \subset U$ for all $l' \in U$ and $\lambda > 0$. For fixed $l' \in U$ we have

$$p(x) = \sup\{l(x) : l \in U\} \geq l'(x) + \sup_{\lambda} \lambda l(x) = +\infty.$$

Hence $\operatorname{dom} p \subset K$ and $U = \operatorname{supp}(p, X^*, K)$. △

Consider now an arbitrary nonempty set $U \subset L = X^*$. Let

$$U_* = \operatorname{cl}(\operatorname{co} U - K^*),$$

where $\operatorname{co} U$ denotes the convex hull of U and cl stands for the w^*-closure. Clearly the set U_* is w^*-closed convex and $-K^*$-conormal. Therefore U_* is (L, X, Q)-convex. In addition

$$\sup_{l \in U_*} l(x) = \sup_{l \in U} l(x) \text{ for all } x \in Q.$$

Therefore U_* is the (L, X, K)-convex hull $\operatorname{co}_L^K U$ of the set U. Thus we have the following result.

PROPOSITION 7.32 *Let $L = X^*$ be the set of all continuous linear functions defined on X and let $K \subset X$ be a closed convex cone. Then*

$$\operatorname{co}_L^K(U) = \operatorname{cl}(\operatorname{co} U - K^*).$$

REMARK 7.6 If U is a w^*-compact convex set, then $\operatorname{co}_L^K = U - K^*$. Moreover, if $K = X$, then U is an (L, X)-convex set and $\operatorname{co}_L^K U = U$.

Let K be a closed convex cone in a Banach space X. We shall now study abstract convex functions with respect to the conjugate cone K^*. Let $Z \subset K$ be a closed convex cone. The following proposition holds:

PROPOSITION 7.33 *A function $p : Z \to \mathbb{R}_{+\infty}$ belongs to $P(K^*, K, Z)$ if and only if there exists a lower semicontinuous sublinear function $\tilde{p} : X \to \mathbb{R}_{+\infty}$ such that*

(i) $\tilde{p}(z) = p(z)$ for all $z \in Z$;

(ii) $\tilde{p}$ is K-increasing: if $x_1 \geq x_2$ with respect to K (that is, $x_1 - x_2 \in K$) then $\tilde{p}(x_1) \geq \tilde{p}(x_2)$.

Proof: Let $p \in P(K^*, K, Z)$. Then there exists a subset U of the cone K^* such that $p(x) = \sup\{l(x) : l \in U\}$ for all $x \in Z$. Let

$$\tilde{p}(x) = \sup\{l(x) : l \in U\} \text{ for all } x \in X.$$

Clearly $p(x) = \tilde{p}(x)$ for all $x \in Z$. If $x_1 \geq x_2$ with respect to K then $l(x_1) \geq l(x_2)$ for all $l \in U$. Therefore, $\tilde{p}(x_1) \geq \tilde{p}(x_2)$.

Assume now that for the function p there exists a lower semicontinuous sublinear function $\tilde{p}$ with properties (i) and (ii). The function $\tilde{p}$ is X^*-convex on X so there exists a subset U of the space X^* such that $\tilde{p}(x) = \sup\{l(x) : l \in U\}$. We now show that $U \subset K^*$. If $x \in K$, then $-x \leq 0$ with respect to K and therefore

$$-l(x) = l(-x) \leq \tilde{p}(-x) \leq \tilde{p}(0) = 0, \qquad l \in U.$$

Thus $U \subset K^*$. Since $p(x) = \tilde{p}(x)$ for all $x \in Z$ we have $p \in P(K^*, K, Z)$. $\triangle$

We now study abstract convex sets, that is, elements of $S(K^*, K, Z)$. We shall need the following definition.

DEFINITION 7.17 Let L_1 and L_2 be cones in X^* with $L_1 \subset L_2$. Then a subset U of the cone L_1 is called (L_1, L_2)-*normal* if $U = \text{cl}\,(U - L_2) \cap L_1$, where cl stands for the w^*-closure. If U is an arbitrary subset of the cone L_1 then the set

$$\text{Nh}\,(U) = \text{cl}\,(U - L_2) \cap L_1$$

is called the (L_1, L_2)-*normal hull* of the set U.

Compare this definition with the definitions of normal sets in various situations, given in Chapter 2 and also with Definition 7.16. Note that (L_1, L_2)-normal sets have been studied in detail in [140].

PROPOSITION 7.34 *$U \in S(K^*, K, Z)$ if and only if U is a w^*-closed convex and (K^*, Z^*)-normal subset of the cone K^*.*

Proof: Let U be a w^*-closed convex and (K^*, Z^*) -normal subset of K^*. Assume that U is proper, that is $U \neq K^*$, $U \neq \emptyset$. Take $l' \in K^*$ such that $l' \notin U$. Since $U = \text{cl}\,(U - Z^*) \cap K^*$ and $l' \in K^*$, it follows that $l' \notin \text{cl}\,(U - Z^*)$. Therefore, there exists $x \in X$ such that

$$l'(x) > \sup\{l(x) : l \in \text{cl}\,(U - Z^*)\}. \tag{7.3.2}$$

It follows from (7.3.2) that

$$\sup\{l(x) : l \in -Z^*\} < +\infty.$$

The set $-Z^*$ is a cone so $l(x) \leq 0$ for all $l \in -Z^*$. Since Z is w^*-closed it follows that $x \in Z^{**} = Z$. Applying (7.3.2), we conclude that $l'(x) > \sup\{l(x) : l \in U\}$. It follows immediately from Lemma 7.1 that $U \in S(K^*, K, Z)$.

Assume now that $U \in S(K^*, K, Z)$, i.e., $U = \text{supp}(p, K^*, Z)$ is the support set of the function p defined on the cone Z by

$$p(x) = \sup\{l(x) : l \in U\} \quad \text{for all } x \in Z.$$

The set U is $(-Z^*)$-conormal. Indeed, if $l \in U$ and $l' \in -Z^*$ then we have for $x \in Z$

$$(l + l')(x) = l(x) + l'(x) \leq l(x) \leq p(x),$$

hence, $l + l' \in \text{supp}(p, X^*, Z) = U$. Thus $U - Z^* \subset U$. We have

$$U - K^* = U - Z^* - K^* = (U - K^*) + (-Z^*),$$

so the set $U - K^*$ is also $(-Z^*)$-conormal. It is easy to check that $\text{cl}\,(U - K^*)$ is $(-Z^*)$-conormal as well. Since the set $\text{cl}\,(U - K^*)$ is w^*-closed, convex and $(-Z^*)$-conormal, it follows from Proposition 7.31 that this set is (X^*, X, Z)-convex. The equality

$$\sup\{l(x) : l \in \text{cl}\,(U - Z^*)\} = p(x) \quad \text{for all } x \in Z$$

implies $\text{cl}\,(U - Z^*) = \text{supp}(p, X^*, Z)$. We have

$$U = \text{supp}(p, K^*, Z) = \text{supp}(p, X^*, Z) \cap K^* = \text{cl}\,(U - Z^*) \cap K^*,$$

hence U is (K^*, Z^*)-normal. △

COROLLARY 7.5 *The K^*-convex hull, $co^Z_{K^*}U$, of an arbitrary set $U \subset K^*$ has the form*

$$\text{co}^Z_{K^*}U = \text{Nh}\,(\text{co}\,U).$$

7.3.3. Affine functions - generating convexity

Let X be a Banach space. Let $L = X^*$ and let $H_L = \{(l, c) : l \in L,\ c \in \mathbb{R}\}$ be the set of all continuous affine functions defined on X. For the sake of simplicity we denote H_L by H in this section. Recall that $h(x) = l(x) - c$ for $h = (l, c) \in L \times \mathbb{R}$. The well-known result of convex analysis (see, for example [84], Section 3.4) asserts that a function $f : X \to \mathbb{R}_+$ is H-convex if and only if f is a lower semicontinuous convex function.

Consider a convex closed subset Z of the space X. It is easy to check, applying the same argument as in Proposition 7.30 that a function f defined on Z is H-convex if and only if this function is lower semicontinuous and convex. Assume now that Z is an arbitrary nonempty subset of X. It is easy to check that a function f defined on Z is H-convex if and only if there exists a lower semicontinuous convex function $\bar{f}$ defined on the closed convex hull, cl co Z, of Z such that, for all $z \in Z$, $f(z) = \bar{f}(z)$. In fact, if $f \in P(H, X, Z)$ then there exists a subset U of the set H such that $f(z) = \sup_{h \in U} h(z)$ for all $z \in Z$. Clearly, the function $\bar{f}$ defined by

$$\bar{f}(x) = \sup_{h \in U} h(x), \qquad (x \in \operatorname{cl} \operatorname{co} Z),$$

is lower semicontinuous convex and coincides with f on the set Z. The inverse assertion is trivial.

Now assume that $Z = X$ and consider the H-convex hull $\operatorname{co}_H f$ of the function f. By definition,

$$\operatorname{co}_H f(x) = \sup\{h(x) : \ h \in \operatorname{supp}(f, H, X)\}.$$

The function $\operatorname{co}_H f$ is usually called the *convex regularization* of the function f.

If f is a convex function then the convex regularization of f coincides with its lower regularization, see Definition 2.2. If follows from the Fenchel-Moreau theorem that for a convex function f

$$f^{**} = \operatorname{cl} f,$$

where $\operatorname{cl} f$ stands for the lower regularization of f and f^{**} stands for the Fenchel-Moreau second conjugate.

Consider now a function $f : X \to \mathbb{R}_{+\infty}$, which is not necessarily convex, its convex regularization $\operatorname{co}_H f$ and its convex hull $\operatorname{co} f$. The function $\operatorname{co} f$ is given by

$$(\operatorname{co} f)(x) = \inf_{x_1, \dots, x_n, \alpha_1, \dots, \alpha_n} \sum_{i=1}^{n} \alpha_i f(x_i)$$

where elements x_i and numbers α_i enjoy the following property

$$\sum_{i=1}^{n} \alpha_i x_i = x, \ \sum_{i=1}^{n} \alpha_i = 1, \ \alpha_i \geq 0, \ i = 1, \dots, n; \ n = 1, 2, \dots .$$

Clearly, $\operatorname{co} f \leq f$. It is easy to check that $\operatorname{co} f$ is a convex function. Assume that the support set $\operatorname{supp}(f, H, X)$ is not empty. If

$h \in \mathrm{supp}(f, H, X)$ and $\sum_{i=1}^n \alpha_i x_i = x$, $\sum_{i=1}^n \alpha_i = 1$ $\alpha_i \geq 0, i = 1, \ldots n$, then

$$h(x) = h(\sum_{i=1}^n \alpha_i x_i) = \sum_{i=1}^n \alpha_i h(x_i) \leq \sum_{i=1}^n \alpha_i f(x_i).$$

Thus

$$h(x) \leq \mathrm{co}\, f(x). \tag{7.3.3}$$

This inequality shows that $h \in \mathrm{supp}(\mathrm{co}\, f, H, X)$ so $\mathrm{supp}(f, H, X) \subset \mathrm{supp}(\mathrm{co}\, f, H, X)$. On the other hand, the inequality $\mathrm{co}\, f \leq f$ implies $\mathrm{supp}(\mathrm{co}\, f, H, X) \subset \mathrm{supp}(f, H, X)$. Thus

$$\mathrm{supp}(\mathrm{co}\, f, H, X) = \mathrm{supp}(f, H, X).$$

The inequality $\mathrm{co}_H\, f \leq cof$ follows immediately from (7.3.3). Since $\mathrm{co}_H\, f$ is a lower semicontinuous function we have $\mathrm{co}_H\, f \leq \mathrm{cl}\, \mathrm{co}\, f$, where cl denotes the lower regularization of a convex function (see Definition 2.2). At the same time $\mathrm{co}_H\, f$ is the greatest lower semicontinuous convex function minored by f. Therefore,

$$\mathrm{co}_H\, f = \mathrm{cl}\, \mathrm{co}\, f.$$

REMARK 7.7 If the space X is finite-dimensional and dom$f$ is a compact set then $\mathrm{co}\, f$ is a lower semicontinuous function. So $\mathrm{co}_H\, f = \mathrm{co}\, f$ in this case (see, for example, [84] Section 3.5 for details).

If f is a positively homogeneous of degree one function then the function $\mathrm{co}\, f$ is positively homogeneous of degree one as well. Applying this fact, we conclude that

$$\mathrm{co}_L\, f = \mathrm{cl}\, \mathrm{co}\, f$$

for a positively homogeneous of degree one function f (recall that $L = X^*$).

We now examine the structure of (H, X, Z)-convex sets, where H is the set of all continuous affine functions defined on a Banach space X, and Z is a closed convex subset of X. We need the notion of inf-convolution of two functions and some results related to inf-convolution.

DEFINITION 7.18 Let f and g be functions defined on a Banach space X and mapping into $\mathbb{R}_{+\infty}$. The function

$$(f \oplus g)(x) = \inf_{y \in Y} (f(y) + g(y - x)) \qquad (x \in X)$$

is called the *inf-convolution* of the functions f and g.

It is well known and easy to see that the inf-convolution of two convex functions is a convex function.

THEOREM 7.6 *(see, for example, [84, 118]) Let X be a Banach space and let $f : X \to \mathbb{R}_{+\infty}$ and $g : X \to \mathbb{R}_{+\infty}$ be proper lower semicontinuous convex functions. Then*

$$(f+g)^* = \mathrm{cl}\,(f^* \oplus g^*).$$

Proof: We have

$$\begin{aligned}(f^* \oplus g^*)^*(x) &= \sup_l [l(x) - \inf_{l_1+l_2=l}(f^*(l_1) + g^*(l_2))] \\ &= \sup_{l_1,l_2 \in X^*} (l_1(x) - f^*(l_1) + l_2(x) - g^*(l_2)) \\ &= f^{**}(x) + g^{**}(x) = (f+g)(x).\end{aligned}$$

Hence

$$\mathrm{cl}\,(f^* \oplus g^*) = (f^* \oplus g^*)^{**} = (f+g)^*. \qquad \triangle$$

DEFINITION 7.19 Let $f : X \to \mathbb{R}_{+\infty}$. The set $\mathrm{epi}^0 f = \{(x,\lambda) \in X \times \mathbb{R} : \lambda > f(x)\}$ is called the *strict epigraph* of the function f.

PROPOSITION 7.35 *(see, for example, [118], Proposition 5.2.1) Let f and g be proper functions defined on a Banach space X and mapping into $\mathbb{R}_{+\infty}$. Let $(f \oplus g)(x) > -\infty$ for all $x \in \mathbb{R}^n$. Then* $\mathrm{epi}^0(f \oplus g) = \mathrm{epi}^0(f) + \mathrm{epi}^0(g)$.

LEMMA 7.2 *1) Let $f : \mathbb{R}^n \to \mathbb{R}_{+\infty}$. Then* $\mathrm{cl}\,(\mathrm{epi}\, f) = \mathrm{cl}\,(\mathrm{epi}^0 f)$*;*

2) Let f_1 and f_2 be functions defined on $\mathbb{R}^n$ and mapping into $\mathbb{R}$. Then $\mathrm{cl}\,(\mathrm{epi}^0 f_1 + \mathrm{epi}^0 f_2) = \mathrm{cl}\,(\mathrm{epi}\, f_1 + \mathrm{epi}\, f_2)$.

Proof: We shall only prove the second statement of Lemma. Since $\mathrm{epi}^0 f_1 + \mathrm{epi}^0 f_2 \subset \mathrm{epi}\, f_1 + \mathrm{epi}\, f_2$ it follows that $\mathrm{cl}\,(\mathrm{epi}^0 f_1 + \mathrm{epi}^0 f_2) \subset \mathrm{cl}\,(\mathrm{epi}\, f_1 + \mathrm{epi}\, f_2)$. We now prove the reverse inclusion. It is sufficient to show that $\mathrm{cl}\,(\mathrm{epi}^0 f_1 + \mathrm{epi}^0 f_2) \supset (\mathrm{epi}\, f_1 + \mathrm{epi}\, f_2)$. Let $(x,\lambda) \in \mathrm{epi}\, f_1 + \mathrm{epi}\, f_2$. Then $x = x_1 + x_2$, $\lambda = \lambda_1 + \lambda_2$, where $\lambda_1 \geq f_1(x_1)$, $\lambda_2 \geq f_2(x_2)$. Let $\varepsilon_k = 1/k$, $k = 1, \ldots$. Then $\lambda_i + \varepsilon_k > f_i(x_i)$, hence $(x_i, \lambda_i + \varepsilon_k) \in \mathrm{epi}^0 f_i$, $i = 1,2$; $k = 1, \ldots$. We have $(x, \lambda + 2\varepsilon_k) \in \mathrm{epi}^0 f_1 + \mathrm{epi}^0 f_2$, so $(x,\lambda) \in \mathrm{cl}\,(\mathrm{epi}^0 f_1 + \mathrm{epi}^0 f_2)$. $\triangle$

Let $Z \subset X$ be a closed convex set. Consider the set

$$K_Z = \{(l,c) \in H : l(x) - c \leq 0 \text{ for all } x \in Z\}.$$

The set K_Z is a closed convex cone.

PROPOSITION 7.36 *Let f be a lower semicontinuous convex function defined on X and $Z \subset X$ be a closed convex set such that $Z \cap \operatorname{dom} f \neq \emptyset$. Then*

$$\operatorname{dom} f \subset Z \iff \operatorname{supp}(f, H, X) \text{ is a } K_Z\text{-conormal set.}$$

Proof: The following holds:

$$\operatorname{dom} f \subset Z \iff f+\delta_Z = f \iff (f+\delta)^* = f^* \iff \operatorname{epi}(f+\delta_Z)^* = \operatorname{epi} f^*. \tag{7.3.4}$$

Here δ_Z is the indicator function of the set Z and * stands for the Fenchel conjugacy operator.

It follows from Theorem 7.6 that

$$(f + \delta_Z)^* = \operatorname{cl}(f^* \oplus \delta_Z^*) \tag{7.3.5}$$

Since $Z \cap \operatorname{dom} f$ is not empty it follows that $\operatorname{dom}(f + \delta_Z) \neq \emptyset$. Hence $(f + \delta_Z)^*(l) > -\infty$ for all l. Note that $\operatorname{cl}(f^* \oplus \delta_Z^*) \leq f^* \oplus \delta_Z^*$, so (7.3.5) implies that $(f^* \oplus \delta_Z^*)(l) > -\infty$ for all l. Using (7.3.5), Lemma 7.2 and Proposition 7.35 we have

$$\begin{aligned} \operatorname{epi}(f + \delta_Z)^* &= \operatorname{epi}(\operatorname{cl}(f^* \oplus \delta_Z^*)) = \operatorname{cl}(\operatorname{epi}(f^* \oplus \delta_Z^*)) \\ &= \operatorname{cl}(\operatorname{epi}^0(f^* \oplus \delta_Z^*)) = \operatorname{cl}(\operatorname{epi}^0 f^* + \operatorname{epi}^0 \delta_Z^*) \end{aligned} \tag{7.3.6}$$

Due to Lemma 7.2, $\operatorname{cl}(\operatorname{epi}^0 f^* + \operatorname{epi}^0 \delta_Z^*) = \operatorname{cl}(\operatorname{epi} f^* + \operatorname{epi} \delta_Z^*)$. Hence

$$\operatorname{epi}(f + \delta_Z)^* = \operatorname{cl}(\operatorname{epi} f^* + \operatorname{epi} \delta_Z^*). \tag{7.3.7}$$

We now calculate $\operatorname{epi} \delta_Z^*$:

$$\begin{aligned} \operatorname{epi} \delta_Z^* &= \{(l, c) : c \geq \delta_Z^*(l)\} = \{(l, c) : c \geq \sup_{z \in Z} l(z)\} \\ &= \{(l, c) : (\forall z \in Z)\ c \geq l(z)\} = K_Z. \end{aligned} \tag{7.3.8}$$

Let $\operatorname{dom} f \subset Z$. Then using (7.3.4), (7.3.7) and (7.3.8) we have

$$\operatorname{cl}(\operatorname{epi} f^* + K_Z) = \operatorname{epi} f^*.$$

and therefore $\operatorname{epi} f^* + K_Z \subset \operatorname{epi} f^*$. On the other hand, since $0 \in K_Z$ we have $\operatorname{epi} f^* + K_Z \supset \operatorname{epi} f^*$. Therefore, $\operatorname{epi} f^* + K_Z = \operatorname{epi} f^*$.

Conversely, let $\operatorname{supp}(f, H, X) = \operatorname{epi} f^*$ is K_Z-normal, that is $\operatorname{epi} f^* + K_Z = \operatorname{epi} f^*$. Since $\operatorname{epi} f^*$ is a w^*-closed set it follows that

$$\operatorname{cl}(\operatorname{epi} f^* + K_Z) = \operatorname{epi} f^*.$$

Applying (7.3.7), (7.3.8) and then (7.3.4), we conclude that $\operatorname{epi}(f + \delta_Z)^* = \operatorname{epi} f^*$ and $\operatorname{dom} f \subset Z$ as required. △

We also need the following result.

PROPOSITION 7.37 *Let H be the set of all continuous affine functions defined on X. Then a nonempty proper set $U \subset H$ is (H, X)-convex (with respect to the entire space X) if and only if U is w^*-closed, convex, $(\{0\} \times \mathbb{R}_+)$-conormal and not $(\{0\} \times \mathbb{R})$-conormal.*

Proof: Let f be a lower semicontinuous convex function and $U = \operatorname{supp}(f, H, X)$. The set U is w^*-closed convex and $(\{0\} \times \mathbb{R}_+)$-conormal. If $U \neq \emptyset$, then (see Proposition 7.8) $\operatorname{supp}(f, H, X) = \operatorname{epi} f^*$. Assume U is $(\{0\} \times \mathbb{R})$-conormal. Since $f^*(l) = \inf\{\lambda : (l, \lambda) \in \operatorname{epi} f^*\}$, we have $f^*(l) = -\infty$ for all $l \in X^*$, so $f(x) = f^{**}(x) = +\infty$ for all $x \in X$ and $U = H$. Hence, if U is a proper set, then U is not $(\{0\} \times \mathbb{R})$-conormal.

Now consider a nonempty w^*-closed convex subset U of $H = L \times \mathbb{R}$. Assume that U is a $(\{0\} \times \mathbb{R}_+)$-conormal set. For $l \in X^*$ define

$$f(l) = \inf\{\lambda \in \mathbb{R} : (l, \lambda) \in U.$$

Since U is $(\{0\} \times \mathbb{R}_+)$ conormal, it follows that $((l, \lambda) \in U, \lambda' \geq \lambda) \implies (l, \lambda') \in U$. Hence $U = \operatorname{epi} f$. Since U is w^*-closed and convex, we deduce that f is w^*-lower semicontinuous and convex. If U is not $(\{0\} \times \mathbb{R})$-conormal then f is a proper function. Applying Proposition 7.8, we conclude that

$$\operatorname{supp}(f^*, H, X) = \operatorname{epi} f^{**} = \operatorname{epi} f = U.$$

So U is an (H, X)-convex set. △

The following result holds:

PROPOSITION 7.38 *Let $Z \subset X$ be a closed convex set. Let U be a proper subset of H which enjoys the following property:*

$$\text{there exists } z \in Z \text{ such that } \sup_{(l,c) \in U} (l(z) - c) < +\infty. \tag{7.3.9}$$

Then U is (H, X, Z)-convex if and only if U is w^*-closed convex, K_Z-conormal and not $(\{0\} \times \mathbb{R})$-conormal.

Proof: It is easy to check that an (H, X, Z)-convex set is w^*-closed convex and K_Z-conormal. Now let U be w^*-closed convex K_Z-conormal and not $(\{0\} \times \mathbb{R})$-conormal. Since $\{0\} \times \mathbb{R}_+ \subset K_Z$, it follows that U is a $(\{0\} \times \mathbb{R}_+)$-conormal set. Proposition 7.37 shows that there is a lower semicontinuous convex function f defined on X such that

$U = \operatorname{supp}(f, H, X)$. It follows from (7.3.9) that $\operatorname{dom} f \cap Z \neq \emptyset$. Applying Proposition 7.36, we conclude that $\operatorname{dom} f \subset Z$. Therefore U is an (H, X, Z)-convex set. $\triangle$

COROLLARY 7.6 *Let H be the set of all continuous affine functions defined on X and let U be an arbitrary nonempty subset of H which is not $(\{0\} \times \mathbb{R})$-conormal and (7.3.9) holds. Then*

$$\operatorname{co}_H^Z U = \operatorname{cl}(\operatorname{co} U + K_Z). \tag{7.3.10}$$

Proof: Note that the set $\operatorname{cl}(\operatorname{co} U + K_Z)$ is w^*-closed convex and K_Z-conormal. Therefore this set is (H, X, Z)-convex. Since, for $z \in Z$,

$$\sup_{h \in U} h(z) = \sup_{h \in \operatorname{cl}(\operatorname{co} U + K_Z)} h(z),$$

it follows that $\operatorname{cl}(\operatorname{co} U + K_Z)$ is the support set of the function $z \mapsto \sup_{h \in U} h(z)$. Hence (7.3.10) follows. $\triangle$

REMARK 7.8 If U is a $(\{0\} \times \mathbb{R})$-conormal set, then for all $x \in X$,

$$\sup_{(l,c) \in U} (l(x) - c) = +\infty$$

and therefore $\operatorname{co}_H U = H$.

PROPOSITION 7.39 *Let I be an arbitrary index set and, for each $i \in I$, let g_i be a lower semicontinuous convex function defined on X and let $U = \operatorname{cone} \bigcup_{i \in I} \operatorname{epi} g_i^*$. Consider assertions*

a) the system

$$g_i(x) \leq 0, \qquad (i \in I)$$

has a solution;

b) the set U is not $\{0\} \times \mathbb{R}$-conormal.

Then a) implies b).

Proof: Let x_0 be a solution to the system and take any $i \in I$. Then, for each $l \in \operatorname{dom} g_i^*$, we have

$$g_i^*(l) = \sup_{x \in X} \{l(x) - g_i(x)\} \geq l(x_0) - g_i(x_0) \geq l(x_0).$$

Thus if $(l, \lambda) \in \operatorname{epi} g_i^*$ then $\lambda \geq g_i^*(l) \geq l(x_0)$. Let $x_0(l) = l(x_0)$. By the above, if $(l, \lambda) \in \operatorname{epi} g_i^*$, then $(l, \lambda) \in \operatorname{epi} x_0$. Thus, for all i, $\operatorname{epi} g_i^* \subset \operatorname{epi} x_0$. Hence

$$\bigcup_{i \in I} \operatorname{epi} g_i^* \subset \operatorname{epi} x_0.$$

Since $\operatorname{epi} x_0$ is a halfspace in $X^* \times \mathbb{R}$ it follows immediately that U cannot be $\{0\} \times \mathbb{R}$-conormal. $\triangle$

7.3.4. Min-type functions - generating convexity-along-rays

Let $X = L = \mathbb{R}^n_{++}$, where $\mathbb{R}^n_{++}$ is the cone of n-dimensional vectors with positive coordinates. Consider a pair (X, L) with the coupling function $\langle l|x\rangle = \langle l, x\rangle$ where $\langle l, x\rangle = \min_{i=1,\dots,n} l_i x_i$. Let H_L (H_X) be the set of L-affine (X-affine) functions, that is, $h \in H_L$ if and only if $h(x) = \langle l, x\rangle - c$ for all $x \in X$ with $l \in L$, $c \in \mathbb{R}$ and $y \in H_X$ if and only if $y(l) = \langle l, x\rangle - c$ for all $l \in L$ with $x \in X$, $c \in \mathbb{R}$.

Recall (see Remark 3.9) that a function $f : X \to \mathbb{R}_{+\infty}$ is H_L-convex if and only if f is a lower semicontinuous ICAR (increasing convex-along-rays) function. Note that $X = L$ and that a coupling function is symmetric ($\langle l, x\rangle = \langle x, l\rangle$). It follows from this observation that a function $g : L \to \mathbb{R}_{+\infty}$ is H_X-convex if and only if g is lower semicontinuous ICAR. Let $f \in F_X$. Since the Fenchel-Moreau conjugate function f^* is H_X-convex it follows that f^* is an ICAR function.

Assume now that $X = L = \mathbb{R}^n_+$, where $\mathbb{R}^n_+$ is the cone of all vectors with nonnegative coordinates. Let the coupling function $\langle l|x\rangle$ be defined by $\langle l|x\rangle = \langle l, x\rangle$, where $\langle l, x\rangle = \min_{i\in I_+(l)} l_i x_i$ with $I_+(l) = \{i : l_i > 0\}$. This coupling function is an extension of the coupling function, which was considered for the cone $\mathbb{R}^n_{++}$. Note that $\langle l, x\rangle \neq \langle x, l\rangle$, so the class of functions generated by L does not coincide with the class of functions generated by X. More precisely, we can identify each element $l \in L$ with the function $x \mapsto \min_{i\in I_+(l)} l_i x_i$ and each element $x \in X$ with the function $l \mapsto \min_{i\in I_+(l)} l_i x_i$. It is known (see Theorem 3.2) that an H_L-convex function defined on $\mathbb{R}^n_+$ is also a lower semicontinuous ICAR function.

We now indicate some simple properties of H_X-convex functions defined on L. For each nonempty subset J of the set $I = \{1, 2, \dots, n\}$ consider the cone

$$\mathbb{R}^J_{++} = \{l \in \mathbb{R}^n_+ : l_i > 0, \ (i \in J), \quad l_i = 0 \ (i \notin J)\}. \tag{7.3.11}$$

The restriction of a function $g : L \to \mathbb{R}_{+\infty}$ to the cone $\mathbb{R}^J_{++}$ is denoted by g_J. Let $f : L \to \mathbb{R}_{+\infty}$ be an H_X-convex function. Then there exists a subset $U \in H_X$ such that

$$f(l) = \sup_{y=(x,c)\in U} \min_{i\in I_+(l)} l_i x_i - c. \tag{7.3.12}$$

The following assertions hold.

1) The function f is CAR. Indeed, it follows from (7.3.12) that

$$f(\lambda l) = \sup_{(x,c)\in U} \min_{i\in I_+(l)} \lambda l_i x_i - c.$$

For each $(x, c) \in U$, the function $\lambda \mapsto \lambda \min_{i \in I_+(l)} l_i x_i - c$ is affine on $[0, +\infty)$, so the function $\lambda \mapsto f(\lambda l)$ is convex as the supremum of affine functions.

2) The restriction of f to the cone R^J_{++} is ICAR for each J. This holds, since the set $I_+(l)$ is constant for $l \in \mathbb{R}^J_{++}$.

3) Let $l \in \mathbb{R}^n_+$, $J \subset I_+(l)$ and let the vector l_J be defined as follows

$$(l_J)_i = \begin{cases} l_i & \text{if } i \in J; \\ 0 & \text{if } i \notin J. \end{cases}$$

Then $f(l_J) \geq f(l)$. Indeed, since $J = I_+(l_J) \subset I_+(l)$, we have

$$f(l_J) = \sup_x \min_{i \in J} l_i x_i \geq \sup_x \min_{i \in I_+(l)} l_i x_i = f(l).$$

It follows from this assertion that for an increasing H_X-convex function f we have $f(l_J) = f(l)$ for all $J \subset N$. The following example shows that an H_L-convex function f is not necessarily increasing and therefore not necessarily ICAR.

EXAMPLE 7.11 Let $n = 2$ and f be a function defined on $\mathbb{R}^2$ by $f(x) = \min(x_1, x_2)$, let $e_1 = (1, 0)$ and $\mathbf{1} = (1, 1)$. It is clear that $\mathbf{1} > e_1$. We have

$$f^*_L(e_1) = \sup_x (x_1 - \min(x_1, x_2)) = \sup_x \max(0, x_1 - x_2) = +\infty$$

and

$$f^*_L(\mathbf{1}) = \sup_x (\min(x_1, x_2) - \min(x_1, x_2)) = 0.$$

Thus $f^*_L(e_1) > f^*_L(\mathbf{1})$.

The following result follows directly from the Fenchel-Moreau theorem.

THEOREM 7.7 *Let $f : \mathbb{R}^n_+ \to \mathbb{R}_{+\infty}$. The equality $f = f^{**}_L$ holds if and only if f is a lower semicontinuous ICAR function.*

7.3.5. Two-step functions - generating quasiconvexity. Abstract convex functions

In this subsection we will study quasiconvex functions as abstract convex with respect to the special classes of two-step functions.

Let X be a Banach space. A function $f : X \to \bar{\mathbb{R}}$ is called *quasiconvex* if its lower level sets $S_c(f) = \{x \in X : f(x) \leq c\}$ are convex for all $c \in \mathbb{R}$. A function f is called *quasiconcave* if its upper level sets $S^c(f) = \{x \in$

$X : f(x) \geq c\}$ are convex for all real c. Clearly, f is quasiconcave if and only if $(-f)$ is quasiconvex. It is known that f is quasiconvex (quasiconcave) if and only if its level sets $T_c(f) = \{x \in X : f(x) < c\}$ ($T^c(f) = \{x \in X : f(x) > c\}$) are convex for all real c. We shall only study two classes of quasiconvex functions: lower semicontinuous quasiconvex functions and evenly quasiconvex functions. First we recall the definition of evenly convex sets. A proper set $Z \subset X$ is called *evenly convex* if for each $x \in Z \setminus X$ there exists a linear function $v \in X^*$ such that $v(x) > v(z)$ for all $z \in Z$. The empty set and the space X are evenly convex by definition. If Z is a closed convex set and $x \notin Z$ then, due to the separation theorem we can find $v \in X^*$ such that $v(x) > \sup_{z \in Z} v(z)$. Thus every closed convex set is evenly convex. It follows from the separation theorem that every open convex set is also evenly convex. The intersection of an arbitrary family of evenly convex sets is also evenly convex.

A quasiconvex function $f : X \to \bar{\mathbb{R}}$ is called *evenly quasiconvex* if its level sets $S_c(f)$ are evenly convex for all c. Since a function $f : X \to \bar{\mathbb{R}}$ is lower semicontinuous if and only if its level sets $S_c(f)$ are closed for all $c \in \bar{\mathbb{R}}$, we conclude that each lower semicontinuous quasiconvex function is evenly quasiconvex. If f is an upper semicontinuous quasiconvex function, then the sets $T_c(f) = \{x : f(x) < c\}$ are convex and open. We know that an open convex set is evenly convex so each upper semicontinuous quasiconvex function is also evenly quasiconvex.

We shall denote by $Q(X)$ the set of all lower semicontinuous quasiconvex functions $f : X \to \bar{\mathbb{R}}$ and by $Q^{ev}(X)$ the set of all evenly quasiconvex functions $f : X \to \bar{\mathbb{R}}$. Thus, $Q(X) \subset Q^{ev}(X)$.

Let us consider a family $(q_i)_{i \in I}$ of quasiconvex functions where I is an arbitrary index set. Set $q(x) = \sup_{i \in I} q_i(x)$. The equality $S_c(q) = \bigcap_{i \in I} S_c(q_i)$ shows that q is a quasiconvex function, so we have $q \in Q(X)$ if $q_i \in Q(X)$ and $q \in Q^{ev}(X)$ if $q_i \in Q^{ev}(X)$. Let $\lambda > 0$. Clearly, $\lambda q \in Q(X)$ if $q \in Q(X)$ and $\lambda q \in Q^{ev}(X)$ if $q \in Q^{ev}(X)$, so both $Q(X)$ and $Q^{ev}(X)$ are conic sets. Note that both $Q(X)$ and $Q^{ev}(X)$ are not closed under summation. Let us give a simple example.

EXAMPLE 7.12 Let $X = \mathbb{R}^2$. Define for $x = (x_1, x_2)$:

$$q_1(x) = \begin{cases} 1 & \text{if } x_1 > 1; \\ 0 & \text{if } x_1 \leq 1. \end{cases}$$

and

$$q_2(x) = \begin{cases} 1 & \text{if } x_2 > 1; \\ 0 & \text{if } x_2 \leq 1 \end{cases}$$

Clearly, q_1 and q_2 are quasiconvex. For $1 < c < 2$ we have

$$S_c(q_1 + q_2) = \mathbb{R}^2 \setminus \{x = (x_1, x_2) : x_1 > 0, x_2 > 0\}.$$

The set $S_c(q_1 + q_2)$ is not convex, therefore $q_1 + q_2$ is not quasiconvex.

We shall consider in this subsection the set Q_0^+ of all nonnegative lower semicontinuous quasiconvex functions $q : X \to \bar{\mathbb{R}}_+$ such that $q(0) = 0$ and the set $Q_0^{ev,+}$ of all nonnegative evenly quasiconvex functions $q : X \to \bar{\mathbb{R}}_+$ with the same property $q(0) = 0$.

Clearly, $Q_0^{ev,+} \supset Q_0^+$. Note that 0 is a global minimizer of each function $q \in Q_0^{ev,+}$. For each $v \in X^*$ and $c \in \mathbb{R}_+$, consider the functions $l_{v,c}$ and $l^{v,c}$ defined by:

$$l_{v,c}(x) = \begin{cases} c & \text{if } v(x) > 1; \\ 0 & \text{if } v(x) \leq 1 \end{cases} \tag{7.3.13}$$

and

$$l^{v,c}(x) = \begin{cases} c & \text{if } v(x) \geq 1; \\ 0 & \text{if } v(x) < 1. \end{cases} \tag{7.3.14}$$

Obviously, $l_{v,c}$ is lower semicontinuous and $l^{v,c}$ is upper semicontinuous. Let

$$L = \{l_{v,c} : v \in X^*, c \in \mathbb{R}_+\} \text{ and } L^{ev} = \{l^{v,c} : v \in X^*, c \in \mathbb{R}^*\}.$$

Then $L \subset Q_0^+$ and $L^{ev} \subset Q_0^{ev,+}$. We shall consider both L and L^{ev} as sets of elementary functions.

The following result holds.

THEOREM 7.8 *A function $q : X \to \bar{\mathbb{R}}_+$ is L-convex if and only if $q \in Q_0^+$.*

Proof: We only need to prove that each function $q \in Q_0^+$ is L-convex. Let $c > 0$. Consider the set $S_c(q) = \{x \in X : q(x) \leq c\}$ and the function q_c, where

$$q_c(x) = \begin{cases} 0 & \text{if } x \in S_c(q); \\ c & \text{if } x \notin S_c(q). \end{cases}$$

The function q_c is quasiconvex and lower semicontinuous, hence $q_c \in Q_0^+$. Since $q_c \leq q$, it follows that $q(x) \geq \sup_{c>0} q_c(x)$ for all $x \in X$. Let $x \in X$ and $\bar{c} = q(x)$. For all $0 \leq c' < \bar{c}$, we have $q_{c'}(x) = c'$. Therefore,

$$\sup_{c>0} q_c(x) \geq \sup_{0 \leq c' < c} q_{c'}(x) = \sup_{0 \leq c' < \bar{c}} \bar{c} = q(x).$$

Thus

$$q(x) = \sup_{c>0} q_c(x). \tag{7.3.15}$$

Consider the set $S_c(q)$ with $c > 0$. Since $S_c(q)$ is convex and closed, it follows that $S_c(q)$ is the intersection of a family of closed halfspaces. In other words, there exist families $(v'_\alpha)_{\alpha\in A} \subset X^*$ and $(c_\alpha)_{\alpha\in A} \subset \mathbb{R}$ such that $S_c(q) = \bigcap_{\alpha\in A} V_\alpha$, where $V_\alpha = \{x \in X : v'_\alpha(x) \le c_\alpha\}$. Since $q \in Q_0$, it follows that $0 \in S_c(q)$, hence $c_\alpha \ge 0$ for all $\alpha \in A$. Note that $\{x : v(x) \le 0\} = \bigcap_{b>0}\{x : v(x) \le b\}$, hence we can assume without loss of generality that $c_\alpha > 0$ for all $\alpha > 0$. Let $v_\alpha = v'_\alpha/c_\alpha$. Then $V_\alpha = \{x : v_\alpha(x) \le 1\}$. Consider the functions

$$l_{v_\alpha,c}(x) = \begin{cases} c & \text{if } x \notin V_\alpha; \\ 0 & \text{if } x \in V_\alpha \end{cases}$$

with $\alpha \in A$. Obviously $l_{v_\alpha,c} \in L$ for each $\alpha \in A$. Let $x \in X$. If $x \in S_c(q)$, then $q_c(x) = 0 = l_{v_\alpha,c}(x)$. Let $x \notin S_c(q)$. Since $S_c(q) \subset V_\alpha$, it follows that $q_c(x) = c \ge l_{v_\alpha,c}(x)$. Hence, $l_{v_\alpha,c} \le q$. Since $S_c(q) = \bigcap_{\alpha\in A} V_\alpha$, it follows that for each point $x \notin S_c(q)$ there exists $\alpha \in A$ such that $l_{v_\alpha,c}(x) = c$. So

$$\sup_{\alpha\in A} l_{v_\alpha,c}(x) = q_c(x) \text{ for all } x \in X. \tag{7.3.16}$$

It follows from (7.3.15) and (7.3.16) that there exists a set $U \subset L$ such that $q(c) = \sup_{l\in L} l(x)$. △

THEOREM 7.9 *A function q is L^{ev}-convex if and only $q \in Q^{ev,+}$.*

Proof: The proof is similar to that of Theorem 7.8 and we omit it. △

REMARK 7.9 Theorem 7.8 asserts that L is a supremal generator for Q_0^+. Theorem 7.9 asserts that L^{ev} is a supremal generator for $Q^{ev,+}$.

Let Z be a closed convex set, $0 \in Z$. Denote by $Q_0^+(Z)$ the set of all lower semicontinuous quasiconvex functions $q : Z \to \mathbb{R}_{+\infty}$ such that $q(0) = 0$. Let $q \in Q_0^+(Z)$ and q' be the extension of q to X defined by

$$q'(x) = \begin{cases} q(x) & \text{if } x \in Z; \\ +\infty & \text{if } x \notin Z. \end{cases} \tag{7.3.17}$$

Then $q' \in Q_0^+$. Let $Q_0^{ev,+}(Z)$ be the set of all evenly quasiconvex functions $q : Z \to \mathbb{R}_{+\infty}$ with $q(0) = 0$. In the same manner, we can extend each function $q \in Q_0^{ev,+}(Z)$ to the function $q' \in Q_0^{ev,+}$.

THEOREM 7.10 *Let $q : Z \to \mathbb{R}_{+\infty}$. The function q is L-abstract convex on Z if and only if $q \in Q_0^+(Z)$. The function q is L^{ev}-abstract convex on Z if and only if $q \in Q_0^{ev,+}(Z)$.*

Proof: Consider the extension q' of the function q defined by (7.3.17). Applying Theorem 7.8 (Theorem 7.9) to the function q', we deduce that $q \in Q_0^+(Z)$ ($q \in Q_0^{ev,+}(Z)$). △

Note that L and L^{ev} are conic set. We now point out (without proofs) some properties of the set L. (See [159] for details.)

PROPOSITION 7.40 *Let $v \in X^*, v \neq 0$, and $c \in \mathbb{R}_+$. Then there exists $x \in X$ such that*

$$l_{v,c}(x) \neq \sup\{l_{v',c'}(x) : l_{v',c'} \leq l_{v,c},\ v' \neq \lambda v \text{ with } \lambda > 0\}.$$

Proposition 7.40 asserts that L is a minimal conic set, which is a supremal generator for Q_0^+.

In spite of Q_0^+ not being a convex cone, we can define extreme rays of this conic set in the usual way. A function $q \in Q_0^+$ is said to be on an extreme ray of $Q(X)$ if the equality $q = q_1 + q_2$ with $q_1, q_2 \in Q_0^+$ implies $q_1 = \lambda_1 q$, $q_2 = \lambda_2 q$ with $\lambda_1, \lambda_2 \in (0, +\infty)$.

THEOREM 7.11 *(see [159]) Each $l \in L$ is on an extreme ray of the conic set Q_0^+.*

Consider now the sets H_L and $H_{L^{ev}}$ of abstract affine functions with respect to L and L^{ev}, respectively. For the sake of simplicity we denote H_L by $\mathcal{J}_0$ and $H_{L^{ev}}$ by $\mathcal{K}_0$. By definition, $j \in \mathcal{J}_0$ if and only if there exists $v \in X^*, d \geq 0$ and $d' \in \mathbb{R}$ such that $j = l_{v,d} - d'$, hence

$$j(x) = \begin{cases} c & \text{if } v(x) > 1; \\ c' & \text{if } v(x) \leq 1, \end{cases} \tag{7.3.18}$$

where $c = d - d'$, $c' = -d'$. A function given by (7.3.18) will be denoted by the ordered triple (v, c, c'). Thus $\mathcal{J}_0$ consists of all functions $j = (v, c, c')$ with $v \in X^*$, c, $c' \in \mathbb{R}$, $c' \leq c$. If $v = 0$, then the set $\{x : v(x) > 1\}$ is empty, so the function $(0, c, c')$ is a constant function equal to c'.

The set $\mathcal{K}_0$ consists of all functions k defined by

$$k(x) = \begin{cases} c & \text{if } v(x) \geq 1; \\ c' & \text{if } v(x) < 1, \end{cases} \tag{7.3.19}$$

where $c, c' \in \bar{\mathbb{R}}$, $c' \leq c$ and $v \in X^*$. Clearly, a function j defined by (7.3.18) is lower semicontinuous, therefore $\mathcal{J}_0 \in Q$. A function k defined by (7.3.19) is upper semicontinuous so $K \in Q^{ev}$. Note that zero is a global minimizer of each function $j \in \mathcal{J}_0$ and of each function $k \in \mathcal{K}_0$.

Let Z be a closed convex subset of the space X with $0 \in Z$. In the rest of this subsection we consider the set $Q_0(Z)$ of all lower semicontinuous quasiconvex functions $q : Z \to \mathbb{R}_{+\infty}$ with the property $q(0) = \inf\{q(x) : x \in Z\}$ and the set $Q_0^{ev}(Z)$ of all evenly quasiconvex functions $q : Z \to \mathbb{R}_{+\infty}$ with the same property.

First, we will describe $\mathcal{J}_0$-convex and $\mathcal{K}_0$-convex functions.

PROPOSITION 7.41 *Let Z be a closed convex subset of X with $0 \in Z$ and let $q : Z \to \mathbb{R}_{+\infty}$. Then*

1) the function q is $\mathcal{J}_0$-convex on Z if and only $q \in Q_0(Z)$;

2) the function q is $\mathcal{K}_0$-convex on Z if and only if $q \in Q_0^{ev}(Z)$.

Proof: We shall only prove the first part of Proposition 7.41. The second part can be proved by the same argument.

Each function $j \in \mathcal{J}_0$ is lower semicontinuous, quasiconvex and $j(0) \leq j(x)$ for all $x \in X$. If there exists a set $U \subset \mathcal{J}_0$ such that $q(z) = \sup_{j \in U} j(z)$ for all $z \in Z$, then the function q enjoys the same properties, hence $q \in Q_0(Z)$. Assume now that $q \in Q_0(Z)$. Let $\tilde{q}(z) = q(z) - q(0)$ $(z \in Z)$. Then $\tilde{q} \in Q_0^+$, hence there exists a set $U \subset L$ such that $\tilde{q}(z) = \sup_{l \in U} l(z)$ for all $z \in Z$. We have $q(z) = \sup_{l \in U}(l(z) + q(0))$ $(z \in Z)$. Since the function $j : x \mapsto l(x) + q(0)$ belongs to $\mathcal{J}_0$, the result follows. $\triangle$

7.3.6. Two-step functions - generating quasiconvexity. Abstract convex sets

We now give a description of $(\mathcal{J}_0, X, Z)$-convex sets. We need the following definition.

DEFINITION 7.20 A subset U of the set $\mathcal{J}_0$ is called *vertically closed* if

$$((v, c_i, c_i') \in U,\ c_{i+1} \geq c_i,\ i = 1, 2, \dots, ; c_i \to c) \implies (v, c, c') \in U.$$

THEOREM 7.12 *Let U be a nonempty subset of $\mathcal{J}_0$ and let $m = \sup\{c' : (v, c, c') \in U\}$. The set U is $\mathcal{J}_0$-convex on a closed convex set Z with $0 \in Z$ if and only if*

(i) U is vertically closed;

(ii) if $(v_i, c_i, c_i')_{i \in I} \in U$, for an arbitrary index set I, and

$$v \in \operatorname{cl}\operatorname{co}\{Z^\circ \cup \{v_i\}_{i \in I}\},\ c \leq \inf_{i \in I} c_i,\ c' \leq m,$$

then $(v,c,c') \in U$. In particular, if I is empty then $(v,c,c') \in U$ for all $v \in Z^\circ$, $c \in \mathbb{R}$ and $c' \leq m$ (as usual we suppose that inf *over the empty set is equal to $+\infty$).*

Here Z° stands for the polar set to Z, that is $Z^\circ = \{v \in X^* : v(x) \leq 1$ for all $x \in Z\}$.

REMARK 7.10 If $Z = X$, then condition (ii) can be represented as follows:

(iii) If $(v_i, c_i, c'_i)_{i \in I} \in U$, for an arbitrary index set I, and

$$v \in \operatorname{cl}\operatorname{co}\{0, \{v_i\}_{i\in I}\}, \; c \leq \inf_{i \in I} c_i, \; c' \leq m,$$

then $(v,c,c') \in U$.

Consider the projection U_1 of the set U with respect to its first argument, that is, $U_1 = \{v : \exists\; c, c'$ such that $(v,c,c') \in U\}$. Assume that all constant functions $(0, -c', -c')$ with sufficiently large $c' > 0$ belong to U. Then $0 \in U_1$ and it follows from (iii) that U_1 is a closed convex set. The set U itself is not convex and the set $\mathcal{J}_0$ is also not convex. Note that instead of convex combinations of the numbers c_i we take the infimum of these numbers.

Proof of Theorem 7.12: Let U be a nonempty set which is $\mathcal{J}_0$-convex on Z and $f(x) = \sup\{j(x) : j \in U\}$ for $x \in Z$. Then

$$U = \{j = (v,c,c') : j(x) \leq f(x) \text{ for all } x \in Z\}.$$

Clearly U is vertically closed. We have

$$m = \sup\{c' : (v,c,c') \in U\} = \sup\{j(0) : j \in U\} = f(0).$$

Since $j(0) \leq j(x)$ for all $x \in Z$ and all $j \in U$, it follows that $m = f(0) \leq f(x)$ for all $x \in Z$. Applying this inequality we shall check condition (ii). Let

$$v \in \operatorname{cl}\operatorname{co}\{Z^\circ \cup \{v_i\}_{i\in I}\}, \; c \leq \inf_{i \in I} c_i, \; c' \leq m,$$

where $j_i = (v_i, c_i, c'_i) \in U$ for all $i \in I$ and $j = (v,c,c')$. If $v(x) \leq 1$ for all $x \in Z$, that is, $v \in Z^\circ$, then $j \in U$. Now assume that the set $Z_v = \{x \in Z : v(x) > 1\}$ is not empty. Let $x \in Z_v$. Assume that $v_i(x) \leq 1$ for all $i \in I$. We also have $w(x) \leq 1$ for all $w \in Z^\circ$. Therefore,

$$v(x) \leq \sup\{v'(x) : v' \in \operatorname{cl}\operatorname{co}(Z^\circ \cup \{v_i\}_{i\in I})\} \leq 1.$$

We have a contradiction which shows that there is an index $i_0 \in I$ such that $v_{i_0}(x) > 1$. For the function $j_{i_0} = (v_{i_0}, c_{i_0}, c'_{i_0})$ we have $j_{i_0}(x) = c_{i_0}$.

Since $j_{i_0} \in U$, it follows that $c_{i_0} \leq f(x)$. Hence $c \leq \inf_{i \in I} c_i \leq c_{i_0} \leq f(x)$. Thus for all $x \in Z$ such that $v(x) > 1$ we have $c \leq f(x)$. If $v(x) \leq 1$, then $j(x) = c' \leq m \leq f(x)$. So $j = (v, c, c') \in U$.

Now let U be a nonempty set with properties (i) and (ii). If $U = \mathcal{J}_0$ then U is a $\mathcal{J}_0$-convex set. Suppose $U \neq \mathcal{J}_0$. Take $j_0 = (v, c_0, c'_0) \notin U$. In order to check that U is a support set we must find $x \in Z$ such that $j_0(x) > \sup\{j(x) : j \in U\}$ (see Lemma 7.1).

Consider the following two cases:

1) Assume $c'_0 > m = \sup\{c' : (v, c, c') \in U\}$. We have

$$\sup_{j \in U} j(0) = \sup\{c' : (v, c, c') \in U\} = m < c'_0 = j_0(0).$$

Therefore, $x = 0$ is an appropriate point.

2) Now assume $c'_0 \leq m$. If also $c_0 \leq m$, then $j_0(x) \leq m$ for all $x \in Z$ and applying (ii) it is easy to check that $j_0 \in U$. So we suppose that $c_0 > m$. Since U is vertically closed it is possible to find $c < c_0$ such that $j = (v, c, c'_0) \notin U$ and $c > m$. Since $j \notin U$ and $c'_0 \leq m$, it follows that the set $Z_v = \{x \in Z : \ v(x) > 1\}$ is not empty.

Assume that for all $x \in Z_v$ there exists a $j_x = (v_x, c_x, c'_x) \in U$ such that $j_x(x) \geq j(x)$.

Take a point $x \in Z_v$. We have

$$j_x(x) \geq j(x) = c > m.$$

However, $c'_x \leq m$. Therefore, $j_x(x) = c_x$ for $x \in Z_v$ and we have

$$c_x \geq c \text{ for all } x \in Z_v. \tag{7.3.20}$$

It follows from the equality $j_x(x) = c_x$ that $v_x(x) > 1$, so

$$Z_v = \{x \in Z : v(x) > 1\} \subset \bigcup_{x \in Z_v} \{x' \in Z : v_x(x') > 1\}.$$

Thus

$$\{x \in Z : v(x) \leq 1\} \supset \bigcap_{x \in Z_v} \{x' \in Z : v_x(x') \leq 1\}. \tag{7.3.21}$$

Denote by M the halfspace $\{x : v(x) \leq 1\}$ and by M_x the halfspaces $\{x' : v_x(x') \leq 1\}$ (where $x \in Z$ and $v(x) > 1$). Thus we have

$$Z \cap M \supset \bigcap_{x \in Z_v} (Z \cap M_x),$$

so

$$(Z \cap M)^\circ \subset (\bigcap_{x \in Z_v} (Z \cap M_x))^\circ.$$

It is easy to check that

$$M^\circ = \mathrm{co}\,\{0, v\}, \qquad M_x^\circ = \mathrm{co}\,\{0, v_x\}.$$

Since M° is compact and $0 \in Z^\circ$ we have

$$(Z \cap M)^\circ = \mathrm{co}\,(Z^\circ \cup M^\circ) = \mathrm{co}\,(Z^\circ \cup \{v\}).$$

and

$$\begin{aligned}
(\bigcap_{x \in Z_v} (Z \cap M_x))^\circ &= (Z \cap (\bigcap_{x \in Z_v} M_x))^\circ \\
&= \mathrm{cl\,co}\,(Z^\circ \cup \mathrm{cl\,co} \bigcup_{x \in Z_v} M_x^\circ) \\
&= \mathrm{cl\,co}\,(Z^\circ \cup \mathrm{cl\,co}\,[\bigcup_{x \in Z_v} \mathrm{co}\,(0, v_x)]) \\
&= \mathrm{cl\,co}\,(Z^\circ \cup \{v_x\}_{x \in Z_v}).
\end{aligned}$$

So it follows from (7.3.21) that

$$\mathrm{co}\,(Z^\circ \cup \{v\}) \subset \mathrm{cl\,co}\,(Z^\circ \cup \{v_x\}_{x \in Z_v}). \tag{7.3.22}$$

Clearly, (7.3.22) holds if and only if

$$v \in \mathrm{cl\,co}\,(Z^\circ \cup \{v_x\}_{x \in Z_v}). \tag{7.3.23}$$

By applying (7.3.20), which shows that $c \leq \inf\{c_x : x \in Z_v\}$, and (7.3.23) we deduce, using (ii), that $j = (v, c, c_0') \in U$. However, this contradicts our assumption. Hence there exists $x' \in Z$ such that $v(x') > 1$ and $j(x') > j'(x')$ for all $j' \in U$. We then have

$$\sup_{j' \in U} j'(x') \leq j(x') = c < c_0 = j_0(x'). \qquad \triangle$$

Now we give a description of the $\mathcal{J}_0$-convex hull of a subset of the set $\mathcal{J}_0$. Let us consider a closed convex set $Z \subset X$ with $0 \in Z$ and an arbitrary subset U of the set $\mathcal{J}_0$. Denote

$$m = \sup\{j(0) : j \in U\} = \sup\{c' : (v, c, c') \in U\}. \tag{7.3.24}$$

Let $\hat{U}$ be the set of all $j = (v, c, c') \in \mathcal{J}_0$ such that there exist an index set I and a family $(j_i)_{i \in I}$ such that $j_i = (v_i, c_i, c_i')$ and

$$v \in \mathrm{cl\,co}\,\{Z^\circ \cup \{v_i\}_{i \in I}\}, \quad c \leq \inf_i c_i, \quad c' \leq m.$$

In particular, when I is an empty index set, $(v, c, c') \in \hat{U}$ for all $v \in Z^\circ$, $c \in \mathbb{R}$, and $c' \leq m$. Let $\bar{U}$ be the vertical closure of the set $\hat{U}$, that is, $(v, c, c') \in \bar{U}$ if and only if there exists a sequence $(v, c_n, c') \in \hat{U}$ such that $c_{n+1} > c_n$ and $c_n \to c$.

THEOREM 7.13 *For $U \subset \mathcal{J}_0$ we have*

$$co^Z_{\mathcal{J}_0} U = \bar{U}.$$

Proof: First we shall show, by applying Theorem 7.12, that $\bar{U}$ is a $\mathcal{J}_0$-convex set. Clearly, $\bar{U}$ is vertically closed. Now let us check that the set $\bar{U}$ possesses property (ii) of Theorem 3.3. Let I be an arbitrary index set, $j_i = (v_i, c_i, c'_i) \in \bar{U}$ for all $i \in I$ and

$$v \in \operatorname{cl} \operatorname{co} \{Z^\circ \cap \{v_i\}_{i \in I}\}, \quad c \leq \inf_i c_i, \quad c' \leq \bar{m},$$

where $\bar{m} = \sup\{c' : (v, c, c') \in \bar{U}\}$.

For a natural number n, let $c_i^n = c_i - \frac{1}{n}$. It follows directly from the definitions of $\hat{U}$ and $\bar{U}$ that $(v_i, c_i^n, c'_i) \in \hat{U}$. Applying again the definition of $\hat{U}$ we find, for each $i \in I$, an index set P_i and a family $(j_{p,i})_{p,i \in P_i}$ with $j_{p,i} = (v_{p,i}, c_{p,i}, c'_{p,i}) \in \mathcal{J}_0$ such that

$$v_i \in \operatorname{cl} \operatorname{co} \{Z^\circ \cap \{v_{p,i}\}_{p,i \in P_i}\}, \quad c_i^n \leq \inf_{p,i \in P_i} c_{p,i}, \quad c'_{p,i} \leq m.$$

Let us consider the index set $T = \bigcup_{i \in I} P_i$ and the family $(j_t)_{t \in T}$. We have for all $i \in I$:

$$v_i \in \operatorname{cl} \operatorname{co} \{Z^\circ \cup \{v_{p,i}\}_{p,i \in P_i}\} \subset \operatorname{cl} \operatorname{co} \{Z^\circ \cup \{v_t\}_{t \in T}\}.$$

Therefore,

$$v \in \operatorname{cl} \operatorname{co} \{Z^\circ \cup \{v_i\}_{i \in I}\} \subset \operatorname{cl} \operatorname{co} \{Z^\circ \cup \{v_t\}_{t \in T}\}.$$

We also have

$$c - \frac{1}{n} \leq \inf_i c_i^n \leq \inf_i \inf_{p,i \in P_i} c_{p_i} = \inf_{t \in T} c_t.$$

It follows from the definition of the set $\hat{U}$ that $(v, c - \frac{1}{n}, c') \in \hat{U}$ for all $n = 1, 2, \ldots$. So $(v, c, c') \in \bar{U}$. Now we apply Theorem 7.12 and deduce that $\bar{U}$ is a $\mathcal{J}_0$-convex set. This fact allows us to conclude that

$$co^Z_{\mathcal{J}_0} \bar{U} = \bar{U}. \tag{7.3.25}$$

By the definition of the $\mathcal{J}_0$-convex hull we have that

$$\mathrm{co}^Z_{\mathcal{J}_0} U = \mathrm{supp}(f, \mathcal{J}_0, Z),$$

where the function f is defined on the set Z by the formula $f(z) = \sup\{j(z) : j \in U\}$. It is easy to check that $\bar{U} \subset \mathrm{supp}(f, \mathcal{J}_0, Z)$. On the other hand, the inclusion $U \subset \bar{U}$ and equality (7.3.25) show that

$$\mathrm{supp}(f, \mathcal{J}_0, Z) = \mathrm{co}^Z_{\mathcal{J}_0} U \subset \mathrm{co}^Z_{\mathcal{J}_0} \bar{U} = \bar{U}.$$

Thus $\bar{U} = \mathrm{supp}(f, \mathcal{J}_0, Z) = \mathrm{co}^Z_{\mathcal{J}_0} U$. $\triangle$

In order to represent the set $\mathrm{co}^Z_{\mathcal{J}_0} U$ in a more convenient form we will introduce algebraic operations and a topology on the set $\mathcal{J}_0$. At first we shall consider the set $\bar{\mathcal{J}}_0$ of two-step functions $j = (v, c, c')$ with $c' = -\infty$. Thus $j \in \bar{\mathcal{J}}_0$ if and only if

$$j(x) = \begin{cases} c & \text{if } v(x) > 1; \\ -\infty & \text{if } v(x) \le 1. \end{cases} \tag{7.3.26}$$

We denote by (v, c) the function given by (7.3.26). Let us introduce the algebraic operation $\oplus$ on the set $\bar{\mathcal{J}}_0$, where

$$(v_1, c_1) \oplus (v_2, c_2) = (v_1 + v_2, \min(c_1, c_2)).$$

Clearly, the operation $\oplus$ is commutative and associative. We can define, in the usual way, an arbitrary finite sum (with respect to $\oplus$) of functions $j_1, j_2, \dots, j_n \in \bar{\mathcal{J}}_0$. We also introduce the multiplication by a nonnegative number $\odot$:

$$\lambda \odot (v, c) = (\lambda v, c) \text{ for } \lambda > 0; \qquad 0 \odot (v, c) = 0.$$

Let $\lambda_i \ge 0$ and $\sum_i \lambda_i = 1$. We shall call the expression

$$\lambda_1 \odot (v_1, c_1) \oplus \lambda_2 \odot (v_2, c_2) \oplus \dots \oplus \lambda_n \odot (v_n, c_n) \tag{7.3.27}$$

a *quasiconvex combination* of the functions $(v_1, c_1), (v_2, c_2), \dots, (v_n, c_n)$ and denote it by $\sum'_i \lambda_i \odot (v_i, c_i)$.

Let $U \in \bar{\mathcal{J}}_0$. The set of all quasiconvex combinations of elements of the set U is called the *quasiconvex hull* of U and is denoted by $\mathrm{co}' U$.

Now we assume that $U \subset \mathcal{J}_0$. Let

$$\tilde{U} = \{(v, c) \in \bar{\mathcal{J}}_0(X) : \text{ there exists } c' \text{ such that } (v, c, c') \in U\}$$

and $m = \sup\{j(0) : j \in U\}$. The set

$$\mathrm{co}' U = \{(v, c, c') \in \mathcal{J}_0 : (v, c) \in \mathrm{co}' \tilde{U},\ c' \le m\}$$

is then called the *quasiconvex hull* of the set U. A subset U of the set $\mathcal{J}_0$ is called *quasiconvex* if $\mathrm{co}'\,U = U$.

Let $U \in \mathcal{J}_0$ with $m = \sup\{j(0) : j \in U\}$. For $W \subset X^*$ denote the following set by $\mathrm{co}'_W U$:

$$\mathrm{co}'_W U = \mathrm{co}'\,(U \cup \{(v, c, c') : v \in W,\ c \in \mathbb{R},\ c' \leq m\}).$$

A subset U of the set $\mathcal{J}_0$ is called ***vertically stable*** if together with a function (v, c, c') this set contains all functions of the form (v, c_1, c'_1) with $c_1 \leq c$, $c'_1 \leq c'$.

PROPOSITION 7.42 *Let U be a $\mathcal{J}_0$-convex set with respect to a closed convex set Z with $0 \in Z$. Then*

1) $\mathrm{co}'_{Z^\circ} U = U$;

2) U is a vertically stable set.

Proof: The proof follows directly from Theorem 7.12. △

Now we introduce a topology τ on the set $\mathcal{J}_0$. Let $j = (v, c, c') \in \mathcal{J}_0$ and let $\mathcal{M}_v = (M_v)$ be a fundamental family of neighbourhoods of the point v. Then a fundamental family of neighbourhoods of the function j under the topology τ consists of all sets of the form

$$N_{M_v,\varepsilon}(j) = \{\tilde{j} = (\tilde{v}, \tilde{c}, \tilde{c}') : \tilde{v} \in M_v,\ |c - \tilde{c}| < \varepsilon,\ |c' - \tilde{c}'| < \varepsilon\}$$

with $M_v \in \mathcal{M}_v$ and $\varepsilon > 0$. In other words, τ is a product-topology induced on the subset $\mathcal{J}_0$ of the product $X^* \times \mathbb{R} \times \mathbb{R}$.

PROPOSITION 7.43 *For an arbitrary closed convex subset Z of the space X, with $0 \in Z$, a $\mathcal{J}_0$-convex (with respect to Z) subset U of $\mathcal{J}_0$ is closed in the topology τ.*

Proof: Let $j_\omega = (v_\omega, c_\omega, c'_\omega)_{\omega \in \Omega}$ be a net of elements of U such that $j_\omega \to j = (v, c, c')$. For $\varepsilon > 0$, take a subnet $(j_\omega)_{\omega \in \Omega'}$ of the given net such that

$$c_\varepsilon =: \inf\{c_\omega : \omega \in \Omega'\} > \lim_{\omega \in \Omega} c_\omega - \varepsilon.$$

We have

$$v \in \mathrm{cl}\,\{v_\omega\}_{\omega \in \Omega'} \subset \mathrm{cl}\,\mathrm{co}\,\{Z^\circ \cup \{v_\omega\}_{\omega \in \Omega'}\}, \qquad c' \leq m = \sup\{j(0) : j \in U\}.$$

It follows from Theorem 7.12 that $(v, c_\varepsilon, c') \in U$. We assume without loss of generality that $c_\varepsilon > c_{\varepsilon'}$ for $\varepsilon < \varepsilon'$. Since the set U is vertically closed and $c_\varepsilon \to c$, it follows that $(v, c, c') \in U$. △

Now we are able to describe the $\mathcal{J}_0$-hull in algebraic and topological terms.

PROPOSITION 7.44 *Let Z be a closed convex subset of the space X with $0 \in Z$. Let U be a vertically stable subset of $\mathcal{J}_0$. Then*

$$co^Z_{\mathcal{J}_0}\, U = cl_\tau\, co'_{Z^\circ}\, U,$$

where cl_τ denotes the closure in the topology τ.

Proof: Since $co^Z_{\mathcal{J}_0}\, U$ is a $\mathcal{J}_0$-convex set, it follows from Proposition 7.42 and Proposition 7.43 that

$$cl_\tau co'_{Z^\circ}\, U \subset co^Z_{\mathcal{J}_0}\, U.$$

Now we will show that the reverse inclusion is true. Let $j = (v, c, c') \in co^Z_{\mathcal{J}_0}\, U$ and $\bar{c} < c$. It follows from Theorem 7.13 that there is a family $(v_i, c_i, c'_i)_{i \in I} \in U$ such that

$$v \in \mathrm{cl}\,\mathrm{co}\,\{Z^\circ \cup \{v_i\}_{i\in I}\}, \quad \bar{c} \leq \inf c_i, \quad c'_i \leq m$$

with $m = \sup\{j(0) : j \in U\}$. Let M_v be a neighbourhood of the element v. Then there exists an element $v_* \in M_v$ such that $v_* \in \mathrm{co}\,\{Z^\circ \cup \{v_i\}_{i\in I}\}$. If $v_* \in Z^\circ$ then $(v_*, c, c') \in co'_{Z^\circ} U$ for all $c' \leq m$. So assume $v_* \notin Z^\circ$. Then there exist

1) a finite subfamily $(v_k, c_k, c'_k)_{k=1}^p$ of the family $(v_i, c_i, c'_i)_{i\in I}$ such that

$$c_* =: \min_{k=1,\dots,p} c_k < \inf_{i \in I} c_i + \varepsilon;$$

2) an element $v_0 \in Z^\circ$;

3) numbers $\alpha_1, \dots, \alpha_p \geq 0$ and $0 \leq \alpha_0 < 1$ with the property $\alpha_0 + \sum_{k=1}^p \alpha_k = 1$,

such that the following representation holds:

$$v_* = \alpha_0 v_0 + \sum_{k=1}^p \alpha_k v_k.$$

Let $c'_* \leq m$. Then we have $(v_*, c_*, c'_*) \in co'_{Z^\circ} U$. Since ε is an arbitrary positive number and M_v is an arbitrary neighbourhood of the point v, it follows that the function $j' = (v, \inf_i c_i, m) \in cl_\tau\, co'_{Z^\circ} U$. Since U is a vertically stable set, it follows that $(v, \bar{c}, c') \in cl_\tau\, co'_{Z^\circ} U$ for all

$c' \le m$. Since $\bar{c}$ is an arbitrary number with the property $\bar{c} < c$, we have $j = (v, c, c') \in \mathrm{cl}_\tau \mathrm{co}'_{Z^\circ} U$. Thus

$$\mathrm{co}^{Z}_{\mathcal{J}_0}\, U \subset \mathrm{cl}_\tau \mathrm{co}'_{Z^\circ} U.$$

$\triangle$

REMARK 7.11 If $Z = X$, then $Z^\circ = \{0\}$ and $\mathrm{co}'_{Z^\circ} U = \mathrm{co}' U$ for a vertically stable set U. Thus if U is vertically stable we have

$$\mathrm{co}^{X}_{\mathcal{J}_0}\, U = \mathrm{cl}_\tau \mathrm{co}'\, U.$$

7.3.7. Two-step functions - generating quasiconvexity. Abstract subdifferentials

We now turn to the subdifferentials with respect to the sets of abstract linear functions L and L^{ev}. Recall that L (L^{ev}) consists of all functions $l_{v,c}$ ($l^{v,c}$) with $v \in X^*$ and $c \in \mathbb{R}_+$, where

$$l_{v,c} = \begin{cases} c & \text{if } v(x) > 1; \\ 0 & \text{if } v(x) \le 1, \end{cases} \qquad l^{v,c} = \begin{cases} c & \text{if } v(x) \ge 1; \\ 0 & \text{if } v(x) < 1. \end{cases}$$

Recall also that $\mathcal{J}_0$ is the set of all L-affine functions and $\mathcal{K}_0$ is the set of all L^{ev}−affine functions. We begin with the L-subdifferential. Consider the sets

$$\partial^*_{\mathcal{J}_0} f(x_0) = \{j \in \mathrm{supp}(f, \mathcal{J}_0) : j(x_0) = f(x_0)\}.$$

It follows from Proposition 7.1 that $\partial_L f(x_0)$ is nonempty if and only if $\partial^*_{\mathcal{J}_0} f(x_0)$ is nonempty.

PROPOSITION 7.45 *The subdifferential $\partial_L f(x_0) \equiv \partial_{L,X} f(x_0)$ is empty for an arbitrary $\mathcal{J}_0$-convex function f if x_0 is not a local minimizer of f over X.*

Proof: Assume that x_0 is not a local minimizer of a $\mathcal{J}_0$-convex function f. Suppose that $\partial_L f(x_0)$ is not empty then $\partial^*_{\mathcal{J}_0} f(x_0)$ is nonempty as well. Let $j = (v, c, c') \in \partial^*_{\mathcal{J}_0} f(x_0)$. We have

$$j(x) = \begin{cases} c & \text{if } v(x) > 1; \\ c' & \text{if } v(x) \le 1. \end{cases}$$

First assume that $v(x_0) \le 1$. Then $f(x_0) = j(x_0) = c'$. Since $c \ge c'$, we have $f(x_0) = c' \le f(x)$ for all $x \in X$. Thus x_0 is a global minimizer of the function f, which contradicts our assumption.

Now suppose $v(x_0) > 1$. Since $f(x_0) = j(x_0)$ we have $f(x_0) = c$. Let N be a neighbourhood of the point x_0 such that $v(x) > 1$ for all $x \in N$. Then $f(x) \geq j(x) = f(x_0)$ for all $x \in N$, so x_0 is a local minimizer of the function f and we have another contradiction. Thus $\partial^*_{\mathcal{J}_0} f(x_0) = \emptyset$. $\triangle$

REMARK 7.12 It follows from Remark 7.4 that $\partial_{L,Z} f(x_0) \neq \emptyset$ if x_0 is a global minimizer of a function $f : Z \to \mathbb{R}_{+\infty}$.

It follows from Proposition 7.41 that f is $\mathcal{J}_0$-convex if and only if f is lower semicontinuous, quasiconvex and $f(0) = \inf_{x \in X} f(x)$.

Assume f is a lower semicontinuous quasiconvex function such that each local minimum of q is its global minimum and $f(x) > f(0)$ for all $x \in X, x \neq 0$. Then the subdifferential $\partial_L f(0) \neq \emptyset$ and $\partial_L f(x)$ is empty at all $x \neq 0$.

We now examine the subdifferential with respect to the set L^{ev}. Consider the set

$$\begin{aligned} \partial^*_{\mathcal{K}_0} f(x_0) &= \{k \in \operatorname{supp}(f, \mathcal{K}_0) : k(x_0) = f(x_0)\} \\ &\equiv \{k \in \mathcal{K}_0 : k \leq f,\ k(x_0) = f(x_0)\}. \end{aligned}$$

The set $\mathcal{K}_0$ consists of all functions k of the form:

$$k(x) = \begin{cases} d & \text{if } v(x) \geq 1; \\ d' & \text{if } v(x) < 1 \end{cases} \tag{7.3.28}$$

with $v \in X^*$, $d, d' \in \mathbb{R}$, $d \geq d'$. Since $\mathcal{K}_0$ is the set of L^{ev}-affine functions we can represent the function (7.3.28) in the following form: $k(x) = l^{v,c}(x) - d'$, where $c = d - d' \geq 0$ and $l^{v,c} \in L^{ev}$.

It has been proved (see Proposition 7.41) that a function $q : X \to \mathbb{R}_{+\infty}$ is $\mathcal{K}_0$-convex if and only if $q \in Q_0^{ev}$, that is, if q is evenly quasiconvex and $q(0) = \min_{x \in X} q(x)$.

It follows from Proposition 7.1 that $l \in \partial_{L^{ev}} f(x_0)$ if and only if

$$k \in \partial^*_{\mathcal{K}_0} f(x_0) \text{ with } k = l(x) - c,\ l \in L^{ev}, c = l(x_0) - f(x_0).$$

Let f be an evenly quasiconvex function $f : X \to \mathbb{R}_{+\infty}$ and $x_0 \in \operatorname{dom} f$. Assume that $k \in \partial^*_{\mathcal{K}_0} f(x_0)$, that is, $k \in \mathcal{K}_0$, $k \leq f$ and $k(x_0) = f(x_0)$.

DEFINITION 7.21 Let X be a Banach space. A function $f : X \to \mathbb{R}$ is called ***regular*** if

$$\{x : f(x) \leq c\} = \operatorname{cl}\{x : f(x) < c\}$$

for all $c > \inf\{f(x) : x \in X\}$.

Let U be a convex set. Recall that the set

$$N(U, x_0) = \{v \in X^* : v(x_0) = \max_{x \in U} v(x)\}$$

is called the *normal cone* to the set U at a point $x_0 \in U$. Set

$$N_1(U, x_0) = \{v \in N(U, x_0) : v(x_0) = 1\}.$$

Note that the conic hull cone $N_1(U, x_0)$ of the set $N_1(U, x_0)$ is equal to $\{v \in N(U, x_0) : v(x_0) > 0\} \cup \{0\}$. Assume that $0 \in \operatorname{int} U$ and $x_0 \in \operatorname{bd} U$. Then $v(x_0) > 0$ for all $v \in N(U, x_0) \setminus \{0\}$, hence $N(U, x_0) \setminus \{0\}$ coincides with the conic hull of the set $N_1(U, x_0)$.

PROPOSITION 7.46 *Let $f \in Q_0^{ev}$ and*

(i) f is regular;

(ii) $0 \in \operatorname{int} \operatorname{dom} f$ and f is continuous on $\operatorname{int} \operatorname{dom} f$.

Then $\partial^*_{\mathcal{K}_0} f(x_0)$ is nonempty for all $x_0 \in \operatorname{dom} f$. If $f(x_0) > f(0)$, then $\partial^*_{\mathcal{K}_0} f(x_0)$ consists of all functions $k \in K$ given by

$$k(x) = \begin{cases} f(x_0) & v(x) \geq 1; \\ c' & v(x) < 1, \end{cases} \tag{7.3.29}$$

where $v \in N_1(U, x_0)$ with

$$U := S_{f(x_0)}(f) \equiv \{x : f(x) \leq f(x_0)\}$$

and $c' \leq f(0)$.

Proof: Since zero is a global minimizer of the function f, it follows that $\partial^*_{\mathcal{K}_0} f(x_0)$ is nonempty for all x_0 such that $f(x_0) - f(0)$. Indeed, $f(x_0)\mathbf{1} \in \tilde{\partial} f(x_0)$, where $\mathbf{1}(x) = 1$ for all x. Assume now that $f(x_0) = \gamma > f(0)$. Then the level set $T_\gamma(f) = \{x : f(x) < \gamma\}$ is nonempty. It follows from (ii) that the interior of the convex set $T_\gamma(f)$ is not empty. Applying the regularity of f we conclude that x_0 is a boundary point of the set $T_\gamma(f)$. Hence the normal cone $N(U, x_0)$ at the point x_0 of the level set $U := S_\gamma(f) = \operatorname{cl} T_\gamma(f)$ is nonempty. Since $0 \in \operatorname{int} U$, it follows that for each $v' \in N(U, x_0)\{0\}$ there exists $v \in N_1(x_0)$ such that $v' = \lambda v$ with $\lambda > 0$. By the definition of the normal cone, we have

$$T_\gamma(f) \subset \{x : v(x) < v(x_0) = 1\}.$$

Therefore, the inequality $v(x) \geq 1$ implies $f(x) \geq f(x_0)$. Thus a function k defined by (7.3.29) is an element of the support set $\operatorname{supp}(f, K, X)$.

Since $k(x_0) = f(x_0)$, it follows that $k \in \partial^*_{\mathcal{K}_0} f(x_0)$. We have proved that the set $\partial^*_{\mathcal{K}_0} f(x_0)$ is nonempty and contains all functions given by (7.3.29).

We now show that an arbitrary function $k \in \partial^*_{\mathcal{K}_0} f(x_0)$ has the form (7.3.29). Since $k \in \mathcal{K}_0(X)$, it follows that there exist $v \in X^*$ and $c, c' \in \mathbb{R}$ with $c \geq c'$ such that

$$k(x) = \begin{cases} c & \text{if } v(x) \geq 1; \\ c' & \text{if } v(x) < 1. \end{cases} \tag{7.3.30}$$

We have $k(x) \leq f(x)$ for all $x \in X$ and $k(x_0) = f(x_0)$. Assume $f(x_0) = c'$. Then,

$$f(x_0) \leq k(x) \leq f(x) \text{ for all } x \in X.$$

Therefore, x_0 is a global minimizer, which contradicts our assumption. Thus $f(x_0) = c > c'$. It follows from (7.3.30) that $(f(x) \geq c) \implies (v(x) \geq 1)$, hence $(v(x) < 1) \implies (f(x) < f(x_0))$. Let $\gamma = f(x_0)$. Then

$$T_\gamma(f) = \{x : f(x) < \gamma\} \subset \{x : v(x) < 1\}.$$

The set $T_\gamma(f)$ is nonempty since $\gamma > f(x_0)$.

Applying the regularity of the function f, we conclude that

$$S_\gamma(f) \subset \{x : v(x) \leq 1\}. \tag{7.3.31}$$

Since $x_0 \in S_\gamma(f)$, it follows from (7.3.31) that $v(x_0) \leq 1$. Assume that $v(x_0) < 1$. The set $T_\gamma(f)$ is open, so there exists a vector $y \in T_\gamma(f)$ such that $v(y) < v(x_0)$. Let

$$\beta = \frac{1 - v(y)}{v(x_0) - v(y)}$$

and $z = y + \beta(x_0 - y)$. Then $v(z) = 1$, so $f(z) \geq \gamma$. We also have

$$x_0 = \frac{1}{\beta} z + \left(1 - \frac{1}{\beta}\right) y.$$

Note that $\beta > 1$, so x_0 is a convex combination of the points z and y. It follows from the quasiconvexity of f that

$$\gamma = f(x_0) \leq \min\{f(z), f(y)\}.$$

However, the inclusion $y \in T_\gamma(f)$ implies $f(y) < \gamma$ and we arrive at a contradiction. Thus we have proved that $v(x_0) = 1$. Applying (7.3.31) we conclude that $v \in N(U, x_0)$, hence $v \in N_1(U, x_0)$. △

THEOREM 7.14 *Assume that all conditions in Proposition 7.46 hold. Then $\partial_{L^{ev}} f(x_0)$ is nonempty for all $x_0 \in \operatorname{dom} f$. If $f(x_0) > f(0)$, then*

$l \in \partial_{L^{ev}} f(x_0)$ if and only if

$$l(x) = \begin{cases} c & \text{if } v(x) \geq 1; \\ 0 & \text{if } v(x) < 1 \end{cases}$$

with $c \geq f(x_0) - f(0)$ and $v \in N_1(U, x_0)$, where $U = S_{f(x_0)}(f)$.

Proof: The proof follows immediately from Proposition 7.1 and Proposition 7.46. △

7.3.8. Infima of families of abstract convex functions

We now consider examples of inf-H-convex functions (see Definition 7.5). Recall that if H coincides with the set of all continuous affine functions defined on a Banach space X, then inf-H-convex functions are called inf-convex.

As it was shown in Example 1.10, an arbitrary proper function $f : X \to \mathbb{R}_{+\infty}$ is inf-convex, namely f is the infimum of a family of convex functions p_z, such that $\operatorname{dom} p_z$ consists of the unique point z:

$$p_z(x) = \begin{cases} f(z) & \text{if } x = z; \\ +\infty & \text{if } x \neq z. \end{cases} \tag{7.3.32}$$

Sometimes we need a representation of a given function f as the infimum of a special family of convex functions (for example, a family of finite convex functions). Let us give some examples.

EXAMPLE 7.13 Let f_1, f_2 be lower semicontinuous convex functions defined on a convex subset Z of a Banach space X. Assume f_2 is finite on Z. Let $f = f_1 - f_2$. Then f can be represented as the infimum of the family of convex functions

$$g(x) = f_1(x) - h(x) \quad \text{with} \quad h \in \operatorname{supp}(f_2, H, Z).$$

Indeed,

$$\begin{aligned} f_1(x) - f_2(x) &= f_1(x) - \sup\{h(x) : h \in \operatorname{supp}(f_2, H, Z)\} \\ &= f_1(x) + \inf\{(-h)(x) : h \in \operatorname{supp}(f_2, H, Z)\} \\ &= \inf\{(f_1 - h)(x) : h \in \operatorname{supp}(f_2, H, Z)\}. \end{aligned}$$

Thus f is the infimum of a family of affine translations of the same function f_1.

A function f defined on a convex subset Z of a Banach space X is called *min-convex* on Z if there exists a family $(p_z)_{z\in Z}$ of convex functions such that $f \le p_z$ and $f(z) = p_z(z)$ for all $z \in Z$. Actually, a function is min-convex if its superdifferential with respect to the set of convex functions is not empty for each $x \in X$. See Section 1.6 for the definition of the superdifferential.

It follows from Example 1.10 that every function defined on Z is min-convex. Sometimes it is possible to find a special convex function from the above mentioned superdifferential. Let us give some examples.

EXAMPLE 7.14 Let Z be a subset of a Banach space X and f be a Lipschitz function on Z with a Lipschitz constant K (i.e., $|f(x)-f(y)| \le K\|x-y\|$ for all $x, y \in Z$). Let $p_z(x) = f(x) + K\|x-z\|$ for $z \in Z$. Then

$$f(x) = \min_{z\in Z} p_z(x) \text{ for all } x \in X.$$

Compare this result with Proposition 7.2.

EXAMPLE 7.15 Let $Z \subset \mathbb{R}^n$ be a compact convex set and let $G \supset Z$ be an open bounded set. Denote by $C^2(G)$ the set of all twice continuously differentiable functions defined on the set G. We shall show that for a function $f \in C^2(G)$ with $G \supset Z$ the following representation holds:

$$f(x) = \min\{p_z(x) : z \in Z\}, \qquad x \in Z,$$

where $p_z(x)$ is a quadratic function of the following form:

$$p_z(x) = a\|x-z\|^2 + [b_z, z] + c_z \quad \text{with } a_z, c_z \in R,\ b_z \in \mathbb{R}^n_+.$$

This assertion is an immediate corollary of the following lemma.

LEMMA 7.3 *Let $f \in C^2(G)$ and $z \in Z$. Then there exists a number k_z such that the quadratic function p_z defined as*

$$p_z(x) = k_z\|x-z\|^2 + [\nabla f(z), x-z] + f(z) \tag{7.3.33}$$

possesses the following properties:

$$p_z(z) = f(z) \text{ and } p_z(x) \ge f(x) \text{ for all } x \in Z.$$

Proof: We have $p_z(z) = f(z)$ for any k_z. Consider the quadratic function (7.3.33) with an arbitrary $k_z \ge \frac{1}{2}\|\nabla^2 f(z)\| + m$, where $m > 0$. Let $F(x) = p_z(x) - f(x)$. We have

$$\nabla F(x) = \nabla p_z(x) - \nabla f(x) = 2k_z(x-z) + \nabla f(z) - \nabla f(x).$$

Thus $\nabla^2 F(x) = 2k_z \,\mathrm{Id} - \nabla^2 f(x)$, where Id stands for the identity mapping. In particular, $\nabla F(z) = 0$ and, for all x

$$\begin{aligned} x^T \nabla^2 F(z) x &= 2k_z \|x\|^2 - x^T \nabla^2 f(z) x \\ &\geq 2k_z \|x\|^2 - \|\nabla^2 f(z)\| \|x\|^2 > m\|x\|^2. \end{aligned}$$

Therefore, for an arbitrary $k_z > \frac{1}{2}\|\nabla^2 f(z)\|$, the function $F(x)$ attains its local maximum at the point z, i.e., there is a neighbourhood V_z of the point z such that $f(x) < p_z(x)$ for all $x \in V_z \setminus \{z\}$. Since the set $Z \setminus V_z$ is compact, it follows that for k_z large enough, $p_z(x) > f(x)$ for all $x \in Z \setminus V_z$. $\triangle$

COROLLARY 7.7 *If $f \in C^2(G)$, then*

$$f(x) = \min_{z \in Z} (k_z \|x - z\|^2 + [\nabla f(z), x - z] + f(z)).$$

Thus every twice continuously differentiable function defined on a compact set Z is min-H-convex, where H is the set of all convex quadratic functions defined on $\mathbb{R}^n$.

7.4. MINKOWSKI DUALITY, C^2-LATTICES AND SEMILINEAR LATTICES

7.4.1. Overview

Let H be a set of finite elementary functions defined on a set X and $Z \subset X$. As it turns out the set $P(H, X, Z)$ of all abstract convex functions and the set $S(H, X, Z)$ of all abstract convex sets are complete lattices. Thus the theory of lattices can be used in the study of abstract convexity. A natural ordered isomorphism between $P(H, X, Z)$ and $S(H, X, Z)$ is the Minkowski duality, which maps each H-convex function on Z to its support set. A well-known example of the Minkowski duality is the relationship between a sublinear function and its support set. Sets of elementary functions H which are conic spaces or semilinear spaces, arise in applications. In such cases, $P(H, X, Z)$ and $S(H, X, Z)$ are conic complete lattices (c^2-lattices) or semilinear lattices, respectively, and the Minkowski duality is an algebraic isomorphism.

The Minkowski duality was introduced and studied in [99, 100]. Examples of c^2-lattices and semilinear lattices from Subsection 7.4.3 can be found in [141]

7.4.2. The Minkowski duality

Let Z be a subset of a set X and H be a set of finite functions defined on X. Let us consider the sets $P(H,X,Z)$ of all H-convex functions on Z and $S(H,X,Z)$ of all (H,X,Z)-convex sets. We assume that both $P(H,X,Z)$ and $S(H,X,Z)$ are equipped with the natural order relations:

$$p_1 \geq p_2 \iff p_1(z) \geq p_2(z) \text{ for all } z \in Z \qquad (p_1, p_2 \in P(H,X,Z);)$$
$$U_1 \geq U_2 \iff U_1 \supset U_2 \qquad U_1, U_2 \in S(H,X,Z).$$

DEFINITION 7.22 The mapping $\varphi : P(H,X,Z) \to S(H,X,Z)$ defined by $\varphi(p) = \mathrm{supp}(p,H,Z)$ is called the *Minkowski duality.*

PROPOSITION 7.47 *The Minkowski duality φ is an isomorphism between the ordered sets $P(H,X,Z)$ and $S(H,X,Z)$.*

Proof: It follows directly from the definition of $S(H,X,Z)$ that $\varphi(p) \in S(H,X,Z)$ for all $p \in P(H,X,Z)$. If $U \in S(H,X,Z)$, then $U = \varphi(p_U)$, where $p_U(x) = \sup_{h\in U} h(x)$ is the support function of the set U. Therefore, φ maps onto $S(H,X,Z)$.

Let $p_1, p_2 \in P(H,X,Z)$ and $p_1 \neq p_2$. Assume for the sake of definiteness that there exists $x_0 \in Z$ such that $p_1(x_0) > p_2(x_0)$. Let $U_1 = \varphi(p_1)$ and $U_2 = \varphi(p_2)$. We have

$$\sup_{h\in U_1} h(x_0) = p_1(x_0) > p_2(x_0) = \sup_{h\in U_2} h(x_0),$$

therefore $U_1 \neq U_2$. Thus φ is a one-to-one mapping. It follows directly from the definition of the support set that

$$p_1 \geq p_2 \iff \varphi(p_1) \supset \varphi(p_2). \qquad \triangle$$

REMARK 7.13 Let L be a set of abstract linear functions defined on a set X and let H_L be the corresponding set of abstract affine functions. Consider an H_L-convex function f. Then (see Proposition 7.8) $\mathrm{supp}(f,H_L) = \mathrm{epi}\, f^*$, where f^* is the Fenchel-Moreau conjugate function to f. Since $\mathrm{supp}(f,H_L) = \varphi(f)$, we conclude that $\varphi(f) = \mathrm{epi}\, f^*$. This equality expresses links between the Minkowski duality and the Fenchel-Moreau conjugacy.

We need the notion of closure in the sense of Moore (see, for example, [23, 114, 175]).

DEFINITION 7.23 Let X be a set and let 2^X be the totality of all subsets of X. A mapping $\Phi : 2^X \to 2^X$ is called a *closure in the sense of Moore* or a *Moore closure operator* if

1) $\Phi(U) \supset U$ for all $U \subset X$;

2) $\Phi(\Phi(U)) = \Phi(U)$ for all $U \subset X$;

3) $U_1 \subset U_2 \implies \Phi(U_1) \subset \Phi(U_2)$.

Let $\Phi : 2^X \to 2^X$ be a closure in the sense of Moore and Y be the image of Φ, that is the totality of all sets of the form $\Phi(U)$ with $U \subset X$. Assume that Y is ordered by inclusion. Then Φ is an isotone mapping, which majorates the identity and coincides with the identity on Y. The following assertion holds (see, for example, [23]):

THEOREM 7.15 (Moore) *The set Y is a complete lattice. If $(U_\alpha)_{\alpha \in A}$ is an arbitrary family of subsets of Y, then*

$$\inf_{\alpha \in A} U_\alpha = \bigcap_{\alpha \in A} U_\alpha; \qquad \sup_{\alpha \in A} U_\alpha = \Phi(\bigcup_{\alpha \in A} U_\alpha).$$

Proof: Let $U = \bigcap_{\alpha \in A} U_\alpha$. It follows from the properties of Φ that $\Phi(U) \supset U$. Let $\alpha \in A$. Since $U \subset U_\alpha$, we have

$$\Phi(U) \subset \Phi(U_\alpha) = U_\alpha,$$

hence $\Phi(U) \subset \bigcap_{\alpha \in A} U_\alpha = U$. Thus $\Phi(U) = U$, that is, $U \in Y$. If $V \in Y$ and $V \subset U_\alpha$ for all $\alpha \in A$, then $V \subset U$, hence $U = \inf_{\alpha \in A} U_\alpha$. Let $U_* = \Phi(\bigcup_{\alpha \in A} U_\alpha)$. Then $U_* \in Y$ and $U_* \supset U_\alpha$ for all $\alpha \in A$. If $V \in Y$ and $V \supset U_\alpha$ for all $\alpha \in A$, then $V \supset \bigcup_{\alpha \in A} U_\alpha$, hence $V = \Phi(V) \subset U_*$. Thus $U_* = \sup_{\alpha \in A} U_\alpha$. △

PROPOSITION 7.48 *Let X be a set, $Z \subset X$ and H be a set of finite functions defined on X. Then the mapping $\mathrm{co}_H^Z : 2^H \to 2^H$ is a closure in the sense of Moore.*

Proof: Let U be a nonempty subset of H. Then by definition,

$$\mathrm{co}_H^Z U = \{h \in H : h(x) \le \sup_{h' \in U} h'(x) \text{ for all } x \in Z\}.$$

If $U = \emptyset$, then $\mathrm{co}_H^Z U = U = \emptyset$. We also have

$$U \subset \mathrm{co}_H^Z U, \ \mathrm{co}_H^Z (\mathrm{co}_H^Z U) = \mathrm{co}_H^Z U \text{ and } U_1 \subset U_2 \implies \mathrm{co}_H^Z U_1 \subset \mathrm{co}_H^Z U_2.$$

Thus the result follows. △

THEOREM 7.16 *The ordered sets $S(H, X, Z)$ and $P(H, X, Z)$ are complete lattices. For an arbitrary family $(U_\alpha)_{\alpha \in A}$ of (H, X)-convex with respect to Z subsets of H, we have*

$$\sup_\alpha U_\alpha = \mathrm{co}_H^Z (\cup_\alpha U_\alpha), \qquad \inf_\alpha U_\alpha = \bigcap_\alpha U_\alpha. \tag{7.4.1}$$

For an arbitrary family of H-convex functions p_α on Z, we have

$$(sup_\alpha^* p_\alpha)(x) = \sup_\alpha p_\alpha(x) \qquad (inf_\alpha^* p_\alpha)(x) = \mathrm{co}_H^Z (\inf_\alpha p_\alpha(x)), \tag{7.4.2}$$

where $\sup_\alpha^* p_\alpha$ and $\inf_\alpha^* p_\alpha$ are boundaries in the lattice $P(H, X, Z)$ and $\sup_\alpha p_\alpha$ and $\inf_\alpha p_\alpha$ are pointwise boundaries.

Proof: It follows from Moore theorem that the ordered set $S(H, X, Z)$ is a complete lattice and (7.4.1) holds. Since the Minkowski duality is an isomorphism between the ordered sets $P(H, X, Z)$ and $S(H, X, Z)$, it follows that $P(H, X, Z)$ is also a complete lattice. The function $x \mapsto \sup_{\alpha \in A} p_\alpha(x)$ is H-convex on Z, hence the first equality in (7.4.2) holds. The validity of the second equality in (7.4.2) follows from the definition of the H-convex hull on the set Z. △

EXAMPLE 7.16 Let H be the space of all affine functions defined on a Banach space X and $(f_\alpha)_{\alpha \in A}$ be a family of H-convex (i.e., lower semicontinuous convex) functions defined on X. Let $Z = X$. Then $\sup_{\alpha \in A}^* f_\alpha$ (in the lattice $P(H, X)$) coincides with the pointwise supremum of this family, the infimum $\inf_{\alpha \in A}^* f_\alpha$ coincides with the closed convex hull of the pointwise infimum.

EXAMPLE 7.17 Let $L = X^*$ be the space of all linear continuous functions defined on X. Let $(U_\alpha)_{\alpha \in A}$ be a family of (L, X)-convex (i.e., w^*-closed convex) sets. Let $Z = X$. Then

$$\sup_\alpha^* U_\alpha = \mathrm{co}_H^Z \bigcup_\alpha U_\alpha = \mathrm{cl}\,\mathrm{co} \bigcup_\alpha U_\alpha,$$

where cl stands for the w^*-closure;

$$\inf_\alpha^* U_\alpha = \bigcap_\alpha U_\alpha.$$

Here sup* and inf* are boundaries in the lattice $S(H, X)$.

7.4.3. Minkowski duality for c^2-lattices and semilinear lattices

In this subsection we shall study connections between the Minkowski duality and some algebraic operations. We begin with the following definitions.

DEFINITION 7.24 A set A is called a *semilinear* space if there exists a binary operation $+$ and the operation of multiplication by a positive scalar defined on the set A such that, for $a, a_1, a_2, a_3 \in A$ and $\lambda, \mu > 0$,

$$a_1 + (a_2 + a_3) = (a_1 + a_2) + a_3, \qquad a_1 + a_2 = a_2 + a_1,$$
$$\lambda(a_1 + a_2) = \lambda a_1 + \lambda a_2, \qquad (\lambda + \mu)a = \lambda a + \mu a,$$
$$(\lambda\mu)a = \lambda(\mu)a, \qquad 1 \cdot a = a.$$

If the cancellation law holds in a semilinear space A, i.e, $a, b, c \in A$, $a + c = b + c \implies a = b$, then there is an isomorphism between A and a convex cone in a vector space. Note that the set of all functions $f : X \to \mathbb{R}_{+\infty}$, where X is an arbitrary nonempty set, is a semilinear space with respect to the usual pointwise definition of addition and multiplication by a positive scalar. However, this set is not a cone in a linear space. The set of all lower semicontinuous convex functions $f : X \to \mathbb{R}_{+\infty}$ defined on a Banach space X is a semilinear space as well.

Convex sets can be defined in a semilinear space in the usual way.

Recall (see Definition 7.9) that a set U is called *conic* if a multiplication by a positive scalar is defined on U with properties as follows:

$$1u = u; \quad \lambda(\mu u) = (\lambda\mu)u \text{ for all } u \in U, \ \lambda, \mu > 0.$$

Each semilinear space is a conic set.

DEFINITION 7.25 A set U is called a *conic complete lattice* (c^2-lattice) if U is a conic set and a complete lattice.

DEFINITION 7.26 Let A be both a semilinear space and a lattice. A is called a *semilinear lattice* if

(a) $x \geq y \implies x + z \geq y + z$ for all $z \in A$;

(b) $x \geq y \implies \lambda x \geq \lambda y$ for all $\lambda > 0$;

(c) for an arbitrary $U \subset A$ and $z \in A$, $\sup(z + U) = z + \sup U$.

An isomorphism between two c^2-lattices and semilinear lattices can be defined in the usual way so we do not provide details.

First, we indicate some algebraic properties of the Minkowski duality and then give some examples of c^2-lattices and semilinear lattices.

Let H be a set of finite functions defined on the set $Z \subset X$ and assume that H is a conic set with respect to the usual (pointwise) multiplication by a positive scalar. It follows directly from the definition of a support set that for a function $f : Z \to \mathbb{R}_{+\infty}$ and $\lambda > 0$ the following holds:

$$\operatorname{supp}(\lambda f, H, Z) = \lambda \operatorname{supp}(f, H, Z) := \{\lambda h : h \in \operatorname{supp}(\lambda f, H, Z).$$

Thus the totality $S(H, X, Z)$ of all H-convex sets on Z is a conic set with respect to the usual multiplication by a positive number ($\lambda U = \{\lambda u : u \in U\}$.) Let p be an H-convex function on Z, i.e., $p(z) = \sup\{h(z) : h \in \operatorname{supp}(p, H, Z)\}$. Then $\lambda p(z) = \sup\{\lambda h(z) : h \in \operatorname{supp}(p, H, Z)\}$, so λp is an H-convex function on Z. Thus $P(H, X, Z)$ is a conic set with respect to the usual pointwise multiplication. Since both $P(H, X, Z)$ and $S(H, X, Z)$ are complete lattices we can consider them as being c^2-lattices. The following assertion has been proved:

PROPOSITION 7.49 *Let $Z \subset X$ and H be a conic (with respect to the pointwise multiplication by a positive number) set of finite functions defined on X. Then the Minkowski duality is an isomorphism between c^2-lattices $P(H, X, Z)$ and $S(H, X, Z)$.*

Assume now that H is a semilinear space with respect to the usual operations of the pointwise addition and the pointwise multiplication by a positive number. Then $P(H, Z, X)$ is also a semilinear space with respect to the pointwise algebraic operations. Indeed, we only need to check that $p_1 + p_2$ is an H-convex function on Z if both p_1 and p_2 are H-convex on Z. For $x \in Z$ we have:

$$\begin{aligned} p_1(x) + p_2(x) &= \sup\{h_1(x) : h_1 \in \operatorname{supp}(p_1, H, Z)\} \\ &+ \sup\{h_2(x) : h_2 \in \operatorname{supp}(p_2, H, Z)\} \\ &= \sup\{h(x) : h = h_1 + h_2 \in \operatorname{supp}(p_1, H, Z) \\ &+ \operatorname{supp}(p_2, H, Z)\}. \end{aligned} \tag{7.4.3}$$

Hence $p_1 + p_2 \in P(H, X, Z)$. Note that $P(H, X, Z)$ is a complete lattice. It follows from the definition of pointwise operations that $P(H, X, Z)$ is a semilinear lattice.

Consider now the lattice $S(H, X, Z)$. It is straightforward to verify that an H-convex set is convex with respect to the algebraic operations on H. We equip $S(H, X, Z)$ with the pointwise multiplication by a positive number and with the binary operation $\oplus$ defined as follows

$$U_1 \oplus U_2 = \operatorname{co}_H^Z (U_1 + U_2) \qquad (U_1, U_2 \in S(H, X, Z)).$$

Let $p_1, p_2 \in P(H, X, Z)$ and $U_1 = \operatorname{supp}(p_1, H, Z), U_2 = \operatorname{supp}(p_2, H, Z)$. It follows from (7.4.3) that

$$\varphi(p_1 + p_2) = \operatorname{supp}(p_1 + p_2) = U_1 \oplus U_2$$

We also have $\varphi(\lambda p) = \lambda\varphi(p)$. Hence the Minkowski duality φ is an algebraic isomorphism between $P(H, X, Z)$ and $S(H, X, Z)$. Note (see Proposition 7.47) that φ is also an ordered isomorphism between these spaces. Thus the following assertion has been proved.

THEOREM 7.17 *Let $Z \subset X$ and H be a semilinear space (with respect to the pointwise operation) of finite functions defined on X. Then both $P(H, X, Z)$ and $S(H, X, Z)$ are semilinear lattices and the Minkowski duality is an isomorphism between $P(H, X, Z)$ and $S(H, X, Z)$.*

As mentioned previously, if H is a conic set (semilinear space), then the sets of H-convex functions and (H, X)-convex sets provide examples of c^2-lattices (semilattices). We consider now some examples of c^2-lattices and semilinear lattices. We examine two situations.

Situation 1. c^2-lattices of nonnegative functions. Let X be a set. Consider a set U of functions $f : X \to \bar{\mathbb{R}}_+$ with the following properties:

1) $0 \in U$;

2) if $u \in U$, $\lambda > 0$, then $\lambda u \in U$, where $(\lambda u)(x) = \lambda u(x)$ for all $x \in X$;

3) if $A \subset U$, then $\sup A \in U$, where $\sup A$ is the pointwise supremum of the set A, i.e., $(\sup A)(x) = \sup_{f \in A} f(x)$ for all $x \in X$.

We consider U as an ordered set with the pointwise order relation.

Let $A \subset U$. Then the function $\sup A$ is the supremum of the set A in the ordered set U. For each function $f : X \to \bar{\mathbb{R}}_+$, we can consider its support set $\mathrm{supp}(f, U)$ and its U-convex hull $(\mathrm{co}_U)(x) = \sup\{u(x) : u \in \mathrm{supp}(f, U)\}$.

Clearly, a function $\mathrm{co}_U \inf A$ is the infimum of the set A in the ordered set U. Here $\inf A$ stands for the pointwise infimum of the set A. Thus U is a complete lattice. We denote the supremum and the infimum in this lattice by $\sup^*$ and $\inf^*$, respectively. It has been shown that $\sup^* A = \sup A$ and $\inf^* A = \mathrm{co}_U \inf A$. Clearly, U is a c^2-lattice with respect to the pointwise multiplication by a positive number.

Here there are examples of c^2-lattices which are sets of functions possessing properties 1) - 3):

1. The set $F(X)$ of all functions defined on a set X and mapping into $\bar{\mathbb{R}}_+$.

2. Sets $M_0(X)$ of all increasing functions and $M^0(X)$ of all decreasing functions defined on an ordered set X and mapping into $\bar{\mathbb{R}}_+$.

Assume now that X is a Banach space. We shall only consider functions defined on X and mapping into $\bar{\mathbb{R}}_+$. Recall that a function f mapping into $\bar{\mathbb{R}}_+$ is called proper if $\operatorname{dom} f = \{x \in X : f(x) < +\infty\}$ is nonempty. We continue with the list of examples.

3. The set $PH(X)$ of all lower semicontinuous nonnegative positively homogeneous of degree one proper functions.

4. The set $SL(X)$ of all lower semicontinuous nonnegative proper sublinear functions.

5. The set $CF(X)$ of all lower semicontinuous nonnegative proper convex functions f such that $f(0) = 0$.

6. The set $Q_0^+(X)$ of all lower semicontinuous nonnegative quasiconvex functions q with the property $q(0) = 0$.

Note that sets in Examples 1 - 5 are semilinear lattices. The set Q_0^+ from Example 6 is only c^2-lattice, but not a semilinear lattice.

Situation 2. c^2- lattices of sets. Let U be a totality of subsets of a vector space X such that

1) $X \in U$;

2) $0 \in \Omega$ for all $\Omega \in U$;

3) if $\Omega \in U$, then $\lambda\Omega = \{\lambda x : x \in \Omega\} \in U$ for all $\lambda > 0$;

4) if $A \subset U$, then $\bigcap\{\Omega : \Omega \in A\} \in U$.

If Ω is a subset of the space X and $0 \in \Omega$, then the U-hull of Ω, denoted by $\operatorname{co}_U \Omega$, is the set

$$\bigcap\{\Omega' : \Omega' \in U, \ \Omega' \supset \Omega\}.$$

Let us consider U as an ordered space: $\Omega_1 \geq \Omega_2$ if and only if $\Omega_1 \supset \Omega_2$. It is easy to check that for an arbitrary subset A of the set U, there exist the supremum and the infimum of A in the ordered set U and

$$\inf A = \bigcap\{\Omega : \Omega \in A\}; \quad \sup A = \operatorname{co}_A \bigcup\{\Omega : \Omega \in A\}.$$

Clearly, U is a c^2-lattice with respect to the natural multiplication by positive numbers. Let us give some examples of c^2-lattices of sets with properties 1) - 4) above. Let X be a Banach space.

1. The set $\operatorname{rad} X$ of all radiant closed subsets of X.

2. The set $CS(X)$ of all convex closed subsets A of X such that $0 \in A$.

3. The set $\mathrm{Cone}(X)$ of all closed convex cones $K \subset X$.

4. The set $\mathrm{Lin}(X)$ of all closed linear subspaces of the space X.

Note that the sets in Examples 2-4 are semilinear lattices.

Let H be a conic set of functions. As it was shown, the Minkowski duality φ is a *homogeneous* ordered isomorphism, that is, $\varphi(\lambda f) = \lambda\varphi(f)$. In some instances it is interesting to consider an *anti-homogeneous* ordered isomorphism, that is, a one-to-one correspondence $\psi : X \to Y$, where X and Y are c^2-lattices, such that $\psi(\lambda x) = \lambda^{-1}\psi(x)$ and $\psi(X) = Y$. Anti-homogeneous ordered isomorphisms can serve for description of various schemes of conjugation (see [141] for details).

Chapter 8

APPLICATION TO GLOBAL OPTIMIZATION: DUALITY

8.1. INTRODUCTION

In the final part of the book we shall discuss possible applications of abstract convexity to global optimization. Some elements of theory of global optimization will be discussed in this chapter.

The main tool in the study of local extrema is the classical calculus and its modern generalizations which allow us to approximate functions in a neighbourhood of a point (in particular, in a neighbourhood of a local extremum). Unfortunately, a local approximation cannot help in the examination of global extrema for many problems. For *global optimization* we need to have *global information* about functions under consideration and we need special tools in order to present such global information in a convenient form. Abstract convexity provides some of these tools.

The support set of an abstract convex function accumulates an essential part of global information about this function in terms of elementary functions. Abstract subdifferentials also have a global nature. Both support set and subdifferentials contain minorants of the function under consideration and the definition of a minorant is *global*. Indeed, a function h is a minorant of a function f if $h(x) \leq f(x)$ *for all* elements x from the domain of these functions.

Necessary and sufficient conditions for local minima play a crucial role in the theoretical study and construction of numerical methods in local optimization. Due to the existence of Calculus, these conditions are efficient in the sense that they can be verified for many classes of functions. In contrast with this, it is very difficult to verify existing necessary and sufficient conditions for global minima. At the current stage of the development of global optimization these conditions play mainly a theoretical role, so they are substantially less important, than conditions in a local theory. It is impossible to obtained verified conditions, which can serve in the study of all global optimization problems. However, we believe that the further development of the theory will allow one to find more constructive conditions for some special classes of optimization problems.

We shall discuss in this chapter two various approaches to the theoretical study of global optima. First of them is based on solvability results for inequality systems. The minimization of an objective function subject to some inequality constraints can be represented in the following form: a certain inequality is a consequence of the given system of inequalities. The dual representations of this assertion are known as solvability results. The solvability approach is a nonlinear extension of the well-known Farkas lemma in linear programming. We shall demonstrate that the Minkowski duality forms a natural framework for the study of the solvability results for optimization problems with abstract convex functions involved. The properties of the Minkowski duality allow us to express the solvability results in terms of the abstract convex hull of some sets. We discuss solvability for many special situations.

The second approach is related to the optimization of the difference of abstract convex functions. The dual characteristic of such problems can be given by the well-known Toland-Singer formula (see, for example, [175] and references therein). In the abstract convex environment this formula can be expressed in terms of support sets and also in terms of abstract subdifferentials. We concider separately two kinds of problems: the minimization of DC (difference of convex) functions and the minimization of the difference of ICAR (increasing convex-along-rays) functions. Note that for both convex and ICAR functions a description of subgradients is known.

Our approach is based on the consideration of maximal elements of support sets. We shall demonstrate that (under some assumptions) the support set can be described in terms of its maximal elements and, in turn, these elements can be described in terms of subgradients. Hence the approach under consideration allows us to use the known description of subgradients for a presentation of optimality conditions given in terms of support sets. A convenient tool for the implementation of this approach is the *excess function*, which is discussed in detail in Subsection 8.3.7.

8.2. GENERAL SOLVABILITY THEOREMS

8.2.1. Overview

Consider the following problem of global optimization

$$\varphi(x) \longrightarrow \min \text{ subject to } g_i(x) \leq 0 \quad (i \in I), \ x \in X, \tag{8.2.1}$$

where I is an arbitrary index set and φ and g_i $(i \in I)$ are functions defined on a set X. Denote the set $\{x \in X : g_i(x) \leq 0 \ (i \in I)\}$ of feasible elements of (8.2.1) by X_*. Let γ be a real number. Consider the following assertion:

$$x \in X, \ (\forall i \in I) \ g_i(x) \leq 0 \implies \gamma - \varphi(x) \leq 0. \tag{8.2.2}$$

It follows from (8.2.2) that the inequality $\gamma \leq \varphi(x)$ holds for each feasible element $x \in X_*$, hence γ is a lower bound of the global minimum $\varphi_* = \min\{\varphi(x) : x \in X_*\}$ of (8.2.1):

$$\gamma \leq \varphi_*. \tag{8.2.3}$$

Similarly we can present the definition of a global minimizer: a point $x_* \in X_*$ is a global minimizer of (8.2.1) if and only if the following holds:

$$x \in X, \ (\forall i \in I) \ g_i(x) \leq 0 \implies \varphi(x_*) - \varphi(x) \leq 0. \tag{8.2.4}$$

Thus in order to give necessary and sufficient conditions for the global minimum φ_* or to describe lower bounds of φ_*, it is enough to examine the following assertion:

$$x \in X, \ (\forall i \in I) \ g_i(x) \leq 0 \implies f(x) \leq 0, \tag{8.2.5}$$

where g_i $(i \subset I)$ and f are functions defined on X. Dual characterizations of (8.2.5) are known as *solvability theorems.* These theorems can be considered as generalizations of the well-known Farkas lemma (see Remark 8.1).

Dual characterizations of the solvability of nonlinear inequality systems are crucial for the development of dual necessary and sufficient conditions for global extrema of some optimization problems. A great deal of attention has been focussed in the recent literature of nonlinear programming and nonsmooth optimization to establish such characterizations in terms of subdifferentials and approximate subdifferentials (see [63, 64, 85, 86, 69, 87, 88, 89]).

Abstract convexity is an appropriate tool in the study of solvability theorems. In particular the Minkowski duality allows us to obtain a *conceptual trivialization* of this problem. Let us explain this statement.

We shall say that we can obtain a conceptual trivialization of a problem by means of a theory if it easy to obtain a clear understanding of the problem in the framework of the theory. It does not mean that we can immediately solve the problem; many obstacles may appear but they are only technical obstacles. In the next section we shall show that the conceptual trivialization of solvability theorems can be obtained by means of the Minkowski duality. The main results from this sections can be found in [155].

8.2.2. Solvability theorems for systems of abstract convex functions

In this subsection we shall develop solvability theorems for infinite systems of inequalities involving abstract convex functions.

We shall consider a set H of elementary functions defined on a set X, the set $P(H,X,Z)$ of abstract convex with respect to H functions, defined on a subset Z of the set X and the set $S(H,X,Z)$ of (H,X)-convex sets on Z. First (Theorem 8.1), we assume that H is a conic set of functions and then (Theorem 8.2) that H is a semilinear space.

THEOREM 8.1 *Let Z be a subset of a set X, let H be a conic set of elementary functions defined on X and let I be an arbitrary index set. Furthermore let f and, for each $i \in I$, g_i be H-convex functions defined on Z. Then the following statements are equivalent:*

(i) $x \in Z,\ (\forall i \in I)\ g_i(x) \le 0 \implies f(x) \le 0$;

(ii) $\operatorname{supp}(f,H,Z) \subset \operatorname{co}_H^Z \operatorname{cone} \bigcup_i \operatorname{supp}(g_i,H,Z)$. *Here* $\operatorname{cone} V$ *is the conic hull of a set* V, *i.e.,* $\operatorname{cone} V = \bigcup_{\lambda>0} \lambda V$.

Proof: Let $g(z) = \sup_{i\in I} g_i(z)$ $(z \in Z)$. Since g_i is H-convex for each $i \in I$, it follows that g is H-convex as well. Consider the level sets $S_0(g) = \{z : g(z) \le 0\}$ and $S_0(f) = \{z : f(z) \le 0\}$. Clearly, statement (i) can be written as follows:

(iii) $S_0(g) \subset S_0(f)$.

Let δ denote the indicator function of the set $S_0(g)$:

$$\delta(z) = \begin{cases} +\infty & \text{if } z \notin S_0(g); \\ 0 & \text{if } z \in S_0(g). \end{cases}$$

It is easy to check that (iii) is equivalent to the inequality $f \le \delta$. Indeed, if $f \le \delta$, then $g(x) \le 0$ implies $f(x) \le 0$. On the other hand, if $S_0(g) \subset S_0(f)$ then $f(x) \le 0$ whenever $x \in S_0(g)$ that is, $f \le \delta$.

The definition of δ implies the equality $\delta = \sup_{\lambda>0} \lambda g$. Since H is a conic set, we have that λg is an H-convex function since g is H-convex. Therefore, δ is a H-convex function as the supremum of H-convex functions. Note that

$$\delta = \sup_{\lambda>0} \sup_{i\in I} g_i = \sup_{\lambda>0,\, i\in I} \lambda g_i.$$

We now compute the support set $\mathrm{supp}(\delta, H, Z)$ of the function δ. Applying Theorem 7.16 and Proposition 7.49, we conclude that

$$\begin{aligned} \mathrm{supp}(\delta, H, Z) &= \mathrm{co}_H^Z \bigcup_{\lambda>0,\, i\in I} \lambda\, \mathrm{supp}(g_i, H, Z) \\ &= \mathrm{co}_H^Z \bigcup_{\lambda>0} \bigcup_{i\in I} \lambda\, \mathrm{supp}(g_i, H, Z) \\ &= \mathrm{co}_H^Z\, \mathrm{cone} \bigcup_{i\in I} \mathrm{supp}(g_i, H, Z). \end{aligned}$$

Since the Minkowski duality is an ordered isomorphism of complete lattices $P(H, X, Z)$ and $S(H, X, Z)$, we conclude that the inequality $f \le \delta$ is equivalent to the inclusion

$$\mathrm{supp}(f, H, Z) \subset \mathrm{supp}(\delta, H, Z) = \mathrm{co}_H^Z\, \mathrm{cone} \bigcup_{i\in I} \mathrm{supp}(g_i, H, Z).$$

Hence the result follows. △

We shall discuss special cases of this solvability result (and Theorem 8.2 to follow) below. It suffices to note that Theorem 8.1 is a very general nonlinear extension of the classical Farkas lemma for finite systems of linear inequalities. Extensions of this lemma have been used extensively in applications to nonsmooth, and most recently, global optimization. This type of solvability theorems provides a dual characterization of inconsistency for a specific inequality system which arises when considering a suitable first order approximation to the programming problem under consideration (see [61, 63, 64, 85, 86]).

PROPOSITION 8.1 *Let I, and for each $i \in I$, g_i be as in Theorem 8.1, then the system*

$$(\forall i \in I)\ g_i(z) \le 0,\ z \in Z \tag{8.2.6}$$

is inconsistent on the set Z if and only if

$$\mathrm{co}_H^Z\, \mathrm{cone} \bigcup_{i\in I} \mathrm{supp}(g_i, H, Z) = H. \tag{8.2.7}$$

Proof: Let $g(z) = \sup_i g_i(z)$ and $\delta(z) = \sup_{\lambda>0} \lambda g(z)$. We have

$$\operatorname{supp}(\delta, H, Z) = \operatorname{co}_H^Z \operatorname{cone} \bigcup_{i \in I} \operatorname{supp}(g_i, H, Z),$$

and

$$\operatorname{supp}(+\infty, H, Z) = H.$$

Clearly, (8.2.6) is inconsistent on the set Z if and only if $g(z) > 0$ for all $z \in Z$. Since

$$(g(z) > 0 \text{ for all } z \in Z) \iff \delta(z) = +\infty \text{ for all } z \in Z,$$

it follows that (8.2.6) is equivalent to (8.2.7). △

8.2.3. Further solvability results

We shall now study characterizations of solvability theorems involving functions expressible as the pointwise infimum of abstract convex functions.

Let $Z \subset X$ and let H be a semilinear space of elementary functions defined on X (see Subsection 7.4.3 for the definition and properties of semilinear spaces). Assume that $0 \in H$. Let f be an inf-H-convex function and $(g_i)_{i \in I}$ be a family of inf-H-convex functions, where I is an arbitrary index set. (See Section 1.6 for the definition of inf-H-convex functions.) Thus for each $i \in I$ there exist a set Δ_i and a family $\{p_{\alpha_i} : \alpha_i \in \Delta_i\}$ with $p_{\alpha_i} \in P(H, X, Z)$ such that,

$$g_i(z) = \inf_{\alpha_i \in \Delta_i} p_{\alpha_i}(z) \text{ for all } z \in Z.$$

In addition, there exist a set Δ and a family $(p_\alpha)_{\alpha \in \Delta} \subset P(H, X, Z)$ such that

$$f(z) = \inf_{\alpha \in \Delta} p_\alpha(z) \text{ for all } z \in Z.$$

We need the following definition.

Let $(A(t))_{t \in T}$ be a family of nonempty sets, then a family a, where

$$a = (a_t)_{t \in T} \in \prod_{t \in T} A(t)$$

is called a *selection function* of the family $(A(t))_{t \in T}$. We shall use the notation

$$(a_t) \in \prod_{t \in T} A(t).$$

The following statement holds.

LEMMA 8.1 *Consider an index set I and a family of sets $(\Delta_i)_{i\in I}$. Let $\mathcal{S}$ be the set of all selection functions a defined on I with $a(i) = a_i \in \Delta_i$ for each $i \in I$. Assume that for each $i \in I$ there exists a function $t_i : \Delta_i \to \mathbb{R}$ such that $\inf_{\alpha_i\in\Delta_i} t_i(\alpha_i) > -\infty$. Then for the function $t : \mathcal{S} \times I \to \mathbb{R}$, $t(a,i) = t_i(a(i))$ we have*

$$\inf_{a\in\mathcal{S}} \sup_{i\in I} t(a,i) = \sup_{i\in I} \inf_{a\in\mathcal{S}} t(a,i).$$

Proof: We have, for all $i \in I$,

$$\inf_{a\in\mathcal{S}} t(a,i) = \inf_{a\in\mathcal{S}} t_i(a(i)) = \inf_{\alpha_i\in\Delta_i} t_i(\alpha_i).$$

Let $\varepsilon > 0$ and, for each $i \in I$, let $\bar{\alpha}_i \in \Delta_i$ be such that

$$t_i(\bar{\alpha}_i) \leq \inf_{\alpha_i\in\Delta_i} t_i(\alpha_i) + \varepsilon.$$

Define the selection function $\bar{a} \in \mathcal{S}$ by $\bar{a}(i) = \bar{\alpha}_i$ for all $i \in I$. Then

$$\begin{aligned}\sup_{i\in I} t(\bar{a},i) &= \sup_{i\in I} t_i(\bar{\alpha}_i)\\ &\leq \sup_{i\in I}\Big(\inf_{\alpha_i\in\Delta_i} t_i(\alpha_i) + \varepsilon\Big) = \sup_{i\in I} \inf_{a\in\mathcal{S}} t(a,i) + \varepsilon.\end{aligned}$$

Consequently,

$$\inf_{a\in\mathcal{S}} \sup_{i\in I} t(a,i) \leq \sup_{i\in I} t(\bar{a},i) \leq \sup_{i\in I} \inf_{a\in\mathcal{S}} t(a,i) + \varepsilon.$$

Since $\varepsilon > 0$ is an arbitrary number, we conclude that

$$\inf_{a\in\mathcal{S}} \sup_{i\in I} t(a,i) \leq \sup_{i\in I} \inf_{a\in\mathcal{S}} t(a,i).$$

The reverse inequality is always true, hence the result follows. Δ

THEOREM 8.2 *Let $Z \subset X$ and let H be a semilinear space of finite elementary functions defined on X. Assume that $0 \in H$. Let f be an inf-H-convex function and $(g_i)_{i\in I}$ be a family of inf-H-convex functions, where I is an arbitrary index set. Then the following statements are equivalent:*

(i) $x \in Z$, $(\forall i \in I)\ g_i(z) \leq 0 \implies f(z) \geq 0$.

(ii) For each $\alpha \in \Delta$ *and* $(\alpha_i) \in \prod_{i\in I} \Delta_i$

$$0 \in \mathrm{co}_H^Z \left(\mathrm{supp}(p_\alpha, H, Z) + \mathrm{co}_H^Z \,\mathrm{cone} \bigcup_{i\in I} \mathrm{supp}(p_{\alpha_i}, H, Z)\right).$$

We need the following assertion in order to prove Theorem 8.2.

LEMMA 8.2 *Let X, Z, H, I, g_i and f be as above. Then the following statements are equivalent:*

(i) $0 \in \mathrm{co}_H^Z\,(\mathrm{supp}(p_\alpha, H, Z) + \mathrm{co}_H^Z \bigcup_i \mathrm{supp}(p_{\alpha_i}, H, Z))$ *for all* $\alpha \in \Delta$ *and for all* $(\alpha_i) \in \prod_{i \in I} \Delta_i)$.

(ii) $f(z) + \sup_i g_i(z) \geq 0$ *for all* $z \in Z$.

Proof: Let $\alpha \in \Delta$ and $a = (\alpha_i) \in \prod_i \Delta_i$. Consider the set

$$V_{\alpha,a} = \mathrm{co}_H^Z\,(\mathrm{supp}(p_\alpha, H, Z) + \mathrm{co}_H^Z \bigcup_i \mathrm{supp}(p_{\alpha_i}, H, Z)),$$

and the function

$$p_{\alpha,a} = p_\alpha + \sup_i p_{\alpha_i}.$$

Since H is a semilinear space it follows that the set $P(H, X, Z)$ of all H-convex functions on Z is a semilinear lattice (see Subsection 7.4.3 for details) and therefore $p_{\alpha,a}$ is an H-convex function. It follows from the properties of the Minkowski duality (see Theorem 7.17) that $V_{\alpha,a}$ is the support set of the function $p_{\alpha,a}$.

Since $V_{\alpha,a} = \{h \in H : h \leq p_{\alpha,a}\}$ and $0 \in H$, we have $0 \in V_{\alpha,a}$ if and only if $p_{\alpha,a}(z) \geq 0$ for all $z \in Z$. Therefore, condition (i) is equivalent to

$$\inf_{\alpha \in \Delta} \inf_{a \in \mathcal{S}} p_{\alpha,a}(z) \geq 0, \quad z \in Z \tag{8.2.8}$$

(here $\mathcal{S}$ is the set of all selections of the indexed family $(\Delta_i)_{i \in I}$). To simplify (8.2.8) we note that for each $z \in Z$:

$$\begin{aligned} \inf_{\alpha \in \Delta} \inf_{a \in \mathcal{S}} p_{\alpha,a}(z) &= \inf_{\alpha \in \Delta} \inf_{a \in \mathcal{S}} (p_\alpha(z) + \sup_i p_{\alpha_i}(z)) \\ &= \inf_{\alpha \in \Delta} p_\alpha(z) + \inf_{a \in \mathcal{S}} \sup_i p_{\alpha_i}(z) \\ &= f(z) + \inf_{a \in \mathcal{S}} \sup_i p_{\alpha_i}(z). \end{aligned}$$

Let $\alpha_i = a(i)$ and $t_i(\alpha_i) = p_{\alpha_i}(z)$. Using Lemma 8.1, we have

$$\inf_{a \in \mathcal{S}} \sup_{i \in I} p_{\alpha_i}(z) = \sup_{i \in I} \inf_{a \in \mathcal{S}} p_{\alpha_i}(z) = \sup_{i \in I} g_i(z).$$

Hence

$$\inf_{\alpha \in \Delta} \inf_{a \mathcal{S}} p_{\alpha,a}(z) = f(z) + \sup_{i \in I} g_i(z).$$

So condition (i) is equivalent to the following:

$$f(z) + \sup_i g_i(z) \geq 0 \text{ for all } z \in Z \tag{8.2.9}$$

as required. △

Proof of Theorem 4.2: In order to prove that (ii) implies (i) let us consider the set $\mathcal{I} = I \times (0, +\infty)$. For each $j = (i, \mu) \in \mathcal{I}$ define the function $\tilde{g}_j$ as follows $\tilde{g}_j(z) = \mu g_i(z)$. Statement (i) can be presented in the following form:

(i′) $x \in Z$, $(\forall j \in \mathcal{I})$ $\tilde{g}_j(z) \leq 0 \implies f(z) \geq 0$.

Each function $\tilde{g}_j$, $j \in \mathcal{I}$, is inf-H-convex. Let $\bar{p}_j = \mu p_{\alpha_i}$ and $\bar{\Delta}_j = \Delta_i$ for each $j = (i, \mu) \in \mathcal{I}$. We have

$$\tilde{g}_j(z) = \mu g_i(z) = \inf_{\alpha_i \in \Delta_i} \mu p_{\alpha_i}(z) = \inf_{\bar{\alpha}_j \in \bar{\Delta}_j} \bar{p}_{\bar{\alpha}_j}(z) \text{ for all } z \in Z.$$

Since the Minkowski duality is an isomorphism of conic sets, we conclude that

$$\operatorname{supp}(\bar{p}_{\bar{\alpha}_j}, H, Z) = \operatorname{supp}(\mu p_{\alpha_i}, H, Z) = \mu \operatorname{supp}(p_{\alpha_i}, H, Z).$$

Let $\alpha \in \Delta$ and let $\bar{\alpha} = (\bar{\alpha}_j)_{j \in \mathcal{I}}$ be a selection from the set $\prod_{j \in \mathcal{I}} \bar{\Delta}_j$. Consider the function $p_{\alpha, \bar{\alpha}} = p_\alpha + \sup_j \bar{p}_{\bar{\alpha}_j}$ and the support set U of this function ($U = \operatorname{supp}(p_{\alpha,\bar{\alpha}}, H, Z)$.) We have

$$\begin{aligned} U &= \operatorname{co}_H^Z \left(\operatorname{supp}(p_\alpha, H, Z) + \operatorname{co}_H^Z \bigcup_{j \in \mathcal{I}} \operatorname{supp}(\bar{p}_{\bar{\alpha}_j}, H, Z)\right) \\ &= \operatorname{co}_H^Z \left(\operatorname{supp}(p_\alpha, H, Z) + \operatorname{co}_H^Z \bigcup_{i \in I, \mu > 0} \mu \operatorname{supp}(p_{\alpha_i}, H, Z)\right) \\ &= \operatorname{co}_H^Z \left(\operatorname{supp}(p_\alpha, H, Z) + \operatorname{co}_H^Z \operatorname{cone} \bigcup_{i \in I} \operatorname{supp}(p_{\alpha_i}, H, Z)\right). \end{aligned}$$

Therefore, the inclusion $0 \in \operatorname{supp}(p_{\alpha,\bar{\alpha}}, H, Z)$ is equivalent to

$$0 \in \operatorname{co}_H^Z \left(\operatorname{supp}(p_\alpha, H, Z) + \operatorname{co}_H^Z \operatorname{cone} \bigcup_{i \in I} \operatorname{supp}(p_{\alpha_i}, H, Z)\right).$$

Thus, by Lemma 8.2, (ii) is equivalent to the inequality

$$f(z) + \sup_{j \in \mathcal{I}} \tilde{g}_j(z) \geq 0 \text{ for all } z \in Z. \tag{8.2.10}$$

Note that

$$\sup_j \tilde{g}_j(z) = \begin{cases} 0 & \text{if } g_i(z) \leq 0, \text{ for all } i \in I \\ +\infty & \text{if there is an } i \in I \text{ such that } g_i(z) > 0 \end{cases}$$

We now check that (8.2.10) implies (i). Suppose that (8.2.10) holds and that $g_i(z) \leq 0$ for all $i \in I$. Then $\sup_{j \in \mathcal{I}} \tilde{g}_j(z) = 0$ and consequently, by (8.2.10), $f(z) \geq 0$. Thus (i) is true. Hence, since (8.2.10) is equivalent to (ii), it follows that (ii) implies (i). Similarly, we can show that (i′) implies (8.2.10). If (8.2.10) does not hold, then there exists $z \in Z$ such that $f(z) + \sup_j \tilde{g}_j(z) < 0$. In particular, $\sup_j \tilde{g}_j(z) < +\infty$ and so $\sup_j \tilde{g}_j(z) = 0$. Hence $f(z) < 0$ and so (i′) is not true. Thus since (i) and (i′) are equivalent the result follows. △

8.2.4. Sublinear inequality systems

We shall now study solvability results with some special classes of functions involved. We begin by considering solvability theorems involving lower semicontinuos sublinear functions f and g_i ($i \in I$) defined on a closed convex cone $K \subset X$, where X is a Banach space. Thus we wish to characterize Theorem 8.1 (i). Consider the set L of all linear continuous functions defined on X as a set of elementary functions. Then a function p defined on K is L-convex on K if and only if this function is lower semicontinuos and sublinear. Proposition 7.32 demonstrates that

$$\begin{aligned} \mathrm{co}_L^K \mathrm{cone} \bigcup_i \partial g_i(0) &= \mathrm{cl}\,(\mathrm{co}\,\mathrm{cone} \bigcup_i \partial g_i(0) - K^*) \\ &= \mathrm{cl}\,(\mathrm{cone}\,\mathrm{co} \bigcup_i \partial g_i(0) - K^*). \qquad (8.2.11) \end{aligned}$$

If $\mathrm{dom}\, g_i \subset K$ then the support set (the subdifferential $\partial g_i(0)$) of the sublinear function g_i is $-K^*$-conormal (see Subsection 7.3.2 for details) and therefore, $\partial g_i(0) - K^* = \partial g_i(0)$. If $\mathrm{dom}\, g_i \subset K$ for all $i \in I$, the (8.2.11) has a very simple form:

$$\mathrm{co}_L^K \mathrm{cone} \bigcup_i \partial g_i(0) = \mathrm{cl}\,\mathrm{cone}\,\mathrm{co} \bigcup_i \partial g_i(0). \qquad (8.2.12)$$

In particular, (8.2.12) holds if $K = X$. In this case statement (ii) of Theorem 8.1 can be written as

$$\partial f(0) \subset \mathrm{cl}\,(\mathrm{cone}\,\mathrm{co} \bigcup_i \partial g_i(0)). \qquad (8.2.13)$$

REMARK 8.1 Assume that $I = \{1, \ldots, m\}$. Let f and g_i ($i \in I$) be linear functions and $K = X$. Then $\partial f(0) = \{f\}$, $\partial g_i(0) = \{g_i\}$. It follows from (8.2.13) and Theorem 8.1 that

$$x \in X, \ (\forall i \in I)\, g_i(x) \leq 0 \implies f(x) \leq 0$$

if and only if there exist numbers $\alpha_i \geq 0$ such that $f = \sum_{i \in I} \alpha_i g_i$. We can consider this assertion as a version of the well-known Farkas lemma.

Suppose now that $(p_{\alpha_i})_{i \in \Delta_i}$ is a family of lower semicontinuous sublinear functions and p_α is a lower semicontinuous sublinear function, then

$$\begin{aligned} & \mathrm{co}_L^K(\partial p_\alpha(0) + \mathrm{co}_L^K \mathrm{cone} \textstyle\bigcup_i \partial p_{\alpha_i}(0)) \\ = \; & \mathrm{cl}\,(\partial p_\alpha(0) + \mathrm{cl}\,[\mathrm{co}\,\mathrm{cone} \textstyle\bigcup_i \partial p_{\alpha_i}(0) - K^*] - K^*). \end{aligned}$$

To simplify this formula we note that, in general, $\mathrm{cl}\,(C + \mathrm{cl}\,D) = \mathrm{cl}\,(C + D)$, and the set $(\mathrm{co}\,\mathrm{cone} \bigcup_i \partial p_{\alpha_i}(0) - K^*)$ is $-K^*$-conormal. Thus we have

$$\begin{aligned} & \mathrm{co}_L^K(\partial p_\alpha(0) + \mathrm{co}_L^K \mathrm{cone} \textstyle\bigcup_i \partial p_{\alpha_i}(0)) \\ = \; & \mathrm{cl}\,(\partial p_\alpha(0) + \mathrm{co}\,\mathrm{cone} \textstyle\bigcup_i \partial p_{\alpha_i}(0) - K^*). \quad . \end{aligned} \tag{8.2.14}$$

If $\mathrm{dom}\, p_\alpha \subset K$ and, for each i, $\mathrm{dom}\, p_{\alpha_i} \subset K$, then we can remove $-K^*$. Thus statement (ii) of Theorem 8.2 can be written as

$$0 \in \mathrm{cl}\,(\partial p_\alpha(0) + \mathrm{co}\, \mathit{cone} \bigcup_i \partial p_{\alpha_i}(0)).$$

8.2.5. Convex inequality systems

We now consider systems involving proper lower semicontinuous convex functions, f and g_i $(i \in I)$ defined on a Banach space X. Assume that Z is a closed and convex subset of X. Consider the set H of all continuous affine functions defined on the space X as a set of elementary functions. Throughout this subsection we assume that the system

$$x \in Z, \; (\forall i \in I) \; g_i(x) \leq 0 \tag{8.2.15}$$

is consistent. Thus, by Proposition 8.1, the set $\mathrm{cone} \bigcup_{i \in I} \mathrm{supp}(g_i, H)$ is not $\{0\} \times \mathbb{R}$-conormal. Note (see Proposition 7.8) that $\mathrm{supp}(f, H) = \mathrm{epi}\, f^*$. Note also that

$$\sup\{l(x) - c : (l, c) \in \mathrm{cone} \bigcup_i \mathrm{supp}(g_i, H)\} < +\infty$$

for a solution x of system (8.2.15). Hence, we can apply Corollary 7.6. Using this corollary, we have

$$\mathrm{co}_H^Z\, \mathrm{cone} \bigcup_i \mathrm{epi}\, g_i^* = \mathrm{cl}\,(\mathrm{co}\,\mathrm{cone} \bigcup_i \mathrm{epi}\, g_i^* + K_Z),$$

where

$$K_Z = \{h = (l, c) \in H : l(z) - c \leq 0 \text{ for all } z \in Z\}.$$

If $\operatorname{dom} g_i \subset Z$ for all $i \in I$, then, by applying Proposition 7.36, we remove K_Z. Therefore, statement (ii) of Theorem 8.1 has the following form:

$$\operatorname{epi} f^* \subset \operatorname{cl}(\operatorname{co}\operatorname{cone} \bigcup_i \operatorname{epi} g_i^*).$$

Now let p_α and p_{α_i}, $(i \in I)$ be lower semicontinuous proper convex functions. We have

$$\begin{aligned} & \operatorname{co}_H^Z (\operatorname{epi} p_\alpha^* + \operatorname{co}_H^Z \operatorname{cone} \textstyle\bigcup_i \operatorname{epi} p_{\alpha_i}^* \\ = \; & \operatorname{cl}\{(\operatorname{epi} p_\alpha^* + \operatorname{cl}[\operatorname{cone}\operatorname{co} \textstyle\bigcup_i \operatorname{epi} p_{\alpha_i}^* + K_Z]) + K_Z\} \\ = \; & \operatorname{cl}(\operatorname{epi} p_\alpha^* + (\operatorname{cone}\operatorname{co} \textstyle\bigcup_i \operatorname{epi} p_{\alpha_i}^*) + K_Z) \end{aligned}$$

and consequently statement (ii) of Theorem 8.2 can be written as

$$0 \in \operatorname{cl}(\operatorname{epi} p_\alpha^* + (\operatorname{co}\operatorname{cone} \bigcup_i \operatorname{epi} p_{\alpha_i}^*) + K_Z). \tag{8.2.16}$$

If $\operatorname{dom} g_i \subset Z$ for all $i \in I$, then we can remove K_Z and present (8.2.16) as

$$0 \in \operatorname{cl}(\operatorname{epi} p_\alpha^* + (\operatorname{co}\operatorname{cone} \bigcup_i \operatorname{epi} p_{\alpha_i}^*)).$$

8.2.6. Positively homogeneous systems

Let L be a conic set of positively homogeneous functions defined on a conic set $K \subset X$, where X is a Banach space. Assume that $0 \in K$. We consider L as a set of abstract linear functions. Let $H_L = \{h = (l, c) : h = l(x) - c,\ h \in L,\ c \in \mathbb{R}\}$ be the set of L-affine functions. In this subsection we consider inequalities involving functions $\bar{p}$ where $\bar{p}(x) = p(x) - 1$ with p being an L-convex function defined on a conic set $Z \subset K$. We begin with the following lemma.

LEMMA 8.3 *Let p be an L-convex function and $\bar{p}(x) = p(x) - 1$. Then*

$$\operatorname{supp}(\bar{p}, H_L, Z) = \operatorname{supp}(p, L, Z) \times [1, +\infty).$$

Proof: If $l(x) \leq p(x)$ for each $x \in Z$ and $c \geq 1$, then

$$l(x) - c \leq p(x) - 1 \text{ for all } x \in Z. \tag{8.2.17}$$

Therefore, $(h, c) \in \operatorname{supp}(\bar{p}, \bar{H}, Z)$. Now let $(h, c) \in \operatorname{supp}(\bar{p}, \bar{H}, Z)$, that is, (8.2.17) holds for all $x \in Z$. If $x = 0$, then (8.2.17) gives $c \geq 1$. If $x \in Z$, $x \neq 0$, then $\mu x \in Z$ with $\mu > 0$ and we have

$$\mu l(x) - c \leq \mu p(x) - 1$$

or equivalently,

$$l(x) - \frac{c}{\mu} \leq p(x) - \frac{1}{\mu} \text{ for all } x \in Z.$$

Letting $\mu \to +\infty$ we conclude that $l(x) \leq p(x)$ for all $x \in Z$, i.e., $l \in \operatorname{supp}(p, L, Z)$. △

PROPOSITION 8.2 *Let g be an L-convex and nonnegative function defined on a conic set Z with $\operatorname{dom} g \subset Z$. Let $\delta : Z \to \mathbb{R}_{+\infty}$ be the indicator function of the level set $S_1(g) = \{x \in Z : g(x) \leq 1\}$, that is*

$$\delta(x) = \begin{cases} 0 & \text{if } g(x) \leq 1; \\ +\infty & \text{if } g(x) > 1. \end{cases}$$

Then

$$\operatorname{supp}(\delta, H_L, Z) = \operatorname{cone}\left(\operatorname{supp}(g, L, Z) \times [1, +\infty)\right) \cup V,$$

where $V = \{(h, 0) \in H_L : h(x) \leq 0$ for all $x \in \operatorname{dom} g\}$.

Proof: By the definition of δ,

$$\operatorname{supp}(\delta, H_L, Z) = \{(l, c) \in H_L : l(x) \leq 1 \text{ if } g(x) \leq 1\}.$$

Let $(l, c) \in \operatorname{supp}(\delta, H_L, Z)$. Since g is a positively homogeneous function we have $g(0) = 0 < 1$. Therefore $c = l(0) \quad c \leq 0$ and $c \geq 0$. First, assume that $c = 0$. Then $l(x) \leq 0$ for all x such that $g(x) \leq 1$. Let $x \in \operatorname{dom} g$ and $g(x) > 0$. Then $g(x/g(x)) = 1$ and therefore $l(x) = g(x) l(x/g(x)) \leq 0$. If $g(x) = 0$, then $l(x) \leq 0$ as well. Thus $l(x) \leq 0$ for all $x \in \operatorname{dom} g$ and so $l \in V$.

Now let $c > 0$. We have $l(x) \leq c$ if $g(x) \leq 1$. In particular, if $g(x) = 1$, then $(1/c) l(x) \leq 1 = g(x)$. Since g is nonnegative and positively homogeneous we have $(1/c) l(x) \leq g(x)$ for all $x \in \operatorname{dom} g$ such that $g(x) > 0$. If $g(x) = 0$, then $g(\lambda x) = 0$ for all $\lambda > 0$ and therefore, $(1/c) l(\lambda x) \leq 1$ for all $\lambda > 0$. We also have $(1/c) l(x) = 0 = g(x)$ in this case and so $(1/c) l(x) \leq g(x)$ for all $x \in Z$. Therefore $(1/c) l \in \operatorname{supp}(g, L, Z)$. Thus

$$(l, c) \in c\left[\operatorname{supp}(g, L, Z) \times [1, +\infty)\right] \subset \operatorname{cone}\left[\operatorname{supp}(g, L, Z) \times [1, +\infty)\right].$$

So far we have proved that

$$\operatorname{supp}(\delta, H_L, Z) \subset \operatorname{cone}[\operatorname{supp}(g, L, Z) \times [1, +\infty)] \cup V.$$

Now we establish the reverse inclusion. By definition, if $(l, 0) \in V$ then $l(x) \leq 0$ for all $x \in \operatorname{dom} g$ and so $(l, 0) \in \operatorname{supp}(\delta, H_L, Z)$. If

$$(l, c) \in \operatorname{cone}[\operatorname{supp}(g, L, Z) \times [1, +\infty)],$$

then there is $\lambda \geq 0$ such that

$$(l, c) \in \lambda \operatorname{supp}(g, L, Z) \times [\lambda, +\infty).$$

We have $c \geq \lambda$ and $l \in \lambda \operatorname{supp}(g, L, Z) = \operatorname{supp}(\lambda g, L, Z)$. Therefore,

$$l(x) \leq \lambda g(x) \leq c g(x).$$

If $g(x) \leq 1$, then $l(x) - c \leq 0$, that is $(l, c) \in \operatorname{supp}(\delta, H_L, Z)$. △

THEOREM 8.3 *Let f and g_i $(i \in I)$ be nonnegative L-convex functions defined on a conic set Z. Then the following statements are equivalent:*

(i) $x \in Z$, $(\forall i \in I)\ g_i(x) \leq 1 \implies f(x) \leq 1$

(ii) $\operatorname{supp}(f, L, Z) \times [1, +\infty) \subset \operatorname{cone}\left(co_L^Z \bigcup_i \operatorname{supp}(g_i, L, Z) \times [1, +\infty)\right)$.

Proof: We apply Theorem 8.1 to show that (i) is equivalent to the following:

(ii′) $\operatorname{supp}(\bar{f}, H_L, Z) \subset \operatorname{supp}(\delta, H_L, Z)$

where $\bar{f} = f - 1$ and

$$\delta(x) = \begin{cases} 0 & \text{if } g(x) \leq 1; \\ +\infty & \text{if } g(x) > 1. \end{cases}$$

where $g(x) = \sup_i g_i(x)$. Applying Proposition 8.2, we have

$$\operatorname{supp}(\delta, H_L, Z) = \operatorname{cone}(\operatorname{supp}(g, L, Z) \times [1, +\infty)) \cup V,$$

where $V = \{(l, 0) \in H_L : l(x) \leq 0 \text{ for all } x \in \operatorname{dom} g\}$. However, by Lemma 8.3, we have

$$\operatorname{supp}(\bar{f}, H_L, Z) = \operatorname{supp}(f, L, Z) \times [1, +\infty).$$

If $(l, \lambda) \in \operatorname{supp}(\bar{f}, H_L, Z)$, then $\lambda \geq 1$; if $(l, \lambda) \in V$, then $\lambda \leq 0$. Therefore, statement (ii′) is equivalent to

$$\operatorname{supp}(f, L, Z) \times [1, +\infty) \subset \operatorname{cone}(\operatorname{supp}(g, L, Z) \times [1, +\infty)).$$

Since

$$\operatorname{supp}(g, L, Z) = \operatorname{supp}(\sup_i g_i, L, Z) = \operatorname{co}_H^Z \bigcup_i \operatorname{supp}(g_i, L, Z),$$

it follows that (ii′) is equivalent to (ii). △

Consider the following example.

EXAMPLE 8.1 Let Z be a closed convex cone in a Banach space X. Let f and g_i $(i \in I)$ be nonnegative lower semicontinuous sublinear functions defined on X. Then the following assertions are equivalent:

(i) $x \in Z,\ (\forall i \in I)\ g_i(x) \le 1 \implies f(x) \le 1$

(ii) $\partial f(0) \times [1, +\infty) \subset \operatorname{cone}([\operatorname{cl}\operatorname{co} \bigcup_i \partial g_i(0) - Z^*] \times [1, +\infty))$.

8.2.7. Twice continuously differentiable systems

In this subsection we apply Theorem 8.2 to obtain a solvability result for systems involving twice continuously differentiable functions. Let Z be a compact convex set in $\mathbb{R}^n$ and let $G \supset Z$ be an open bounded set. Denote by $C^2(G)$ the set of all twice continuously differentiable functions defined on G.

Consider a semilinear space P of all quadratic functions p defined on $\mathbb{R}^n$ by

$$p(x) = a\|x\|^2 + [l, x] + c,$$

where $a \ge 0$, $l \in \mathbb{R}^n$, and $c \in \mathbb{R}$. Let I be an arbitrary index set and let $f \in C^2(G)$, $g_i \in C^2(G)$ $(i \in I)$. It follows from Lemma 7.3 that there exist numbers $k_z > 0$ and $k_z^i > 0$ $(z \in Z,\ i \in I)$ such that

$$f(x) = \min_{z \in Z} p_z(x), \quad g_i(x) = \min_{z \in Z} p_z^i(x), \text{ for all } x \in Z,$$

where

$$p_z(x) = k_z\|x - z\|^2 + [\nabla f(z), x - z] + f(z), \tag{8.2.18}$$

$$p_z^i(x) = k_z^i\|x - z\|^2 + [\nabla g_i(z), x - z] + g_i(z) \tag{8.2.19}$$

Clearly $p_z \in P$, $p_z^i \in P$. Now let H be the set of all continuous affine functions defined on $\mathbb{R}^n$. Since $k_z > 0$ and $k_z^i > 0$ the functions p_z and p_z^i are convex and therefore, H-convex.

Let $p \in P$, $p(x) = a\|x-z\|^2 + [l, x-z] + c$. It is straightforward to compute affine functions $h(x) = [k, x-z] + b$ such that $h(x) \leq p(x)$ for all $x \in \mathbb{R}^n$. We have

$$\begin{aligned} \operatorname{supp}(p, H, \mathbb{R}^n) &= \{(k,b) : [k, x-z] + b \leq a\|x-z\|^2 + [l, x-z] + c\} \\ &= \{(k,b) : c - b \geq \frac{1}{4a}\|k-l\|^2\}. \end{aligned}$$

Corollary 7.6 shows that

$$\operatorname{supp}(p, H, Z) = \operatorname{cl}(\operatorname{supp}(p, H, \mathbb{R}^n) + K_Z)$$

where $K_Z = \{h = (l,c) \in H : l(x) - c \leq 0 \text{ for all } x \in Z\}$. We can apply Theorem 8.2 to yield necessary and sufficient conditions for the following statement:

$$x \in Z,\ (\forall i \in I)\ g_i(x) \leq 0 \Longrightarrow f(x) \geq 0.$$

8.2.8. Quasiconvex inequality systems

In this subsection we present a solvability result for quasiconvex inequality systems. We shall exploit the notation, definitions and results from Subsection 7.3.5, in particular we shall use the description of the abstract convex hull with respect to the set J_0 of two-step functions (see Proposition 7.44). It follows from Proposition 7.44 and Theorem 8.1 that the following assertion holds.

PROPOSITION 8.3 *Let Z be a closed convex subset of a Banach space X with $0 \in Z$ and let I be an arbitrary index set. Let $f : Z \to \mathbb{R}_{+\infty}$ and, for each $i \in I$, let $g_i : Z \to \mathbb{R}_{+\infty}$ be lower semicontinuous quasiconvex functions which attain their minimum over Z at the point zero. Then the following statements are equivalent:*

(i) $x \in Z,\ (\forall i \in I)\ g_i(x) \leq 0 \Longrightarrow f(x) \leq 0$;

(ii) $\operatorname{supp}(f, \mathcal{J}_0, Z) \subset cl_\tau \operatorname{co}'_{Z^\circ} \operatorname{cone} \bigcup_i \operatorname{supp}(g_i, \mathcal{J}_0, Z)$.

We now assume that $Z = X$ and we compare results obtained for convex and quasiconvex functions. For lower semicontinuous convex functions f and g_i we represent statement (ii) of Theorem 8.1 in the following form:

$$\operatorname{supp}(f, \mathcal{A}(X), X) \subset \operatorname{cl}\operatorname{co}\operatorname{cone} \bigcup_i \operatorname{supp}(g_i, \mathcal{A}(X), X),$$

where $\mathcal{A}(X)$ is the space of all continuous affine functions defined on X.

For lower semicontinuous quasiconvex functions f and g_i, which attain their minimum over X at the origin, statement (ii) has the following form (see Proposition 8.3):

$$\operatorname{supp}(f, \mathcal{J}_0, X) \subset \operatorname{cl}_\tau \operatorname{co}' \operatorname{cone} \bigcup_i \operatorname{supp}(g_i, \mathcal{J}_0, X).$$

8.2.9. Systems involving functions expressible as the infima of families of convex functions

In this subsection we shall assume that f and g_i (for $i \in I$) are inf-convex functions defined on a subset Z of a Banach space X. Then there exist families $(p_\alpha)_{\alpha\in\Delta}$ and, for each $i \in I$, $(p_{\alpha_i})_{\alpha_i\in\Delta_i}$, of proper lower semicontinuous convex functions such that, for each $z \in Z$,

$$f(z) = \inf_{\alpha\in\Delta} p_\alpha(z), \quad g_i(z) = \inf_{\alpha_i\in\Delta_i} p_{\alpha_i}(z). \tag{8.2.20}$$

Theorem 8.4 *Let $Z \subset X$ be a convex set and let I be an arbitrary index set. Let $f : X \to \mathbb{R}_{+\infty}$ and, for each $i \in I$, $g_i : Z \to \mathbb{R}_{+\infty}$ be inf-convex, that is there exist families $(p_\alpha)_{\alpha\in\Delta}$ and $(p_{\alpha_i})_{\alpha_i\in\Delta_i}$ such that (8.2.20) holds. Furthermore suppose that for each selection (α_i), the system*

$$z \in Z, \ (\forall i \in I) \ p_{\alpha_i}(z) \le 0,$$

is consistent. Then the following statements are equivalent:

(i) $z \in Z, (\forall i \in I)\ g_i(z) \le 0 \implies f(z) \ge 0.$

(ii) For each $\alpha \in \Delta$ and $(\alpha_i) \in \prod_{i\in I} \Delta_i$

$$0 \in \operatorname{cl}(\operatorname{epi} p_\alpha^* + (\operatorname{co} \operatorname{cone} \bigcup_{i\in I} \operatorname{epi} p_{\alpha_i}^*) + K_Z),$$

where $K_Z = \{(l, c) \in X^* \times \mathbb{R} : \ (\forall x \in Z)\ l(x) - c \le 0\}$.

If f and, for each i, g_i are continuous convex functions with $\operatorname{dom} f$ *and* $\operatorname{dom} g_i$ *contained in Z then (ii) becomes*

$$0 \in \operatorname{cl}(\operatorname{epi} f^* + (\operatorname{co} \operatorname{cone} \bigcup_{i\in I} \operatorname{epi} g_i^*)). \tag{8.2.21}$$

Proof: The proof follows directly from Theorem 8.2, Proposition 7.8 and Corollary 7.6. △

As a special case of Theorem 8.1 let f and, for each $i \in I$, let g_i be continuous convex functions defined on X. Then the following holds:

COROLLARY 8.1 *Let f and, for each $i \in I$, g_i be continuous convex functions defined on X and suppose that the system*

$$(\forall i \in I)\ g_i(x) \le 0$$

is consistent. Then the following statements are equivalent:

(i) $x \in X;\ (\forall i \in I)\ g_i(x) \le 0 \implies f(x) \le 0.$

(ii) $\operatorname{epi} f^* \subset \operatorname{cl}\operatorname{co}\operatorname{cone} \bigcup_{i\in I} \operatorname{epi} g_i^*.$

The next result is a direct corollary of Corollary 8.1.

COROLLARY 8.2 *Let I and J be nonempty index sets and let, for each $j \in J$ and $i \in I$, f_j and g_i be continuous convex functions defined on X. Suppose that the system*

$$x \in X,\ (\forall i \in I)\ g_i(x) \le 0$$

is consistent. Then the following statements are equivalent:

(i) $x \in X,\ (\forall i \in I)\ g_i(x) \le 0 \implies (\forall j \in J)\ f_j(x) \le 0.$

(ii) $\bigcup_{j\in J} \operatorname{epi} f_j^* \subset \operatorname{cl}\operatorname{co}\operatorname{cone} \bigcup_{i\in I} \operatorname{epi} g_i^*.$

Proof: The proof follows immediately from Corollary 8.1 with $f = \sup_j f_j$. △

Particular cases of this results were obtained by Gwinner [69] and Ha [70].

It is straightforward, using Theorem 8.2, to establish a solvability theorem for systems involving DC (difference convex) functions.

8.2.10. Convex maximization

As it was mentioned in Overview, solvability theorems lead to necessary and sufficient conditions for global extrema. We illustrate this assertion by the following example.

EXAMPLE 8.2 Consider the convex maximization problem

$$f(x) \longrightarrow \max \quad \text{subject to } g_i(x) \le 0,\ i \in I,$$

where I is an arbitrary set of indices, f and g_i, $(i \in I)$ are convex functions. Assume that the system $g_i(x) \le 0$ for all $i \in I$ is consistent.

A point x_0 is a global minimizer of this problem if and only if $g_i(x_0) \leq 0$ for all $i \in I$ and

$$x \in X, \ (\forall i \in I) \ g_i(x) \leq 0 \implies f(x) - f(x_0) \leq 0.$$

The function $\phi(x) = f(x) - f(x_0)$ is convex. Let H be the set of continuous affine functions. We have

$$\begin{aligned}
\text{supp}(\phi, H) &= \{(l, c) : (\forall x \in X)\, l(x) - c \leq f(x_0) \leq f(x)\} \\
&= \{(l, c) : (\forall x \in X)\, l(x) - (c - f(x_0)) \leq f(x)\} \\
&= \{(l, c' + f(x^0)) : (\forall x \in X)\, l(x) - c' \leq f(x)\} \\
&= \text{supp}(f, H) + (0, f(x_0) = \text{epi}\, f^* + (0, f(x_0).
\end{aligned}$$

Thus by a direct application of Corollary 8.1, x_0 is a global maximizer if and only if the following holds:

$$\text{epi}\, f^* + (0, f(x_0)) \subset \text{cl}\,\text{co}\,\text{cone} \bigcup_{i \in I} \text{epi}\, g_i^*. \tag{8.2.22}$$

This condition provides a characterization of global optimality for the convex maximization problem. Using Proposition 7.8 and Proposition 7.9, we can express (8.2.22) in terms of ε-subdifferentials. This approach, using the ε-subdifferential, to characterizing global optimality has been used in [74, 75, 87, 89]. It should be noted that convex maximization problems have recently received considerable attention in the literature (see, for example, [189] and references therein).

It is straightforward to obtain characterizations of global optimality for a range of related problems such as DC (difference convex) programming problems by a suitable application of Theorem 8.4. (See [89] for details.)

8.3. MAXIMAL ELEMENTS OF SUPPORT SETS AND TOLAND-SINGER FORMULA

8.3.1. Overview

The support set $\text{supp}(f, H)$ of an abstract convex function with respect to a set of elementary functions H possesses the following property:

$$(h \in \text{supp}(f, H),\ h' \leq h) \implies h' \in H.$$

Hence in some instances the support set is too large. We shall demonstrate in this section, that under some fairly strong assumptions this set can be completely described by its maximal elements and, in turn, maximal elements can be described by elements of the subdifferentials at points of the domain. The established links between support sets and subdifferentials are very useful in some applications (see next section for details).

The well-known Toland-Singer formula (see [175] and references therein) is one of the main results of abstract convex analysis. Roughly speaking, this formula demonstrates that the infimum of the difference of abstract convex functions f and g is equal to the infimum of the difference of the Fenchel-Moreau conjugate functions g^* and f^*. We describe some links between this formula and support sets. We also describe, through links between support sets and subdifferentials, some relations between subdifferentials of f and g, which follow from the Toland-Singer formula.

The results obtained in this section will be used in the next section in the study of optimization problems with differences of abstract convex functions involved. In particular we shall consider the differences between convex functions and between ICAR functions. We shall examine the concept of *excess function* as a valuable tool in the study of these optimization problems. Some results from this section can be found in [152, 154].

8.3.2. Maximal elements

Let L be a set of elementary functions defined on a set X and let H_L be a set of L-affine functions. Recall that the subdifferential $\partial_{L,Z}f(x_0)$ of an H_L-convex function f on a set Z was defined by

$$\partial_{L,Z}f(x_0) = \{l \in L : l(x) - l(x_0) \leq f(x) - f(x_0) \text{ for all } x \in Z\}.$$

Later on we will use the following set:

$$\mathcal{D}f(x_0) = \{h \in H_L : h(x) = l(x) - l(x_0),\ l \in \partial_{L,Z}f(x_0)\}.$$

Thus $h \in \mathcal{D}f(x_0)$ if and only if $h = (l, l(x_0))$ with $l \in \partial_{L,Z}f(x_0)$.

We will examine connections between the maximal elements of the support set, $\operatorname{supp}(f, H_L, Z)$, and the maximal elements of the set $\mathcal{D}f(x_0)$ $(x_0 \in Z)$.

Consider a set U of functions defined on a set Z. As usual, we assume that U is equipped with the natural (pointwise) order relation. Recall that a function f is called a *maximal element* of the set U, if $f \in U$ and

$$\bar{f} \in U,\ \bar{f}(x) \geq f(x) \text{ for all } x \in Z \Longrightarrow \bar{f} = f.$$

PROPOSITION 8.4 *Let $Z \subset X$ and let $f \in P(H_L, X, Z)$. Furthermore, let $h_0 \in (l, l(x_0))$ be a maximal element of $\mathcal{D}f(x_0)$. Then $h = (l, l(x_0) - f(x_0))$ is a maximal element of the set* $\operatorname{supp}(f, H_L, Z)$.

Proof: Let $c = l(x_0) - f(x_0)$. Applying Proposition 7.1 we see that $h = (l, c) \in \text{supp}(f, H, Z)$ and $h(x_0) = f(x_0)$. Take $h' = (l', c') \in \text{supp}(f, H, Z)$, $h' \geq h$. In order to establish the result, it is enough to show that $h' = h$. Since $l'(x_0) - c' = h'(x_0) = h(x_0) = f(x_0)$, we have $c' = l'(x_0) - f(x_0)$. Applying Proposition 7.1 again, we conclude that $l' \in \partial_{L,Z} f(x_0)$, so $h'_0 = (l', l'(x_0)) \in \mathcal{D}f(x_0)$. The inequality $h' \geq h$ shows that

$$\begin{aligned} h'_0(x) &= l'(x) - l'(x_0) = l'(x) - c + f(x_0) \\ &= h'(x) + f(x_0) \geq h(x) + f(x_0) = l(x) - l(x_0) = h_0(x). \end{aligned}$$

Since h_0 is a maximal element of $\mathcal{D}f(x_0)$ it follows that $h_0 = h'_0$. Therefore, $h = h'$. △

We now show that under some additional assumptions the converse assertion is also valid.

THEOREM 8.5 *Let $X = Y^*$ be a Banach space conjugate to a Banach space Y. Let Z be a w^*-closed subset of X, L be a set of abstract linear functions and H_L be a set of L-affine functions. Let f be an H_L-convex function defined on Z such that* $\text{dom} f \neq \emptyset$ *and*

1) the function $f - l$ is w^*- lower semicontinuous on Z for all $l \in L$;

2) (for Z unbounded)

$$\lim_{\|x\| \to +\infty, x \in Z} (f(x) - l(x)) = +\infty. \qquad (8.3.1)$$

Then for $l \in L$ the following assertions are equivalent :

(i) There exists a number c such that $h = (l, c)$ is a maximal element of the support set $\text{supp}(f, H_L, Z)$.

(ii) There exists a point $x_0 \in Z$ such that $\bar{h} = (l, l(x_0))$ is a maximal element of the set $\mathcal{D}f(x_0)$ and $\bar{h}(x_0) = f(x_0)$.

Proof: (ii) $\Longrightarrow$ (i) follows from Proposition 8.4.

(i) $\Longrightarrow$ (ii). Let $h = (l, c)$ be a maximal element of $\text{supp}(f, H, Z)$ and $g(x) = f(x) - h(x) = f(x) - l(x) + c$. Clearly, $g(x) \geq 0$ for all $x \in Z$ and g is w^*-lower semicontinuous on Z.

Assume that Z is unbounded. Applying (8.3.1), we find a number $\mu > 0$ such that

$$\inf_{x \in Z} g(x) = \inf_{x \in Z \cap B_\mu} g(x),$$

where $B_\mu = \{x : \|x\| \le \mu\}$. Since Z is w^*-closed it follows that $Z \cap B_\mu$ is w^*-compact, hence the function g achieves its minimum on Z. The inequality $g(x) \ge 0$ for all $x \in Z$ shows that

$$\min_{x \in Z} g(x) \ge 0. \tag{8.3.2}$$

If Z is bounded, then (8.3.2) trivially holds. Assume that $\min_{x\in Z} g(x) = \varepsilon > 0$ and take the function $h' = h + \varepsilon \mathbf{1}$ (with $\mathbf{1}(x) = 1$ for all $x \in X$). Clearly, $h' \in \text{supp}(f, H, Z)$ and $h' > h$. Thus h is not a maximal element of $\text{supp}(f, H, Z)$ and we have a contradiction. We have proved that $\min_{x\in Z} g(x) = 0$, hence there exists a point $x_0 \in Z$ such that

$$\min_{x \in Z} g(x) = g(x_0) = 0.$$

Equivalently, $h(x_0) = f(x_0)$. Since $h = (l, c)$ we have, by applying Proposition 7.1, that $l \in \partial_{L,Z} f(x_0)$ and $c = l(x_0) - f(x_0)$.

We now show that $\bar{h} = (l, l(x_0))$ is a maximal element of the set $\mathcal{D}f(x_0)$. Assume that there exists $l' \in \partial_{L,Z} f(x_0)$ such that $l'(x) - l'(x_0) \ge l(x) - l(x_0)$. Then, for $h'(x) = l'(x) - l'(x_0) + f(x_0)$, we have $h' \ge h$. Since $l' \in \partial_{L,Z} f(x_0)$, it follows (from Proposition 7.1) that $h' \in \text{supp}(f, H_L, Z)$. Thus $h' = h$ and $(l', l'(x_0)) = (l, l(x_0))$. △

REMARK 8.2 Let $f : Z \to \mathbb{R}_{+\infty}$ be a *coercive* function, that is

$$\lim_{\|z\| \to +\infty} \frac{f(z)}{\|z\|} = +\infty.$$

Then (8.3.1) holds.

Both the w^*-lower semicontinuity of $f - l$ and condition (8.3.1) are essential in Theorem 8.5. To show that (8.3.1) is essential consider the following example.

EXAMPLE 8.3 Let $X = Z = \mathbb{R}$ and

$$L = \{l_{a,b} : (\forall x \in \mathbb{R})\, l_{a,b}(x) = ax^2 + bx,\ a \le 0,\ b \in \mathbb{R}\}.$$

Then

$$H_L = \{h : (\forall x \in \mathbb{R})\, h(x) = ax^2 + bx - c,\ a \le 0,\ b,\ c \in \mathbb{R}\}.$$

For $x \in \mathbb{R}$, let $f(x) = -x^2 + 1/(1 + x^2)$. The function f is H_L-convex. It is easy to check that the function $h(x) = -x^2$ is a maximal element of $\text{supp}(f, H, Z)$ but there is no point x_0 such that $h \in \partial_L f(x_0)$.

Consider the classical situation. Let X be a Banach space, and let $L = X^*$, where X^* is the space conjugate to X, then H_L is the space of continuous affine functions defined on X. Let f be a lower semi-continuous convex (that is, H_L-convex) function defined on X and let $x_0 \in X$. Assume $\partial f(x_0) = \partial_L f(x_0) \neq \emptyset$. It is easy to check that each element of the set $\mathcal{D}f(x_0) = \{(l, l(x_0)) : l \in \partial f(x_0)\}$ is a maximal element of this set. Indeed, if $(l, l(x_0)) \in \mathcal{D}f(x_0)$, $(l', l'(x_0)) \in \mathcal{D}f(x_0)$ and $l(x) - l(x_0) \geq l'(x) - l'(x_0)$ then the linear function $l - l'$ is bounded from below over the space X. Therefore $l = l'$. The following statement is valid.

PROPOSITION 8.5 *Let Y be a Banach space and f be a w^*-lower semi-continuous convex coercive function defined on the Banach space $X = Y^*$. Then the following assertions are equivalent:*

(i) There exists a number c such that $h = (l, c)$ is a maximal element of the support set $\text{supp}(f, H_L, X) = \text{epi}\, f^*$.

(ii) There exists a point x_0 such that $l \in \partial_L f(x_0)$.

Here f^* is the Fenchel-Moreau conjugate to f.

8.3.3. Existence of maximal elements of support sets

Consider a set X and a set H of functions $h : X \to \mathbb{R}$. Assume that $Z = X$, so we shall study abstract convex functions defined on the set X. We assume in this subsection that H enjoys the following property (B):

(B): let $h(x) = \lim_{\alpha \in \Delta} h_\alpha(x)$, $x \in X$, where Δ is a directed set and $(h_\alpha)_{\alpha \in \Delta} \subset H$ is a net, which is bounded from below. Then either $h \subset H$ or $h \equiv +\infty$.

Recall that an ordered set Δ with an order relation $\succ$ is called a *directed set* if for each $\alpha, \beta \in \Delta$ there exists $\gamma \in \Delta$ such that $\gamma \succ \alpha$ and $\gamma \succ \beta$. A function $(x_\alpha)_{\alpha \in \Delta}$ mapping Δ into a set X is called a net (generalized sequence). If X is an ordered set with an order relation $>$, then a net $(x_\alpha)_{\alpha \in \Delta}$ is called bounded from below if there exist $\gamma \in \Delta$ and $y \in X$ such that $x_\alpha \geq y$ for all $\alpha \succeq \gamma$. If X is a topological space, then the limit of a net can be defined in the usual way.

It follows from the property (B) that for the support set, $\text{supp}(f, H) \equiv \text{supp}(f, H, X)$, of a proper function $f : X \to \mathbb{R}_{+\infty}$ the following holds:

If $h(x) = \lim_{\alpha \in \Delta} h_\alpha(x)$, where Δ is a directed set and $(h_\alpha)_{\alpha \in \Delta} \subset H$ is an increasing net, then $h \in \text{supp}(f, H)$.

Indeed, we have $h \leq f$. Since f is proper, it follows that also h is proper, hence $h \in H$. Thus $h \in \{h \in H : h \leq f\} = \text{supp}(f, H)$.

The support set, $\text{supp}(f, H)$, also possesses the following property. If $h \in \text{supp}(f, H)$ and $h' \in H$, $h'(x) \leq h(x)$ for all $x \in X$, then $h' \in \text{supp}(f, H)$. It follows from this observation that support sets are often very large. Sometimes it is possible to describe these sets by means of maximal elements.

We now show, by applying the well-known Zorn's lemma (see for example [168], Subsections 2.2.2 and 2.2.3), that the following statement is valid.

PROPOSITION 8.6 *Let H be a set of finite functions and assume that (B) holds. Let $f : X \to \mathbb{R}_{+\infty}$ be a proper function. Then for each $h \in \text{supp}(f, H)$ there exists a maximal element h' of* $\text{supp}(f, H)$ *such that $h' \geq h$.*

Proof: Let $h \in \text{supp}(f, H)$ and $U = \{h' \in \text{supp}(f, H, X) : h' \geq h\}$. Let V be an infinite chain of elements U, i.e., for all $h', h'' \in V$ we have either $h' \geq h''$ or $h'' \geq h'$. We can consider V as a net $(h_\alpha)_{\alpha \in \Delta}$, where the index set Δ is equal to V. Obviously this net is bounded from below. Let $\bar{h} = \sup\{h' : h' \in V\}$. Clearly, $\bar{h}(x) = \lim_{\alpha \in \Delta} h_\alpha(x)$ for all $x \in X$. Since f is proper and $\bar{h} \leq f$ it follows that $\bar{h}$ is proper as well, hence due to (B), $\bar{h} \in H$. The inequalities $h \leq h' \leq f$ for all $h' \in V$ imply $h \leq \bar{h} \leq f$. Thus $\bar{h} \in U$. It follows from Zorn's lemma that there exists a maximal element $\hat{h}$ of the set U. Clearly, $\hat{h} \geq h$. Let us check that $\hat{h}$ is a maximal element of the support set $\text{supp}(f, H)$. Indeed, if $h' \geq \hat{h}$ and $h' \in \text{supp}(f, H)$, then $h' \in U$ and therefore, $h' = \hat{h}$. $\triangle$

Proposition 8.6 shows that we can use maximal elements to describe support sets. In particular, the following simple assertion holds.

PROPOSITION 8.7 *Let H enjoy the property (B) and let f and g be functions defined on X. Then the following statements are equivalent:*

(i) $\text{supp}(f, H) \subset \text{supp}(g, H)$;

(ii) for each maximal element h_1 of $\text{supp}(f, H)$ *there exists a maximal element h_2 of* $\text{supp}(g, H)$ *such that $h_1(x) \leq h_2(x)$ for all $x \in X$.*

Proof: (i) $\Longrightarrow$ (ii). Let $\text{supp}(f, H) \subset \text{supp}(g, H)$ and h_1 be a maximal element of $\text{supp}(f, H)$. Then $h_1 \in \text{supp}(g, H)$. Applying Proposition 8.6 we find a maximal element of the set $\text{supp}(g, H)$ such that $h_1 \leq h_2$.

(ii) $\Longrightarrow$ (i). Let $h \in \text{supp}(f, H)$. Proposition 8.6 shows that there exists a maximal element h_1 of $\text{supp}(f, H)$ such that $h \leq h_1$. Let h_2 be a maximal element of $\text{supp}(g, H)$ such that $h_1 \le h_2$. We have $h \le h_2$, therefore $h \in \text{supp}(g, H)$. $\triangle$

We now give some examples of sets, which possess the property (B). First, we prove the following simple result.

PROPOSITION 8.8 *Let L be a set of finite functions defined on a set X. Assume that L possesses the property (B) and there exists an element $0 \in X$ such that $l(0) = 0$ for all $l \in L$. Then the set $H_L = \{h : (\forall x \in X)\ h(x) = l(x) - c,\ l \in L, c \in \mathbb{R}\}$ also possesses the property (B).*

Proof: Let Δ be a directed set, $h_\alpha \in H_L$ for all $\alpha \in \Delta$ and the net (h_α) is bounded from below. Let $h(x) = \lim_{\alpha\in\Delta} h_\alpha(x)$. Assume that h is a proper function, then $h \in H$. The function h is finite, in particular $h(0) < +\infty$. Let $h_\alpha = l_\alpha - c_\alpha$ with $l_\alpha \in L$ and $c_\alpha \in \mathbb{R}$. Then $c_\alpha = -h_\alpha(0) \to -h(0) > -\infty$, so the net $(c_\alpha)_{\alpha\in\Delta}$ is bounded from below. Since both $(h_\alpha)_{\alpha\in\Delta}$ and $(c_\alpha)\alpha \in \Delta$ are bounded from below, it follows that $(l_\alpha)_{\alpha\in\Delta}$ is also bounded from below. There exists $\lim_{\alpha\in\Delta} l_\alpha(x) = l(x)$, where $l(x) = h(x) - h(0)$. Since L enjoys the propety (B), the function l belongs to L, hence $h \in H_L$. $\triangle$

Let (X, L) be a pair of conic sets with the coupling function $\langle\cdot|\cdot\rangle : L \times X \to \mathbb{R}$. Assume that $0 \in X$ and $\langle l|0\rangle = 0$ for all $l \in L$. As usual we consider L as a set of abstract linear functions on X. Let H_L be the corresponding set of abstract affine functions. It follows from Proposition 8.8 that H_L enjoys the property (B) if and only L enjoys this property.

We now describe some sets of functions, which enjoys or does not enjoy the property (B).

PROPOSITION 8.9 *Consider the pair (X, L), where $X = L = \mathbb{R}^n$, with a coupling function $\langle l|x\rangle = [l, x] := \sum_{i=1}^n l_i x_i$. Then L enjoys the property (B).*

Proof: The proof is straightforward. $\triangle$

Consider the pair (X, L), where $X = L = \mathbb{R}^n_+$, with a coupling function $\langle l|x\rangle = \langle l, x\rangle := \min_{i\in I_+(l)} l_i x_i$. (Here $I_+(l) = \{i : l_i > 0\}$.) Then the set L does not possess (B).

The following example confirms this assertion.

EXAMPLE 8.4 Let $n = 3$ and let

$$l^k = (k, 1, 1), \qquad k = 1, 2, \dots . \tag{8.3.3}$$

Then $\langle l^k, x\rangle \to f(x)$ as $k \to +\infty$, where

$$f(x) = \begin{cases} \min(x_2, x_3) & \text{if } x_1 > 0; \\ 0 & \text{if } x_1 = 0. \end{cases}$$

We now check that f is not a min-type function. Indeed, assume that there exists $l = (l_1, l_2, l_3) \in \mathbb{R}^3_+$ such that $f(x) = \min_{i \in I_+(l)} l_i x_i$. Clearly $l \neq 0$, so $I_+(l) \neq \emptyset$. If $1 \notin I_+(l)$, then for the vector $x = (0, 1, 1)$ we have

$$f(x) = 0, \qquad \langle l, x \rangle > 0.$$

Hence $1 \in I_+(l)$. Let $x = (\varepsilon, 1, 1)$. Then for small enough $\varepsilon > 0$ we have $\min_{i \in I_+(l)} l_i x_i < \min(x_2, x_3) = 1$ and we arrive at a contradiction.

Assume now that $X = L = \mathbb{R}^n_{++}$ with the same coupling function, that is $\langle l, x \rangle = \min_{i=1,\dots,n} l_i x_i$. It is easy to see that L also does not enjoy (B). Indeed consider the sequence l^k defined by (8.3.3). Then $\langle l^k, x \rangle \to \min(x_2, x_3)$ for all $x \in \mathbb{R}^n_{++}$. It is easy to see that there exists no $l \in \mathbb{R}^n_{++}$ such that $\langle l, x \rangle = \min(x_2, x_3)$ for all $x \in \mathbb{R}^n_{++}$.

However, if we consider the pair $(\mathbb{R}^n_{++} \cup \{0\}, \mathbb{R}^n_+)$, then we obtain a set of elementary functions, which enjoy the property (B).

PROPOSITION 8.10 *Let $X = \mathbb{R}^n_{++} \cup \{0\}$ and $L = \mathbb{R}^n_+$. Assume that the pair of conic sets (X, L) is equipped with the coupling function $\langle l, x \rangle = \min_{i \in I_+(l)} l_i x_i$. Consider L as a set of functions defined on X. Then the set L enjoys the propery (B).*

Proof: Let Δ be a directed index set and let $(l^\alpha)_{\alpha \in \Delta} \subset L$ be a bounded from below net. Let $q(x) = \lim_{\alpha \in \Delta} \langle l^\alpha, x \rangle$. Extracting a subnet, if necessary, we can find a set $I_+ \subset I = \{1, \dots, n\}$ such that $I_+(l_\alpha) = I_+$ for all $\alpha \in \Delta$. If $I_+ = \emptyset$ then $\langle l^\alpha, x \rangle = 0$ for all $\alpha \in A$, so $f(x) = 0 = \langle 0, x \rangle$. Hence $q \in L$. Assume that I_+ is nonempty. If the nets $(l_i^\alpha)_{\alpha \in \Delta}$ are unbounded for all $i \in I_+$ then $\langle l^\alpha, x \rangle \to +\infty$ for all $x \in \mathbb{R}^n_{++}$. Assume now that the set

$$I_* = \{i \in I_+ : \text{ the net } (l_i^\alpha)_{\alpha \in \Delta} \text{ is bounded } \}$$

is nonempty. Let l_*^α be a vector such that

$$(l_*^\alpha)_i = \begin{cases} l_i^\alpha & \text{if } i \in I_*; \\ 0 & \text{if } i \notin I_*. \end{cases}$$

We have for $x \in \mathbb{R}^n_{++} \cup \{0\}$

$$\begin{aligned} q(x) &= \lim_{\alpha \in \Delta} \langle l^\alpha, x \rangle = \lim_{\alpha} \min_{i \in I_+} l_i^\alpha x_i \\ &= \lim_{\alpha} \min_{i \in I_*} l_i^\alpha x_i = \lim_{\alpha} \langle l_*^\alpha, x \rangle \end{aligned}$$

The net $(l_*^\alpha)_{\alpha \in \Delta}$ is bounded and we can assume without loss of generality that there exists $\lim l_*^\alpha := l_*$. Let $l_* = (l_1, \dots, l_n)$. Obviously $I_+(l_*) \subset I_*$. If $I_+(l_*) = I_*$, that is $l_i > 0$ for each $i \in I_*$, then

$$q(x) = \lim_{\alpha \in \Delta} \langle l_*^\alpha, x \rangle = \langle l_*, x \rangle, \qquad x \in \mathbb{R}^n_{++} \cup \{0\},$$

hence $f \in L$. Assume now that $I_+(l_*) \neq I_*$. Then there exists $j \in I_*$ such that $l_j = 0$. We have for $x \in \mathbb{R}^n_{++}$:

$$0 \leq q(x) = \lim_{\alpha \in \Delta} \min_{i \in I_*} l_i^\alpha x_i \leq \lim_{\alpha \in \Delta} l_j^\alpha x_j = l_j x_j = 0.$$

Hence $q(x) = 0 = \langle 0, x \rangle$ for all $x \in \mathbb{R}^n_{++} \cup \{0\}$ and $f \in L$. △

8.3.4. Positively homogeneous elementary functions

In this subsection we assume that a set of elementary functions L consists of positively homogeneous functions defined on a conic set X with $0 \in X$. We shall examine maximal elements of support sets. Let Z be a conic subset of the set X. We now describe connections between maximal elements of the support set $\text{supp}(f, L, Z)$ and maximal elements of the subdifferential $\partial_{L,Z} f(x_0)$ for $x_0 \in Z$.

PROPOSITION 8.11 *Let f be an L-convex function defined on Z and $x_0 \in Z$. Then a maximal element of the subdifferential $\partial_{L,Z} f(x_0)$ is a maximal element of the support set* $\text{supp}(f, L, Z)$.

Proof: Let l be a maximal element of the subdifferential $\partial_{L,Z} f(x_0)$. Assume $l' \in \text{supp}(f, L, Z)$ and $l' \geq l$. Then $l'(x_0) = l(x_0)$ that is $l' \in \partial_{L,Z} f(x_0)$. So $l' = l$ and l is a maximal element of the support set $\text{supp}(f, L, Z)$. △

Since $\text{supp}(f, L, Z) = \partial_{L,Z} f(0)$ (see (7.2.19)), it follows that for each maximal element l of the support set there exists $x \in Z$ (for example $x = 0$) such that l is a maximal element of $\partial_{L,Z} f(x)$. In some instances, each element of $\partial_{L,Z} f(0)$ is a maximal element of this set. For example, if $Z = X$ is a linear space and L is a set of linear functions defined on X, then the inequality $l(x) \geq l'(x)$ for all $x \in X$ implies $l = l'$. Therefore, all elements of the set $\text{supp}(f, L, Z) = \partial_{L,Z} f(0)$ are maximal elements of this set. If Z is not a linear space, this is no longer true without some additional assumptions. We now give some sufficient conditions which guarantee that for each maximal element l of the support set, there exists a point $x_0 \neq 0$ such that l is a maximal element of the subdifferential $\partial_{L,Z} f(x_0)$.

PROPOSITION 8.12 *Let X be a finite-dimensional normed space and let Z be a closed conic subset of X. Suppose that L is a subset of $PH(Z)$ with the following properties:*

1) L consists of continuous functions;

2) L is a convex cone;

3) there is $l' \in L$ such that $l' \neq 0$ and

$$0 \leq l'(x) \leq \|x\| \text{ for all } x \in Z. \tag{8.3.4}$$

Let f be an L-convex function on Z such that $\operatorname{dom} f \neq \emptyset$. *Then for a maximal element l of the support set* $\operatorname{supp}(f, L, Z)$ *there exists a point $x_0 \neq 0$ such that l is a maximal element of the subdifferential $\partial_{L,Z} f(x_0)$.*

Proof: Let $g(x) = f(x) - l(x)$ $(x \in Z)$. Clearly, $g(x) \geq 0$ for all $x \in Z$. Since f is lower semicontinuous (as an L-convex function) and l is continuous, it follows that g is lower semicontinuous. So g attains its minimum over the compact set $S = \{x \in Z : \|x\| = 1\}$. Let $\min_{x \in S} g(x) = \varepsilon$ and suppose $\varepsilon > 0$. Since g is positively homogeneous we have, for non-zero $x \in Z$,

$$g(x) = g\left(\frac{x}{\|x\|}\right) \|x\| \geq \varepsilon \|x\|.$$

Let $l' \in L$ be a non-zero function such that (8.3.4) holds. Then $f(x) \geq l(x) + \varepsilon l'(x)$. Since L is a convex cone, $l + \varepsilon l' \in L$. Thus $l + \varepsilon l' \in \operatorname{supp}(f, L, Z)$ and $l \leq l + \varepsilon l'$. Since l is a maximal element of the support set, we arrive at a contradiction. Thus $\varepsilon = 0$. Let x_0 be a minimizer of the function g over the set S. Then $f(x_0) = l(x_0)$ i.e., $l \in \partial_{L,Z} f(x_0)$. Since l is a maximal element of the support set we conclude that l is also a maximal element of the subdifferential $\partial_{L,Z} f(x_0)$. △

We now give one more sufficient condition which is weaker for L but stronger for f.

PROPOSITION 8.13 *Let X and Z be as in Proposition 8.12, and let a conic set L consist of continuous nonnegative positively homogeneous functions defined on Z. Let*

1) f be an L-convex function such that $\operatorname{dom} f \neq \emptyset$;

2) f be continuous;

3) $f(x) > 0$ for all non-zero $x \in Z$.

Then for a maximal element $l \neq 0$ of the set $\operatorname{supp}(f, L, Z)$ *there exists a point $x_0 \neq 0$ such that l is a maximal element of the subdifferential $\partial_{L,Z} f(x_0)$.*

Proof: Since $f(x) > 0$ for $x \neq 0$ and f is continuous it follows that the set $S_1 = \{x \in Z : f(x) = 1\}$ is compact. Let $g(x) = f(x) - l(x)$ and $\varepsilon = \min\{g(x) : x \in S_1\}$. Clearly, $\varepsilon \geq 0$. If $\varepsilon > 0$, then we have for non-zero $x \in Z$,

$$f(x) - l(x) = g(x) = f(x)g\left(\frac{x}{f(x)}\right) \geq \varepsilon f(x).$$

Therefore, $f(x) \geq \frac{1}{1-\varepsilon} l(x)$. Since l is nonnegative and $l \neq 0$, it follows that $\dfrac{1}{1-\varepsilon} l(x) > l(x)$ for all x with $l(x) \neq 0$. We have a contradiction which shows that $\varepsilon = 0$. If x_0 is a minimizer of the function g over S_1, then $l(x_0) = f(x_0)$ i.e., $l \in \partial_{L,Z} f(x_0)$. Clearly, l is a maximal element of the subdifferential $\partial_{L,Z} f(x_0)$. △

8.3.5. Maximal elements of support sets with respect to conic sets of positively homogeneous functions

Let X and L be a pair of conic sets with a coupling function $\langle \cdot | \cdot \rangle : L \times X \to \mathbb{R}$ such that

$$\langle \lambda l | x \rangle = \langle l | \lambda x \rangle = \lambda \langle l | x \rangle \text{ for all } l \in L,\ x \in X.$$

As usual, we consider L as a set of functions defined on X and X as a set of functions defined on L. Denote by H_L the set of L-affine functions defined on X. We indentify an H_L-convex function $h(x) = \langle l | x \rangle - \lambda$ with the pair (l, λ). We also assume that H_L is equipped with the pointwise order relation

$$h_1 \geq h_2 \iff h_1(x) \geq h_2(x) \text{ for all } x \in X. \tag{8.3.5}$$

In particular, we have, for $l_1, l_2 \in L$,

$$l_1 \geq l_2 \iff \langle l_1 | x \rangle \geq \langle l_2 | x \rangle \text{ for all } x \in X. \tag{8.3.6}$$

Recall that (see Proposition 7.8)

$$\text{supp}(f, H_L) = \text{epi } f^* := \{(l, \lambda) \in L \times \mathbb{R} : \lambda \geq f^*(l)\} \tag{8.3.7}$$

and (see Proposition 7.29) the function f^* is increasing with respect to the order relation $\geq$.

We now describe maximal elements of the support set $\text{supp}(f, H_L)$ in terms of the Fenchel-Moreau conjugate function.

PROPOSITION 8.14 *Let $f : X \to \mathbb{R}_{+\infty}$. A point (l, λ) of the support set* $\text{supp}(f, H_L)$ *is a maximal element of this set if and only if l is a* *maximal element of the level set* $S_c(f^*) = \{l' : f^*(l') \le c\}$ *with* $c = f^*(l)$ *and* $\lambda = f^*(l)$.

Proof: Let $h = (l, \lambda) \in L$ and $h' = (l', \lambda') \in L'$. First, we show that

$$h' \ge h \iff (l' \ge l,\ \lambda' \le \lambda).$$

Let $h'(x) = \langle l'|x\rangle - \lambda' \ge \langle l|x\rangle - \lambda$ for all $x \in X$ and let $x \in X$ and $t > 0$. Then

$$t\langle l'|x\rangle - \lambda' = \langle tl'|x\rangle - \lambda' \ge \langle tl|x\rangle - \lambda = t\langle l|x\rangle - \lambda. \tag{8.3.8}$$

Passing $t \to +0$ we obtain $\lambda' \le \lambda$. We also have $\langle l'|tx\rangle - \langle l, tx\rangle \ge \lambda' - \lambda > -\infty$ for all $x \in X$ and $t \in \mathbb{R}_+$, hence

$$\langle l'|x\rangle - \langle l, x\rangle \ge \frac{1}{t}(\lambda' - \lambda) \to_{t\to+\infty} 0.$$

Thus $l' \ge l$. We have proved that $h' \ge h \implies (l' \ge l,\ \lambda' \le \lambda)$. The reverse statement is trivial.

Let $h = (l, \lambda)$ be a maximal element of $\mathrm{supp}(f, H_L)$. Let $h' = (l', \lambda') \in \mathrm{supp}(f, H_L)$ and $h' \ge h$. These relations are equivalent to

$$\lambda \ge f^*(l), \quad \lambda' \ge f^*(l'), \quad l' \ge l, \quad \lambda' \le \lambda. \tag{8.3.9}$$

Note that the first two inequalities follows from (8.3.7). Assume that $\lambda > f^*(l)$. Let $l' = l$ and $\lambda' = f^*(l)$. Applying (8.3.9) we conclude that $(l, \lambda') \ge (l, \lambda)$, which contradicts the maximality of (l, λ). Thus $\lambda = f^*(l)$.

Assume now that $f^*(l') \le f^*(l)$, that is, $l' \in S_c(f^*)$ with $c = f(l)$. Let $\lambda = f^*(l)$ and $\lambda' = f^*(l')$. If $l' \ge l$, then, as it follows from (8.3.9), $(l', \lambda') \ge (l, \lambda)$, so the maximality of (l, λ) implies $l = l'$. Thus l is a maximal element of the set $S_c(f^*)$.

To prove the reverse statement, take $\lambda = f^*(l)$ and a maximal element l of the set $S_c(f^*)$ with $c = f^*(l)$. Let $(l', \lambda') \ge (l, f^*(l))$. It follows from (8.3.9) that $l' \ge l$ and $f^*(l') \le \lambda' \le \lambda \equiv f^*(l)$, hence $l' \in S_c(f^*)$. Since l is a maximal element of $S_c(f^*)$, we have $l = l'$. Applying (8.3.9) again, we conclude that $\lambda' = \lambda$. Thus $(l, f^*(l))$ is a maximal element of the support set. △

EXAMPLE 8.5 Let X be a Banach space, $L = X^*$ and let $\langle l|x\rangle = l(x)$. Then $l' \ge l$ if and only if $l' = l$. Hence (l, λ) is a maximal element of the support set $\mathrm{supp}(f, H_L)$ if and only if $\lambda = f^*(l)$.

8.3.6. The Toland-Singer formula and maximal elements of support sets

Let X and L be a pair of sets and let $\langle l|x\rangle$ be a coupling function defined on $L \times X$. We consider L as a set of abstract linear functions defined on X. Let H_L be the set of abstract affine functions with respect to L.

We assume that $(+\infty) - (+\infty) = +\infty$.

Consider proper H_L-convex functions $p, q : X \to \mathbb{R}_{+\infty}$ such that $\operatorname{dom} p \supset \operatorname{dom} q$. Let $f = q - p$, that is,

$$f(x) = \begin{cases} q(x) - p(x) & \text{if } x \in \operatorname{dom} q; \\ +\infty & \text{if } x \notin \operatorname{dom} q. \end{cases}$$

Clearly $\operatorname{dom} f = \operatorname{dom} q$. We consider the following problem of global minimization:

$$f(x) \longrightarrow \min \text{ subject to } x \in X. \tag{8.3.10}$$

Note that $\operatorname{dom} p \not\supset \operatorname{dom} q$ implies $\inf(q(x) - p(x)) = -\infty$, so in this case the problem (8.3.10) is trivial. Thus $\inf_{x\in X} f(x) > -\infty$ implies $\operatorname{dom} q \subset \operatorname{dom} p$.

We now give an outline of the approach we propose for the study of global optimization problems involving functions that can be represented as the difference of H_L-convex functions. Assume $f = q - p$, where $q, p : X \to \mathbb{R}_{+\infty}$ are H_L- convex functions and $\operatorname{dom} p \supset \operatorname{dom} q$. Let $\gamma' = \inf\{f(x) : x \in \mathbb{R}^n\} > -\infty$ and $\gamma \leq \gamma'$. Then $f(x) \geq \gamma$ for all $x \in X$, that is,

$$q(x) \geq p(x) + \gamma \text{ for all } x \in X. \tag{8.3.11}$$

Let $\tilde{p}(x) = p(x) + \gamma$. The inequality (8.3.11) is equivalent to the inclusion

$$\operatorname{supp}(q, H_L) \supset \operatorname{supp}(\tilde{p}, H_L).$$

Taking into account the obvious equality

$$\operatorname{supp}(\tilde{p}, H_L) = \operatorname{supp}(p, H_L) - (0, \gamma),$$

we conclude that the following statement holds.

PROPOSITION 8.15 *Let $f = q - p$ with proper H_L-convex functions $p, q : X \to \mathbb{R}_{+\infty}$. Let $\gamma' = \inf_{x\in X} f(x) > -\infty$. Then $\gamma \leq \gamma'$ if and only if*

$$\operatorname{supp}(q, H_L) \supset \operatorname{supp}(p, H_L) - (0, \gamma). \tag{8.3.12}$$

The dual characterization of problem (8.3.10) is given by the well-known Toland-Singer formula (see [185, 173] for the classical convex case, and [175] and references therein for the abstract convex situation). We shall prove this formula only for a particular case.

THEOREM 8.6 (Toland-Singer formula). *Let $p, q : X \to \mathbb{R}_{+\infty}$ be proper H_L-convex functions such that $\inf_{x \in X}(p(x) - q(x)) > -\infty$. Then*

$$\inf\{q(x) - p(x) : x \in X\} = \inf\{p^*(l) - q^*(l) : l \in L\}. \tag{8.3.13}$$

Proof: Let $\gamma \leq \gamma' := \inf\{q(x) - p(x)\}$. Then (8.3.12) holds. Since $\text{supp}(q, H_L) = \text{epi}\, q^*$ and $\text{supp}(p, H_L) = \text{epi}\, p^*$ we deduce that (8.3.12) is equivalent to the inequality $q^* \leq (\tilde{p})^* = p^* - \gamma\mathbf{1}$. Since $\gamma' > -\infty$, it follows that $\text{dom}\, p^* \subset \text{dom}\, q^*$ and $p^* - q^* \geq \gamma'\mathbf{1}$. Thus $\gamma'' := \inf_{l \in L} p^*(l) - q^*(l) \geq \gamma'$. We apply this result to functions p^* and q^*. Since $q^{**} = q$ and $p^{**} = p$, we also have $\gamma' \geq \gamma''$. Thus $\gamma' = \gamma''$. $\triangle$

REMARK 8.3 The Toland-Singer formula is equivalent to (8.3.12), so we can say that Proposition 8.15 is a geometric form of the Toland-Singer formula.

REMARK 8.4 Here we only consider a particular case of the Toland-Singer formula, when both p and q are proper H_L- functions. The original Toland-Singer formula holds true assuming only that p is H_L-convex while q is arbitrary and without the assumption $\inf_{x \in X} q(x) - p(x) > -\infty$.

We now express the Toland-Singer formula in terms of maximal elements of support sets. We assume that the set H_L of abstract L-affine functions possesses the property (B) (see Subsection 8.3.3).

PROPOSITION 8.16 *Let H_L possesses the property (B). Let $f : X \to \mathbb{R}_{+\infty}$ and $g : X \to \mathbb{R}_{+\infty}$ be H_L-convex functions. Then the following statements are equivalent:*

1) $\text{supp}(f, H_L) \subset \text{supp}(g, H_L)$;

2) for each maximal element h_1 of $\text{supp}(f, H_L)$ *there exists a maximal element h_2 of* $\text{supp}(g, H_L)$ *such that $h_1 \leq h_2$.*

Proof: 1) $\Longrightarrow$ 2) Let $\text{supp}(f, H_L) \subset \text{supp}(g, H_L)$. Let h_1 be a maximal element of $\text{supp}(f, H_L)$. Then $h_1 \in \text{supp}(g, H_L))$. Since H_L enjoys the property (B), we can apply Proposition 8.6 to deduce that there exists a maximal element h_2 of $\text{supp}(g, H_L)$ such that $h_2 \geq h_1$.

2) $\implies$ 1) Let $h \in \text{supp}(f, H_L)$. Applying Proposition 8.6 we find a maximal element h_1 of $\text{supp}(f, H_L)$ such that $h \leq h_1$. Let $h_2 \in \text{supp}(g, H_L)$, $h_2 \geq h_1$. Then $h_2 \geq h$, hence $h \in \text{supp}(g, H_L)$. △

We can exploit Proposition 8.16 in order to describe inclusion (8.3.12), which is a geometric form of the Toland-Singer formula in terms of the maximal elements of support sets. We can also apply subdifferentials in this description. More precisely, we can take advantage of the structure of the set $\mathcal{D}f(x_0)$.

As an example of this approach we shall prove the following assertion related to functions defined on a conic set X.

PROPOSITION 8.17 *Let X be a conic subset of the space $\mathbb{R}^n$ and L be a set of continuous positively homogeneous of degree one functions defined on X. Let H_L be the set of L-affine functions. Assume that H_L possesses the property (B). Let p and q be continuous real-valued H_L-convex functions such that*

$$\lim_{\|x\|\to+\infty}(q(x) - l(x)) = \lim_{\|x\|\to+\infty}(p(x) - l(x)) = +\infty \tag{8.3.14}$$

for all $l \in L$. Let x_0 be a global minimizer of the function $f = q - p$ over the set X. Then the inclusion (8.3.12) with $\gamma = f(x_0)$ holds if and only if for each $y \in X$ with the nonempty subdifferential $\partial_L f(y)$ and for each maximal element $(l_y, l_y(y))$ of the set $\mathcal{D}p(y)$, there exist an element $z \in X$ and a maximal element $(l_z, l_z(z))$ of the set $\mathcal{D}q(z)$ such that

$$l_y \leq l_z \text{ and } l_y(y) - l_z(z) + p(y) - q(z) + f(x_0) \geq 0. \tag{8.3.15}$$

Proof: We consider a pair of conic sets (X, L) with a coupling function $\langle l|x\rangle = l(x)$. It is clear that (l, c) is a maximal element of $\text{supp}(\tilde{p}, H_L)$ with $\tilde{p}(x) = p(x) + f(x_0)$ if and only if $(l, c - f(x_0))$ is a maximal element of $\text{supp}(p, H_L)$. It follows from Theorem 8.5 and Proposition 8.16 then the inclusion (8.3.12) with $\gamma = f(x_0)$ holds if and only if for each $y \in X$ and each maximal element $h_y = (l_y, c_y)$ of the set $\mathcal{D}p(y)$ (with $c_y = l_y(y) - p(y) + f(x_0)$), there exist an element $z \in X$ and a maximal element $h_z = (l_z, c_z)$ of the set $\mathcal{D}q(z)$ (with $c_z = l(z) - q(z)$) such that $l_y(x) - c_y \leq l_z(x) - c_z$ for all $x \in X$. Thus we have obtained the following inequality:

$$l_y(x) - l_z(x) \leq d \text{ for all } x \in X, \tag{8.3.16}$$

where $d = l_y(y) - l_z(z) + p(y) - q(z) + f(x_0)$. Let $x \in X$ and $t > 0$. It follows from (8.3.16) that $t(l_y(x) - l_z(x)) \leq d$ for all $t > 0$, so $l_y(x) - l_z(x) \leq 0$. Applying (8.3.16) again, we have $d \geq 0$. Thus the result follows. △

8.3.7. Excess functions

In order to examine some problems in global optimization involving H_L-convex functions, it is convenient to apply the concept of an excess function. Let $f : X \to \mathbb{R}$ be a function defined on a conic set X. Assume that the following limits exist:

$$f'(x,x) = \lim_{t \to +0} \frac{f((1+t)x) - f(x)}{t},$$

$$f'(x,-x) = \lim_{t \to +0} \frac{f((1-t)x) - f(x)}{t}.$$

These limits are called the right (left) derivative of f at the point x. If X is a vector space, then $f'(x,x)$ ($f'(x,-x)$) coincides with the directional derivative of the function f at the point x in the direction x $(-x)$.

Let f be a positively homogeneous of degree $k > 0$ function, that is $f(\lambda x) = \lambda^k f(x)$ for all $\lambda > 0$ and $x \in X$. Then the right and left derivatives exist. We have

$$\begin{aligned} f'(x,x) &= \lim_{t \to +0} \frac{1}{t}(f((1+t)x - f(x)) \\ &= \lim_{t \to +0} \frac{1}{t}((1+t)^k f(x) - f(x)) = kf(x). \end{aligned} \quad (8.3.17)$$

The same argument shows that

$$f'(x,-x) = -(1/k)f(x). \quad (8.3.18)$$

We now recall Euler's theorem for positively homogeneous of degree k functions.

PROPOSITION 8.18 (Euler's theorem). *Let X be a conic set and $f : X \to \mathbb{R}$ be a positively homogeneous of degree $k > 0$ function. Then for all $x \in X$ we have $f(x) = (1/k)f'(x,x) = -(1/k)f'(x,-x)$.*

Proof: The proof follows from (8.3.17) and (8.3.18). △

It follows from Euler's theorem that the function $x \mapsto f(x) - f'(x,x)$ is zero for a positively homogeneous of degree one function f. Thus, for $k = 1$, Euler's theorem motivates the following definition.

DEFINITION 8.1 Let f be a function defined on set X. Assume that for all $x \in X$ there exists the right derivative $f'(x,x)$. Then the function E_f defined on X by

$$E_f(x) = f(x) - f'(x,x)$$

is called the *excess function* of f.

It follows from Euler's theorem that $E_f = -(k-1)f$ for a positively homogeneous of degree k function f. Assume that $X \subset \mathbb{R}^n$ and f is a smooth function at a point $x \in X$, that is the gradient of f, $\nabla f(x)$, exists. Then $E_f(x) = f(x) - [\nabla f(x), x]$, where $[\cdot,\cdot]$ stands for the inner product.

REMARK 8.5 If x is a stationary point of a differentiable function f, that is, $\nabla f(x) = 0$, then $E_f(x) = f(x)$.

The following proposition indicates the link between the excess function E_f and the positively homogeneous extension $\tilde{f}$ for a differentiable function f defined on a conic set $X \subset \mathbb{R}^n$. Recall (see Subsection 7.2.8) that the *positively homogeneous extension* $\tilde{f}$ is defined on the set $X \times (0, +\infty)$ by $\tilde{f}(x, \lambda) = \lambda f(x/\lambda)$.

PROPOSITION 8.19 *Let f be a differentiable function and $\tilde{f}$ be its positively homogeneous extension. Then*

$$E_f(x) = \frac{\partial \tilde{f}}{\partial \lambda}(x, 1).$$

Proof: We have for $\lambda > 0$:

$$\frac{\partial \tilde{f}}{\partial \lambda}(x, \lambda) = \frac{\partial}{\partial \lambda}\left(\lambda f\left(\frac{x}{\lambda}\right)\right) = f\left(\frac{x}{\lambda}\right) + \lambda\left[\nabla f\left(\frac{x}{\lambda}\right), \frac{x}{\lambda}\right]\left(-\frac{1}{\lambda^2}x\right).$$

Thus the desired result follows. △

PROPOSITION 8.20 *Let f be a twice continuously differentiable function. Then*

$$\nabla E_f(x) = -x^T \cdot \nabla^2 f(x).$$

Proof: We have

$$\begin{aligned} \nabla E_f(x) &= \nabla(f(x) - [\nabla f(x), x]) = \nabla f(x) - x^T \cdot \nabla^2 f(x) - \nabla f(x) \\ &= -x^T \cdot \nabla^2 f(x). \end{aligned}$$

Thus the result follows. △

There are at least two situations, where the excess function E_f can be expressed in terms of the Fenchel -Moreau conjugate function f^*. We present these situations as examples.

EXAMPLE 8.6 Let $X \subset \mathbb{R}^n$ be a convex cone and let $f : X \to \mathbb{R}_{+\infty}$ be a convex function. Assume that f is differentiable at a point $x \in \operatorname{int} X$. Then $\partial f(x) = \{\nabla f(x)\}$ and using Proposition 7.7, we deduce that $[\nabla f(x), x] = f(x) + f^*(\nabla f(x))$. Thus

$$E_f(x) = f(x) - [\nabla f(x), x] = -f^*(\nabla f(x)). \tag{8.3.19}$$

EXAMPLE 8.7 Let f be an ICAR function defined on $\mathbb{R}^n_{++}$. Then f is H_L-convex with respect to the set L of all min-type functions $l(x) := \langle l, x\rangle$ with $l \in \mathbb{R}^n_{++}$ where $\langle l, x\rangle = \min_{i=1,\dots,n} l_i x_i$ $(x \in \mathbb{R}^n_{++})$. Let $x \in \mathbb{R}^n_{++}$. Proposition 7.7 implies that $l \in \partial_L f(x)$ if and only if $\langle l, x\rangle = f(x) + f^*(l)$. We now consider the element

$$f^{\nabla}(x) = \frac{f'(x,x)}{x}, \tag{8.3.20}$$

where

$$\frac{f'(x,x)}{x} = \left(\frac{f'(x,x)}{x_1}, \dots, \frac{f'(x,x)}{x_n}\right).$$

It has been proved (see (3.3.16)) that $f^{\nabla}(x) \in \partial_L f(x)$. So

$$f'(x,x) = \langle f^{\nabla}(x), x\rangle = f(x) + f^*(f^{\nabla}(x)). \tag{8.3.21}$$

It follows from (8.3.21) that

$$E_f(x) = -f^*(f^{\nabla}(x)). \tag{8.3.22}$$

REMARK 8.6 Excess functions have already appeared in mathematical programming in the study of some kinds of duality. For example, the so-called symmetric duality for twice differentiable functions can be formulated in terms of excess functions (see [34]). V.L. Levin [102] has used excess functions in the study of some classes of convex-along-rays functions.

8.4. OPTIMIZATION OF THE DIFFERENCE OF ABSTRACT CONVEX FUNCTIONS

8.4.1. Overview

Recently several authors have discussed the theoretical development of optimality conditions for specific classes of global optimization problems, see for example [74, 191, 89, 88, 189] and the references therein.

In many cases, the optimization problems have involved differences of convex functions (DC functions).

In this section we apply results obtained in Section 8.3 to study the optimization of the difference of abstract convex functions. In particular, we apply these results to functions expressible as the difference of either convex or ICAR functions with some additional properties. This approach is also useful in the study of the global minimization of the quotient of coercive differentiable convex functions.

This section contains several examples, which provide an important insight into the theoretical global optimality conditions we have developed.

We outline a dual approach to the study of global optimization problems involving functions which can be represented as the difference of H-convex functions. Assume for the sake of definiteness that H coincides with the set of all affine functions defined on the space $\mathbb{R}^n$, so a function f is H-convex if and only if f is lower semicontinuous and convex.

Our approach is based on the Toland-Singer formula. For the problem

$$q(x) - p(x) \longrightarrow \inf \text{ subject to } x \in \mathbb{R}^n \tag{8.4.1}$$

with lower semicontinuous convex functions q and p, the dual problem can be formulated in terms of conjugate functions q^* and p^*. It is well known that some results of the theory of convex duality and convex optimization can be made more precise for coercive differentiable convex functions by using the Legendre transformation (see, for example, [134], Scction 26). Applying this transformation, we formulate dual problems in terms of gradients of the functions involved without a formal use of duality. Similarly, we make more precise the Toland-Singer formula for the differences of the form $q - p$, where q and p are convex functions and, in addition, q is coercive differentiable and p is also either coercive differentiable or sublinear. If p is coercive differentiable, we present the dual results by means of the gradients of corresponding functions without a formal use of duality. If p is sublinear, we use methods of nonsmooth optimization. In both cases, we construct an auxiliary function, say g, that describes a dual information in primal terms. We show that a stationary point of the given function $f = q - p$ is a global minimizer of f if and only if this point is a global minimizer of the auxiliary function g. Thus we can try to distinguish a global minimizer by looking for stationary points of f, which are not stationary points of the auxiliary function g. We show that this is impossible: stationary points of both functions coincide under very weak assumptions. Thus dual methods based on the Toland-Singer formula cannot help us to obtain sufficient conditions that distinguish a global minimizer among stationary points.

We also discuss a similar presentation of dual information in primal terms for the difference of ICAR differentiable-along-rays functions.

The main results of this section can be found in [154, 152].

8.4.2. Minimization of the difference of coercive convex functions

Let $p, q : \mathbb{R}^n \to \mathbb{R}_{+\infty}$ be proper lower semicontinuous convex functions, that is, the functions p and q are H_L-convex, where L is the set of all affine functions defined on $\mathbb{R}^n$. Let $f(x) = q(x) - p(x)$. We assume that $\inf_{x \in \mathbb{R}^n} f(x) > -\infty$. Consider the following problem of global optimization:

$$q(x) - p(x) \longrightarrow \min \text{ subject to } x \in \mathbb{R}^n. \tag{8.4.2}$$

The dual characterization of problem (8.4.2) is given by the Toland-Singer formula:

$$\inf\{q(x) - p(x) : x \in \mathbb{R}^n\} = \inf\{p^*(l) - q^*(l) : l \in \mathbb{R}^n\}. \tag{8.4.3}$$

We now apply Proposition 8.15 to study (8.4.2) where p and q are lower semicontinuous convex coercive functions. Recall that a function $f : \mathbb{R}^n \to \mathbb{R}_{+\infty}$ is called coercive if

$$\lim_{\|z\| \to +\infty} \frac{f(z)}{\|z\|} = +\infty.$$

It follows from coercivity that we can apply Proposition 8.5 to describe the maximal elements of the support sets of p and q.

PROPOSITION 8.21 *A point x_0 is a global minimizer of the function f if and only if for each y with the nonempty subdifferential $\partial p(y)$ and $v \in \partial p(y)$, there exists z such that $v \in \partial q(z)$ and*

$$(q(z) - [v, z]) - (p(y) - [v, y]) - (q(x_0) - p(x_0)) \geq 0. \tag{8.4.4}$$

Proof: A point x_0 is a global minimizer of the function $f = q - p$ if and only if $f(x) \geq f(x_0)$ for all $x \in X$, that is $q(x) \geq p(x) + f(x_0)$. This inequality is equivalent to the inclusion $\text{supp}(q, H_L) \supset \text{supp}(p, H_L) - (0, \gamma)$, where $\gamma = f(x_0)$ and H_L coincides with the set of all affine functions defined on $\mathbb{R}^n$. The desired result follows now from Proposition 8.17 .$\triangle$

Using a different approach, A. Strekalovsky (see, for example [177]) has obtained the same result.

REMARK 8.7 Assume that p and q are differentiable functions. Then $\partial p(y) = \{\nabla p(y)\}$, $\partial q(z) = \{\nabla q(z)\}$. Hence the statement "for an element y we can find z such that there exists $v \in \partial p(y)$ and $v \in \partial q(z)$" can be expressed in the following form: " for an element y the equation $\nabla p(y) = \nabla q(z)$ has a solution".

We now present necessary and sufficient conditions for the minimum of the difference of coercive differentiable convex functions. We say that a function f is differentiable if $\operatorname{dom} f$ is an open set and there exists the gradient $\nabla f(x)$ for each $x \in \operatorname{dom} f$.

PROPOSITION 8.22 *Let p and q be coercive differentiable convex functions with* $\operatorname{dom} p \supset \operatorname{dom} q$. *Let x_0 be a stationary point of the function f. Then x_0 is a global minimizer of the function $f = q - p$ if and only if for each $y \in \operatorname{dom} p$ there exists $z \in \operatorname{dom} q$ such that $\nabla p(y) = \nabla q(z)$ and*

$$E_q(z) - E_p(y) \geq E_q(x_0) - E_p(x_0), \qquad (8.4.5)$$

where E_q and E_p are the corresponding excess functions.

Proof: The following necessary and sufficient optimality condition follows from Proposition 8.21 and Remark 8.7: a point x_0 is a global minimizer of the function $f = q - p$ if and only if for each vector y there exists a vector $z = z(y)$ such that $\nabla p(y) = \nabla q(z)$ and (8.4.4) holds with $v = \nabla p(y) = \nabla q(z)$. Since x_0 is a stationary point of f, we conclude that $\nabla p(x_0) = \nabla q(x_0)$, that is, $z(x_0) = x_0$. It follows from Remark 8.5 that $E_f(x_0) = f(x_0)$. Let $y \in \mathbb{R}^n$ and $v = \nabla p(y)$. Since $v = \nabla q(z)$ with $z = z(y)$, we can express (8.4.4) in the following form

$$E_q(z) - E_p(y) \geq E_f(x_0) = E_q(x_0) - E_p(x_0). \qquad \triangle$$

Consider the problem:

$$f(x) \equiv q(x) - p(x) \longrightarrow \min \text{ subject to } x \in \mathbb{R}^n, \qquad (8.4.6)$$

where q and p are as in Proposition 8.22. The problem

$$p^*(l) - q^*(l) \longrightarrow \min \text{ subject to } l \in \mathbb{R}^n \qquad (8.4.7)$$

is called *dual* to problem (8.4.6). Due to the Toland-Singer formula, the following zero duality gap property holds.

PROPOSITION 8.23 *Let q and p be functions as in Proposition 8.22. Then*

$$\inf_{x \in \mathbb{R}^n} (q(x) - p(x)) = \inf_{l \in \mathbb{R}^n} (p^*(l) - q^*(l)).$$

We now establish links between a solution of problem (8.4.6) and a solution of dual problem (8.4.7).

THEOREM 8.7 *Let x_0 be a solution of problem (8.4.6). Then the vector $l_{x_0} = \nabla q(x_0) = \nabla p(x_0)$ is a solution of problem (8.4.7).*

Proof: Since a global minimizer is a stationary point we have $\nabla f(x_0) = 0$. Thus $\nabla q(x_0) = \nabla p(x_0)$ so the vector l_{x_0} is well defined. Remark 8.5 shows that $E_f(x_0) = f(x_0)$. Then, by applying (8.3.19), we have

$$E_f(x_0) = E_q(x_0) - E_p(x_0) = p^*(l_{x_0}) - q^*(l_{x_0}).$$

On the other hand,

$$\inf_l(p^*(l) - q^*(l)) = \inf_x(q(x) - p(x)) = f(x_0).$$

Thus the desired result follows. △

Assume that q and p are finite coercive differentiable convex functions defined on $\mathbb{R}^n$ and, in addition, the function q is strictly convex. Then (see Theorem 26.6 in [134]) the mapping $x \mapsto \nabla q(x)$ is a one-to-one mapping from $\mathbb{R}^n$ onto itself. Let $g(y) = E_q((\nabla q)^{-1}\nabla p(y)) - E_p(y)$. Consider the problem

$$g(y) \longrightarrow \min \text{ subject to } y \in \mathbb{R}^n. \tag{8.4.8}$$

It follows from (8.4.5) that a stationary point x_0 is a global minimizer of the function f if and only if x_0 is a solution of problem (8.4.8), that is, a global minimizer of the function g. Hence in order to distinguish a global minimizer from a choice of stationary points we can look for stationary points of the initial function f, which are not stationary points of the function g. Unfortunately, as the following statement shows, this is impossible at least for twice continuously differentiable functions.

THEOREM 8.8 *Let p, q be coercive twice continuously differentiable convex functions and, in addition, q be strictly convex. Then a stationary point x_0 of the function f is simultaneously a stationary point of the function g. If x_0 is a stationary point of g and the Hessian $\nabla^2 p(x_0)$ is a non-singular matrix, then x_0 is a stationary point of f.*

Proof: First, we calculate the gradient of the function g. Since $g(y) = E_q(z) - E_p(y)$ with $\nabla p(y) = \nabla q(z)$, we have

$$\nabla g(y) = \nabla E_q(z)\nabla_y z - \nabla E_p(y) \text{ and } \nabla^2 p(y) = \nabla^2 q(z)\nabla_y z.$$

Using Proposition 8.20 we conclude that

$$\begin{aligned}\nabla g(y) &= -z^T\nabla^2 q(z)\nabla_y z + y^T\nabla^2 p(y) \\ &= -z^T\nabla p^2(y) + y^T\nabla^2 p(y) = (y-z)^T\nabla^2 p(y) \\ &= (y - (\nabla q)^{-1}p(y))^T\nabla^2 p(y).\end{aligned}$$

Let x_0 be a stationary point of the function f. Then $\nabla p(x_0) = \nabla q(x_0)$ so $x_0 - (\nabla q)^{-1}p(x_0) = 0$. Thus $\nabla g(x_0) = 0$. If the matrix $\nabla^2 p(x_0)$ is not singular then a stationary point x_0 of the function g is a stationary point of the function f. $\triangle$

REMARK 8.8 Both the Toland-Singer formula and Proposition 8.22 are analytic expressions of inclusion (8.3.12) (see Remark 8.3). Thus Theorem 8.8 demonstrates that at least for twice continuously differentiable functions p and q with q strictly increasing, the dual approach based on (8.3.12), can not help to distinguish a global minimizer from the totality of stationary points.

Let x_0 be a global minimizer of the function f, that is the inequality $f(x) \geq f(x_0)$ holds. Let $f = q - p$, where q and p as in Theorem 8.8. It follows from Proposition 8.22, that this inequality is equivalent to the inequality

$$E_q(z) - E_p(y) \geq E_q(x_0) - E_p(x_0) \qquad y \in \mathbb{R}^n,\ z = (\nabla q)^{-1}p(y).$$

In some cases one of these inequalities is stronger then the other. (The inequality $a(x)+b(x) \geq 0$ is said to be stronger than $a(x) \geq 0$ at a point x if $b(x) > 0$.) So we can claim that the stronger inequality follows from the weaker one. Let us present a corresponding example.

EXAMPLE 8.8 Let A be a positive definite operator, i.e $[Ax, x] > 0$ for all $x \neq 0$. Here $[y, x]$ is the inner product of vectors y and x. Let $f = q-p$ with p being coercive, differentiable and convex and $q(x) = [Ax, x]$. Furthermore, let x_0 be a global minimizer of f, that is,

$$f(y) - f(x_0) \geq 0 \text{ for all } y \in \mathbb{R}^n. \tag{8.4.9}$$

Applying (8.4.5) we deduce that

$$[2Az, y - z] + [Az, z] - p(y) \geq f(x_0).$$

for each $y \in \mathbb{R}^n$ and $z = (1/2)A^{-1}(\nabla p(y))$. Since

$$[2Az, y-z]+[Az, z] = [Ay, y]-[A(y-z), y-z] = q(y)-[A(y-z), y-z]$$

we assert that the inequality $f(y) - f(x_0) \geq 0$ implies that

$$f(y) - f(x_0) \geq [A(y-z), y-z] \text{ for all } y \in \mathbb{R}^n, \quad z = \frac{1}{2}A^{-1}(\nabla p(y)). \tag{8.4.10}$$

Clearly, $[A(y-z), y-z] = 0$ if and only if $y = z$, that is $(1/2)Ay = \nabla p(y)$. This equality holds if and only if y is a stationary point of f. Thus inequality (8.4.10)is stronger than inequality (8.4.9) for all regular (non stationary) points of the function f.

Now assume that q is coercive differentiable and strictly convex and $p(x) = [Ax, x]$ with A being positive definite. If $A = \alpha\mathrm{Id}$ where $\alpha > 0$ and Id is the identity operator, then $f = q - p$ is a paraconvex (or weakly convex) function ([138]).

The same reasoning as above shows that the inequality $f(y) - f(x_0) \geq 0$ is equivalent to the inequality

$$f(x_0) \leq f(z) + [A(z-y), z-y], \tag{8.4.11}$$

where y is an arbitrary element and z is a solution of the equation $\nabla q(z) = 2Ay$. Properties of strictly convex functions (see [134], Section 26) imply that this equation has only one solution for each $y \in \mathbb{R}^n$ and for each z there exists y such that $\nabla q(z) = 2Ay$. Thus we can consider the inequality (8.4.11) for all $z \in \mathbb{R}^n$ with $y = (2A)^{-1}\nabla q(z)$. If the inequality (8.4.11) holds for all z, then the inequality $f(x_0) \leq f(z)$ is also valid for all z and the latter is stronger than the former for all regular points of the function f.

8.4.3. Minimization of the difference of coercive convex and sublinear functions

In this subsection we shall study the following problem:

$$f(x) \equiv q(x) - p(x) \longrightarrow \min \quad \text{subject to} \quad x \in X, \tag{8.4.12}$$

where $X \subset \mathbb{R}^n$ is a convex cone, q and p are lower semicontinuous functions with q being finite coercive, strictly convex and differentiable and p being finite and sublinear.

PROPOSITION 8.24 *Let $U = (\nabla q)^{-1}(\partial p(0))$, where $\partial p(0) = \{l \in \mathbb{R}^n : [l, x] \leq p(x)\}$ for all $x \in X$. Then*

$$\inf_{x \in X}(q(x) - p(x)) = \inf_{z \in U} E_q(z).$$

Proof: Since p is a positively homogeneous function, it follows (see Proposition 7.20) that p^* coincides with the indicator function $\delta_{\partial p(0)}(l)$ of the subdifferential $\partial p(0)$ (which is equal to the support set of p with respect to the set of linear functions). Using the Toland-Singer formula we infer that

$$\begin{aligned}\inf_{x\in X}(q(x)-p(x)) &= \inf_{l\in\mathbb{R}^n}(p^*(l)-q^*(l)) \\ &= \inf_{l\in\partial p(0)}(-q^*(l)) = \inf_{z\in U}(-q^*(\nabla q(z))) = \inf_{z\in U} E_q(z).\end{aligned}$$

Thus the result follows. △

THEOREM 8.9 *Let x_0 be a local minimizer of problem (8.4.12). Then x_0 is a global minimizer of this problem if and only if*

$$E_q(x_0) = \min_{z\in U} E_q(z). \tag{8.4.13}$$

Proof: Since x_0 is a *local* minimizer of f over $\mathbb{R}^n$, we can apply the well-known necessary conditions for a local minimum in terms of directional derivatives (see, for example, [43], Section 5.1):

$$f'(x_0,u) \equiv [\nabla q(x_0),u] + p'(x_0,u) \geq 0$$

for all $u \in \mathbb{R}^n$. In particular

$$[\nabla q(x_0),x_0] + p'(x_0,x_0) \geq 0$$

and

$$[\nabla q(x_0),-x_0] + p'(x_0,-x_0) \geq 0.$$

It follows from Euler's theorem that $p'(x_0,x_0) = -p'(x_0,-x_0)$. So

$$[\nabla q(x_0),x_0] + p'(x_0,x_0) = 0.$$

Since $E_p(x_0) = 0$ for the positively homogeneous of degree one function p we have

$$\begin{aligned}f(x_0) &= q(x_0)-p(x_0) \\ &= (q(x_0)-[\nabla q(x_0),x_0]) - (p(x_0)-p'(x_0,x_0)) \\ &= E_q(x_0)-E_p(x_0) = E_q(x_0).\end{aligned}$$

Thus, applying Proposition 8.24, we can conclude that

$$f(x_0) = E_q(x_0) = \min\{E_q(z) : z \in U\}$$

if and only if x_0 is a global minimizer of the function f. $\triangle$

Theorem 8.9 is an analogue of Proposition 8.22 for the case under consideration and the problem

$$E_q(z) \longrightarrow \min \text{ subject to } z \in U = (\nabla q)^{-1}(\partial p(0)). \tag{8.4.14}$$

is an analogue of the problem (8.4.8). We now demonstrate that problem (8.4.14) also cannot disctinguish a global minimizer from the totality of stationary points. For this purpose we shall compare inf-stationary points of the problems (8.4.12) and (8.4.14). First we define such points.

DEFINITION 8.2 A point x_0 is called an *inf-stationary point* of problem (8.4.12) if $x_0 \in \operatorname{int} X$ and the following necessary condition for a local minimum holds:

$$[\nabla q(x_0), u] - p'(x_0, u) \geq 0 \text{ for all } u \in \mathbb{R}^n. \tag{8.4.15}$$

LEMMA 8.4 *The function p is differentiable at an inf-stationary point x_0 and $\nabla p(x_0) = \nabla q(x_0)$.*

Proof: It follows from (8.4.15) that

$$[\nabla q(x_0), u] \geq [l, u] \text{ for all } u \in \mathbb{R}^n \text{ and } l \in \partial p(x_0).$$

This inequality allows us to conclude that $l = \{\nabla q(x_0)\}$ for each $l \in \partial p(x_0)$, so the directional derivative

$$p'(x_0, u) = \max_{l \in \partial p(x_0)} [l, u]$$

coincides with the linear function $[\nabla q(x_0), u]$. Thus p is differentiable at the point x_0 and $\nabla p(x_0) = \nabla q(x_0)$. $\triangle$

Lemma 8.4 shows that a point x_0 is inf-stationary for the function $f = q - p$ if and only if x_0 is stationary in the usual sense, i.e., there exists the gradient $\nabla p(x_0)$ and $\nabla q(x_0) = \nabla p(x_0)$.

Let x_0 be a stationary point of the problem (8.4.12). Since $\nabla p(x_0) \in \partial p(0)$ and $\nabla q(x_0) = \nabla p(x_0)$, it follows that $x_0 \in U = (\nabla q^{-1})\partial p(0)$.

Let z_0 be a local minimizer of problem (8.4.14). Then the following well-known necessary condition for the minimum holds (see, for example [43], Section 5.1):

$$[\nabla E_q(z_0), h] \geq 0 \text{ for all } h \in \Gamma, \tag{8.4.16}$$

where $\Gamma = \Gamma(U, z_0)$ is the *Bouligand cone* of the set U at the point z_0 (see Definition 5.16). A point x_0 is called an *inf-stationary point* of the problem (8.4.14) if (8.4.16) holds. We now show that under some additional assumptions an arbitrary inf-stationary point of the initial problem (8.4.12) is an inf-stationary point of problem (8.4.14).

THEOREM 8.10 *Let q and p be as above and, in addition, the function q be twice continuously differentiable. Furthermore, let x_0 be a stationary (in particular, inf-stationary) point of problem (8.4.12). Then x_0 is an inf-stationary point of problem (8.4.14).*

Proof: We should check that $[\nabla E_q(x_0), h] \geq 0$ for all $h \in \Gamma$. Applying Proposition 8.20 we represent this inequality in the following form

$$x_0^T \nabla^2 q(x_0)h \equiv [\nabla^2 q(x_0)h, x_0] \leq 0 \text{ for all } h \in \Gamma. \tag{8.4.17}$$

By the definition of the Bouligand cone, for each $h \in \Gamma$ there exist a sequence $\alpha_k \to +0$ and a sequence $h_k \to h$ such that $x_0 + \alpha_k h_k \in U$. In other words, $\nabla q(x_0 + \alpha_k h_k) \in \partial p(0)$. Exploiting the differentiability of the mapping ∇q, we have

$$\nabla q(x_0) + \alpha_k \nabla^2 q(x_0)h_k + o(\alpha_k) \in \partial p(0). \tag{8.4.18}$$

Since $\nabla q(x_0) = \nabla p(x_0)$ we can present (8.4.18) as

$$\nabla p(x_0) + \alpha_k \nabla^2 q(x_0)h_k + o(\alpha_k) \in \partial p(0). \tag{8.4.19}$$

It follows from the definition of the subdifferential and (8.4.19) that

$$[\nabla p(x_0), x_0] + \alpha_k[\nabla^2 q(x_0)h_k, x_0] + [o(\alpha_k), x_0] \leq p(x_0).$$

By Euler's theorem $[\nabla p(x_0), x_0] = p(x_0)$, so $[\nabla^2 q(x_0)h, x_0] \leq 0$. △

REMARK 8.9 Inclusion (8.4.19) demonstrates that $\nabla^2 q(x_0)h$ belongs to the tangent cone at the point $\nabla p(x_0)$ with respect to the convex set $\partial p(0)$.

We now apply Theorem 8.9 to examine the problem

$$f(x) \equiv [Ax, x] - p(x) \longrightarrow \min \text{ subject to } x \in \mathbb{R}^n \tag{8.4.20}$$

where A is a positive definite operator and p is a finite non-negative sublinear function p. We denote by p^* the polar function to p:

$$p^*(l) = \max\{[l, x] : p(x) \leq 1\}.$$

It is well-known (see [134], Section 15) that $\partial p(0) = \{l : p^*(l) \leq 1\}$.

First, we describe the excess function E_q of the function $q(x) = [Ax, x]$ and the set $U = \{z : \nabla q(z) \in \partial p(0)\}$. We have:

$$E_q(x) = q(x) - [\nabla q(x), x] = [Ax, x] - 2[Ax, x] = -[Ax, x] \quad (8.4.21)$$

and

$$U = \{z : 2Az \in \partial p(0)\} = \left\{z : p^*(Az) \le \frac{1}{2}\right\}. \quad (8.4.22)$$

Combining (8.4.21), (8.4.22) and (8.4.13), we have:

$$[Ax_0, x_0] \ge [Az, z] \text{ for all } z \text{ such that } p^*(Az) \le \frac{1}{2}.$$

In other words (8.4.13) holds if and only if the vector x_0 can be expressed as $x_0 = Au_0$ where u_0 is a solution of the problem

$$[u, A^{-1}u] \longrightarrow \max \quad \text{subject to} \quad p^*(u) \le \frac{1}{2}. \quad (8.4.23)$$

Applying Theorem 8.9 we conclude that the following is valid.

PROPOSITION 8.25 *Let $f(x) = [Ax, x] - p(x)$ where A is a positive definite operator and p is a finite nonnegative sublinear function. If a vector x_0 is a global minimizer of problem (8.4.20), then $x_0 = Au_0$, where u_0 is a solution of problem (8.4.23). If a vector u_0 is a solution of problem (8.4.23) and x_0 is a local minimizer of problem (8.4.20), then x_0 is a global minimizer of (8.4.20).*

Proposition 8.25 allows us to investigate problem (8.4.20). We now give two examples.

EXAMPLE 8.9 Consider the following problem:

$$[Ax, x] - \|x\| \longrightarrow \min \quad \text{subject to} \quad x \in \mathbb{R}^n, \quad (8.4.24)$$

where A is a positive definite matrix and $\|\cdot\|$ is the Euclidean norm. In order to apply Proposition 8.25 we have to consider the following problem

$$[A^{-1}u, u] \longrightarrow \max \quad \text{subject to} \quad \|u\| \le \frac{1}{2}. \quad (8.4.25)$$

It is well known that the solution of the problem (8.4.25) is an eigenvector u_0 of the matrix A^{-1} corresponding to its greatest eigenvalue (that is, the least eigenvalue of the matrix A with the eigenvector u_0 such that $\|u_0\| = \frac{1}{2}$). It follows from Proposition 8.25 that a vector $x_0 = Au_0$ is a solution of the problem (8.4.24).

EXAMPLE 8.10 Consider the following problem:

$$[Ax, x] - \max_{i \in I} [l_i, x] \longrightarrow \min \quad \text{subject to} \quad x \in \mathbb{R}^n, \tag{8.4.26}$$

where I is a finite index set. Assume the function $p(x) = \max_{i \in I}[l_i, x]$ is nonnegative on $\mathbb{R}^n$. It is well known that the subdifferential $\partial p(0)$ of the function p is equal to the convex hull, $V := \operatorname{co}\{l_i : i \in I\}$, of the vectors l_i. Since $\partial p(0) = \{l : p^*(l) \leq 1\}$, we can express problem (8.4.23) in the following form:

$$[A^{-1}u, u] \longrightarrow \max \quad \text{subject to} \quad u \in \frac{1}{2}V. \tag{8.4.27}$$

The convex function $x \mapsto [A^{-1}u, u]$ attains its maximal value on the convex set V at one of the extreme points of this set. Hence, a solution u_0 of problem (8.4.27) is one of the points l_i, $i \in I$. It follows from Proposition 8.25 that a solution x_0 of problem (8.4.26) has the form $x_0 = Au_0$.

8.4.4. Dual optimality conditions for the difference of ICAR functions

Consider the pair of conic sets X, L, where $X = L = \mathbb{R}^n_{++}$ and a coupling function $\langle \cdot, \cdot \rangle : \mathbb{R}^n_{++} \times \mathbb{R}^n_{++} \to \mathbb{R}_+$ defined by $\langle l, x \rangle = \min l_i x_i$. We consider the set $L = \mathbb{R}^n_{++}$ as a set of abstract linear functions defined on the set $X = \mathbb{R}^n_{++}$. Let H_L be the set of all L-affine functions. Tnen a function $f : \mathbb{R}^n_{++} \to \mathbb{R}$ is H_L-convex if and only if f is an ICAR function. Thc coupling function $\langle \cdot, \cdot \rangle$ allows us to define the Fenchel-Moreau conjugacy operator. For a function $f : \mathbb{R}^n_{++} \to \mathbb{R}_{+\infty}$ we have

$$f^*(l) = \sup_{x \in \mathbb{R}^n_{++}} (\min_i l_i x_i - f(x)).$$

Recall (see Definition 3.6) that function $f : \mathbb{R}^n_+ \to \mathbb{R}$ is called differentiable-along-rays if, for all $x \in \mathbb{R}^n_{++}$, the function of one variable $f_x(\alpha) = f(\alpha x)$ is differentiable. Let f be a differentiable-along-rays ICAR function. Then (see Proposition 3.11) the subdifferential $\partial_L f(x)$ for each $x \in \mathbb{R}^n_{++}$ is nonempty and consists of the unique element $f^\nabla x$, where

$$f^\nabla x = f'(x, x)/x. \tag{8.4.28}$$

Consider the following extremal problem:

$$q(x) - p(x) \longrightarrow \min \text{ subject to } x \in \mathbb{R}^n_{++}, \tag{8.4.29}$$

where p and q are ICAR functions. Problem (8.4.29) can be considered as a certain analogue of problem (8.4.6) relating to the minimization of the difference of convex functions. The problem

$$p^*(l) - q^*(l) \longrightarrow \min \text{ subject to } l \in \operatorname{dom} q^* \tag{8.4.30}$$

is called the dual problem with respect to (8.4.29). The Toland-Singer formula (see Theorem 8.6) allows us to establish the zero duality gap property:

PROPOSITION 8.26 *Let q and p be ICAR functions. Then*

$$\inf\{q(x) - p(x) : x \in \mathbb{R}^n_{++}\} = \inf\{p^*(l) - q^*(l) : l \in \mathbb{R}^n_{++}\}.$$

Proposition 8.26 is similar to Proposition 8.23. We now consider an analogue of Theorem 8.7 and establish links between solutions of problem (8.4.29) and problem (8.4.30). We need the following assertion.

LEMMA 8.5 *Let p and q be convex-along-rays functions defined on $\mathbb{R}^n_+$ and let $y \neq 0$ be a local minimizer of the function $f = q - p$ on the ray $\tilde{R}_y = \{\alpha y : \alpha > 0\}$. Assume that the function of one variable $q_y(\alpha) = q(\alpha y)$ is differentiable at the point $\alpha_0 = 1$. Then the function $p_y(\alpha) = p(\alpha y)$ is also differentiable at this point and $q'(y, y) = p'(y, y)$.*

Proof: It follows from the necessary conditions for a local minimum that

$$f'(y, y) \geq 0 \text{ and } f'(y, -y) \geq 0. \tag{8.4.31}$$

In other words, $p'(y, y) \leq q'(y, y)$ and $p'(y, -y) \leq q'(y, -y)$. Since the function $p_y(\alpha)$ is convex on $(0, +\infty)$, its right derivative at the point $\alpha = 1$ is greater than or equal to its left derivative at this point. So $-p'(y, y) \leq p'(y, -y)$. Differentiability of q_y implies $-q'(y, -y) = q'(y, y)$. Since

$$-q'(y, y) \leq -p'(y, y) \leq p'(y, -y) \leq q'(y, -y) = -q'(y, y),$$

it follows that $p'(y, y) = -p'(y, y)$. This equality shows that the function p_y is differentiable at the point $\alpha_0 = 1$. It follows from (8.4.31) that $f'(x_0, x_0) = 0$. So $p'(y, y) = q'(y, y)$. △

REMARK 8.10 Let q be a differentiable-along-rays function and y be a local minimizer of the function $f = q - p$ on the ray $\{\alpha y : \alpha > 0\}$. Applying Lemma 8.5 we conclude that $p^{\nabla} y = q^{\nabla}(y)$.

THEOREM 8.11 *Let p be an ICAR function and let q be a differentiable-along-rays ICAR function. Let $y \neq 0$ be a local minimizer of the function $f = q - p$ on the ray $\{\alpha y : \alpha > 0\}$ and $l_y = p^{\nabla}(y) = q^{\nabla}(y)$. Then y is a global minimizer of the function f on $\mathbb{R}^n_+$ if and only if l_y is a global minimizer of the function $p^*(l) - q^*(l)$ on* $\operatorname{dom} q^*$.

Proof: Let y be a global minimizer of the function f. Proposition 8.26 demonstrates that

$$f(y) = \inf\{(p^*(l) - q^*(l)) : l \in \mathbb{R}^n_{++}\}. \tag{8.4.32}$$

It follows from (8.3.22) that $E_p(y) = -p^*(p^{\nabla}(y))$, $E_q(y) = -q^*(q^{\nabla}(y))$. Applying (8.4.32) and the equality $p'(y, y) = q'(y, y)$, we conclude that

$$\begin{aligned} p^*(p^{\nabla}(y)) - q^*(q^{\nabla}(y)) &= E_q(y) - E_p(y) = q(y) - p(y) \\ &= f(y) = \inf_{l \in \operatorname{dom} \mathbb{R}^n_{++}} p^*(l) - q^*(l). \end{aligned}$$

Thus $l_y = p^{\nabla}(y) = q^{\nabla}(y)$ is a global minimized of $p^* - q^*$.

We now assume that l_y is a global minimizer of the function $p^* - q^*$. It follows from Proposition 8.26 that

$$p^*(p^{\nabla}(y)) - q^*(q^{\nabla}(y)) = \inf\{q(x) - p(x) : x \in \mathbb{R}^n_{++}\}.$$

We have

$$\begin{aligned} p^*(p^{\nabla}(y)) - q^*(q^{\nabla}(y)) &= E_q(y) - E_p(y) \\ &= q(y) - p(y) - (q'(y, y) - p'(y, y) = f(y). \end{aligned}$$

Thus y is a global minimizer of the function f. △

8.4.5. Necessary and sufficient conditions for the minimum of the difference of coercive strictly ICAR functions

In this subsection we present necessary and sufficient conditions for the minimum of the difference $q - p$ of coercive strictly ICAR functions q, p defined on $\mathbb{R}^n_{++}$ in terms of abstract subgradients $p^{\nabla}(x)$ and $q^{\nabla}(x)$, where for an ICAR function f we have $f^{\nabla}(x) = f'(x, x)/x$.

Consider the pair of conic sets (X, L), where $X = \mathbb{R}^n_{++} \cup \{0\}$ and $L = \mathbb{R}^n_+$, with the coupling function $\langle l, x\rangle = \min_{i \in I_+(l)} l_i x_i$. Here $I_+(l) = \{i : l_i > 0\}$. Let

$$H_L = \{h : (\forall x \in X)\ h(x) = \langle l, x\rangle - c,\ l \in L, c \in \mathbb{R}\} \tag{8.4.33}$$

be the set of all L-affine functions. It follows easily from Theorem 3.2 that a function $f : X \to \mathbb{R}$ is H_L-convex if and only if f is ICAR. Proposition 8.8 and Proposition 8.10 imply that the set H_L possesses the property (B) (see Subsection 8.3.3 for the definition of this property).

PROPOSITION 8.27 *Let p and q be coercive strictly increasing convex-along-rays continuous functions defined on $\mathbb{R}^n_{++}$ and $f = q - p$. Assume that there exists a point $x \in \mathbb{R}^n_{++}$ such that $(q_x)'_+(0) = 0$. Then a point $x_0 \in \mathbb{R}^n_{++}$ is a global minimizer of the function f on $\mathbb{R}^n_{++}$ if and only if for each $y \in \mathbb{R}^n_{++}$ there exists $z \in \mathbb{R}^n_{++}$ such that*

$$\frac{p'(y,y)}{y_i} \leq \frac{q'(z,z)}{z_i} \quad \textit{for all } i = 1, \dots, n \tag{8.4.34}$$

and

$$E_q(z) - E_p(y) \geq f(x_0). \tag{8.4.35}$$

Proof: Since the set H_L enjoys the property (B), we can apply Proposition 8.17. Due to this proposition, x_0 is a global minimizer if and only if for each $y \in \mathbb{R}^n_{++} \cup \{0\}$ and for each maximal element $(l_y, \langle l_y, y\rangle)$ of the set $\mathcal{D}p(y)$ there exist $z \in \mathbb{R}^n_{++} \cup \{0\}$ and a maximal element $(l_z, \langle l_z, z\rangle)$ of the set $\mathcal{D}q(z)$ such that:

$$\langle l_y, x\rangle \leq \langle l_z, x\rangle \quad \text{for all } x \in \mathbb{R}^n_{++} \tag{8.4.36}$$

and

$$\langle l_y, y\rangle - \langle l_z, z\rangle + q(z) - p(y) - f(x_0) \geq 0. \tag{8.4.37}$$

If $y = 0$ then both (8.4.36) and (8.4.37) hold with $z = 0$. For a strictly increasing function g and a point u the set $\mathcal{D}g(u)$ contains the unique maximal element $(g^{\nabla}(u), g'(u,u))$. Since g is strictly increasing it follows easily that $g'(u,u) > 0$. Let $y \neq 0$. Then $l_y = p^{\nabla}(y)$ is not equal to zero. Assume that $z = 0$. Applying Corollary 3.3, we conclude that $l_z = 0$, which contradicts (8.4.36. Thus we can consider only $y \in \mathbb{R}^n_{++}$ and $z \in \mathbb{R}^n_{++}$.

Let x_0 be a global minimizer. Applying (8.4.36) and (8.4.37) we can deduce that for each $y \in \mathbb{R}^n_{++}$ there exists $z \in \mathbb{R}^n_{++} \cup \{0\}$ such that

(a) $\langle p^{\nabla}(y), x\rangle \leq \langle q^{\nabla}(z), x\rangle$ for all $x \in \mathbb{R}^n_{++}$;

(b) $\langle p^{\nabla}(y), y\rangle - \langle q^{\nabla}(z), z\rangle + (q(z) - p(y) - f(x_0) \geq 0$.

Clearly (b) is equivalent to $p'(y,y) - q'(z,z) + (q(z) - p(y) - f(x_0) \geq 0$, which is in turn equivalent to (8.4.35). We now check that the inequality

(a) is valid if and only if (8.4.34) holds. We can present (a) as

$$p'(y,y)\min_{k\in I}\frac{x_k}{y_k}\le q'(z,z)\min_{k\in I}\frac{x_k}{z_k} \tag{8.4.38}$$

where $I=\{1,\dots,n\}$. (Clearly, $I_+(y)=I_+(z)=I$.) Let (a) be true. Take an index $i\in I$ and consider a vector x such that $x_i=z_i>0$ and x_j are sufficiently large numbers for $j\ne i$. Then

$$\min_{k\in I}\frac{x_k}{z_k}=1;\qquad \min_{k\in I}\frac{x_k}{y_k}=\frac{x_i}{y_i}=\frac{z_i}{y_i}.$$

It follows from (8.4.38) that $p'(y,y)(z_i/y_i)\le q'(z,z)$ so $p'(y,y)/y_i\le q'(z,z)/z_i$ for all $i\in I$. Thus (a) implies (8.4.34). Assume now that (8.4.34) holds. Then

$$p'(y,y)\frac{1}{y_i}\le q'(z,z)\frac{1}{z_i}\ \text{for all } i\in I$$

so for all $x\in \mathbb{R}^n_+$ we have

$$p'(y,y)\min_i\frac{x_i}{y_i}\le p'(y,y)\min_i\frac{x_i}{y_i}\le q'(z,z)\min_i\frac{x_i}{z_i},$$

that is (a) holds. Thus if x_0 is a global minimizer then (8.4.34) is valid; it was mentioned above that (8.4.35) holds as well.

Consider now a point x_0 such that (8.4.34) and (8.4.35) hold. It has been shown that (8.4.34) implies (a) and (8.4.35) implies (b). It follows from (a) and (b) that for each $y\ne 0$ and each maximal element of the set $\mathcal{D}p(y)$ there exist $z\ne 0$ and a maximal element of the set $\mathcal{D}q(z)$ such that (8.4.36) and (8.4.37) hold. Thus x_0 is a global minimizer. △

THEOREM 8.12 *Let p and q be coercive strictly increasing convex-along-rays functions defined on $\mathbb{R}^n_{++}\cup\{0\}$. Moreover, let q be differentiable-along-rays and there exists $x\in\mathbb{R}^n_{++}$ such that $(q_x)'_+(0)=0$. Let $f=q-p$ and $x_0\in\mathbb{R}^n_{++}$ be a local minimizer of the restriction of the function f on the ray $\{\alpha x_0:\alpha>0\}$. Then x_0 is a global minimizer of f on $\mathbb{R}^n_{++}$ if and only if for each $y\in\mathbb{R}^n_{++}$ there exists $z\in\mathbb{R}^n_{++}$ such that*

i) $\frac{p'(y,y)}{y_i}\le\frac{q'(z,z)}{z_i}$ *for all* $i=1,\dots,n$;

ii) $E_q(z)-E_p(y)\ge E_q(x_0)-E_p(x_0)$.

Proof: Applying Lemma 8.5 we have $E_q(x_0)-E_p(x_0)=E_f(x_0)=f(x_0)-f'(x_0,x_0)=E_f(x_0)$. Thus the result follows from Proposition 8.27. △

8.4.6. Minimization of the quotient of convex functions

Results obtained for the difference of abstract convex functions can often be applied in the study of the quotient of such functions. As an example, we shall consider optimization problems involving a function f of the form $f(x) = q(x)/p(x)$ with lower semicontinuous nonnegative convex functions p and q, such that $\operatorname{dom} q \subset \operatorname{dom} p$ and $p(x) = 0$ if and only if $q(x) = 0$. It is assumed that $+\infty/+\infty = +\infty$.

We shall study the following problem:

$$\frac{q(x)}{p(x)} \longrightarrow \min \quad \text{subject to} \quad p(x) \neq 0. \tag{8.4.39}$$

Assume that the value β of problem (8.4.39) is positive. Then

$$\inf_{x \in \mathbb{R}^n} (q(x) - \beta p(x)) = 0.$$

Applying the Toland -Singer formula we obtain the following result:

$$\inf_{l \in \mathbb{R}^n} ((\beta p)^*(l) - q^*(l)) = 0. \tag{8.4.40}$$

Consider a function f and a number α. It follows directly from the definitions of the Fenchel-Moreau congjugate function and the excess function that

$$(\alpha f)^*(l) = \alpha f^* \left(\frac{l}{\alpha}\right), \tag{8.4.41}$$

$$E_{\alpha f} = \alpha E_f. \tag{8.4.42}$$

Using (8.4.41), we can present (8.4.40) in the following form

$$\inf_{l \in \mathbb{R}^n} \left(\beta p^* \left(\frac{l}{\beta}\right) - q^*(l)\right) = 0. \tag{8.4.43}$$

Thus we should calculate p^* and q^* at different points. It is more convenient to have a dual problem, which contains the values of both functions p^* and q^* at the same points. Due to (8.4.42) we can try to use excess functions for constructing such a problem.

Assume that q and p are finite coercive differentiable functions and q is strictly convex. Then we can apply the excess functions to examine problem (8.4.39). Note that we can reformulate this problem in the following way: find $\beta > 0$ and x_0 such that

$$q(x) \geq \beta p(x) \text{ for all } x \in \operatorname{dom} p \text{ and } q(x_0) = \beta p(x_0). \tag{8.4.44}$$

We also consider the following problem: find $\beta > 0$ and x_0 such that

$$E_q(z(y)) \geq \beta E_p(y) \text{ for all } y \in \operatorname{dom} p \text{ and } E_q(x_0) = \beta E_p(x_0). \quad (8.4.45)$$

where $z(y) = (\nabla q)^{-1}(\nabla p(y))$.

PROPOSITION 8.28 *Let q and p be coercive differentiable functions and let q be strictly convex. Then a pair (β, x_0) is a solution of problem (8.4.44) if and only if this pair is a solution of problem (8.4.45).*

Proof: A pair (β, x_0) is a solution of problem (8.4.44) if and only if

$$\min(q(x) - \beta p(x)) = q(x_0) - \beta p(x_0) = 0.$$

Applying Proposition 8.26 and (8.4.42), we conclude that (β, x_0) is a solution of problem (8.4.44) if and only if for each $y \in \operatorname{dom} p$,

$$E_q(z(y)) - \beta E_p(y) \geq E_q(x_0) - \beta E_p(x_0) = 0. \qquad \triangle$$

Under some additional assumptions problem (8.4.45) can be presented in the form symmetric to (8.4.39). We shall assume that $E_p(y) \leq 0$ for all y. This inequality certainly holds if $p(0) = 0$. Indeed, we have for all l:

$$p^*(l) = \sup_x([l, x] - p(x)) \geq [l, 0] - p(0) = 0,$$

so $E_p(y) = -p^*(\nabla p(y)) \leq 0$. We now consider the following problem:

$$\frac{E_q(z(y))}{E_p(y)} \longrightarrow \max \text{ subject to } E_p(y) \neq 0. \quad (8.4.46)$$

PROPOSITION 8.29 *Let p and q be as above and $E_p(y) \leq 0$ for all y. Then problems (8.4.39) and (8.4.46) have the same values and the same solutions.*

Proof: The proof follows directly from Proposition 8.28. $\triangle$

Chapter 9

APPLICATION TO GLOBAL OPTIMIZATION: NUMERICAL METHODS

9.1. INTRODUCTION

Numerical methods for global optimization are becoming an increasingly important field of study. There is a recognized practical need for methods which will efficiently solve global optimization problems (see, for example, Horst and Tuy [81]). However, in general, such problems are, by their very nature, extremely difficult to solve. This is primarily due to the lack of tools which provide *global* information about the objects (sets and functions) under study in contrast with local optimization, where the classical calculus and its modern generalization provide necessary tools. It should also be noted that standard nonlinear programming methods are not generally applicable as solution methods due to the intrinsic multi-extremality of global optimization problems. Finally, the computational difficulties of global optimization through its essentially combinatorial nature make the development of general efficient solution methods unlikely. However, despite these difficulties it is possible to provide methods for the solution of some specific global optimization problems, which cover many practical needs. In this chapter we shall use techniques from abstract convexity for developing some numerical methods for global optimization.

The simplest example of a global optimization problem is that of minimizing a convex function over a convex set. There is a variety of methods for solving such problems based either on global techniques or on local techniques exploiting the fact that the local extrema are also global ones. The local techniques are essentially gradient or (local) subgradient approaches in which the method developed uses a local approximation to the function (as provided by the subdifferential). The development of methods for the search for a local extremum for general nonsmooth optimization problems is often based on an extension of a subgradient method. However, there are techniques for solving convex minimization problems which are essentially global in nature. The prime example is the cutting plane method which is based on the representation of a convex function as the upper envelope of its affine minorants. The conceptual scheme of this approach is suitable for the minimization of an abstract convex function which is the upper envelope of a set of elementary functions. Two problems arise in the implementation of this approach. The first is to find an abstract subgradient (an element of the abstract subdifferential) of an abstract convex function at a point. The other problem is to solve the subprobem (auxiliary optimization problem): minimize the finite maximum of elementary functions over the feasible set. As it was shown in the previous chapters, the first problem can be easily solved for some important classes of elementary functions. The solution of the auxiliary problem is a difficult task, however this problem can be solved in some special cases.

Since the maximum of a finite number of elementary functions is a minorant of an abstract convex function, the minimization of this maximum (that is, solving the subproblem) provides a lower estimate for the global minimum of this function. Lower estimates of such kinds are very useful in the implementation of many methods in global optimization, such as branch and bound or tabu search. We discuss conceptual schemes of some of these approaches based on abstract convexity in Section 9.2. The main focus is on the generalized cutting plane method. The numerical implementation of this method for abstract convexity based on min-type functions is examined in details in Section 9.3. We consider various approaches for solving the subproblem (auxiliary problem) and report results of numerical experiments for IPH (increasing positively homogeneous), ICAR (increasing convex-along-rays), ICR (inceasing coradiant) and Lipschitz functions. Some numerical results related to the branch-and-bound method based on abstract convexity are discussed as well.

Abstract quasiconvexity can also be useful in the design and the study of numerical methods for global optimization. Level sets of an abstract quasiconvex function g are abstract convex, hence a point which does not belong to the level set $S_c(g) = \{x : g(x) \leq c\}$ of g can be separated from $S_c(g)$ by an elementary function. Such a kind of separation leads to the *outer approximation* of the level set. This observation can be useful for construction of numerical methods for solving optimization problems based on the outer approximation of the feasible set of this problem. In

Section 9.5 we consider a very fruitful example based on this approach, namely, monotone optimization.

Sections 9.2 and 9.3 are written jointly with Mikhail Andramonov and Section 9.4 written jointly with Adil Bagirov.

9.2. CONCEPTUAL SCHEMES OF NUMERICAL METHODS

9.2.1. Overview

In this section we present conceptual schemes of some methods of global optimization based on abstract convexity. The main idea behind this application is the following. If f is an abstract convex function with respect to the set of elementary functions H, then the maximum of finite collections of functions $h_1, \dots, h_n \in \text{supp}(f, H)$ is a minorant of the function f. We get the best minorants of such a kind if we consider elements of H-subdifferentials of the function f at some points $x_1, \dots, x_k$. We show that such minorants approximate, in a certain sense, the function f. The minimization of minorants over the feasible set provides a lower estimate of the global minimum f over this set. These estimates can be useful while constructing some numerical methods for the search for global minima.

We do not discuss here how to find an element of the H-subdifferential and how to minimize the maximum of finite collections of elementary functions. These questions will be addressed, for some special classes of functions, in the next two sections.

The main results from this section can be found in [7, 6, 4, 5].

9.2.2. Special minorates of abstract convex functions

Let H be a set of elementary functions defined on a set X. Consider an H-convex function $f : X \to \mathbb{R}$ and a point $x \in X$. We are interested in the set $\partial_H^* f(x)$ defined by (7.2.7):

$$\partial_H^* f(x) = \{h \in \text{supp}(f, H) : h(x) = f(x)\}. \tag{9.2.1}$$

If H is closed under vertical shifts that is $(h \in H, c \in \mathbb{R}) \implies h - c\mathbf{1} \in H\}$, then $\partial_H^* f(x) = \partial_H f(x)$. If H coincides with the set H_L of all abstract affine functions with respect to a set of abstract linear functions L, then (see Proposition 7.1)

$$\partial_H^* f(x) = \{h_{l,c} : l \in \partial_L f(x),\ c = l(x) - f(x)\}. \tag{9.2.2}$$

where $h_{l,c}(y) = l(y) - c, \ \ y \in X$. A function $h \in \partial_H^* f(x)$ is a minorant of the function f, which coincides with f at a point x. We can consider h as a special kind of approximation from below of the function f at the point x. Of course, this approximation can be very distant from f at points different from x. In order to get a more precise approximation, consider a finite collection $x^1, \dots, x^k$ of points of the set X. Assume that $\partial_H^* f(x^i)$ is nonempty for all $i = 1, \dots, k$ and let $h_i \in \partial_H^* f(x^i)$. The function

$$p(x) = \max_{i=1,\dots,k} h_i(x) \tag{9.2.3}$$

is again a minorant of f and p coincides with f at the points $x^1, \dots, x^k$, so we can assume that the function p approximates from below the function f more accurately than $h_1, \dots, h_k$. Sometimes we will use the notation p_k instead of p in order to underline the quantity of elementary functions which form p. The function $p \equiv p_k$ defined by (9.2.3) is called a *k-minorant* of the function f by the set of elementary functions H.

The following simple statement shows that k-minorants with large k, really approximate the function f.

PROPOSITION 9.1 *1) Let f be an H-convex function defined on X and assume that $\partial_H^* f(x) \neq \emptyset$ for all $x \in X$. Let T be the set of all finite collections $t = (x^1, \dots, x^k)$, where $x^i \in X$, $i = 1, \dots, k$, $k = 1, 2, \dots$. Assume that T is equipped with the order relation by the inclusion. Consider a net (generalized sequence) $(p^t)_{t \in T}$ where, for $t = (x^i)_{i=1}^k$, we have $p^t(x) = \max_{i \in t} h_i(x)$ with an arbitrary element h_i of the set $\partial_H^* f(x^i)$. Then*

$$\lim_{t \in T} p^t(x) = f(x) \text{ for each } x \in X.$$

2) Let X be a compact metric space and let the function f be continuous. Furthermore, let H consist of continuous functions. Then $p^t \to f$ uniformly, that is, for each $\varepsilon > 0$ there exists $t \in T$ such that $\max_{x \in X}(f(x) - p^t(x) < \varepsilon$.

Proof: 1) Since the net $(p^t)_{t \in T}$ is increasing,

$$\lim_{t \in T} p^t(x) = \sup_{t \in T} p^t(x) = \sup_{x \in X} h_x(x)$$

where h_x is an arbitrary element of $\partial_H^* f(x)$. Since $h_x(x) = f(x)$ and $h_x(y) \le f(y)$ for all $y \in X$ it follows that $\lim_{t \in T} p^t(x) = f(x)$.

2) This assertion follows from 1) and the well-known Dini theorem (see, for example, [168]): if $p^t(x) \to f(x)$ for each $x \in X$, the net p^t is increasing and functions p^t and f are continuous, then $p^t \to f$ uniformly. $\triangle$

Various methods for global minimization of a function f on a set X are based on the minimization of k-minorants. Thus the problem

$$\max_{i=1,\dots,k} h_i(x) \longrightarrow \min \text{ subject to } x \in X \tag{9.2.4}$$

becomes very important. This problem is very complicated and we shall examine it carefully for some classes of elementary functions. We now present some examples of problems of such a kind. In these examples we indicate classes of abstract linear functions L, having in mind that the class H of elementary functions coincides with the set H_L of all abstract affine functions. We use (9.2.2) for the calculation of elements of $\partial^* f_{H_L} f(x)$.

1) Let L be the set of all linear functions defined on $X = \mathbb{R}^n$. Then a finite function f is H_L-convex if and only if f is convex; the L-subdifferential coincides with the usual subdifferential of a convex function and is nonempty for all $x \in \mathbb{R}^n$. Moreover, problem (9.2.4) can be represented as follows: find

$$\min_{x\in\mathbb{R}^n} \max_{i=1,\dots,k} ([l_i, x] - c_i),$$

where $l_i \in \mathbb{R}^n, c_i \in \mathbb{R}$. Here $[l, x]$ stands for the inner product of vectors l and x. It is well known that this problem can be transformed into a linear programming problem.

2) Let L be the set of all quadratic functions l of the form $l(x) = a\|x\|^2 + [b, x]$ with $a \le 0$ and $b \in \mathbb{R}^n$, defined on a compact set $X \subset \mathbb{R}^n$. Then a function f is H-convex if and only if f is lower semicontinuous. If f is twice continuously differentiable then the L-subdifferential is nonempty for all $x \in X$. Moreover, the problem (9.2.4) can be written as follows: find

$$\min_{x\in X} \max_{i=1,\dots,k} (a_i\|x\|^2 + [b_i, x] - c_i)$$

with $a_i \le 0, c_i \in \mathbb{R}, b_i \in \mathbb{R}^n$. This is a very complicated problem, which can be transformed into the problem of DC programming (see, for example, [189]).

3) Let X be a Banach space and L be the set of all functions of the form $l(x) = a\|x - y\|$ with $y \in X$ and $a \le 0$. Then the function f is H_L-convex if and only if f is lower semicontinuous. Moreover, the L-subdifferential is nonempty for each Lipschitz function f. Problem (9.2.4) can be formulated as follows: find

$$\min_{x\in X} \max_{i=1,\dots,k} (a_i\|x - x^i\| - c_i)$$

with $a_i \leq 0$ and $x^i \in X$. Since functions $x \mapsto a_i\|x - x^i\|$ are concave, this problem can again be transformed into a DC problem.

4) Let $X \subset \mathbb{R}^n_+$ and let L be the set of all min-type functions $l(x) = \min_{i \in I_+(l)} l_i x_i$ with $l \in \mathbb{R}^n_+$, here $I_+(l) = \{i : l_i > 0\}$. A function f is H_L-convex if and only if f is an ICAR (increasing convex-along-rays) function. The L-subdifferential is nonempty for each finite ICAR function. Problem (9.2.4) can be represented as follows: find

$$\min_{x \in X} \max_{i=1,\dots,k} \min_{j \in I_+(l)} (l^i_j x_j - c_i),$$

where $l^i = (l^i_j)^n_{j=1} \in \mathbb{R}^n_+$, $c_i \in \mathbb{R}$. This is a complicated *min-max-min* problem, however the functions under the sign *min-max-min* are very simple. We shall discuss some approaches to this problem later (see Section 9.3 and Section 9.4).

9.2.3. Generalized cutting plane method

There are many methods of convex minimization based on the notion of the subdifferential. Almost all of them use the local definition of the subdifferential, hence they use the directional derivative of a convex function. However, there is a method, which only uses the global definition of the subdifferential, i.e., its support properties. This is the so-called cutting plane method (see, for example, [92, 131]). The cutting plane method can be easily generalized for the abstract convex setting. This method was studied in detail by D. Pallaschke and S. Rolewicz [118] in a very general situation. They use the term *Φ-bundle method.* (In our notation it should be H-bundle method.) We are mainly interested in a computational realization of this approach, so we shall study a less general situation. However, this situation is more adaptable to discussions, relating to the implementation of this approach. We consider the corresponding method as the *generalized cutting plane method.*

Let H be a set of elementary functions defined on a compact subset X of a Banach space. We shall consider an H-convex function f defined on X. Assume that the set

$$\partial^*_H f(x) = \{h \in \text{supp}(f, H_L) : h(x) = f(x)\} \tag{9.2.5}$$

is not empty for all $x \in X$. We now consider a numerical method for the solution of the problem

$$f(x) \longrightarrow \min \quad \text{subject to} \quad x \in X. \tag{9.2.6}$$

The algorithm

Step 0. Set $k := 0$. Choose an arbitrary initial point $x^0 \in X$.

Step 1. Calculate $h_k \in \partial^*_H f(x^k)$. Let

$$f_k(x) = \max_{i=0,\dots,k} h_i(x) \text{ for all } x \in X \tag{9.2.7}$$

be a k-minorant of the function f.

Step 2. Find a global optimum of the problem

$$f_k(x) \longrightarrow \min \text{ subject to } x \in X. \tag{9.2.8}$$

Step 3. Let y^* be a solution of problem (9.2.8). Set $k := k+1, x^k = y^*$ and go to Step 1.

A convergence of this algorithm has been proved in [118] (see Theorem 9.1.1) under very weak assumptions, however the proof in [118] is not simple. We provide the statement from [118] for the case under consideration without any proof and then give a short proof of convergence under some additional assumptions.

THEOREM 9.1 (Pallaschke and Rolewicz [118]) *Assume that the set H consists of continuous functions and let f be a continuous H-convex function. Then each limit point of the sequence (x^k) produced by the generalized cutting plane algorithm is a global minimizer of the function f over the set X.*

We shall prove a convergence under some additional assumptions using a different approach than in [118]. In order to study the convergence, we introduce the following quantities (for the classical convex case they were considered in 1965 by A. M. Rubinov in [139]) :

$$\mu_k = f_k(x^k) \ \ (k = 0, 1, 2, \dots); \qquad \lambda_k = f_{k-1}(x^k) \ \ (k = 1, 2 \dots), \tag{9.2.9}$$

where f_k is defined by (9.2.7). Let us point out some properties of the functions f_k and the numbers μ_k and λ_k:

1) $f_o(x) \le f_1(x) \le \dots \le f_k(x) \dots \le f(x)$ for all $x \in X$.

2) $\mu_k = f(x^k) = h_k(x^k)$. Indeed,

$$f_k(x^k) \le f(x^k) = h_k(x^k) \le \max_{0 \le i \le k} h_i(x^k) = f_k(x^k).$$

Thus $\mu_k = f_k(x^k) = f(x^k) = h_k(x^k)$.

3) The inequality $\lambda_k \leq \lambda_{k+1}$ holds for all k. In fact,

$$\lambda_k = f_{k-1}(x^k) = \min_{x\in X} f_{k-1}(x) = \min_{x\in X} \max_{0\leq i\leq k-1} h_i(x)$$
$$\leq \min_{x\in X} \max_{0\leq i\leq k} h_i(x) = \min_{x\in X} f_k(x) = \lambda_{k+1}.$$

4) Monotonicity of the sequence $\{\lambda_k\}$ implies the existence of the limit $\lim_{k\to+\infty} \lambda_k$.

5) $\lambda_k \leq \min_{x\in X} f(x) \leq \mu_k$ for all $k = 1, 2, \ldots$. Indeed, for a solution u of the problem (9.2.6) we have

$$\min_{x\in X} f(x) = f(u) \geq f_{k-1}(u) \geq \min_{x\in X} f_{k-1}(x) = \lambda_k.$$

On the other hand, $\mu_k = f(x^k) \geq \min_{x\in X} f(x)$.

It follows immediately from these properties that if for some k we have $x^{k+1} = x^k$, then x^k is a solution of problem (9.2.6). In this case, we can take $h_{k+1} = h_k$. So

$$\begin{aligned} \lambda_{k+1} &= f_k(x^{k+1}) = \max_{0\leq i\leq k} h_i(x^{k+1}) = \max_{0\leq i\leq k+1} h_i(x^{k+1}) = f_{k+1}(x^{k+1}) \\ &= \mu_{k+1}. \end{aligned}$$

It follows from 5) above that $f(x^k) = \min_{x\in X} f(x)$.

REMARK 9.1 Let $\nu_k = \min_{i\leq k} \mu_i$. The sequence ν_k is decreasing and

$$\nu_k = \min_{i\leq k} \mu_i \geq \max_{i\leq k} \lambda_i = \lambda_k.$$

Thus the algorithm produces the decreasing sequence (ν_k) and the increasing sequence (λ_k) such that

$$\nu_k \geq \min_{x\in X} f(x) \geq \lambda_k \quad \text{and} \quad \lim_{k\to\infty} \nu_k = \min_{x\in X} f(x) = \lim_{k\to\infty} \lambda_k.$$

REMARK 9.2 Let f_* be the value of the global minimum of a function f over a set X. The precision δ_r of the current point x^j is defined as follows:

$$\delta_r(x^j) = \min\left(f(x^j) - f_*, \ \frac{f(x^j) - f_*}{f_*} \right).$$

Very often the quantity f_* is unknown, in such a case we can use the lower estimate λ_k and consider the following number

$$\delta(x^j) = \min\left(f(x^j) - \lambda_j, \ \frac{f(x^j) - \lambda_j}{\lambda_j} \right)$$

as an estimate of the precision. Note that $\delta(x^j) > \delta_r(x^j)$. Numerical experiments show that $\delta_r(x^j)$ is substantially less than $\delta(x^j)$ in many instances. Thus very often, we have a substantially more precise solution than is indicated by the estimate δ of the precision.

We shall describe two sets of conditions providing the convergence of this algorithm. We assume that a set of elementary functions H coincides with the set H_L of abstract affine functions with respect to a set of abstract linear functions L. Then (see Proposition 7.1) $h \in \partial^*_{H_L} f(x^0)$ if and only if $h(x) = l(x) - (l(x^0) - f(x^0))$ with $l \in \partial_L f(x^0)$. Hence functions h_k from the generalized cutting plane algorithm have the following form:

$$h_k(x) = l^k(x) - (l^k(x^k) - f(x^k). \tag{9.2.10}$$

PROPOSITION 9.2 *Assume that the set L consists of concave functions defined on an open convex set, which contains the compact set X. Let the sequence $\{x^k\}$, produced by the algorithm, be infinite. Assume that the directional derivatives $u \mapsto h'_k(x,u)$ of the concave functions h_k defined by (9.2.10) at points $x \in X$ are uniformly bounded on the set X:*

$$\|h'_k(x,\cdot)\| = \max_{\|u\|\le 1} |h'_k(x,u)| \le r < +\infty \quad \textit{for all} \quad x \in X, k = 0,1,\ldots. \tag{9.2.11}$$

Then each limit point x of the sequence (x^k) is a solution of the problem (9.2.6).

Proof: Let $x = \lim_{j\to\infty} x^{k_j}$. For each j, consider the function h_i with $i \le k_{j-1}$ and find a supergradient a_i of the concave function h_i at the point x^{k_j}. Let $b_i(x) = h_i(x^{k_j}) + [a_i, x - x^{k_j}]$. Then $b_i(x) \ge h_i(x)$ for all $x \in X$ and $b_i(x^{k_j}) = h_i(x^{k_j})$. We have

$$\begin{aligned}
\mu_{k_{j-1}} &= f_{k_{j-1}}(x^{k_{j-1}}) = \max_{0\le i\le k_{j-1}} h_i(x^{k_{j-1}}) \le \max_{0\le i\le k_{j-1}} b_i(x^{k_{j-1}}) \\
&= \max_{0\le i\le k_{j-1}} (h_i(x^{k_j}) + [a_i, x^{k_{j-1}} - x^{k_j}]) \\
&\le \max_{0\le i\le k_j-1} h_i(x^{k_j}) + \max_{0\le i\le k_{j-1}} \|a_i\|\|x^{k_{j-1}} - x^{k_j}\|.
\end{aligned}$$

Let $\bar\partial h_i(x^{k_j})$ be the superdifferential of the concave function h_i at the point x^{k_j}. Since $a_i \in \bar\partial h_i(x^{k_j})$, we have

$$h'_i(x^{k_j}, u) = \min\{[l,u] : l \in \bar\partial h_i(x^{k_j})\} \le [a_i, u] \text{ for all } u \in \mathbb{R}^n.$$

Hence

$$-h'_i(x^{k_j}, -u) \ge -[a_i, -u] = [a_i, u].$$

It follows from (9.2.11) that

$$\|a_i\| = \max_{\|u\|\le 1} [a_i, u] \le \max_{\|u\|\le 1} -h_i'(x^{k_j}, -u) \le \max_{\|u\|\le 1} |h_i'(x^{k_j}, u| \le r.$$

Let $\lambda = \lim_{k\to\infty} \lambda_k$. Since $\|x^{k_j-1} - x^{k_j}\| \to 0$, we have $\limsup \mu_{k_j} \le \lambda$. On the other hand, $\mu_{k_j} \ge \lambda_{k_j-1}$ implies $\liminf \mu_{k_j} \ge \lambda$. Thus $\mu_{k_j} = f(x^{k_j}) \to \lambda$. Clearly, $\lambda = \min\{f(x') : x' \in X\}$. △

We now prove one more convergence result.

PROPOSITION 9.3 *Assume that*

1) *the set L consists of continuous functions;*

2) *the set $M(X) = \bigcup_{x\in X} \partial^*_{H_L} f(x)$ is uniformly sequentially compact (for each sequence $h_k \in M(X)$ there exists a uniformly convergent subsequence h_{k_i});*

3) *the subdifferential mapping $x \to \partial_L f(x)$ is closed.*

If the sequence $\{x^k\}$ produced by the algorithm is infinite, then each limit point of this sequence is a solution of (9.2.6).

Proof: Let $h_k \in \partial^*_{H_L}(x^k)$ be a sequence of L-affine functions used for the construction of the sequence x^k. Let x be a limit point of this sequence. It follows from conditions 2) and 3) that there exists a subsequence k_j such that $x^{k_j} \to x$ and h_{k_j} converges uniformly to a function $h \in M(x)$. Let ε be an arbitrary positive number. Since h is a continuous function, we have, for sufficiently large j:

$$h(x) \le h(x^{k_j}) + \frac{\varepsilon}{2} \le h_{k_{j-1}}(x^{k_j}) + \varepsilon \le \max_{0\le i\le k_j-1} h_i(x^{k_j}) + \varepsilon = \lambda_{k_j} + \varepsilon, \tag{9.2.12}$$

where (λ_k) is defined by (9.2.9). It follows from (9.2.12) and the properties of the sequence (λ_k) that $h(x) \le \min_{x'\in X} f(x')$. Since $x \in X$ it follows that $h(x) \ge \min_{x'\in X} f(x')$. △

REMARK 9.3 1) The subdifferential mapping can be very large in some cases. If this happens, we can only choose a special subset $m(x)$ of the set $\partial^*_{H_L} f(x)$ for the construction of the sequence (x^k), however the closedness of the mapping m is required.

2) We can consider our approach within the framework of successive underestimation methods, studied in [81]. This method is related to outer approximation techniques. Proposition 9.3 is similar to, although independent of, the theorem of convergence of outer approximation algorithms considered in [82].

There are two major difficulties with the numerical implementation of the generalized cutting plane algorithm. The first is the calculation of an element l^k from $\partial_L f(x^k)$. In general, it is very difficult to numerically find an element of the L−subdifferential, however it is possible for many important particular cases. The second difficulty relates to the subprolem (auxiliary problem) at Step 2. It is straightforward in the case of affine or convex functions h_i, but if the functions h_i are minima of affine functions or quadratic non-convex functions, the subproblem is essentially of a combinatorial nature and is difficult to solve.

In order to accelerate the convergence of the generalized cutting plane method, it can be productive to use a hybridization, that is, to use a local search in order to obtain a local minimum and then to try to find another local minimum with a smaller value of the objective function by the generalized cutting plane techniques. Also, if the current iteration is close enough to the global solution, the local search, in most cases, will find this global optimum within a good precision. Thus the purpose of the generalized cutting plane method would be to find a point which is close enough to a global solution.

Generalized cutting plane method with a local search

Step 0. Set $k := 0; x^0 \in X$ and let $S = \emptyset$.

Step 1. Calculate $h_k \in \partial_H^* f(x^k)$. Set $f_k(x) = \max_{i=0,\dots,k} h_i(x)$.

Step 2. Solve the problem

$$f_k(x) \longrightarrow \min \text{ subject to } x \in X.$$

Let y^* be its solution.

Step 3. Find a local minimum $\overline{y}$ by any method, starting from y^*.

Step 4. If $\overline{y} \notin S$, then $k := k+1$, $x^k := y$, $S := S \cup \{\overline{y}\}$ and go to Step 1.

Step 5. Otherwise, set $k := k+1$, $x^k := y^*$, $S := S$ and go to Step 1.

PROPOSITION 9.4 *For $\varepsilon > 0$, let $f_* = \min_{x \in X} f(x)$ and*

$$L_f^*(\varepsilon) = \{x \in X : f(x) \leq f_* + \varepsilon\}.$$

Suppose that the number of local minima is finite and that there exists $\varepsilon > 0$ such that the local method converges to a global optimum provided *the initial point is in $L_f^*(\varepsilon)$. Then a global minimum is found in a finite number of iterations.*

Proof: Let N be the number of local minima. Suppose that we know r $(0 \le r < N)$ local minima but not the global one. Then after a finite number of iterations, we either find a new local minimum, or we perform the iterations of the generalized cutting plane method with additional functions h_i. In the second case it follows from Theorem 9.1 that an ε-optimal point will be found in a finite number of iterations for any $\varepsilon > 0$. The number of local minima is finite, so when ε becomes small enough, we arrive at a global optimum. $\triangle$

9.2.4. Branch-and-bound methods

One of the main approaches to global optimization is based on branch-and-bound methods. The feasible set is being partitioned and then the upper and lower bounds are being computed for each element of the partition. After the comparison of the bounds and the values of the objective function, the elements of partition, which cannot contain an optimal point, are deleted. Convergence to a global optimum can be proved under very weak assumptions.

In this subsection we consider branch-and-bound algorithms for optimizing abstract convex functions.

The most straightforward form of such methods is based on a partition of the feasible set. For each element of the partition, an element of the abstract subdifferential is computed at some point, which gives a lower estimate of the optimal value of the objective function.

Let H be a set of continuous elementary functions defined on a set $X \subset \mathbb{R}^n$. Consider the following problem

$$f(x) \longrightarrow \min \text{ subject to } x \in X, \tag{9.2.13}$$

where f is a continuous H-convex function defined on X.

Best estimate algorithm

Step 0. Take an initial partition P_0 of the feasible set X. Set $k := 0$.

Step 1. Let $P_k = \{S_1^k, S_2^k, \dots, S_{N_k}^k\}$, where $X \supset \bigcup_{i=1}^{N_k} S_i^k$. For each set S_i^k choose a finite set of points $X_i^k \subset S_i^k$.

Step 2. For each set X_i^k compute the elements $h_i^k[y] \in \partial_H^* f(y)$, where $y \in X_i^k$. Let

$$\hat{h}_i^k(x) = \max_{y \in X_i^k} h_i^k[y](x) \qquad (x \in X).$$

Step 3. Minimize each function $\hat{h}_i^k$ over the set S_i^k. Let

$$\underline{h}_i^k = \min_{x \in S_i^k} \hat{h}_i^k(x).$$

Step 4. Find an index j such that $\underline{h}_j^k = \min_i \underline{h}_i^k$.

Step 5. Partition the set S_j^k. Let

$$S_j^k = \hat{S}^k \cup \tilde{S}^k.$$

Then

$$P_{k+1} := (P_k \backslash \{S_j^k\}) \cup \{\hat{S}^k\} \cup \{\tilde{S}^k\}.$$

Step 6. For each set X_i^k and all $y \in X_i^k$ compute $f(y)$. Let $\hat{f} = \min_y f(y)$.

Step 7. Delete all sets S_i^k, for which $\hat{h}_i^k \geq \hat{f}$ from the partition.

Step 8. Set $k := k + 1$ and go to Step 1.

The efficiency of this algorithm depends on the choice of sets X_i^k, elements $h_i^k[y]$ and partition in Step 5. In order to determine the sets X_i^k, we suggest to perform several iterations of the generalized cutting plane method (a starting point is being chosen arbitrarily from S_i^k). This gives lower estimates, which are as close as necessary to the optimal value. Instead of partitioning the set S_j^k into two parts, it is possible to divide it into an arbitrary number of parts. The method does not change essentially.

There are at least two possibilities to partition the set S_j^k. Assume that the feasible set X is convex and all sets S_j^k are convex as well.

a) Consider the problem:

$$\begin{aligned} & \lambda \longrightarrow \max \\ \text{subject to} \quad & x \pm \lambda e_i \in S_j^k, \quad i = 1, \dots, n, \\ & x \in S_j^k. \end{aligned}$$

Let $(\overline{x}, \overline{\lambda})$ be a solution of this problem. We consider the hyperplane $\{x : [a, x] = [a, \overline{x}]\}$, where $a \in \mathbb{R}^n_+, \|a\| = 1$. Then the partition is defined as follows:

$$\hat{S}^k = S_j^k \cap \{x : [a, x] \leq [a, \overline{x}]\},$$

$$\tilde{S}^k = S_j^k \cap \{x : [a, x] \geq [a, \overline{x}]\}.$$

b) Let S_j^k be a polygon, that is $S_j^k := \mathrm{co}\,\{v_1, \dots, v_m\}$. Find indices r and s such that

$$\|v_r - v_s\| = \max_{i,j} \|v_j - v_i\|.$$

Set $w = (v_r + v_s)/2$ and choose a vector b such that $\|b\| = 1$ and $[b, v_r - v_s] = 0$. Let

$$\hat{S}^k = S_j^k \cap \{x : [b, x] \le [b, w]\},$$

$$\tilde{S}^k = S_j^k \cap \{x : [b, x] \ge [b, w]\}.$$

We now discuss conditions, which provide the convergence of this method. Assume that the following holds:

At each iteration k the inequalities

$$\mathrm{diam}\,(\hat{S}^k) \le \gamma_1 \mathrm{diam}\,(S_j^k), \tag{9.2.14}$$

$$\mathrm{diam}\,(\tilde{S}^k) \le \gamma_2 \mathrm{diam}\,(S_j^k) \tag{9.2.15}$$

hold for $0 < \gamma_1, \gamma_2 < 1$.

Then we have that $\mathrm{diam}\,(S_j^k) \le \varepsilon$ for some k. The minimum value of the maximum of abstract subgradients over S_j^k does not exceed the value of problem (9.2.13). Hence, due to the continuity and the abstract convexity of f, we can find an approximate global solution within any required precision. Indeed, for some point $y \in S_j^k$ we have

$$f(y) = h_j^k(y) \le h(\tilde{x}) + \delta \le f^* + \delta,$$

where $\tilde{x}$ is the point from S_j^k, at which $\hat{h}_j^k$ attains its minimum and $\delta > 0$ is small (δ can be made arbitrarily small by decreasing ε.) Thus y is an approximate global solution.

It is possible to keep the diameters of the sets S_i^k constant and to perform iterations of generalized cutting plane method in each set. Then convergence of the algorithm to a global optimum follows directly from the results on the convergence of the generalized cutting plane method.

In order to improve the efficiency (the number of sets in P_k can become very high), it is beneficial to use hybrid methods, combining branching with local optimization. An arbitrary local method can be used which guarantees convergence to a stationary point, starting from a point, which is sufficiently close to this stationary point (see [5] for details).

9.2.5. Tabu search

In order to find a global optimum of a function with multiple extrema, we should be able to avoid local extrema. One efficient way to do this is the method of tabu search (see, for example, [66]). This method is typically considered as a meta-heuristics applied to another heuristics. However, it is possible to give a proof of its convergence under certain assumptions (see [4]).

Let H be a set of elementary functions defined on a set X. Consider the problem

$$f(x) \longrightarrow \min \text{ subject to } x \in X, \tag{9.2.16}$$

where f is an H-convex function. Denote by X^* the set of global minimizers problem (9.2.16), that is,

$$X^* = \{x^* \in X : f(x^*) \leq f(x) \text{ for all } x \in X\}.$$

We consider the optimization process as a sequence of iterations. Assume that we have at each iteration the current partition of the feasible set:

$$P_k = \{V_1^k, V_2^k, \ldots, V_{N_k}^k\},$$

where $X \supset \bigcup_{i=1}^{N_k} V_i^k$.

Assume that each set V_i^k is not partitioned any further, whenever its diameter becomes smaller than $\varepsilon > 0$ (this means stopping with ε-optimality by norm). Assume that we use special rules for partitions, which allow us to have only a finite number of partitions for each given set. The total number of partitions (for each initial partition P_0) is finite in this case.

Assume that the kth iteration of a numerical method is characterized by a partition $P_k = \{V_1^k, V_2^k, \ldots, V_{N_k}^k\}$. Let x_{ij}^k be a finite subset of a set V_i^k, which carries out a "good" (in a certain sense) approximation of the set V_i^k . The behaviour of the objective function f on the set V_i^k can be described by its values at points x_{ij}^k and abstract subgradients $h_{ij}^k \in \partial_H^* f(x_{ij}^k)$. Let f_r^k be the minimal known value of the objective function, which was calculated for iterations $1, \ldots, k$. We need to compare sets V_i^k in the partition P_k in order to find the "best" (in a certain sense) from these sets. For this purpose, we introduce the notion of the quality Q_i^k of the set V_i^k in the partition P_k. The quality Q_i^k is a number, which shows how promising the set V_i^k is for finding a global optimum. For each set V_i^k, we have a lower bound of the minimum value of the objective function

$$\underline{h}(V_i^k) = \min_{y \in V_i^k} \max_j h_{ij}^k(y).$$

Then the quality is determined by the following values:

1) $A_i^k = f_r^k - \underline{h}(V_i^k)$.

2) $B_i^k = \sum_{j,r=1}^{|V_i^k|} |f(x_{ij}^k) - h_{ir}^k(x_{ij}^k)|$. This is the precision of the approximation of the objective function by abstract subgradients.

3) $\mathcal{V}_i^k = \int_{V_i^k} dx$. This is the n–dimensional volume of the set V_i^k.

4) S_i^k is the number of stationary (in a certain sense) points in V_i^k.

5) $\mathcal{M}_i^k = \min_{y \in V_i^k} f(y)$ is the best value of the objective function on V_i^k.

It seems reasonable to increase the quality of each set, when the first four criteria increase, and the last criterion decreases. Thus we have

$$Q_i^k = \Phi_1(A_i^k) + \Phi_2(B_i^k) + \Phi_3(\mathcal{V}_i^k) + \Phi_4(S_i^k) - \Phi_5(\mathcal{M}_i^k),$$

where Φ_s are increasing functions for $s = 1, \dots, 5$. If we put a threshold on the quality value, say, choose an element with $Q_j^k > \bar{Q}$, where $\bar{Q}$ is a given lower bound, then all other possibilities are prohibited (tabu), and thus we have a tabu search method. There is still a freedom of choice, which leaves the space for different heuristics.

Let us consider a conceptual scheme of a method based on tabu search for minimizing an H-convex function.

The algorithm

Step 0. Set $k := 0$. Take the initial partition $P_k = \{V_1^k, V_2^k, \dots, V_{N_k}^k\}$.

Step 1. For each set V_i^k, compute one or more points $x_{ij}^k \in V_i^k$. For each x_{ij}^k compute $f(x_{ij}^k)$ and $h_{ij}^k \in \partial_H^* f(x_{ij}^k)$.

Step 2. Compute the quality Q_i^k of each set V_i^k.

Step 3. Delete all sets V_i^k which cannot contain a global optimum.

Step 4. Choose an element $V^* \subset P_k$ with the quality Q_j^k which is sufficiently high.

Step 5. Let $V^* = V_1^* \cup V_2^*, P_{k+1} := P_k \backslash \{V^*\} \cup \{V_1^*, V_2^*\}$. Set $k := k+1$ and go to Step 1.

Note that all sets with $A_i^k < 0$ can be deleted as they cannot contain a global optimum (see Step 3). If we can compute a function $\overline{h}_i^k \in \partial_H^* f(z)$

of f, where z is an arbitrary point from V_i^k, then we can check the following inequality for each m:

$$\min_{y \in V_m^k} \max_j h_{mj}^k(y) \geq \min_{y \in V_i^k} \overline{h}_i^k(y).$$

If it holds, the set V_m^k can be deleted as it does not contain global optima.

The convergence of a conceptual method of such a kind has been studied in [4]. There are some different versions of the tabu search with abstract convexity, for example, a scheme with memory (see [4] for details).

9.2.6. External centres method

In this subsection we present a conceptual scheme of a version based on abstract convexity of the external centres method. This version was proposed by Yu. Evtushenko and V. Zhadan (private communication). The external centres method uses a lower estimate of the optimal value of the problem under consideration, which is updated during the iterative process. Some versions of the external centres method have been discussed in [117, 53, 68, 97, 200].

Let H be a set of elementary functions defined on a set $X \subset \mathbb{R}^n$ and let $X_1 \subset X$ be a compact set. Consider the following optimization problem:

$$f(x) \longrightarrow \min \text{ subject to } x \in X_1, \tag{9.2.17}$$

where f is an H-convex function defined on X. Let f_* be the optimal value of (9.2.17). Assume that a lower estimate η of f_* is known. Let P be a nonnegative function defined on X with the property

$$x \in X_1 \iff P(x) = 0.$$

Compose the auxiliary function M for problem (9.2.17):

$$M(x, \eta) = (f(x) - \eta)_+^2 + P(x), \tag{9.2.18}$$

where $a_+ := \max(a, 0)$ is a positive part of a number a. Auxiliary functions of such a kind have been studied in detail in [197].

Let $\mu(\eta) = \min_{x \in X} M(x, \eta)$. It can be proved under weak assumptions (see [117]) that

$$\eta \leq f_* \implies \eta + \sqrt{\mu(\eta)} \leq f_*. \tag{9.2.19}$$

Assume that (9.2.19) holds.

External centres algorithm

Step 0. Let $J_0 = \{-r, -r+1, \dots, -1\}$. Take a set $G_0 \subset X$ of initial points $x^{-r}, x^{-r+1}, \dots, x^{-1}$; choose $\eta_0 \le f_*$. Set $k := 0$.

Step 1. Find a function $h_i \in \partial_H^* M_{\eta_i}(x^i)$, $i \in J_k$, where $M_\eta(x) = M(x, \eta)$.

Step 2. Compose the function

$$h^k(x) = \max_{i \in J_k} h_i(x)$$

and solve the problem

$$h^k(x) \longrightarrow \min \text{ subject to } x \in X. \qquad (9.2.20)$$

Step 3. Let x^k be a solution of problem (9.2.20). If $h^k(x^k) = 0$, then stop, otherwise update the lower estimate η_k by

$$\eta_{k+1} = \eta_k + \sqrt{h_k(x^k)}. \qquad (9.2.21)$$

Let $J_{k+1} = J_k \cup \{k\}$, $G_{k+1} = G_k \cup \{x^k\}$.

Step 4. Set $k := k+1$ and go to Step 1.

In order to implement this conceptual scheme we need to solve problem (9.2.20). It can be done for some broad classes of problems using the results of the next section.

9.2.7. Lipschitz programming via abstract convexity

The problem of Lipschitz programming is very important for a number of applications. In many problems which arise in various areas of human activity, the objective function and/or constraints are Lipschitz. The problem of Lipschitz programming has been studied thoroughly in the book by J. Pinter ([130]) who presents main known approaches to its solution. The following property of a Lipschitz function is crucial for global optimization: if we know the value of a Lipschitz constant of a function and its value at a point, then we can find a lower and an upper bound of this function at any point.

We shall look at the Lipschitz optimization problem from the point of view of abstract convexity.

There exist several known methods of global optimization of Lipschitz functions which are based on abstract convexity (see [112, 195]). In these methods, the graph of a Lipschitz function is supported by some cones. The algorithms have shown high efficiency for small dimensions and are related to the cutting angle method (for this version of the generalized cutting plane method, see Section 9.3 and Section 9.4), as they use a similar scheme with a special kind of min-type elementary functions.

Let L be a set of elementary functions l defined on $\mathbb{R}^n$ by

$$l(x) = a\|x - y\| \qquad x \in \mathbb{R}^n.$$

Here $y \in \mathbb{R}^n$ and $a \leq 0$. We consider L as a set of abstract linear functions. Then the set H_L of abstract affine functions consists of functions h of the form $h(x) = a\|x - y\| - c$ with $y \in \mathbb{R}^n$, $a \leq 0$ and $c \in \mathbb{R}$. As it was mentioned in Subsection 7.2.3 (see Proposition 7.2), for an arbitrary Lipshitz function $f : X \to \mathbb{R}$ with $X \subset \mathbb{R}^n$ the L-subdifferential $\partial_L f(x)$ at $x \in X$ is not empty, hence the set $\partial_{H_L} f(x)$ is also nonempty. Thus the general cutting plane methods with the set of elementary functions H_L can be applied for global minimization of Lipschitz functions. This approch was suggested in [52, 27]. H. Tuy [186] discussed possible applications of DC programming to generalized cutting plane methods in Lipschitz programming.

The first method which actually uses abstract convexity for multiextremal optimization is the Pijavski-Shubert algorithm for optimization on the real line.([129, 180]). This algorithm approximates the hypograph of the objective function by angles and is essentially a version of the cutting angle method with min-type elementary functions (see Subsection 9.3). The Pijavski-Shubert algorithm has been extended to an n-dimensional optimization with $n > 1$ in two different ways. In the approach by R. Mladineo ([112]) the graph of the objective function is supported by a union of cones, whose intersections with horizontal hyperplanes are spheres . If f is a Lipschitz function, then the set $\partial^*_{H_L} f(y)$ contains the function $g(x) = f(y) - L\|x - y\|$. Therefore, if we use the Euclidean norm and minimize at each iteration k the function

$$f_k(x) = \max_{0 \leq i \leq k} (f(x^i) - L\|x - x^i\|) \tag{9.2.22}$$

over the feasible set X, we obtain a version of the generalized cutting plane method for Lipschitz functions. The methods for solving the subproblems of minimizing the functions (9.2.22) are proposed in [112] and they have shown practical applicability in small dimensions. Obviously, it is possible to use the norm $\|\cdot\|_\infty$ or the norm $\|\cdot\|_1$ instead, thus f_i are becoming min-type functions.

Another approach, developed by G. Wood (see [194, 195] and also [201]), uses polyhedral cones for the support of the graph of the objective function. The intersection of each cone with a horizontal plane in this case is a simplex. The author is reasoning in terms of such simplices. Those of them which cannot contain the global optimum are deleted. For other simplices a procedure of bisection is used. Each simplex corresponds to a min-type function which is an abstract subgradient of the objective function, so the method is based on abstract convexity as well. Numerical experiments have shown the efficiency of the approach for a number of multi-extremal problems.

9.3. CUTTING ANGLE METHOD

9.3.1. Overview

The conceptual schemes from Section 9.2 can be implemented for some special classes of elementary functions. In this and the next sections, we shall study a version of the generalized cutting plane method for various sets of min-type elementary functions. Namely, we consider the following three sets of functions defined on the cone of $\mathbb{R}^n_+$:

- $L = \{l : l(x) = \langle l, x\rangle \ (x \in \mathbb{R}^n_+), l \in \mathbb{R}^n_+\}$; (f is L-convex if and only if f is IPH);
- $H_L = \{h : h(x) = \langle l, x\rangle - c, \ (x \in \mathbb{R}^n_+) \ l \in \mathbb{R}^n_+, c \in \mathbb{R}\}$; ($f$ is H_L-convex if and only if f is ICAR);
- $H = \{h : h(x) = \min(\langle l, x\rangle, c) \ (x \in \mathbb{R}^n_+), l \in \mathbb{R}^n_+, c > 0\}$; ($f$ is H-convex if and only if f is ICR),

where $\langle l, x\rangle = \min_{i\in I_+(l)} l_i x_i$. We shall also study the minimization of a Lipschitz function which can be reduced to the minimization of ICAR (and even IPH) function.

Since the level sets $\{x \in \mathbb{R}^n_+ : \langle l, x\rangle \geq 0\}$ of the function $l \mapsto \langle l, x\rangle$ are *right angles* (more precisely, right cones, that is shifts of the cone $\mathbb{R}^n_+$), the generalized cutting plane method is reduced, in this case, to the *cutting angle method.*

The cutting angle method is suitable for the search of an *approximate* solution of many problems of global optimization. Very often, it is enough to have such a solution since it can be made more precise by an appropriate local method.

Since the abstract subdifferentials of abstract convex functions in the cases under consideration are known (see Chapter 3 for details) we concentrate here on the solution of the subproblem (auxiliary problem) of the cutting angle method. We propose and discuss various approaches for solving the subproblem. In this section, we examine separately the cutting angle method for three classes of functions (ICAR, ICR and Lipschitz), describe methods for solving the subproblem and report results of numerical experiments. We also consider a branch-and-bound method for Lipschitz functions based on applications of the cutting angle approach. The main results from this section can be found in [7, 6, 147, 148].

9.3.2. Cutting angle method for ICAR functions

In this subsection we apply the scheme of generalized cutting plane method for the minimization of increasing convex-along-rays (ICAR) functions.

Let L be the set of all min-type functions $l(x) = \langle l, x\rangle$ with $l \in \mathbb{R}^n_+$, where

$$\langle l, x\rangle = \min_{i\in I_+(l)} l_i x_i \text{ with } I_+(l) = \{i : l_i > 0\}. \tag{9.3.1}$$

Let H_L be the set of all L-affine functions. Then a function $f : \mathbb{R}^n_+ \to \mathbb{R}$ is H_L-convex if and only if f is ICAR. (See Section 3.3 for details). Let f be an ICAR function. It follows from Theorem 3.4 that the L-subdifferential $\partial_L f(x)$ is not empty and contains the function l^x, where

$$l^x(y) = \langle f^{\nabla}(x), y\rangle \text{ for all } y \in \mathbb{R}^n_+ \text{ with } f^{\nabla}(x) = \frac{f'(x,x)}{x}. \tag{9.3.2}$$

In other words

$$l^x(y) = f'(x,x) \min_{i\in I_+(x)} \frac{y_i}{x_i}.$$

It follows from Proposition 7.1 that the set $\partial^*_{H_L} f(x)$ is not empty for all $x \in \mathbb{R}^n_+$ and this set contains the function h_x where

$$\begin{aligned} h_x(y) &= \langle f^{\nabla}(x), y\rangle - (f'(x,x) - f(x)) \\ &= \min_{i\in I_+(x)} \frac{f'(x,x)y_i}{x_i} - (f'(x,x) - f(x)). \end{aligned} \tag{9.3.3}$$

Note that $f'(x,x) - f(x)$ coincides with the value of the excess function E_x (see Subsection 8.3.7).

Let $x \in \mathbb{R}^n_+$. We can consider the sets $\{y \in \mathbb{R}^n_+ : h_x(y) = c\}$ as analogues of "right angles". Hence applying the functions h_x for cuts, we use the cuts by right angles instead of cuts by hyperplanes in the classical version of the cutting plane method. Thus the corresponding version of the generalized cutting plane method can be called the *cutting angle method.*

By using h_x as an element of the set $\partial^*_{H_L} f(x)$, we can represent the cutting angle method in the following form.

The algorithm

Step 0. Set $k := 0$. Choose an arbitrary point $x^0 \in X$.

Step 1. Calculate a vector l^k with the coordinates l_i^k :

$$l_i^k = \frac{f'(x^k, x^k)}{x_i^k}, \quad \text{if} \quad x_i^k \neq 0 \text{ and } l_i^k = 0 \quad \text{if} \quad x_i^k = 0,$$

where x_i^k is the ith coordinate of the vector x^k .

Step 2. Define the function h_k by

$$h_k(x) = \min_{i \in I_+(x^k)} l_i^k x_i - ((f'(x^k, x^k) - f(x^k)),$$

where l_i^k is the ith coordinate of the vector l^k .

Step 3. Solve the problem:

$$\max_{0 \leq i \leq k} h_i(x) \longrightarrow \min \quad \text{subject to} \quad x \in X. \tag{9.3.4}$$

Step 4. Let y^* be a solution of the problem (9.3.4) . Set $k := k+1, x^k := y^*$ and go to Step 1.

REMARK 9.4 We can construct a version of the cutting angle method in order to solve the following problem:

$$f(x) \longrightarrow \max \text{ subject to } x \in X, \tag{9.3.5}$$

where X is a compact set and f is a function represented as the infimum of a family of L^+-affine functions. Here L^+ is the set of all max-type functions of the form $l : l(x) = \max_{i=1,2,\dots,n} l_i x_i$ with nonnegative vectors l.

A function f of this type is increasing and concave-along-rays. The subproblem now has the form:

$$\min_{i=0,\dots,k} (\max_{j=1,\dots,n} l_j^i x_j + b_i) \longrightarrow \max \text{ subject to } x \in X.$$

We can apply the approach described in the next section to study this problem as well.

9.3.3. The subproblem

The main difficulty in implementation of the cutting angle method for ICAR functions is to solve the subproblem in Step 3. This subproblem can be represented as

$$\max_{i=0,\dots,k}\{\langle l^i,x\rangle + b_i\} \longrightarrow \min \text{ subject to } x\in X. \tag{9.3.6}$$

Various approaches can be applied to find the solution of (9.3.6). Here we shall discuss two of these approaches. One of them which we shall discuss in detail is based on the scheme of dynamic programming and the other uses mixed integer linear programming with 0-1 variables.

Let us consider the first approach. We can rewrite problem (9.3.6) in the following equivalent form:

$$t \longrightarrow \min \text{ subject to } \langle l^i,x\rangle + b_i \le t, \quad i=0,\dots,k, \quad x\in X. \tag{9.3.7}$$

Assume for the sake of simplicity that $I_+(l^i) = I := \{1,\dots,n\}$ for all i. If we denote the coordinates of each vector l^i by l_{ij}, (9.3.7) can be represented as

$$t \longrightarrow \min \text{ subject to } \min_{j=1,\dots,n} l_{ij}x_j + b_i \le t, \quad i=0,\dots,k, \quad x\in X. \tag{9.3.8}$$

If we can find a feasible point of system (9.3.7), then an optimal t can be found by a simple dichotomy. Thus it is enough to be able to solve (9.3.8) or determine that there are no solutions at all.

We consider two kinds of feasible subsets of $\mathbb{R}^n_+$.

1) X is a compact convex set;

2) X is the intersection of a conormal set with a rectangle $\{x\in\mathbb{R}^n_+ : x\le c\}$. Recall that the set Y is called conormal if $(y\in Y, y'\ge y) \Longrightarrow y'\in Y$.

Problem (9.3.8) can be represented in the following form:

$$t \longrightarrow \min \text{ subject to } \min_{j=1,\dots,n} l_{ij} \le t-b_i,\ i=0,\dots,k, \quad x\in X.$$

We propose to apply to this system a method similar to dynamic programming. Note that if X is a polygon, then we have a special problem of disjunctive programming for which there are general results and methods (see [13, 14]).

Suppose first that $k,n>1$. We denote by $P(k,n)$ the problem of finding a solution of (9.3.7) with n variables and k min-type functions.

Let us show how we can reduce $P(k,n)$ to a sequence of lower dimension problems.

Consider the values:

$$\theta_1 = \frac{t-b_1}{l_{11}};\ \theta_2 = \frac{t-b_2}{l_{21}};\ldots\theta_n = \frac{t-b_k}{l_{k1}}.$$

Let us sort them out in the increasing order (t is now a fixed number so we can do this):

$$\theta_{i_1} \le \theta_{i_2} \le \ldots \theta_{i_k}.$$

Assume that $\theta_{i_s} < x_1 \le \theta_{i_{s+1}}$. For homogeneity, let $\theta_{i_0} = 0$ and $\theta_{i_{k+1}} = c$, where c is an upper bound for the variables (if X is convex and bounded, we can find it numerically using convex programming). All the constraints with numbers $i_{s+1},...,i_k$ are satisfied and all other constraints are violated. Then we obtain a family of subproblems:

$$t \longrightarrow \min \text{ subject to}$$

$$\min\{l_{i_p 2}x_2;\ldots;l_{i_p n}x_n\} \le t - b_{i_p} \quad p = 1,\ldots,s,$$
$$\theta_{i_s} < x_1 \le \theta_{i_{s+1}},$$
$$x \in X.$$

Thus the number of min-type constraints does not increase and the number of variables in these constraints decreases by at least one in each case. Therefore, applying this procedure consecutively for $x_2, x_3, ..., x_n$ we finally arrive at one of the three situations: $P(1,q)$, $P(0,q)$ or $P(s,0)$. Let us consider all three cases ($1 \le q \le n$, $1 \le s \le k$).

a) $P(1,q)$: this is a problem of the form

$$t \longrightarrow \min \text{ subject to}$$

$$\min\{l_1x_1;...;l_qx_q\} \le t - b \qquad x \in \tilde{X},$$

where $\tilde{X}$ is obtained by adding to X some new upper and lower bounds for $n-q$ variables.

We can simply consider all problems:

$$t \longrightarrow \min \text{ subject to } \ l_jx_j \le t - b, \qquad x \in \tilde{X}$$

for $j = 1,\ldots,q$ which are equivalent to the problems

$$l_jx_j \longrightarrow \max \text{ subject to } x \in \tilde{X}.$$

b) $P(0, q)$: this is a problem of the form: to find a point x such that

$$a_i \leq x_i \leq b_i, \ i = 1, \ldots, n \qquad x \in X.$$

If X is convex, we can apply any known method for finding a feasible point of a convex set. If X is the intersection of a conormal set with a rectangle, we just take $x_i = b_i, \ i = 1, \ldots, n$. If the point $(b_1, \ldots, b_n)$ does not belong to X, then there are no solutions at all.

c) $P(s, 1)$: here we have the following:

$$t \longrightarrow \min \text{ subject to } l_{i_p s} x_s \leq t - b_{i_p}, \quad p = 1, \ldots, s, \qquad x \in \tilde{X}.$$

These constraints can be presented in the following form:

$$x_s \leq \min_{p=1,\ldots,s} \frac{t - b_{i_r}}{l_{i_p} s}, \qquad x \in \tilde{X}$$

and we should again seek a point in X with additional upper and lower bounds on the variables.

Assume that a feasible set X has the following form:

$$X = \{x \in \mathbb{R}^n_+ : \ g(x) \geq 0, \ x_i \leq c, \ i = 1, \ldots, n\},$$

where g is increasing. (If X is the intersection of a conormal set with a rectangle, then X can be represented in such a form with an appropriate function g.) In this case the subproblem can be reduced to problems with the constraints:

$$g(x) \geq 0, \quad 0 \leq \alpha_i \leq x_i \leq \beta_i \leq c, \quad i = 1, \ldots, n.$$

In order to solve these problems, we take $x_i = \beta_i$ for all $i = 1, \ldots, n$. If $g(\beta_1, \beta_2, ..., \beta_n) \geq 0$, the solution exists, otherwise it does not exist.

REMARK 9.5 If some l_{ij} are equal to zero, the corresponding θ is equal to $+\infty$, and we consider the constraint number i as violated. In the next step, the dimension of the subproblem decreases anyway and the procedure indicated above remains essentially the same.

We now consider briefly a different approach for finding a solution of the subproblem. We show that the subproblem can be reduced to a problem of mixed integer linear programming with 0-1 variables (see [13, 14] and the references therein). This is a well-known technique which is based on the introduction of a large positive parameter M. Then each constraint $\langle l^j, x\rangle \leq \tilde{t}$ can be represented as $n+1$ linear constraints of the form:

$$l_{jk}x_1 - My_{jk} \leq \tilde{t}, \qquad k = 1, \dots, n,$$
$$\sum_{i=1}^{n} y_{ji} \leq M - 1,$$

where $y_{ji} \in \{0,1\}$ for all i, j. This technique increases the dimension of the problem by n binary variables at each iteration. However, it allows us to apply existing software packages for mixed integer linear programming in order to solve the subproblem.

9.3.4. Numerical results - ICAR objective functions

In order to verify the practical efficiency of the cutting angle algorithm, a number of numerical experiments have been carried out. We consider the problem of the form:

$$f(x) \longrightarrow \min \text{ subject to } g(x) \geq 0,\ x_i \leq c_i,\ i = 1, \dots, n.$$

Most experiments involved increasing convex-along-rays objective functions belonging to one of the three classes:

1) $f(x) = [Qx, x]$, where Q is a matrix not necessarily positive semidefinite. Q was chosen as a nonnegative matrix, which guarantees that the corresponding quadratic function is increasing and convex-along-rays.

2) A Cobb-Douglas type function:

$$f(x) = \prod_{i=1}^{s} x^{\alpha_i}, \qquad \sum_{i=1}^{s} \alpha_i \geq 1.$$

(The problems of minimizing Cobb-Douglas type functions were studied in [179], where a different algorithm was proposed.)

3) The maximum of a finite number either quadratic non-convex functions or Cobb-Douglas type functions.

The constraint function g was taken either as a quadratic concave function with positive coefficients or as the maximum of linear functions or as the maximum of quadratic concave and linear functions. Thus the feasible set was taken as in the problem of reverse convex programming (see [81, 79, 130]). Various values of the upper bound c for the variables were taken. Usually it was preferable to take a smaller c. The initial point was chosen either randomly or as a local minimum of the objective function found by some local search method. The number of variables varied between two and twenty. Both the cutting angle method and the hybrid method were examined.

We used the technique for solving the subproblem, which was discussed in the previous section. The code has been written in TurboC for DOS/Windows and has also been implemented for the Unix operating system.

In some instances a local search can be used, in particular, in a neighbourhood of a global minimum. Such an improvement has been implemented as a hybrid method and proved to be much more efficient than the direct application of the general algorithm.

The local improvement has been implemented as Shor's subgradient method with the step chosen as a divergent series (see [178]). This method is preferable, since each iteration requires very little time and it also works with non-smooth functions. However, it is sensitive to the choice of the initial step and the performance significantly depended on it. In all cases the minimum was found within a precision $\varepsilon = 0.001$. If we started from a local minimum, the next iterate proved to be far from it and from the subsequent iterate the descent to a global minimum was done.

Let us consider some concrete examples. They are, in fact, multi-extremal problems (that is, problems with local minima, which are not global).

EXAMPLE 9.1 Consider the problem

$$x_1x_2x_3x_4x_5x_6x_7 + \sqrt{x_1x_2x_3x_4x_5x_6x_7} \longrightarrow \min$$

subject to

$$x_1^2 + 4x_2^2 + 2x_3^2 + 3x_4^2 + 2x_5^2 + 3x_6^2 + x_7^2 \geq 24$$

$$1 \leq x_i \leq 10, \quad i = 1, \dots, 10.$$

The optimal solution of this problem is the point $(1, \sqrt{3}, 1, 1, 1, 1, 1)$ The best solution, found by the hybrid algorithm, is:

$$x_1 = 1.000083,\ x_2 = 1.732257, x_3 = 1.000035,\ x_4 = 1.000007,$$

$$x_5 = 1.000042,\ x_6 = 1.000031,\ x_7 = 1.000018.$$

In order to find this point, the hybrid method required 21 iterations. The absolute precision by the maximum norm is 0.000206 and the relative precision by the objective function is 10^{-7}.

EXAMPLE 9.2 Consider the problem

$$x_1^2x_2^2 + 3x_1x_2 + x_1 + 5x_2 + 2 \longrightarrow \min$$

subject to

$$x_1^2 + x_2^2 \geq 1,$$

$$0 \leq x_1 \leq 1, \quad 0 \leq x_2 \leq 1.$$

The optimal solution of this problem is the point (1,0). The best solution, found by the hybrid algorithm, is

$$x_1 = 0.99981000, \; x_2 = 0.001000.$$

In order to find this point, the cutting angle method required 12 iterations.

EXAMPLE 9.3 Consider the problem

$$\sum_{i=1}^{19} x_i x_{i+1} \longrightarrow \min$$

subject to

$$\sum_{i=1}^{n} x_i = 1,$$

$$0 \leq x_i, \; i = 1, \ldots, 20.$$

The optimal solution of this problem is the point z such that $z_{20} = 1$ and $z_i = 0$ for $i \neq 20$. Cutting angle method has found this solution in exactly 20 iterations.

EXAMPLE 9.4 Consider the problem

$$x_1^2 + x_2^2 + x_3^2 + 2x_1^2x_2^2 + 2x_1^2x_3^2 \longrightarrow \min$$

subject to

$$x_1 + x_2 + x_3 \geq 1,$$

$$0 \leq x_i \leq 1, \; i = 1, 2, 3.$$

The best solution, found by the hybrid algorithm, is

$$x_1 = 0.26369, \; x_2 = 0.35694, \; x_3 = 0.37936.$$

The value of the objective function at this point is 0.37858. In order to find this point, the cutting angle method required 80 iterations.

EXAMPLE 9.5 Consider the problem

$$\max\{2x_1^2x_2; 5x_2x_3^2; 2x_1^2 + x_2^2 + 3x_3^2\} \longrightarrow \min$$

subject to

$$x_1 + x_2 + x_3 \geq 1,$$
$$0 \leq x_i \leq 1, \; i = 1, 2, 3.$$

The best solution, found by the hybrid algorithm, is

$$x_1 = 0.25253, \; x_2 = 0.57021, \; x_3 = 0.17725.$$

The value of the objective function is 0.5469449. In order to find this point, the hybrid method required 80 iterations.

The cutting angle method has shown the following behaviour. It works efficiently for non-smooth max-type functions and Cobb-Douglas type functions (especially in combination with local optimization). The reason for this is, may be, that min-type functions are good approximations to these classes of functions. For quadratic functions the performance has been satisfactory though the required precision was not obtained for several examples (more iterations were necessary) probably due to the fact that they were ill-conditioned. The worst case for the cutting angle method occurs with a linear objective function, since min-type functions cannot provide a good approximation in this case.

Finally, we can say that the hybrid method (cutting angle together with local search by Shor's method) proved to be very efficient for minimizing increasing convex-along-rays functions when the number of variables is relatively small. It is not sensitive to the constraints and gave similar results for both linear and non-linear constraints.

9.3.5. Cutting angle method for increasing co-radiant functions

The function $f : \mathbb{R}^n_+ \to \mathbb{R}$ is called increasing co-radiant (ICR) if $x \geq y$ implies $f(x) \geq f(y)$ and $f(\lambda x) \geq \lambda f(x)$ for all $x \in \mathbb{R}^n_+$ and $\lambda \in [0, 1]$. ICR functions have been studied in Section 3.2, where it was shown that ICR functions are $\mathcal{H}^-$-convex. Here $\mathcal{H}^-$ is the set of functions $h : \mathbb{R}^n_+ \to \mathbb{R}$ of the form $h(x) = \min(\langle l, x\rangle, c)$ with $l \in \mathbb{R}^n_+$ and $c > 0$. It has also been shown that the $\mathcal{H}^-$-subdifferential $\partial_{\mathcal{H}^-} f(y)$ of an ICR function f at a point $y \in \mathbb{R}^n_+$ contains the function l, where

$$l(x) = \min\left(\min_{i \in I_+(y)} \frac{f(y)x_i}{y_i}, f(y)\right). \tag{9.3.9}$$

Since $\mathcal{H}^-$ is a conic set, we conclude, by applying Proposition 1.4.6, that $\partial^*_{\mathcal{H}^-} f(x) = \partial_{\mathcal{H}^-} f(x)$ hence $l \in \partial^*_{\mathcal{H}^-} f(x)$, where l is defined by (9.3.9).

Consider the following optimization problem

$$f(x) \longrightarrow \min \quad \text{subject to} \quad x \in X, \tag{9.3.10}$$

where f is an ICR function (in particular, it can be an IPH function), $X \subset \mathbb{R}^n_+$ is a closed convex set.

We propose for the solution of this problem a variant of the generalized cutting plane method.

The algorithm

Step 0. Set $k := 0$. Choose an arbitrary initial point $x^0 \in X$.

Step 1. Calculate the function l^k defined by (9.3.9) with $y = x^k$.

Step 2. Solve the following subproblem

$$\max_{0 \le i \le k} h_i(x) \longrightarrow \min \text{ subject to } x \in X. \tag{9.3.11}$$

Step 3. Let y^* be a solution of the problem (9.3.11). Set $k := k+1, x^k = y^*$ and go to Step 1.

REMARK 9.6 If we have a maximization problem,

$$f(x) \longrightarrow \max \quad \text{subject to} \quad x \in X, \tag{9.3.12}$$

where f is an ICR function and $X \subset \mathbb{R}^n_{++}$, the situation is similar. Instead of an element of the subdifferential we need to take an element of the superdifferential and instead of minimizing the maximum of $f_k(x)$ we need to maximize the minimum of $f_k(x)$.

The main advantage of ICR functions (as well as ICAR functions) is that it is possible to calculate explicitly an element of the abstract subdifferential. This allows us to construct a fully implementable algorithm for minimization.

The crucial part of our method is finding the solution of the auxiliary subproblem in Step 2. This subproblem is similar to the subproblem for ICAR functions which was discussed in Subsection 9.3.5. The approaches from Subsection 9.3.5 can be directly applied to the case under consideration.

9.3.6. Numerical results - ICR objective functions

A number of numerical experiments have been carried out for test examples, most of which have multiple local minima which are not global ones. Some practical problems arising in engineering have an ICR objective function and a convex feasible set. In many cases it is possible to transform the initial problem in order to reduce it to problem (9.3.10).

Let us consider now the following test examples taken from [58, 77].

EXAMPLE 9.6 [77] Consider the problem

$$0.01x_1^2 + x_2^2 \longrightarrow \min$$

subject to

$$x_1x_2 \geq 25,$$
$$x_1^2 + x_2^2 \geq 25,$$
$$2 \leq x_1 \leq 50, \quad 0 \leq x_2 \leq 50.$$

The objective function in this (and also, in the next) example is not ICR, but it is a nonnegative quadratic function so its square root was minimized instead. Of course, the square root of this function is an IPH (hence, also an ICR) function. The optimal solution of this problem is the point $(\sqrt{250}, \sqrt{2.5})$. The best found solution is

$$x_1 = 15.811402, \, x_2 = 1.581137.$$

The absolute precision by the maximum norm is 0.000014 and the relative precision by the objective function is 0.000003. In order to find this solution, the algorithm required 18 iterations.

EXAMPLE 9.7 [77] Consider the problem

$$x_1^2 + x_2^2 + x_3^2 \longrightarrow \min$$

subject to

$$x_1^2 + x_2^2 - 1 \geq 0,$$
$$1 \leq x_1 \leq 10, \quad -10 \leq x_2 \leq 10, \quad -10 \leq x_3 \leq 10.$$

As the feasible domain for this test problem does not belong to $\mathbb{R}^3_+$, it is necessary to shift the feasible set into $\mathbb{R}^3_+$ and minimize the objective function over the received set. The optimal solution of this problem is the point $(1, 0, 0)$. The best found solution is

$$x_1 = 1.000000, \, x_2 = 0.000000, \, x_3 = 0.000000.$$

The absolute precision by the maximum norm is 0 and the relative precision by the objective function is 0. In order to find this solution, the algorithm required 20 iterations. Note that 12 iterations were necessary in order to find the optimal solution within the precision 0.001.

In the next examples, which are taken from practical engineering problems, the objective function is the sum of an ICR function and a linear function. It has the following form:

$$f(x) = g(x) + [c, x],$$

where g is an ICR function and $c \in \mathbb{R}^n$. In this case we introduced a new variable $v = [c, x]$ for the linear part, thus the objective function becomes ICR on $\mathbb{R}^{n+1}_+$. However, as the optimal solution v could be negative, it was necessary to slightly adjust the subproblem, including v inside the min-type functions. Namely, we generated subproblems of the following type:

$$t \longrightarrow \min \text{ subject to } \min(\langle h_i, x\rangle + v, c_i + v) \leq t, \quad i = 0, \dots, k, \quad x \in X,$$

which can be reduced to mixed integer programming problems or solved as disjunctive programming problems (see [13, 14]). It is important to note that the algorithm allows us to check optimality of the best found point. If two subsequent iterates coincide, it means that they both are a global minimum point of the objective function. The optimality has been confirmed for Example 9.8 and Example 9.9 in this way.

EXAMPLE 9.8 [58] Consider the problem

$$x_1^{0.6} + x_2^{0.6} - 6x_1 - 4u_1 + 3u_2 \longrightarrow \min$$

subject to

$$x_2 - 3x_1 - 3u_1 = 0, \quad x_1 + 2u_1 \leq 4, \quad x_2 + 2u_2 \leq 4,$$

$$x_1 \leq 3, \quad u_2 \leq 1.$$

The optimal solution of this problem is the point $(4/3, 4, 0, 0)$ and

$$x_1 = 1.333333, \ x_2 = 3.999999, \ u_1 = 0.000000, \ u_2 = 0.000000$$

is the best found solution. The absolute precision by the maximum norm is 0.000001 and the relative precision by the objective function is $3 \cdot 10^{-7}$. In order to find this solution, the algorithm required 12 iterations.

EXAMPLE 9.9 [58] Consider the problem

$$x_1^{0.6} + 2x_2^{0.6} + 2u_1 - 2x_2 - u_2 \longrightarrow \min$$

subject to

$$x_2 - 3x_1 - 3 = 0, \quad x_1 + 2u_1 \le 4, \quad x_2 + u_2 \le 4,$$

$$x_1 \le 3, \quad u_2 \le 2.$$

The algorithm found the following optimal solution:

$$x_1 = 0.000000,\ x_2 = 2.999999,\ u_1 = 0.000000,\ u_2 = 0.999999.$$

The absolute precision by the maximum norm is 0.000001 and the relative precision by the objective function is $2 \cdot 10^{-7}$. In order to find this solution, the algorithm required 11 iterations.

EXAMPLE 9.10 [58] Consider the problem

$$x_1^{0.6} + x_2^{0.6} + x_3^{0.4} + 2u_1 + 5u_2 - 4x_3 - u_3 \longrightarrow \min$$

subject to

$$x_2 - 3x_1 - 3u_1 = 0,$$

$$x_3 - 2x_2 - 2u_2 = 0,$$

$$4u_1 - u_3 = 0, \quad x_1 + 2u_1 \le 4,$$

$$x_2 + u_2 \le 4, \quad x_3 + u_3 \le 6,$$

$$x_1 \le 3, \quad u_2 \le 2, \quad x_3 \le 4.$$

The optimal solution is the point $(2/3, 2, 4, 0, 0, 0)$. The best found solution is the following point:

$$x_1 = 0.666666,\ x_2 = 2.000000,\ x_3 = 3.999999,$$

$$u_1 = 0.000000,\ u_2 = 0.000000,\ u_3 = 0.000000.$$

The absolute precision by the maximum norm is 0.000001 and the relative precision by the objective function is 10^{-7}. In order to find this solution, the algorithm required 27 iterations.

EXAMPLE 9.11 [77] Consider the problem

$$2 - x_1x_2x_3 \longrightarrow \max$$

subject to

$$x_1 + 2x_2 + 2x_3 - x_4 = 0,$$

$$0 \le x_i \le 1, \quad i = 1, 2, 3, \quad 0 \le x_4 \le 2.$$

The optimal solution is the point $(2/3, 1/3, 1/3, 2)$. The best found solution is:

$$x_1 = 0.667039,\ x_2 = 0.333240,\ x_3 = 0.333240,\ x_4 = 2.000000.$$

As the product of variables is not an ICR function, we reduce this problem to the one of minimizing the cubic root of the function $x \mapsto x_1 x_2 x_3$ (which is an IPH function). The absolute precision by the maximum norm is 0.000372 and the relative precision by the objective function is $7 \cdot 10^{-8}$. In order to find this solution, the algorithm required 57 iterations.

The subproblem has been solved by the software package CPLEX MIP 4.0-5.0 (after its reduction to a problem of mixed integer programming). The code has been written in Borland C++ for Windows 95 and has also been implemented for the AIX 3.2 operating system on IBM RS 6000. The initial point was chosen by convex optimization.

Optimization problems with IPH objective functions can be solved by the cutting angle method for ICAR functions or by the cutting angle method for ICR functions. The experiments have shown that the latter usually outperforms the cutting angle method for ICAR functions, probably due to the fact that some constraints in the subproblem are redundant and this reduces its dimension. It is important that the algorithm is applicable not only to minimization but also to maximization problems with increasing positively homogeneous functions.

9.3.7. Cutting angle method for Lipschitz functions

As it was shown in Subsection 3.3.6, a positive Lipschitz function f defined on the unit simplex $S = \{x \in \mathbb{R}^n_+ : \sum_i x_i = 1\}$ can be extended to a finite ICAR function g defined on $\mathbb{R}^n_+$. For example,

$$g(x) = \begin{cases} f\left(\frac{x}{\sum_i x_i}\right)(\sum_{i=1}^n x_i)^p & \text{if } x \neq 0; \\ 0 & \text{if } x = 0, \end{cases} \tag{9.3.13}$$

where

$$p \geq \frac{2L}{\min_{x \in S} f(x)},$$

is an ICAR extension of the function f. Here

$$L = \inf_{x,y \in S, x \neq y} \frac{|f(x) - f(y)|}{\|x - y\|_1}$$

is the Lipschitz constant of the function f with respect to the norm $\|\cdot\|_1$: $\|x\|_1 = \sum_i |x_i|$. Applying this extension, we can present the problem

$$f(x) \longrightarrow \min \text{ subject to } x \in S \tag{9.3.14}$$

in the form

$$g(x) \longrightarrow \min \text{ subject to } x \in S \tag{9.3.15}$$

with the ICAR objective function g and use the cutting angle method to find the solution of this problem.

In order to solve a Lipschitz programming problem with some constraints, we should transform it so that the feasible set belong to the unit simplex. This is an easy task for problems with linear constraints.

Let us consider the following mathematical programming problem:

$$f(x) \longrightarrow \min \quad \text{subject to } Ax = b,\ x \geq 0, \tag{9.3.16}$$

where f is a Lipschitz function on $\mathbb{R}^n$. Note that any problem with a Lipschitz objective function and linear constraints can be reduced to this form by introducing auxiliary variables.

Moreover, if the initial problem contains nonlinear Lipschitz constraints and has the form:

$$f(x) \longrightarrow \min \quad \text{subject to } g(x) \leq 0,\ Ax = b,\ x \geq 0,$$

then, using methods of sequential unconstrained minimization such as the penalty function methods, we can remove nonlinear constraints so that the objective function preserve the Lipschitz property. For instance, we can solve the problem:

$$f(x) + Rg(x)^+ \longrightarrow \min \quad \text{subject to } Ax = b,\ x \geq 0,$$

where $R > 0$ is a large positive number. Obviously, the objective function of this problem is a Lipschitz function, if f and g are Lipschitz functions.

Now, for problem (9.3.16) a projective transformation can be used, similar to the transformation introduced by N. Karmarkar for his polynomial method of linear programming (see [91]) and later applied successfully to a number of interior point methods.

Problem (9.3.16) can be rewritten in the following way:

$$f(x) \longrightarrow \min$$

subject to

$$Ax - bx_{n+1} = 0,\ x_{n+1} = 1,\ (x, x_{n+1}) \geq 0.$$

Next, we use the projective transformation ([91]):

$$y_i = \frac{x_i}{\sum_{j=1}^{n+1} x_j}, \quad i = 1, \dots, n+1, \quad x_1, \dots, x_n \geq 0, \quad x_{n+1} = 1.$$

Then $\sum_{i=1}^{n+1} x_i = 1/y_{n+1}$, so the inverse transformation exists and has the form:

$$x_i = \frac{y_i}{y_{n+1}}, \quad i = 1, \dots, n\,.$$

Finally, problem (9.3.16) takes the following form:

$$f\left(\frac{y_1}{y_{n+1}}, \frac{y_2}{y_{n+1}}, \dots, \frac{y_n}{y_{n+1}}\right) \longrightarrow \min \tag{9.3.17}$$

subject to

$$\tilde{A}y = 0, \ \sum_{i=1}^{n+1} y_i = 1, \ y \geq 0. \tag{9.3.18}$$

Here matrix $\tilde{A} = [A, b]$ consists of matrix A with vector b added as the last column. When y_{n+1} has a lower positive bound (that is, the feasible set of the initial problem is bounded), the objective function of this problem is Lipschitz. Recalling the tranformation, it is easy to prove that problem (9.3.17 – 9.3.18) has a Lipschitz objective function, if the distance between the feasible set of the initial problem (9.3.16) and the origin is positive.

Note that the feasible set of problem (9.3.17– 9.3.18) belongs to the unit simplex. This means that the techniques of reduction to a problem of minimizing an ICAR function, described in the beginning of this subsection, can be applied to it. Thus it is possible to apply the cutting angle method to a wide class of Lipschitz programming problems.

There is at least one more possible transformation of the feasible set of Lipschitz programming problem. Suppose the optimization problem is as follows:

$$f(x) \longrightarrow \min \quad \text{subject to } g(x) \leq 0, \ 0 \leq x \leq C, \tag{9.3.19}$$

where $C > 0$ is an upper bound for the variables. Then, introducing auxiliary variables:

$$y_i = \frac{x_i}{Cn + q}, \ i = 1, \dots, n,$$

where $q > 0$ and adding a slack variable y_{n+1}, it is possible to rewrite (9.3.19) as a problem with the feasible set contained in the unit simplex.

We now consider the scheme of the cutting angle method for minimizing Lipschitz functions. Consider the problem:

$$f(y) \longrightarrow \min$$

subject to

$$Ay = 0, \ \sum_{i=1}^{n+1} y_i = 1, \ y \geq 0.$$

Let $M > 0$ and $p \geq 1$. Then we can present this problem in the following form:

$$g(y) := \left(M + f\left(\frac{y}{\sum_{i=1}^{n+1} y_i}\right)\right)(\sum_{i=1}^{n+1} y_i)^p \longrightarrow \min \tag{9.3.20}$$

subject to

$$Ay = 0, \ \sum_{i=1}^{n+1} y_i = 1, \ y \geq 0. \tag{9.3.21}$$

Suppose that the parameters $M > 0$ and $p \geq 1$ are chosen in such a way that the objective function g of problem (9.3.20–9.3.21) is ICAR. (It follows from the discussion above that it is possible to find such M and p.) The ICAR function g is H_L-convex, where H_L is the set of all abstract affine functions with respect to the set L of all min-type functions. As it was shown in Subsection 3.3.6, for each $y \in S$, the L-subdifferential $\partial_L g(y)$ contains the element

$$g^\nabla(y) := (pf(y))/y$$

(and does not contain another element, if $p \min_{z \in S} f(z) > 2L$,where L is a Lipschitz constant of f). Hence the set $\partial^*_{H_L} f(y)$ is not empty, moreover this set contains the element h_y such that

$$h_y(z) = \langle g^\nabla(y), z\rangle - (\langle g^\nabla(y), y\rangle - g(y)) = \langle g^\nabla(y), z\rangle - (1-p)(f(y)+M).$$

The algorithm

Step 0. Set $k := 0$. Choose an arbitrary point $y^0 \in S$ such that $Ay^0 = 0$ and parameters $M > 0$ and $p > 1$ such that the function g is ICAR.

Step 1. Calculate a vector $g^\nabla(y^k) := l^k$ with the coordinates l_i^k given by

$$l_i^k = \frac{p(f(y^k) + M)}{y_i^k}, \quad \text{if} \quad y_i^k \neq 0; \quad l_i^k = 0 \quad \text{if} \quad y_i^k = 0,$$

where y_i^k is the ith coordinate of the vector y^k. Next, construct the function h_k :

$$h_k(z) = \langle l^k, z\rangle - (p-1)(f(y^k) + M).$$

Step 2. Find a global optimum of the problem:

$$\max_{0\le i\le k} h_i(y) \longrightarrow \min$$
$$\text{subject to} \quad y \in S,\ Ay = 0. \tag{9.3.22}$$

Step 3. Let y^* be a solution of the problem (9.3.22). Set $k := k+1, y^k := y^*$ and go to Step 1.

There are two parameters in problem (9.3.20– 9.3.21). In order to find a value of the parameter M, an initial lower estimate of the optimal value of the objective function should be calculated. The second parameter $p \geq 1$ depends on the Lipschitz constant and the value of M.

9.3.8. Numerical results - Lipschitz functions

The applicability of the proposed approach was checked by solving a number of test problems with Lipschitz objective functions. We consider some examples from books of test problems and published papers [133, 51, 26].

The feasible set in these examples is a polygon, which allow us to apply the CPLEX 5.0 software package for solving a subproblem of minimizing a maximum of min-type functions. If the polygon did not belong to $\mathbb{R}^n_+$, lower bounds for variables were found and the feasible set was moved to $\mathbb{R}^n_+$ by a simple linear transformation.

The code has been written in C++ for the Unix and AIX 3.2 operating systems. The parameters M and p were adjusted manually. It was important that both parameters not assume 'large values'. The value of p must be in the interval between 2 and 8 and the value of M should not exceed several thousands in order to guarantee finding an optimal solution (otherwise numerical instability appears). We chose a fixed initial value of p which was increased if the optimal solution was not found after a given number of iterations.

In most cases the minimum was found within the precision $\varepsilon = 0.0001$. Probably, the efficiency of the approach depends very much on particular properties of the objective function.

Let us consider three examples of problems of Lipschitz programming which were solved by the cutting angle method. In these examples objective functions have local minima which are not global.

EXAMPLE 9.12 [133] Consider the problem

$$x_1^2 + x_2^2 - \cos(18x_1) - \cos(18x_2) \longrightarrow \min$$

subject to

$$-1 \leq x_1 \leq 1, \quad -1 \leq x_2 \leq 1.$$

The optimal solution of this problem is the point $(0, 0)$. There are about 50 local minima. The best point, found by the cutting angle algorithm is $(-0.001336, -0.001336)$. In order to find this point 18 iterations and 10 seconds were required. The absolute precision by the objective function is 0.0004. The best performance was obtained with $p = 7$.

EXAMPLE 9.13 [51] Consider the problem

$$x_1^2 + x_2^2 \longrightarrow \max$$

subject to

$$4x_1 + 7x_2 \geq 28,$$
$$x_1 - 5x_2 \leq 5,$$
$$1 \leq x_1 \leq 3, \ 0 \leq x_2 \leq 2$$

The optimal solution (up to six digits after decimal point) of this problem is the point $(6.481481; 0.296296)$. The best solution, found by the cutting angle algorithm, is

$$x_1 = 6.481481, \quad x_2 = 0.296296.$$

The precision by the maximum norm is at least 10^{-7}. The algorithm required 3 iterations and 0.01 seconds in order to find the optimal solution.

EXAMPLE 9.14 [26] Consider the problem

$$(x_2 - \frac{5.1}{4\pi^2}x_1^2 + \frac{5}{\pi}x_1 - 6)^2 + 10(1 - \frac{1}{8\pi})\cos(x_1) + 10 \longrightarrow \min,$$

subject to

$$-5 \leq x_1 \leq 10, \quad 0 \leq x_2 \leq 15.$$

The global optimum is attained at three points:

$$(-3.142, 12.275), \ (3.142, 2.275), \ (9.425, 2.425).$$

The minimum value of the objective function is 0.3979. The best point found by the the cutting angle method, is

$$x_1 = 3.104749, \quad x_2 = 2.353947$$

with the value of the objective function equal to 0.3995. It required 18 iterations with the value of M equal to 25 and the value of p equal to 7. Note that by applying a local search method, starting from best found point, it was possible to find the solution with the precision 10^{-6} fast. Thus for such examples it is beneficial to use a hybrid method, using the cutting angle method for finding an approximate solution. For this example the best value of the lower estimate was equal to 0.3946 and it was obtained with $M = 1$ and $p = 8$. However, often the speed of convergence of the lower estimates λ_k to the optimal value could be slow. The speed of convergence depends on the particular properties of an objective function and constraints.

9.3.9. Branch-and-bound method for Lipschitz functions

A conceptual scheme of branch-and-bound techniques with abstract convexity was discussed in Subsection 9.2.4. This conceptual scheme was implemented for problems with Lipschitz objective functions, using the technique, described in Subsection 9.3.7. This technique has allowed us to compute approximate solutions. Using these solutions as starting points, the local search methods find the global optimizer with any required precision. The abstract subgradients based branch-and-bounds technique allows us to find a sufficiently small neighbourhood of a global optimum. Thus, there is no need to apply complicated global optimization techniques, as the standard methods of nonlinear optimization find the solution. We now present some examples which have been solved by using branch-and-bound techniques.

EXAMPLE 9.15 Consider the problem

$$2(x_1 - 1)^2 + 3(x_2 - 2)^2 + 5(x_3 - 3)^2 + (x_4 - 2)^2 + 2(x_5 - 3)^2 \longrightarrow \max$$

subject to

$$0 \le x_i \le 5, \; i = 1, \dots, 5.$$

The optimal solution is $(5, 5, 0, 5, 0)$. The best solution, found by the branch-and-bound algorithm, is:

$$x_1 = 4.98413, \; x_2 = 4.99756, \; x_3 = 0.00244, \; x_4 = 4.99756, \; x_5 = 0.00244.$$

The relative precision by the objective function is 0.0025. The algorithm required 200 iterations in order to find the optimal solution.

EXAMPLE 9.16 Consider the problem

$$\frac{(x_1 + x_2 - 2)^2}{x_1 + 3x_2} + \frac{(3x_1 - x_2)^2}{2x_1 + x_2} \longrightarrow \max$$

subject to

$$0.5 \leq x_i \leq 5, \; i = 1, 2.$$

The optimal solution is $(0.5, 1.5)$. The best solution, found by the branch-and-bound algorithm, is

$$x_1 = 0.4828101, \quad x_2 = 1.468550.$$

The absolute precision by the objective function is 0.0007. The algorithm required 100 iterations in order to find the optimal solution.

EXAMPLE 9.17 Consider the problem

$$2\sin^2 2x + 3\sin^2 3x \longrightarrow \max$$

subject to

$$1 \leq x_i \leq 5, \; i = 1, 2.$$

The optimal solution is $\left(\frac{3\pi}{2}, \pi\right)$. The best solution, found by the branch-and-bound algorithm, is

$$x_1 = 4.723145, \quad x_2 = 3.145508.$$

The absolute precision by the objective function is 0.001339.

9.4. CUTTING ANGLE METHOD (CONTINUATION)

9.4.1. Overview

In this subsection we discuss the cutting angle method in a special situation: we consider the minimization of an IPH function over the unit simplex. Note that many problems of Lipschitz programming can be reduced to problems of such a kind by using the technique described in Subsection 9.3.7. The subproblem of the cutting angle method in this case has the following form:

$$h(x) \longrightarrow \min \text{ subject to } x \in S, \tag{9.4.1}$$

where S is the unit simplex and

$$h(x) = \max_{k \leq j} \min_{i \in I_+(l^k)} l_i^k x_i,$$

$$l^k = (l_1^k, \ldots, l_n^k) \in \mathbb{R}_+^n, \; I_+(l^k) = \{i : l_i^k > 0\}, \; k = 1, \ldots, j, \; j \geq n.$$

Some methods of finding a solution of this (and even a more general) problem based on dynamic programming and integer programming were discussed in Subsection 9.3.3. These methods allow us to successfully apply the cutting angle method when the dimension n of the space is not very high.

In this section we propose an alternative approach for solving problem (9.4.1). This approach is based on a convenient description of all local minima of the function h, then a global minimizer can be found by sorting out local minima.

The cutting angle method based on the proposed approach allows us to solve some problems with up to 50 variables, using a conventional PC. An appropriate combination of the cutting angle method with a local search can also be applied to the search for a global minimizer.

The results from Subsections 9.4.2- 9.4.5 can be found in [25] and the results from Subsection 9.4.6 can be found in [11].

9.4.2. Cutting angle method for the minimization of IPH functions over the unit simplex

We now present an algorithm for finding a global minimizer of an IPH function f over the unit simplex

$$S = \{x \in \mathbb{R}^n_+ : \sum_{i=1}^{n} x_i = 1,\ x_i \geq 0,\ i = 1, \dots, n\}.$$

Note that an IPH function is nonnegative on $\mathbb{R}^n_+$, since $f(x) \geq f(0) = 0$. We assume that $f(x) > 0$ for all $x \in S$. It follows from the positiveness of f that $I_+(l) = I_+(x)$ for all $x \in S$ and $l_i = f(x)/x_i$ for all $i = 1, \dots, n$.

Subsequently, we will the use the unit orths of the space $\mathbb{R}^n$. Consider the mth orth $e_m = (0, \dots, 0, 1, 0, \dots, 0)$. Since $I_+(e_m) = \{m\}$, it follows that the vector $l = f(e_m)/e_m$ can be represented in the form $l = f(e_m)e_m$. Clearly, $\langle f(e_m)e_m, x\rangle = f(e_m)x_m$.

The algorithm

Step 0. Take points $x^m = e_m$ for $m = 1, \dots, n$. Choose arbitrary points $x^{n+i} \in S$, $x^{n+i} \gg 0$, $i = 1, ..q$. Let $l^k = f(x^k)/x^k$, $k = 1, \dots, n+q$. Define the function h_{n+q} by

$$h_{n+q}(x) = \max_{k=1,\dots,n+q} \min_{i \in I_+(x^k)} l^k_i x_i.$$

Step 1. Let the point x^j have been constructed. Let $l^j = f(x^j)/x^j$ and

$$h_j(x) = \max(h_{j-1}(x), \min_{i \in I_+(l^j)} l^j_i x_i) \equiv \max_{k=1,\dots,j} \min_{i \in I_+(l^k)} l^k_i x_i.$$

Step 2. Solve the problem

$$h_j(x) \longrightarrow \min \quad \text{subject to} \quad x \in S. \tag{9.4.2}$$

Step 3. Let y^* be a solution of the problem (9.4.2). Set $j := j + 1$, $x^j = y^*$ and go to Step 1.

This algorithm can be considered as a version of the cutting angle method.

REMARK 9.7 Let

$$\lambda_j = \min_{x \in S} h_j(x) \tag{9.4.3}$$

be the value of problem (9.4.2). It has been shown (see Subsection 9.2.3) that λ_j, $j = 1, \dots$ is an increasing sequence, $\lambda_j \le f_*$ and $\lambda_j \to f_*$, where $f_* = \min_{x \in S} f(x)$ is the value of the problem under consideration. Set

$$\delta_r(x^j) = \min \left(f(x^j) - f_*, \frac{f(x^j) - f^*}{f_*} \right), \tag{9.4.4}$$

$$\delta(x^j) = \min \left(f(x^j) - \delta_j, \frac{f(x^j) - \lambda_j}{\lambda_j} \right). \tag{9.4.5}$$

It was mentioned in Remark 9.2 that $\delta_r(x^j)$ can be considered as the precision of the current point x^j and $\delta(x^j)$ as an estimate of this precision. We shall use $\delta(x_j)$ in order to estimate the current state x^j. However, $\delta_r(x_j)$ is substantially less than $\delta(x_j)$ in many instances, so actually we have a more precise solution than it is shown by the estimate δ.

REMARK 9.8 Consider a Lipschitz function ϕ defined on the simplex S. Let $f(x) = \phi(x) + M$, where M is a number such that $c = \min_{x \in S} f(x) > 0$ and $(2K/c) \le 1$, where K is a Lipschitz constant of the function ϕ in the $\| \cdot \|_1$ norm. As it was mentioned in Subsection 9.3.7, there is an IPH function g, defined on $\mathbb{R}^n_+$ which coincides with f on S. Since g and ϕ have the same minimizers over the simplex S, we can apply the proposed algorithm for the minimization of the function g over S in order to find a global minimizer of the function ϕ. It is easy to express the cutting angle method in terms of the function ϕ itself. Indeed, since $g(x) = \phi(x) + M$ for $x \in S$, we should make some changes only while constructing the vectors l^k (see Step 0 and Step 1). These vectors should be defined in the following way:

$$l^k = \frac{\phi(x^k) + M}{x^k}.$$

REMARK 9.9 The computation time for the solution of the subproblem sharply increases with the number j of vectors l^k. One possible way to reduce the computation time is to use a renewal process: adding a new vector we exclude one of the vectors $l^{n+1}, \dots, l^j$. Hence, the maximal number $j = n+m$ of vectors l^k is fixed. Starting with the iterate $n+m+1$, the $(n+i)$th vector is replaced by the $(n+i+1)$th vector $(i = 2, \dots, m+1)$ and the $(n+1)$th vector is removed. Here the value of m depends on the concrete problem. It should be noted that the convergence of the main algorithm is proved without any renewal process, so we can not guarantee that this process leads to a global minimizer. However, having the lower bound λ_j, defined by (9.4.3), we can easily evaluate how far from the global minimum we are. Numerical experiments demonstrate that in some instances this method allows us to find an approximate solution quickly enough.

9.4.3. The subproblem: local minima

Step 2 (i.e., finding the global minimum of problem (9.4.2)) is the most difficult part of the algorithm. This problem can be represented in the following form (we omit the index j for the sake of simplicity):

$$h(x) \longrightarrow \min \text{ subject to } x \in S, \tag{9.4.6}$$

where

$$h(x) = \max_{k \le j}\left(\min_{x \in I_+(l^k)} l_i^k x_i\right) \text{ for } j \ge n+q, \tag{9.4.7}$$

$l^k = f(x^k)/x^k$ are given vectors $(k = 1, \dots, j)$. Note that $x^k = e_k$ for $k = 1, \dots, n$. Let

$$\phi_k(x) := \langle l^k, x\rangle = \min_{i \in I_+(l^k)} l_i^k x_i, \tag{9.4.8}$$

then $h(x) = \max_{k \le j} \phi_k(x)$.

PROPOSITION 9.5 *Let $j \ge n$, let $l^k = l_k^k e_k$ for $k = 1, \dots, n$, and $l^k \gg 0$ for $k = n+1, \dots, j$. Then each local minimizer of the function h, defined by (9.4.7) over the simplex S, is a strictly positive vector.*

Proof: It is sufficient to show that for each nonstrictly positive $x \in S$ and for each $\varepsilon > 0$ there exists $x' \in S$ such that $x' \gg 0$, $\|x' - x\| < \varepsilon$ and $h(x') < h(x)$. For $x \in S$, consider the set $I_0(x) = I \setminus I_+(x) = \{i \in I : x_i = 0\}$. Assume that $I_0(x)$ is nonempty. Let us calculate the value of functions ϕ_k (defined by (9.4.8)) at the point x. We have

$$\phi_k(x) = \min_{i \in I_+(l^k)} l_i^k x_i = l_k^k x_k, \qquad k = 1, \dots, n. \tag{9.4.9}$$

In particular,

$$\phi_k(x) = 0, \qquad k \in I_0(x). \tag{9.4.10}$$

We also have

$$\phi_k(x) = \min_{i \in I} l_i^k x_i = 0, \qquad k = n+1, \dots, j. \tag{9.4.11}$$

It follows from (9.4.9), (9.4.10) and (9.4.11) that $\phi_k(x) > 0$ if and only if $k \leq n$ and $k \notin I_0(x)$, that is, $k \in I_+(x)$. Hence

$$h(x) = \max_k \phi_k(x) = \max_{k \in I_+(x)} l_k^k x_k. \tag{9.4.12}$$

Let $\varepsilon > 0$ be a small number and $m = |I_0(x)|$. Consider the point $x(\varepsilon)$, where

$$x(\varepsilon)_i = \begin{cases} x_i - \varepsilon & \text{if } i \in I_+(x); \\ \dfrac{n-m}{m}\varepsilon & \text{if } i \in I_0(x). \end{cases}$$

We have $x(\varepsilon) \in S$ and $x(\varepsilon) \gg 0$ for sufficiently small ε. Let us calculate $h(x(\varepsilon)) = \max_{k \leq j} \phi_k(x(\varepsilon))$. Let $k \geq n+1$. For sufficiently small ε we have:

$$\phi_k(x(\varepsilon)) = \min_{i \in I} l_i^k x(\varepsilon)_i = \min_{i \in I_0(x)} l_i^k \frac{n-m}{m}\varepsilon = \frac{n-m}{m}\varepsilon \min_{i \in I_0(x)} l_i^k. \tag{9.4.13}$$

Let $k \in I_0(x)$. Then

$$\phi_k(x(\varepsilon)) = \min_{i \in I_+(l^k)} l_i^k x(\varepsilon)_i = l_k^k \frac{n-m}{m}\varepsilon. \tag{9.4.14}$$

Let $k \in I_+(x)$. Then

$$\phi_k(x(\varepsilon)) = \min_{i \in I_+(l^k)} l_i^k x(\varepsilon)_i = l_k^k (x_k - \varepsilon). \tag{9.4.15}$$

It follows from (9.4.13), (9.4.14) and (9.4.15) that for very small ε

$$h(x(\varepsilon)) = \max_{k \leq j} \phi_k(x) = \max_{k \in I_+(x)} l_k^k (x_k - \varepsilon). \tag{9.4.16}$$

Using (9.4.12) and (9.4.16) we deduce that $h(x(\varepsilon)) < h(x)$. △

COROLLARY 9.1 *Let $(x_1, \dots, x_j, \dots$ be a sequence generated by the algorithm. Then $x_j \gg 0$ for all $j > n$, hence $l^j \gg 0$ for all $j > n$.*

Proof: The proof is by induction on j. △

Let

$$\operatorname{ri} S = \{x \in S : x_i > 0 \text{ for all } i \in I\}$$

(ri S is the relative interior of the simplex S).

Applying Proposition 9.5 and Corollary 9.1 we can solve problem (9.4.6) by sorting out local minima of the function h over the set ri S. We now describe some properties of local minima of h on ri S.

It is well known that functions ϕ_k and h are directionally differentiable. In order to show it we can use the well-known results related to the directional derivative of the functions

$$x \mapsto \max_{i \in I} g_i(x), \qquad x \mapsto \min_{i \in I} g_i(x)$$

where I is a finite set (see, for example, [43], Corollary 3.2). We present the corresponding result without a proof. Set

$$R(x) = \{k : \phi_k(x) = h(x)\} \text{ and } Q_k(x) = \{i \in I_+(l^k) : l_i^k x_i = \phi_k(x)\}. \tag{9.4.17}$$

PROPOSITION 9.6 *For $x \gg 0$*

$$\phi_k'(x, u) = \min_{i \in Q_k(x)} l_i^k u_i;$$

$$h'(x, u) = \max_{k \in R(x)} \phi_k'(x, u) = \max_{k \in R(x)} \min_{i \in Q_k(x)} l_i^k u_i.$$

Let $x \in S$. The cone

$$K(S, x) = \{u \in \mathbb{R}^n : \ \exists \alpha_0 > 0 \text{ such that } x + \alpha u \in S \ \forall \alpha \in (0, \alpha_0)\}$$

is called the tangent cone at the point x with respect to the simplex S (this cone coincides with the Booligand cone $\Gamma(S, x)$).

The following necessary conditions for a local minimum are well known (see, for example, [43]).

PROPOSITION 9.7 *Let $x \in S$ be a local minimizer of the function h over the set S. Then $h'(x, u) \geq 0$ for all $u \in K(S, x)$.*

PROPOSITION 9.8 *Let $x \in$ ri S. Then*

$$K(S, x) = \{u : \sum_{i \in I} u_i = 0\}.$$

Proof: The proof follows directly from the definition of the tangent cone. △

Applying Proposition 9.6, Proposition 9.7 and Proposition 9.8 we can obtain the next result.

PROPOSITION 9.9 *Let x be a local minimizer of the function h over the set $\operatorname{ri} S$, such that $h(x) > 0$. Then there exists a subset $\{l^{k_1}, l^{k_2}, \ldots, l^{k_n}\}$ of the set $\{l^1, \ldots, l^j\}$ such that*

1)

$$x = \left(\frac{d}{l_1^{k_1}}, \ldots, \frac{d}{l_n^{k_n}}\right), \text{ where } d = \frac{1}{\sum_{i \in I} (l_i^{k_i})^{-1}} \tag{9.4.18}$$

2)

$$\max_{k \leq j} \min_{i \in I_+(l^k)} \frac{l_i^k}{l_i^{k_i}} = 1; \tag{9.4.19}$$

3) either $k_i = \{i\}$ for all $i \in I$ or there exists $m \in I$ such that $k_m \geq n+1$; if $k_m \leq n$, then $k_m = m$;

4) if $k_m \geq n+1$, then $l_i^{k_m} > l_i^{k_i}$ for all $i \in I$, $i \neq m$.

Proof: Let $x \gg 0$ be a local minimizer. Then for each $u \in K(S, x)$ there exists $k \in R(x)$ such that

$$\phi_k'(x, u) \equiv \min_{i \in Q_k(x)} l_i^k u_i \geq 0. \tag{9.4.20}$$

Let $m \in I$. For each $i \neq m$ choose a number $\lambda_i > 0$ such that $\sum_{i \neq m} \lambda_i = 1$. Consider the vector u, where

$$u_i = \begin{cases} 1 & \text{if } i = m; \\ -\lambda_i & \text{if } i \neq m. \end{cases}$$

It follows from Proposition 9.8 that $u \in K(S, x)$. Hence there exists $k \in R(x)$ such that (9.4.20) holds. If $Q_k(x) \neq \{m\}$, then

$$\phi_k'(x, u) = \min_{i \in Q_k(x)} l_i^k u_i < 0$$

which contradicts (9.4.20). Hence $Q_k(x) = \{m\}$. Thus for each $m \in I$ there exists $k_m \in R(x)$ such that $Q_{k_m}(x) = \{m\}$. Let $m_1 \neq m_2$ and let k_{m_i} be an index such that $Q_{k_{m_i}} = \{m_i\}$ for $i = 1, 2$. Since $m_1 \neq m_2$ it follows that $k_{m_1} \neq k_{m_2}$.

Let $m \in I$ and $k_m \in R(x)$ be an arbitrary index such that $Q_{k_m}(x) = \{m\}$. Since $k_m \in R(x)$ and $m \in Q_{k_m}(x)$, we have

$$h(x) = \phi_{k_m}(x) = l_m^{k_m} x_m.$$

Hence

$$x_m = \frac{h(x)}{l_m^{k_m}} \text{ for all } m \in I. \tag{9.4.21}$$

It follows from (9.4.21) that

$$x = \left(\frac{h(x)}{l_1^{k_1}}, \dots, \frac{h(x)}{l_n^{k_n}}\right). \tag{9.4.22}$$

Since $x \in S$, we have

$$1 = \sum_{i \in I} x_i = h(x) \sum_{i \in I} \frac{1}{l_i^{k_i}}.$$

Hence

$$h(x) = \frac{1}{\sum_{i \in I} (l_i^{k_i})^{-1}}. \tag{9.4.23}$$

Thus we have shown that 1) in Proposition holds. We now demonstrate that 2) holds as well. Using (9.4.22) we infer that

$$h(x) = \max_{k \le j} \min_{i \in I_+(l^k)} l_i^k x_i = \max_{k \le j} \min_{i \in I_+(l^k)} l_i^k \frac{h(x)}{l_i^{k_i}}.$$

Thus

$$1 = \max_{k \le j} \min_{i \in I_+(l^k)} \frac{l_i^k}{l_i^{k_i}}$$

and 2) has been proved. Set

$$K = \{k_1, \dots, k_n\}, \quad K_1 = \{k_m \in K : k_m \le n\}, \quad K_2 = K \setminus K_1.$$

Let $k_m \in K_1$, that is, $k_m \le n$. It follows from (9.4.21) that $l_m^{k_m} \neq 0$. Since the unique non-zero coordinate of the vector l^{k_m} is $l_{k_m}^{k_m}$, we have $k_m = m$ and

$$x_m = \frac{h(x)}{l_m^m}, \qquad (k_m = m \in K_1). \tag{9.4.24}$$

If $K_2 = \emptyset$, then $K_1 = \{1, 2, \dots, n\}$ and $x = \bar{x}$, where

$$\bar{x} = h(\bar{x}) \left(\frac{1}{l_1^1}, \dots, \frac{1}{l_n^n}\right)$$

with

$$h(\bar{x}) = \frac{1}{\sum_{i \in I} \frac{1}{l_i^i}}.$$

Thus, if $x \neq \bar{x}$ then $K_2 \neq \emptyset$. Hence 3) has been proved.

Let $k_m \in K_2$. Then $I_+(l^{k_m}) = I$ and $Q_{k_m} = \{i \in I : l_i^{k_m} x_i = \phi_{k_m}(x)\}$. Since $k_m \in R(x)$ and $Q_{k_m} = \{m\}$, we have $l_i^{k_m} x_i > \phi_{k_m}(x) = h(x)$ for $i \neq m$. It follows from (9.4.22) that

$$l_i^{k_m} \frac{h(x)}{l_i^{k_i}} > h(x) \qquad i \neq m.$$

Hence

$$l_i^{k_m} > l_i^{k_i} \text{ for all } k_m \in K_2, \; i \neq m.$$

and 4) holds. △

REMARK 9.10 Consider equality (9.4.19). Let

$$v = \max_k \min_{i \in I_+(l^k)} \frac{l_i^k}{l_i^{k_i}}.$$

It follows from (9.4.19) that $v = 1$. Since $I_+(l^m) = \{m\}$ for $m \leq n$ and $I_+(l^k) = I$ for $k \geq n+1$, we have

$$v = \max\left(\max_{m \leq n} \frac{l_m^m}{l_m^{k_m}}, \; \max_{k \geq n+1} \min_{i \in I} \frac{l_i^k}{l_i^{k_i}}\right)$$

Assume that $K_1 \neq \emptyset$ and $m \in K_1$. Then $k_m = m$, hence $l_m^m / l_m^{k_m} = 1$. Thus $v = 1$ if and only if

$$l_m^m \leq l_m^{k_m} \text{ for all } m \leq n, \; m \notin K_1, \text{ and } \min_{i \in I} \frac{l_i^k}{l_i^{k_i}} \leq 1 \text{ for all } k \geq n+1.$$

9.4.4. The subproblem: global minima

Proposition 9.5 and Proposition 9.9 imply that we can find a global minimizer of the function h, defined by (9.4.7), over the unit simplex using the following procedure:

- Sort out all subsets $(l^{k_1}, \ldots, l^{k_n})$ of the given set of vectors $l^1, \ldots, l^j$, such that (9.4.19) holds and the following holds: 1) $l_i^{k_m} > l_i^{k_i}$ for indices $k_m \geq n+1$, 2) $k_m = m$ for $k_m \leq n$.
- For each such a subset, find the vector x defined by (9.4.18).

- Choose the vector (from all vectors described above) which minimizes the function h.

Thus, the search for a global minimizer is reduced to sorting out some subsets containing n elements of the given set $\{l^1, \dots, l^j\}$ with $j > n$. Fortunately, Proposition 9.9 allows us to substantially decrease the number of sorted out subsets. The following simple example will clarify this point.

EXAMPLE 9.18 Assume we have the following set of three-dimensional vectors:

$$l^1 = (4,0,0); \; l^2 = (0,3,0); \; l^3 = (0,0,2)$$

$$l^4 = (3,4,5); \; l^5 = (4,2,3); \; l^6 = (8,3,1).$$

We would like to find all subsets $(l^{k_1}, l^{k_2}, l^{k_3})$ of this set such that

1)

$$\max\left(\max\left(\frac{4}{l_1^{k_1}}, \frac{3}{l_2^{k_2}}, \frac{2}{l_3^{k_3}}\right), \max_{k=4,5,6} \min_{i=1,2,3} \frac{l_i^k}{l_i^{k_i}}\right) = 1; \tag{9.4.25}$$

2)

$$\text{either } k_i = i, \; i = 1,2,3 \text{ or } \exists \; m \in \{1,2,3\} \text{ such that } k_m \geq 4;$$

$$\text{if } k_m \leq 3, \text{ then } k_m = m, \; m = 1,2,3; \tag{9.4.26}$$

3)

$$\text{if } k_m \geq 4, \text{ then } l_i^{k_m} > l_i^{k_i} \; i = 1,2,3, \; i \neq m. \tag{9.4.27}$$

One of these subsets, $I_1 = \{1,2,3\}$, is known. All other subsets should contain at least one vector with the index greater than 3. It follows from (9.4.26) that $k_1 \neq 2,3$. Since $l_1^1/l_1^4 > 1$, (9.4.25) does not hold if $k_1 = 4$. Hence there are no subsets I' with the required properties which contain l^4 as the first vector. There exist only two subsets such that (9.4.26) and (9.4.27) hold:

$$I_2 = \{1,6,3\}, \quad I_3 = \{1,6,5\}.$$

However, (9.4.25) is not valid for both I_2, I_3. Thus there exists only one subset I_1 satisfying (9.4.25)– (9.4.27).

The algorithm for the search for a global minimizer of the auxiliary function based on Proposition 9.9 consists of two parts. First, we sort out subsets consisting of n different elements from the given set $\{l^1, \dots, l^j\}$, excluding those which do not provide the global minimum. We also include in the algorithm the renewal procedure, described in Remark 9.9. These steps allow us to significantly reduce the computation time. Second, we choose a subset with the least value of the function h.

We now describe a version of the algorithm for sorting out all subsets of the set

$$\{l^1, \dots, l^j\}. \tag{9.4.28}$$

Let $\varepsilon > 0$ be a given tolerance. It should be noted that the choice of the number $\varepsilon > 0$ depends on the concrete problem. For a broad class of problems a suitable tolerance is $\varepsilon = 10^{-4} \div 10^{-7}$.

Sorting subsets of the set given by (9.4.28)

Step 0. Set $m = 1$ and $i = 1$.

Step 1. Take a vector l^i.

Step 2. If $l^i_m \leq \varepsilon$ or $l^m_m / l^i_m > 1$, then set $i = i + 1$ and go to Step 1. Otherwise, set $k_m = i$.

Step 3. Set $m = m + 1$, $i = 1$. If $m > n$, then stop.

Step 4. Take a vector l^i. If $l^i_{m-1} > l^{k_{m-1}}_{m-1}, l^i_m > \varepsilon, l^i_m \geq l^m_m$ and $l^i_m < l^{k_m}_m$, then set $k_m = i$ and go to Step 3.

Step 5. Otherwise, set $i = i + 1$. If $i \leq j$ then go to Step 4, otherwise stop.

In order to find local minima we should choose subsets such that (9.4.19) holds and then using (9.4.18) calculate x. Various improvements of this algorithm which allow us to significantly reduce the computational time, have been done by A. M. Bagirov. In particular, we can use the recursion relation:

$$h_{j+1}(x) = \max(h_j(x), \phi_{j+1}(x)), \tag{9.4.29}$$

where $\phi_{j+1}(x) = \min_{i=1,\dots,n} l^{j+1}_i x_i$, and take into account only local minima, which arise due to the new function ϕ_{j+1}. In such a case we should use schemes with memory. We do not discuss other versions of these algorithms and their numerical implementations.

9.4.5. Results of numerical experiments

A number of numerical experiments have been carried out in order to verify the practical efficiency of the algorithm, proposed in Subsection 9.4.4. We consider five different classes of problems. Three of them are problems with IPH objective functions. We also consider two problems with special kinds of Lipschitz functions.

The codes have been written in Microsoft Fortran-90. Numerical experiments have been carried out on an IBM Pentium-S CPU 150 MHz.

We present the results of numerical experiments for some concrete typical problems from each of these classes. In all examples we use the following notations:

- $f = f(x)$ is the objective function;
- n is the number of variables;
- k is the number of iterations;
- t is the computation time.

We present the computational time t in the relative form. If n is the number of variables, then the corresponding value of t is calculated as $t = t_n/t_5$, where t_n is time required for the calculation with n variables. Hence for $n = 5$, we have $t = 1$ for all examples.

The estimate $\delta(x^k)$ of the precision $\delta_r(x^k)$ of the current point x^k is defined as follows:

$$\delta(x^k) = \min\left(f(x^k) - \lambda_k, \frac{f(x^k) - \lambda_k}{\lambda_k}\right),$$

where λ_k is the lower estimate of the global minimum, defined by (9.4.3). See Remark 9.7 for discussions related to this estimate.

The results of numerical experiments with the estimates of the precision $\delta = 10^{-2}$ are presented below.

EXAMPLE 9.19

$$f_1(x) = \max\{[a^i,\ x] :\ i = 1, 2, ..., 40\} + \min\{[b^j, x] : j = 1, 2, ..., 20\},$$

where

$$a^i_k = \frac{20i}{k(1 + |i - k|)},\quad k = 1, 2, ..., n;\ \ i = 1, 2, ..., 40,$$

$$b^j_k = 5|\sin(j)\ \sin(k)|,\ k = 1, 2, ..., n,\ \ j = 1, 2, ..., 20.$$

Results of numerical experiments.

Table 1

n	k	t
5	18	1.0
10	35	88.0
20	35	1055.1
30	30	2546.7
50	26	6571.1

EXAMPLE 9.20

$$f_2(x) = \max_{1 \le i \le 20} \min_{1 \le j \le n} [a^{ij}, x],$$

where

$$a_k^{ij} = \frac{10j}{k(1 + |k - j|)} |\cos(i - 1)|$$

$$i = 1, \dots, 20, \; j = 1, \dots, n, \; k = 1, \dots, n.$$

Results of numerical experiments.

Table 2

n	k	t
5	108	1.0
10	107	122.1
15	167	510.8

EXAMPLE 9.21

$$f_3(x) = \Big(\sum_{i=1}^{n} \sum_{j=1}^{n} a_{ij} x_i x_j \Big)^{1/2},$$

where

$$a_{ij} = \begin{cases} 12 + n/i & \text{if } i = j, \\ 0 & \text{if } i = j + 1, \\ 0 & \text{if } j = i + 2, \\ 15/(i + 0.1j) & \text{otherwise.} \end{cases}$$

Results of numerical experiments.

Table 3

n	k	t
5	20	1.0
10	44	47.2
20	49	980.3
30	34	2546.1
50	22	3208.4

EXAMPLE 9.22

$$f_4(x) = \sum_{i=1}^{n} \min\{0,\ 15\|x - a^i\|^2 - b_i\},$$

where a^i, $i = 1, 2, ..., n$ are n-vectors with coordinates:

$$a^i_j = \begin{cases} (n+1)/2n & \text{if } \ j = i, \\ 1/2n & \text{if } \ j \neq i. \end{cases}$$

$$b = (b_1, b_2, ..., b_n),\ b_1 = 4,\ b_i = b_{i-1} - 2/(n-1), i > 1.$$

Results of numerical experiments.

Table 4

n	k	t
5	29	1.0
10	21	8.8
15	34	86.8

EXAMPLE 9.23

$$f_5(x) = (x_1 - \frac{r_1}{r_2}x_2)^2 + (x_2 - \frac{r_2}{r_3}x_3)^2 + \ldots + (x_{n-1} - \frac{r_{n-1}}{r_n}x_n)^2 - 1,\ x \in S.$$

Set

$$r_1 = \frac{1}{2n},\ r_i = r_1 + \frac{i}{(n+2)(n-1)} \text{ for } i \neq 1.$$

Clearly,

$$r_i > 0 \text{ and } \sum_{i=1}^{n} r_i = 1.$$

Results of numerical expriments.

Table 5

n	k	t
5	48	1.0
10	53	4.2
20	63	14.0

Comments

Example 9.20 is a linear maxmin problem (see [47] for discussions related to the problems of such kind). Example 9.19 is a special case of this maxmin problem. Example 9.21 is equivalent to the minimization of the quadratic function $f(x) = \sum_{i=1}^{n} \sum_{j=1}^{n} a_{ij}x_i x_j$. In contrast with previous problems, the objective functions in Example 9.22 and Example 9.23 are Lipschitz but not IPH.

The function f_4 is a nonsmooth Lipschitz function with n local minima, which attains its global minimum over the unit simplex at the point a^1. The version of the cutting angle method described in Remark 9.8 was applied for the solution of this problem. We consider $g(x) = f_4(x) + M$ with $M = 10$ in order to apply the algorithm. Experiments show that numbers $M' > 10$ do not improve the convergence. Since the value of the global minimum of the function f_4 is known, we can compare the number of iterates which we need in order to get an approximate solution with the precision $\delta_r(x^k)$ and with its estimate $\delta(x^k)$ for Example 9.22. Experiments show that we need almost half the iterates indicated in Table 4 in order to obtain the approximate solution with the prescribed precision $\delta_r(x^k)$.

The function f_5 is a quadratic function which was studied by P. Pardalos, Y. Ye and Ch.-G. Han in [120].

The global minimizer x^* of this function is

$$x^* = (r_1, r_2, \dots, r_n), \quad f(x^*) = -1.$$

The algorithm allows us to calculate all local and global minima of the objective function h_j of the subproblem. As it turns out, the number of local minimizers increases very quickly with the dimension n of the problem and with the number j of min-type functions. This number significantly depends on the objective function f of the global optimization problem under considertion. For instance, for quadratic functions (more exactly, for IPH functions, which are the square root of nonnegative quadratic functions) the number of local minimizers is significantly less than the number of those for the functions of max-min type. Consequently, the number of global minimizers may also be large enough, however this number does not necessarily increase with n and j. Note that for max-min type objective functions the number of global minimizers of subproblem can be very large (up to several hundreds).

The cutting angle method is based on the following procedure: we should compute exactly one global minimizer of the subproblem and use it at the next iteration. The proposed method allows us to compute *all* global minimizers. Thus the following question arises. How do we choose a global minimum of the subproblem which can be used for the next iteration of the cutting angle method?

If the number of global minimizers of subproblems is large enough, than a very slow convergence is possible. The lower estimate λ_k can keep its value for many iterations and then significantly reduce it. A version of the cutting angle method can be considered, where all global minimizers of the subproblem at the current iteration are used for the next iteration. This provides a strict increase of λ_k. This approach leads to acceleration of the convergence of the algorithm. On the other hand, a sharp increase of the number of support vectors can also slow down this convergence. Numerical experiments show that it is not necessarily to compute new support vectors for all points of the global minimum.

Another important question arises in the study of the subproblem. Many local minimizers of the objective function in the previous iteration remain local minimizers for the new objective function at the current iteration. Hence modifications of the algorithm are required which are based only on calcalutation of new local minima corresponding to new support vectors. One such modification was successfuly used by A. M. Bagirov (private communication).

Finally, we can take into account not only global minimizers of the subproblem but also points of local minima, which are *approximate global minimizers.* (Compare with the notion of ε-subdifferential in convex optimization.) This approach guarantees that λ_k increases.

The results of numerical experiments show that a slow increase of λ_k is the main drawback of the cutting angle method. Actually this method often computes the global minimizer quickly enough but a slow convergence of λ_k does not allow us to assert that a global minimizer has exactly been computed. The above mentioned modifications allow us to surpass this drawback in some instances.

9.4.6. An exact method for solving the subproblem

In this subsection we discuss an approach for the solution of the subproblem, which was developed by D. Babayev [11]. Under some additional assumptions this approach allows us to obtain the "almost explicit" solution. It will be convenient to use the following notation. Let $\ell = (l_i^k)$ be a $m \times n$ matrix with m rows l^k $(k = 1, \dots, m)$. Let $I = \{1, \dots, n\}$. We assume that

1) the first n rows of the matrix ℓ form a diagonal matrix with the positive diagonal, that is, $l_i^k > 0$ if $i = k$ and $l_i^k = 0$ if $i \neq k$, $i, k = 1, \dots, n$;

2) $l_i^k > 0$ for $k > n$ and $i \in I$.

Let

$$h(x) = \max_{k=1,\dots,m} \min_{i \in I_+(l^k)} l_i^k x_i,$$

where $I_+(l^k) = \{k\}$ for $k \leq n$ and $I_+(l^k) = I$ for $k > n$. We shall study the problem:

$$h(x) \longrightarrow \min \text{ subject to } x \in S \tag{9.4.30}$$

where S is again the unit simplex. This problem is completely determined by the matrix ℓ and we shall call it (if necessary) *problem (9.4.30) for the matrix ℓ*.

The following assertion will be useful later.

LEMMA 9.1 *Let $a_i > 0$ for $i = 1, \dots, n$ and let $u = (u_1, \dots, u_n) \in S$ be a vector such that*

$$\max_{i \in I} a_i u_i = \min_{x \in S} \max_{i \in I} a_i x_i.$$

Then $a_1 u_1 = \dots = a_n u_n$.

Proof: Let $P = \{p \in I : a_p u_p = \max_{i \in I} a_i u_i\}$ and assume that $P \neq I$. Then $a_p u_p > a_q u_q$ for each $p \in P$ and $q \notin P$. Let $\varepsilon_i > 0$ be numbers such that

$$a_p(u_p - \varepsilon_p) > a_q(u_q + \varepsilon_q) \text{ for all } p \in P,\ q \notin P, \qquad \sum_{p \in P} \varepsilon_p = \sum_{q \notin P} \varepsilon_q$$

and v be the vector with the coordinates

$$v_p = u_p - \varepsilon_p \ (p \in P), \qquad v_q = u_q + \varepsilon_q \ (q \notin P).$$

Then $v \in S$ and

$$\max_{i \in I} a_i v_i = \max_{p \in P} a_p v_p < \max_{p \in P} a_p u_p = \max_{i \in I} a_i u_i$$

and we arrive at a contradiction. △

COROLLARY 9.2 *It follows from the equalities $a_1u_1 = \ldots = a_nu_n$ and $u_1 + \ldots + u_n = 1$ that*

$$u_i = \frac{d}{a_i} \text{ with } d = \frac{1}{\sum_{p=1}^{n} \frac{1}{a_p}}. \qquad (9.4.31)$$

Lemma 9.1 allows us to solve the subproblem if $m = n$ quite easily.

PROPOSITION 9.10 *If $m = n$, then problem (9.4.30) has a unique solution*

$$x = (x_i), \text{ where } x_i = \frac{d}{l_i^i} \text{ with } d = \frac{1}{\sum_{p=1}^{n} \frac{1}{l_p^p}}. \qquad (9.4.32)$$

Proof: If $m = n$, then ℓ is a diagonal matrix. We have $I_+(l^k) = \{k\}$, hence

$$h(x) = \max_k \min_{i \in I_+(l_i^k)} l_i^k x_i = \max_i l_i^i x_i$$

and problem (9.4.30) has the form

$$\max_{i \in I} l_i^i x_i \longrightarrow \min \text{ subject to } x \in S.$$

The desired result follows directly from (9.4.31). △

Consider now the case $m > n$. First, we assume that

$$\text{for each } k > n \text{ there exists } i \in I \text{ such that } l_i^k \le l_i^i. \qquad (9.4.33)$$

PROPOSITION 9.11 *If (9.4.33) holds, then (9.4.30) has a unique solution (9.4.32).*

Proof: We have

$$\max_{k \le m} \min_{i \in I_+(l^k)} l_i^k x_i = \max(\max_{k \le n} l_k^k x_k, \max_{n < k \le m} \min_{i \in I} l_i^k x_i).$$

It follows from (9.4.33) that

$$\min_{i \in I} l_i^k x_i \le l_i^i x_i \le \max_{p \in I} l_p^p x_p.$$

Hence

$$\max_{k \le m} \min_{i \in I_+(l^k)} l_i^k x_i = \max_{i \in I} l_i^i x_i.$$

The result follows now from Proposition 9.10. △

REMARK 9.11 A. M. Bagirov (private communication) pointed out that (9.4.33) can occur in the cutting angle method only at its final iteration.

The most interesting case is when (9.4.33) does not hold, that is, there exists k such that $l_i^k > l_i^i$ for all $i \in I$. Let

$$K = \{k : \min_{i \in I}(l_i^k - l_i^i) > 0\} \tag{9.4.34}$$

and

$$\tilde{l}_i = \max_{k \in K} l_i^k \text{ for } i \in I, \qquad Q_i = \{q \in K : l_i^q = \tilde{l}_i\}.$$

The following assertion holds:

PROPOSITION 9.12 *Suppose that $m > n$, the set K defined by (9.4.34) is not empty and there exist an index $j \in I$ and an index $q(j) \in Q_j$ such that*

$$\min_{i \in I}\left(\frac{1}{l_i^i} - \frac{1}{l_i^{q(j)}}\right) = \frac{1}{l_j^j} - \frac{1}{l_j^{q(j)}}. \tag{9.4.35}$$

Let $\ell_n = (l_i^k)$ be the diagonal $n \times n$ matrix formed by the first n given vectors $l^1, \dots l^n$ and let $\bar{\ell}_n = (\bar{l}_i^k)$ be the matrix obtained from ℓ_n by replacing the element l_j^j with $l_j^{q(j)}$. Then the solution of problem (9.4.30) for the matrix $\bar{\ell}_n$ is a solution of this problem for the initial matrix ℓ.

Proof: Let $k \in K$ and $1 \le j \le n$. Let $\max_{k \in K} l_j^k = \tilde{l}_j$ and consider the $m \times n$-matrix $\ell'_j = (l_i'^k)$ which is obtained from the original matrix ℓ by changing only one element, namely by replacing l_j^j with $\tilde{l}_j$. Let $k \in K$. Observe that

$$l_j'^j = \tilde{l}_j = \max_{p \in K} l_j^p \ge l_j^k.$$

Thus for each $k \in K$ there exists an index i (namely $i = j$) such that $l_i'^k \le l_i'^i$. It follows from the definition of the set K that for each $k \notin K$ there exists an index i such that $l_i^k \le l_i^i$. Hence $l_i'^k \le l_i'^i$. We have shown that conditions in Proposition 9.11 hold for the matrix ℓ'_j.

Let j be an index such that (9.4.35) holds. Applying Proposition 9.11 we find the solution x^* of problem (9.4.30) for the matrix ℓ'_j. This solution is given by

$$x_i^* = \frac{d_j}{l_i'^i} \text{ with } d_j = \frac{1}{\sum_{p=1}^n \frac{1}{l_p'^p}}. \tag{9.4.36}$$

It follows from (9.4.36) (see also Lemma 9.1) that $h(x^*) = l_i'^i x_i^*$ for all $i \in I$, where

$$h(x) = \max_{k \le m} \min_{i \in I} l_i^k x_i$$

is an objective function of problem (9.4.30) for the initial matrix ℓ. Since

$$l'^i_i = \begin{cases} l^i_i & \text{if } i \neq j; \\ \tilde{l}_j & \text{if } i = j \end{cases}$$

we have

$$h(x^*) = \tilde{l}_j x^*_j = l^i_i x^*_i \text{ for all } i \neq j. \tag{9.4.37}$$

We claim that x^* is a global minimizer of (9.4.30) for the initial matrix ℓ. Thus, we should check that $h(y) \geq h(x^*)$ for all $y \in S$.

Suppose that $y \in S$. Assume that there exists $i \neq j$ such that $y_i > x^*_i$. Applying (9.4.37) we conclude that

$$h(y) = \max_{k=1,\dots,m} \min_i l^k_i y_i \geq \max_{k=1,\dots,n} l^k_k y_k \geq l^i_i y_i \geq l^i_i x^*_i = h(x^*).$$

Hence it is sufficient to consider only vectors y such that $y_i \leq x^*_i$ for all $i \neq j$ and $y \neq x^*$. Let

$$\delta_i = x^*_i - y_i \geq 0 \text{ for } i \neq j \text{ and } \delta = \sum_{i \neq j} \delta_i. \tag{9.4.38}$$

Note that $\delta > 0$. Since $x^* \in S$ and $y \in S$,

$$y_j = x^*_j + \delta. \tag{9.4.39}$$

Let $q \equiv q(j) \in Q_j$ be an index such that (9.4.35) holds. Note that $l^q_j = \max_{k \in K} l^k_j = \tilde{l}_j$. Then (see (9.4.37))

$$h(x^*) = \tilde{l}_j x^*_j = l^q_j x^*_j. \tag{9.4.40}$$

We need to check the inequality $h(y) \geq h(x)$ for $y \in S$ such that $y_i \leq x^*_i$ for all $i \neq j$. If $l^j_j y_j \geq h(x^*)$ then

$$h(y) = \max_{k=1,\dots,m} \min_{i \in I_+(l^k)} l^k_i y_i \geq \min_{i \in I_+(l^j)} l^j_i y_i = l^j_j y_j \geq h(x^*).$$

Thus it is sufficient to consider vectors y such that $l^j_j y_j < h(x^*)$. Since $y_j = x^*_j + \delta$, it follows that

$$l^j_j (x^*_j + \delta) < h(x^*).$$

Combining the last inequality with (9.4.40) we have

$$l^j_j x^*_j + l^j_j \delta < l^q_j x^*_j$$

which implies

$$\delta < \frac{l_j^q - l_j^j}{l_j^j} x_j^* = l_j^q \left(\frac{1}{l_j^j} - \frac{1}{l_j^q} \right) x_j^*. \tag{9.4.41}$$

Recall (see (9.4.38)) that $\delta = \sum_{i \neq j} \delta_i$, where $\delta_i = x_i^* - y_i$. We also have (see (9.4.40)) $\tilde{l}_j = l_j^q$. Therefore, (9.4.41) demonstrates that

$$\delta_i \leq l_j^q \left(\frac{1}{l_j^j} - \frac{1}{l_j^q} \right) x_j^* = l_j^q \left(\frac{1}{l_j^j} - \frac{1}{\tilde{l}_j} \right) x_j^*.$$

It follows from (9.4.35) that

$$\delta_i \leq l_j^q \left(\frac{1}{l_i^i} - \frac{1}{l_i^q} \right) x_j^*.$$

Since $q \in K$ and $\tilde{l}_i = \max_{k \in K} l_i^k$ we have $\tilde{l}_i \geq l_i^q$. Combining this inequality with (9.4.37), we have

$$x_i^* - y_i = \delta_i \leq l_i^i x_i^* \left(\frac{1}{l_i^i} - \frac{1}{l_i^q} \right) = x_i^* \left(1 - \frac{l_i^i}{l_i^q} \right).$$

Due to (9.4.40), we get

$$y_i \geq \frac{l_i^i}{l_i^q} x_i^* = \frac{h(x^*)}{l_i^q}.$$

Hence $l_i^q y_i \geq h(x^*)$ for all $i \neq j$. Combining (9.4.39) and (9.4.40) we have $l_j^q y_j > l_j^q x_j^* = h(x^*)$. Thus we conclude that

$$h(y) = \max_k \min_i l_i^k y_i \geq \min_i l_i^q y_i \geq h(x^*).$$

The proof is completed. △

REMARK 9.12 The main idea of the approach, proposed by D. Babaeyv, follows from Proposition 9.11 and Proposition 9.12: if (9.4.33) is valid then there exists a unique global minimizer, which depends only on the first n vectors, and remaining vectors have no influence on the global minimum. If (9.4.33) does not hold and there exists an index j such that (9.4.35) is valid, then we should 'touch up' the initial matrix. In order to do this, we can change only one number from the given n numbers. We take into account the remaining vectors only for this purpose: to define this number and determine its change.

From the point of view of the cutting angle method, we have the following situation: the first n vectors, which form a diagonal matrix ℓ_n, appear at the first iteration of this method. Then we shall add a new vector at each iteration. If the number j exists, such that (9.4.35) is valid, then the family of all obtained vectors allows us to change one of the numbers l_i^i and then to find a global minimizer of the subproblem.

9.5. MONOTONE OPTIMIZATION

9.5.1. Overview

The generalized cutting plane method and, in particular, the cutting angle method are based on the special approximation of the objective function without any approximation of constraints. The different approach is possible where the feasible set is approximated. The outer approximation of a feasible set [81, 95, 187, 188, 189] is a general approach for solving global optimization problems. Its successful implementation is based on specific properties of the problem under consideration. In particular, the outer approximation method for concave minimization [81, 95, 189] exploits separation properties of convex sets and linear approximation properties of convex functions. Let H be a set of elementary functions defined on a set X. Recall that a function $g : X \to \mathbb{R}$ is called abstract quasiconvex with respect to H if its level sets $S_c(g) = \{x \in X : g(x) \leq c\}$ are (X, H)-convex, that is each point $x \notin S_c(g)$ can be separated from $S_c(g)$ by an elementary function. This kind of separation property can be useful in applications of the outer approximation approach. We apply this idea to problems with an increasing objective function and an increasing constraint function defined on the cone $\mathbb{R}^n_+$.

A function f defined on $\mathbb{R}^n_+$ is abstract quasiconvex with respect to the class of all min-type functions L if and only if this function is increasing and lower semicontinuous (see Proposition 2.22). In other words, a function f is lower semicontinuous and increasing if and only if its level sets $S_c(f) = \{x \in \mathbb{R}^n_+ : f(x) \leq c\}$ are closed and normal for all $c \in \mathbb{R}$. Since closed normal sets are abstract convex with respect to L (see Proposition 2.19) it follows that each point which does not belong to a normal closed set G can be separated from G by a min-type function $l(x) = \langle l, x\rangle \equiv \min_{i=1,\dots,n} l_i x_i$, that is, there exists a vector $l \in \mathbb{R}^n_{++}$ such that

$$\sup_{y\in G}\langle l, y\rangle = c < \langle l, x\rangle.$$

Since

$$\{z : \langle l, z\rangle > c\} = \{z : l_i z_i > c \text{ for all } i\} = \frac{c}{l} + \mathbb{R}^n_{++},$$

it follows that geometrically we can separate each point $x \notin G$ by a shift of the cone $\mathbb{R}^n_{++}$. Thus we again exploit cutting cones (*cutting angles* for $n = 2$), however in this situation they are used in order to discard unfit elements of the feasible set.

The resulting algorithm, which is extremely simple conceptually and very easy to implement, turns out to be practically quite efficient as well. In fact, preliminary numerical experiments using conventional PC computers have shown that this algorithm can solve problems of dimension over ten fairly quickly (while it is known that general concave minimization problems of this dimension are already very hard to solve by the standard outer approximation method).

The results from this section were published in [163]. The further developments of this approach have been carried out by Hoang Tuy (see [190] for the survey of these results). Many examples of problems of the global optimization which can be converted to a problem of monotonic optimization can be found in [190].

9.5.2. Preliminaries

We begin by introducing some notations and concepts.

If $a, b \in \mathbb{R}^n_+$ and $a \le b$ we define the rectangle $[a, b]$ to be the set of all x such that $a \le x \le b$. We also write $(a, b] := \{x : a \ll x \le b\}$. As usual $\mathbf{1} = (1, \dots, 1)$ and e_i is the ith orth of $\mathbb{R}^n$.

Recall (see (2.3.10)), that the normal hull $\operatorname{Nh} U$ of a set $U \subset \mathbb{R}^n_+$ is defined by

$$\operatorname{Nh} U = (U - \mathbb{R}^n_+) \cap \mathbb{R}^n_+.$$

Properties of the normal hull are discussed in detail in Subsection 2.3.5.

Let G be a normal set in $\mathbb{R}^n_+$. A point $y \in G$ is said to be a *Pareto point* of G if $(y' \in G, y' \ge y) \implies y' = y$. In other words, a Pareto point is a maximal element of the set G with respect to the order relation $\ge$. Since $y \le x$ if and only if $y \in [0, x]$ it follows that $y \in G$ is a Pareto point of G if and only if the set $G \setminus \{y\}$ is still normal.

PROPOSITION 9.13 *A compact normal set G is the normal hull of the set of its Pareto points V.*

Proof: Suppose that $y \in G$ and $U = \{y' \in G : y' \ge y\}$. Let W be an infinite chain of elements of U that is, for all $y', y'' \in W$ either $y' \ge y''$ or $y'' \ge y'$. Since G is compact, W is bounded. Let $w = \sup_{y \in W} y$. It follows from the compactness of G that $w \in G$. By Zorn's lemma the set U contains a maximal element, say v. It follows from the definition of U that $v \geq y$. Let us check that v is a maximal element of G. Indeed, if $y' \geq v$, then $y' \geq y$, hence $y' \in U$. Since $y' \geq v$ and v is a maximal element of U it follows that $y' = v$. Thus the set V of Pareto points (maximal elements) of G is not empty and $y \in \mathrm{Nh}\, V$. Hence, $G \subset \mathrm{Nh}\, V$. The reverse inclusion is trivial. △

A normal set $M = \bigcup\{[0, z] : z \in Z\}$ where Z is a finite set, is called a *polyblock* generated by Z. (The notion of polyblock was introduced by H.Tuy (private communication).) It is easy to see, that $z \in Z$ is a Pareto point of M if and only if $(z' \in Z, z' \geq z) \implies z' = z$.

It follows directly from the definition that the following assertions are equivalent:

1) M is a polyblock.

2) M is the normal hull of a finite set.

3) M is a compact normal set with a finite number of Pareto points.

A Pareto point of a polyblock will be referred to as a *vertex*.

Recall (see Definition 2.13) that a point $x \in G$ is said to be a *weak Pareto point* (briefly w.P. point) of G if $\lambda x \notin G$ for all $\lambda > 1$.

The set of weak Pareto points of G is called the *upper boundary* of G and will be denoted by $\mathrm{bd}^+ G$. Of course any Pareto point of G is a weak Pareto point, but not conversely.

EXAMPLE 9.24 Let $G = \{(x_1, x_2) \in \mathbb{R}^2_+ : 0 \leq x_1 \leq 1, 0 \leq x_2 \leq 1\}$. The set G has the unique Pareto point $(1, 1)$. The upper boundary of this set consists of segments $\{(x_1, 1) : 0 \leq x_1 \leq 1\}$ and $\{(1, x_2) : 0 \leq x_2 \leq 1\}$.

PROPOSITION 9.14 *Let G be a normal set and $x \in G \cap \mathrm{int}\, \mathbb{R}^n_+$. Then x is a w.P. point of G if and only if*

$$(x + \mathbb{R}^n_{++}) \cap G = \emptyset. \tag{9.5.1}$$

Proof: Let x be a w.P. point of G and $u \gg 0$. There exists $\varepsilon > 0$ such that $\varepsilon x \leq u$. Since $(1 + \varepsilon)x \notin G$ and G is normal, we conclude that $x + u \notin G$. Now assume that (9.5.1) holds. Take $\lambda > 1$. Since $\lambda x \in x + \mathbb{R}^n_{++}$, we have $\lambda x \notin G$, so x is a w.P. point. △

Each normal closed set G can be represented as $G = \{x \in \mathbb{R}^n_+ : g(x) \leq 1\}$ where g is a lower semicontinuous increasing function. Indeed, a normal set is a radiant subset of the cone $\mathbb{R}^n_+$, hence (see Subsection 5.2.2), for a closed normal set G the following equality holds:

$$G = \{x \in \mathbb{R}^n_+ : \mu_G(x) \le 1\}$$

where μ_G is the Minkowski gauge of G. Since G is normal and closed, it follows that μ_G is increasing and lower semicontinuous. Let f be a lower semicontinuous increasing function defined on $\mathbb{R}^n_+$. Then its level sets $S_c(f) = \{x \in \mathbb{R}^n_+ : f(x) \le c\}$ are closed and normal for all $c \in \mathbb{R}$. Consider the set L of all min-type functions of the form $x \mapsto \langle l, x\rangle$ with $l \in \mathbb{R}^n_+$, where $\langle l, x\rangle = \min_{i \in I_+(l)} l_i x_i$. As it was mentioned before (see Proposition 2.22), a function $f : \mathbb{R}^n_+ \to \mathbb{R}_{+\infty}$ is L-quasiconvex if and only if f is lower semicontinuous and increasing.

9.5.3. Problems of monotonic optimization

We will consider the following nonconvex global optimization problem

$$(P) \qquad f(x) \longrightarrow \max \text{ subect to } g(x) \le 1,\ x \in \mathbb{R}^n_+,$$

where $f, g : \mathbb{R}^n_+ \to \mathbb{R}$ are increasing functions, f is upper semicontinuous, g is lower semicontinuous and the following holds:

1) $g(0) < 1$;

2) for each $x \in \mathbb{R}^n_+ \setminus \{0\}$ there exists $\lambda > 0$ such that $g(\lambda x) \le 1$.

Note that 2) holds if g is continuous. Indeed, if there exists x such that $g(\lambda x) > 1$ for all $\lambda > 0$, then

$$1 > g(0) = \lim_{\lambda \to 0} g(\lambda x) \ge 1,$$

which is impossible. Denote by G the feasible set of problem (P):

$$G = \{x \in \mathbb{R}^n_+ : g(x) \le 1\}.$$

Since g is increasing and lower semicontinuous, it follows that G is normal and closed. For each $x \ne 0$ there exists $\lambda > 0$ such that $\lambda x \in G$, hence the cone hull $\operatorname{cone} G$ of the set G coincides with $\mathbb{R}^n_+$. We assume that G is bounded, therefore an optimal solution of problem (P) exists.

Note that any problem of the form

$$(P') \qquad f(x) \longrightarrow \max \text{ subject to } x \in U$$

where $U \subset \mathbb{R}^n_+$ is an arbitrary compact set can easily be reduced to (P). In fact, since f is increasing, it follows that any optimal solution of (P') is also an optimal solution of the problem:

$$(P'') \qquad f(x) \longrightarrow \max \text{ subject to } x \in \operatorname{Nh}(U).$$

Let g be the Minkowski gauge of the set $\mathrm{Nh}(U)$. Since $\mathrm{Nh}(U)$ is normal and closed it follows that g is increasing and lower semicontinuous. We have $\mathrm{Nh}(U) = \{x \in \mathbb{R}^n_+ : g(x) \leq 1\}$, hence the problem (P') with U being an arbitrary compact set is equivalent to problem (P) with g being equal to the Minkowski gauge of $\mathrm{Nh}(U)$.

Many nonconvex optimization problems can be rewritten as monotonic optimization problem (P'), and hence as problem (P). For instance, if $f : S \to R_+$ is a positive Lipschitz function defined on the unit simplex $S = \{x \in \mathbb{R}^n_+ : \sum_{i=1}^n x_i = 1\}$, then one can find an increasing function F on $\mathbb{R}^n_+$, which coincides with f on S (see Subsection 3.3.6). Consequently, the problem of minimizing $f(x)$ on a subset U of S is reduced to the problem of minimizing F on U, which in turn is problem (P'), and can be reduced to problem (P).

Consider the following simple example of a monotonic optimization problem.

EXAMPLE 9.25 Let $n = 2$ and let $G = [0, x^1] \cup [0, x^2]$ be a polyblock with the vertices $x^1 = (1, 3)$ and $x^2 = (2, 1)$. Let $f(x) = x_1 + x_2$ for $x \in \mathbb{R}^2_+$. Clearly, x^2 is a local maximum and x^1 is a global maximum of the function f over the set G. We can represent G as $G = \{x \in \mathbb{R}^2_+ : g(x) \leq 1\}$ with

$$g(x) = \min\left(\max\left(x_1, \frac{1}{3}x_2\right), \max\left(\frac{1}{2}x_1, x_2\right)\right).$$

We shall show that by using the monotonic structure, it is possible to propose a variant of the outer approximation method for coping with this problem.

9.5.4. Basic properties

We first present some properties of normal sets and monotonic functions which play a basic role in the analysis of the problem.

PROPOSITION 9.15 *For any $x \in \mathbb{R}^n_+ \setminus G$ the segment $\{\lambda x : 0 \leq \lambda \leq 1\}$ contains a uniquely defined weak Pareto point of G.*

Proof: Since G is normal and bounded and the conic hull $\mathrm{cone}\, G$ coincides with $\mathbb{R}^n_+$, it follows that the set $G \cap \{tx : t \geq 0\}$ is a segment. Let y be the right endpoint of this segment, i.e. $y = \theta x$, where

$$\theta := \theta(x) = \max\{t > 0 : tx \in G\}. \tag{9.5.2}$$

Since $x \notin G$ it follows that $\theta < 1$. Clearly y is the unique weak Pareto point of G on the segment $\{\lambda x : 0 \leq \lambda \leq 1\}$. △

REMARK 9.13 We have

$$\begin{aligned}\theta(x) &= \max\{t > 0 : tx \in G\} = \max\left\{\frac{1}{\lambda} : x \in \lambda G\right\} \\ &= \frac{1}{\min\{\lambda > 0 : x \in \lambda G\}} = \frac{1}{\mu_G(x)}.\end{aligned}$$

PROPOSITION 9.16 *Let G be a compact normal set contained in a polyblock M with the vertex set $V \subset \mathbb{R}^n_{++}$. Let $z \in V \setminus G$, and $y = \theta z$ where $\theta = \theta(z)$ is defined by (9.5.2). Consider the set $Z' = (V \setminus \{z\}) \cup \{x^1, \ldots, x^n\}$, where*

$$x^i = z - (z_i - y_i)e_i, \quad i = 1, \ldots, n. \tag{9.5.3}$$

Then $x^i \gg 0$ and the polyblock M' generated by Z' satisfies

$$M \supset M' \supset G, \quad z \in M \setminus M'. \tag{9.5.4}$$

Proof: We have

$$x^i_j = \begin{cases} \theta z_i & \text{if } j = i; \\ z_j & \text{if } j \neq i. \end{cases}$$

Since $\theta > 0$ it follows that $x^i \gg 0$. We have $z \in y + \mathbb{R}^n_{++} \subset \mathbb{R}^n_+ \setminus G$. But $y + \mathbb{R}^n_{++} = \cap_{i=1}^n \{x : x_i > y_i\}$, so

$$\begin{aligned}[0, z] \setminus (y + \mathbb{R}^n_{++}) &= \bigcup_{i=1}^n \{x \in [0, z] : \ x_i \leq y_i\} \\ &= \bigcup_{i=1}^n \{x : 0 \leq x_i \leq y_i, \ 0 \leq x_j \leq z_j \ \text{ for all } j \neq i\} \\ &= \bigcup_{i=1}^n [0, x^i].\end{aligned}$$

Consequently, the polyblock $M' = \mathrm{Nh}\,(Z')$ satisfies (9.5.4) (the right-hand inclusion in (9.5.4) holds because z is a vertex of M and so $z \notin \mathrm{Nh}\,(V \setminus \{z\})$). △

Proposition 9.16 shows that the feasible set G (a compact normal set) can be approximated by a polyblock $M \supset G$, that is, an approximate optimal solution of problem (P) can be obtained by solving an approximate problem

$$\max\{f(x) : x \in M\} \tag{9.5.5}$$

where M is a suitable polyblock containing G. Since the function f is increasing, its maximum over a polyblock is attained at a vertex of M. Hence problem (9.5.5) can be solved just by enumeration of the vertices of M. Therefore, we only should find a polyblock $M \supset G$ sufficiently close to G in some neighbourhood of a global minimizer. This can be done by using a procedure similar to the outer approximation method for concave programs (see [81], [95] and also [189]), namely:

Start from an initial polyblock M_1. If the maximizer of $f(x)$ over M_1 belongs to G, the procedure terminates; otherwise, using Proposition 9.16 we construct a polyblock M_2 still containing G but excluding x^1. Then repeat this procedure, until a polyblock M_k is obtained such that the maximizer x^k of $f(x)$ over M_k is sufficiently close to G.

9.5.5. Proposed solution method

We now describe the algorithm in precise terms. Let ε be a given tolerance. Consider the set

$$G_\varepsilon = \{x \in G : x \geq \varepsilon \mathbf{1}\}.$$

We assume that ε is so small that $G_\varepsilon \neq \emptyset$. A global optimal solution of the problem

$$(P_\varepsilon) \qquad f(x) \longrightarrow \max \text{ subject to } x \in G_\varepsilon$$

will be called an *ε-optimal solution* of (P).

A solution $\bar{x} \in G$ such that $f(\bar{x})$ differs from the optimal value in (P_ε) by at most $\eta > 0$ will be referred to as an *(ε, η)-approximate optimal solution* of (P).

For the initialization of the outlined outer approximation procedure, we need a simple polyblock containing G. This initial polyblock is taken to be a box $[0, b] \supset G$ constructed as follows. For each $i = 1, \ldots, n$ find an upper estimate $b_i \geq \max\{x_i : x \in G\}$. Since $G \cap \mathbb{R}^n_{++} \neq \emptyset$, it follows that $\max\{x_i : x \in G\} > 0$, so $b \gg 0$. We have

$$G \subset [0, b]$$

because for any $x \in G$ and for all i we have $x_i \leq b_i$. We shall assume that the tolerance $\varepsilon > 0$ is so small that $x^1 \geq \varepsilon \mathbf{1}$.

Algorithm for monotonic optimization

Step 0. Compute $b_i \geq \max\{x_i : x \in G\}$, $i = 1, \ldots, n$. Let $M_1 = [0, b], V_1 = \{b\}, x^1 = b, \bar{y}^1 = y^1$ (the intersection point of the upper boundary of G with the segment between 0 and x^1). Set $k = 1$.

Step 1. Compute

$$x^k \in \text{Argmax}\,\{f(x):\ x \in V_k,\ x \geq \varepsilon\mathbf{1}\},$$

where $\text{Argmax}\{f(x) : x \in V\} = \{x \in V : f(x) = \max_{y \in V} f(y)\}$. If $x^k \in G$ then the algotithms terminates: x^k is an ε-optimal solution.

Step 2. Compute the intersection point y^k of the upper boundary of G with the segment between 0 and x^k. Set $\bar{y}^k = \text{Argmax}\{f(\bar{y}^{k-1}), f(y^k)\}$. If $f(\bar{y}^k) \geq f(x^k) - \eta$ then the algorithm terminates: $\bar{y}^k$ is an (ε, η)-approximate solution of (P).

Step 3. Compute n extreme points of the rectangle $[y^k, x^k]$ that are adjacent to x^k :

$$x^{k,i} = x^k - (x_i^k - y_i^k)e_i, \quad i = 1, \ldots, n \tag{9.5.6}$$

(where, as usual, e_i is the ith orth of $\mathbb{R}^n$). Set

$$Z_{k+1} = (V_k \setminus \{x^k\}) \cup \{x^{k,1}, \ldots, x^{k,n}\}.$$

Let V_{k+1} be the set obtained from Z_{k+1} after dropping out all those $z \in Z_{k+1}$ that are not vertices (i.e., such that z is dominated by some other $z' \in Z_{k+1}$).

Step 4. Set $k := k + 1$ and go to Step 1.

REMARK 9.14 In outer approximation algorithms for concave minimization, the number of vertices of the enclosing polytope increases exponentially at each iteration, with the new vertices becoming more and more difficult to compute accurately (see e.g. [189]). By contrast, in the present algorithm the vertex set V_k increases at most by $n - 1$ at each iteration and the new vertices are extremely easy to compute. In spite of that, the vertex set of the current polyblock may reach a prohibitively large size, creating storage problems. Should this happen, it is recommended to break off the procedure and to restart it from the last x^k. Specifically, let S be the critical size for the vertex set (so difficulties may arise if there are more than S vertices to be recorded). Step 4 should be modified as follows:

Step 4. If $|V_{k+1}| \leq S$, then set $k := k + 1$ and go to Step 1. Otherwise go to Step 5.

Step 5. Redefine $V_{k+1} = \{b - (b_i - y_i^k)e_i,\ i = 1, \ldots, n\}$, (i.e., $M_{k+1} = [0, b] \setminus (x^k, b])$, then set $k := k + 1$ and go to Step 1.

This restarting procedure enables us to overcome memory space limitations and sometimes also to speed up the convergence.

Note that the sequence generated by the algorithm consists of strictly positive vectors.

9.5.6. Convergence

To establish the convergence of the above algorithm we first prove the following result.

PROPOSITION 9.17 *For each* $k = 1, \ldots$ *we have* $M_k \supset G$ *and* $x^k \in \mathrm{Argmax}\{f(x) : x \in M_k, x \geq \varepsilon e\mathbf{1}\}$. *Hence*

$$f(\bar{y}^k) \leq \max\{f(x) : x \in G - \varepsilon\} \leq f(x^k). \qquad (9.5.7)$$

Proof: By Proposition 9.16 we have $M_k \supset G$ for all k. Furthermore, since V_k is the vertex set of M_k it follows that $x^k \in \mathrm{Argmax}\{f(x) : x \in M_k, x \geq \varepsilon e\mathbf{1}\}$. Thus the right-hand side inequality in (9.5.7) holds. Since $y^1, \ldots, y^k \in G_\varepsilon$ and $\bar{y}^k$ is one of these points, we conclude that the left-hand side inequality in (9.5.7) holds as well. △

As a consequence, the answer given by the algorithm when it terminates is correct (x^k is an ε-optimal solution if the termination occurs at Step 1, while $\bar{y}^k$ is an (ε, η)-approximate solution if the termination occurs at Step 2). It remains to prove the finiteness of the algorithm.

THEOREM 9.2 *The algorithm terminates after finitely many steps, yielding an* ε*-optimal solution or an* (ε, η)*-approximate solution.*

Proof: We first prove that for any $\delta > 0$ there exists k such that $\min_{i=1,\ldots,n}(x_i^k - y_i^k) \leq \delta$. Suppose the contrary, that $\min_{i=1,\ldots,n}(x_i^k - y_i^k) > \delta$ for fairly large k. Observe that for any x^k generated by the algorithm there exists a sequence

$$x^1 = x^{h_1} \geq x^{h_2} \geq \ldots \geq x^{h_p} = x^k$$

such that

$$x^{h_{q+1}} = x^{h_q} - (x_{i_q}^{h_q} - y_{i_q}^{h_q})e_{i_q}, \qquad q = 1, \ldots, p-1. \qquad (9.5.8)$$

Hence, setting

$$\alpha_q = x_{i_q}^{h_q} - y_{i_q}^{h_q} > \delta$$

and adding $p-1$ equalities in (9.5.8), we obtain

$$x^k = x^1 - \sum_{q=1}^{p-1} \alpha_q e_{i_q}. \tag{9.5.9}$$

Let $i_q = j$ for N_j values of $q = 1, \ldots, p-1$. Since $x^1 = b$, (9.5.9) implies that, for every $j = 1, \ldots, n$,

$$b_j \geq b_j - x_j^k = \sum_{q:i_q=j} \alpha_q \geq N_j \delta,$$

so that $N_j \leq b_j/\delta$ for all $j = 1, \ldots, n$. Hence, $p = 1 + N_1 + \ldots + N_n \leq 1 + \sum_{j=1}^n (b_j/\delta)$, conflicting with p becoming arbitrarily large as $k \to +\infty$.

We have thus proved that for any $\delta > 0$ there exists k such that $\min_{i=1,\ldots,n}(x_i^k - y_i^k) \leq \delta$. Noting that $y^k = \lambda_k x^k$ and $x_i^k \geq \varepsilon\ \forall i$, we then have

$$\min_{i=1,\ldots,n}(x_i^k - y_i^k) = \min_{i=1,\ldots,n}(1-\lambda_k)x_i^k \geq (1-\lambda_k)\varepsilon,$$

hence $1 - \lambda_k \leq \frac{1}{\varepsilon}\min_{i=1,\ldots,n}(x_i^k - y_i^k) \leq \delta/\varepsilon$, i.e.,

$$\|x^k - y^k\| = (1-\lambda_k)\|x^k\| \leq (\delta/\varepsilon)\|b\|.$$

Thus, if $\delta > 0$ is chosen so that $f(x^k) - f(y^k) \leq \eta$ whenever $\|x^k - y^k\| \leq (\delta/\varepsilon)\|b\|$ then for some sufficiently large k we have $f(x^k) - f(\bar{y}^k) \leq f(x^k) - f(y^k) \leq \eta$ and the algorithm will stop. $\triangle$

9.5.7. Computational experience

The algorithm has been tested on problems with 2, 4, 8 and 16 variables. The following types of increasing functions were chosen for both the objective function $f(x)$ and the constraint function $g(x)$:

- Cobb-Douglas functions $f(x) = \prod x_i^{\alpha_i}$.
- Quadratic functions $f(x) = x^T Ax$, where A is a random matrix with nonnegative entries.
- Polynomials in x with nonnegative cofficients.
- Min-max functions.

In many instances, we found that, whilst the algorithm itself proved to be efficient (in terms of number of iterations required to reach the optimum), much of the computational time was spent on locating the weak Pareto points of the feasible set.

We have used two approaches for finding the weak Pareto points. One is a version of Newton's technique for solving equations, the other is Bolzano's bisection procedure. For small problems (of dimension two or four) the two approaches produced roughly similar results, but for larger problems Bolzano's bisection procedure was found to perform considerably better. For example, for an 8-variable problem where the constraint function was of the minmax or $\sum_i(x_i^a)$ type and the objective function was either quadratic or of Cobb-Douglas type, the computational time when using this procedure never exceeded five minutes (even for very small values of the tolerance parameters ε and η), while it exceeded 1 hour when using Newton's method and in the worst case, the process had to be prematurely aborted after 2 hours of unsuccessful computation. This definitely confirms that for the implementation of our algorithm the bisection technique, rather than Newton's method, should be used to locate the weak Pareto points.

Execution times were also sensitive to the values of ε and η. Typically, we set each of these values to be 0.01 and, when a solution was obtained, we reset these values depending upon the relative sizes of the x_i values and the final value of the function. Typically, a reduction in the values of ε and η by the factor of 10 led to an increase in calculation times by the factor of 3 to 4. The final important point was the dimension of the problem. Because the number of vertices to be examined at each iteration increases at a linear rate, the execution time for each iteration also increases linearly and so the total execution time to locate a solution increases at a greater rate. For example, the total execution time for an 8-dimensional problem would be about 4 times for a 4-dimensional problem of the same type.

References

[1] Abasov, T. M. and Rubinov, A. M., On the class of H-convex functions, *Russian Acad. Sci. Dokl. Math.* **48** (1994), 95-97.

[2] Abasov, T. M. and Rubinov, A. M., Subdifferentials of some classes of nonsmooth functions, in: *Mathematical Models of Analysis of Nonsmooth models* (V. F. Demyanov (ed.)), St. Petersburg University Press, St Petersburg (1996) (In Russian).

[3] Andramonov, M. Yu., An approach to constructing generalized penalty functions, Research Report 21/97, SITMS, University of Ballarat (1997).

[4] Andramonov, M. Yu., Some conditions of convergence of tabu search, Research Report 14/99, SITMS, University of Ballarat (1999).

[5] Andramonov, M. Yu., Some general approaches to abstract convex programming, Research Report 38/99, SITMS, University of Ballarat (1999).

[6] Andramonov, M. Yu, Rubinov, A. M. and Glover, B. M., Cutting angle methods in global optimization. *Applied Mathem. Letters*, **12** (1999), 95-100.

[7] Andramonov, M. Yu., Rubinov, A. M. and Glover, B. M., Cutting angle method for minimizing increasing convex-along-rays functions, Research Report 7/97, SITMS, University of Ballarat (1997).

[8] Aubin, J.-P. and Ekeland, I., *Applied Nonlinear Analysis*, Wiley & Sons (1984).

[9] Auslender, A., Cominetti, R. and Haddou, M., Asymptotic analysis for penalty and barrier methods in convex and linear programming, *Mathematics of Operations Research*, **22** (1997), 43-62.

[10] Avriel, M., Diewert, W. E., Schaible, S. and Zang, I., *Generalized Concavity*, Mathematical Concepts and Methods in Science and Engineering, Plenum Press, New York **36** (1988).

[11] Babayev, Dj. A., An exact method for solving the subproblem of the cutting angle method for global optimization, submitted for publication.

[12] Bagirov, A. M. and Rubinov, A. M., Global minimization of increasing positively homogeneous functions over the unit simplex, *Annals of Operation Research*, to appear.

[13] Balas, E., Disjunctive programming, *Annals of Discrete Mathematics* **5** (1979), 3-51.

[14] Balas, E., Tama, J. and Tind, J., Sequential convexification in reverse convex and disjunctive programming, *Mathematical Programming* **44, 3** (1989), 337-350.

[15] Balder, E. J., An extension of duality-stability relations to non-convex optimization problems, *SIAM J. Contr. Optim.* **15** (1977), 329-343.

[16] Barbara, A. and Crouzeix, J.-P., Concave gauge functions and applications, *Zeitschrift fur Operations Research*, **40** (1994), 43 -74.

[17] Bauschke, H. H. and Borwein, J. M., On projection algorithms for solving convex feasibility problems, *SIAM Rev.* **38** (1996), 367-426.

[18] Bartels, S. G., Kuntz, L. and Sholtes, S., Continuous selections of linear functions and nonsmooth critical point theory, *Nonlinear Analysis, TMA* **24** (1995), 385 -407.

[19] Beer, G. A., Starshaped sets and the Hausdorff metrics, *Pacific Journal of Mathematics*, **61** (1955), 21-27.

[20] Beer, G and Klee, V., Limits of starshaped sets, *Arch. Math.* **48**(1987), 241-249.

[21] Bertsekas, D. P., *Constrained Optimization and Lagrange Multiplier Methods*, Academic Press (1982).

[22] Bertsekas, D. P., Nondifferentiable optimization via approximation, *Math. Prog.*, **3** (1976), 1-25.

[23] Birkhoff, G., *Lattice Theory*, third ed., American Mathematical Society Colloqiuim Publications, vol.25. AMS, 1967.

[24] Bragard, L., Ensembles ètoilès et irradiès dans un espace vectoriel topologique, *Bull. Soc. Roy Sci. Liège*, **37** (1968), 276-285.

[25] Bragard, L., Propriètès des ensembles irradiès, *Bull. Soc. Roy Sci. Liège*, **39** (1970), 264 -268.

[26] Branin, F. H. and Hoo, S. K., A method for finding multiple extrema of a function of n variables, in: *Lootsma, F. A. (ed.), Numerical Methods of Nonlinear Optimization*, Academic Press, London (1972), 231-237.

[27] Bulatov, V. P., *Embedding methods in optimization problems*, Nauka, Novosibirsk (1977). (In Russian).

[28] Burke, J. V., Calmness and exact penalization, *SIAM J. Control and Optimization* **29** (1991), 493-497.

[29] Burke, J. V., An exact penalization viewpoint of constrained optimization, *SIAM J. Control and Optimization*, **29** (1991), 968 -998.

[30] Clarke, F. H., *Optimization and Nonsmooth Analysis*, John Wiley, New York (1983).

[31] Chen, C. and Mangasarian, O. L., Smoothing methods for convex inequalities and linear complementarity problems, *Mathematical Programming*, **71** (1995), 51-69.

[32] Churilov, L. and Sniedovich, M., Optimization in engineering design: composite concave programming perspective, in *The Role of Mathematics in Modern Engineering*, Eaton and Stainer (eds.), Chartwell-Brat (1995), 559 -564.

[33] Goh, C. J. and Yang, X. G., A sufficient and necessary condition for nonconvex constrained optimization, *Applied Mathematics Letters*, **10** (1997), 9-12.

[34] Craven, B. D., *Mathematical Programming and Control Theory*, Chapman and Hall, London (1978).

[35] Crouzeix, J.-P., Conjugacy in quasiconvex analysis, in: *Convex Analysis and its Applications*, Auslender, A. (ed.), Springer-Verlag, Berlin (1977), 66-99.

[36] Crouziex, J.-P., About differentiability of order one of quasiconvex functions on $\mathbb{R}^n$, *Journal of Optimization Theory and Applications*, **36(3)** (1982), 367-385.

[37] Crouziex, J.-P., Characterizations of generalized convexity and generalized monotonicity, a survey. [38], 237-256.

[38] Crouzeix, J.-P., Martinez-Legaz, J. E. and Volle, M. (Eds.), *Generalized Convexity, Generalized Monotonicity*, Proceedings of the Fifth International Symposium on Generalized Convexity, held at the Centre International de Rencontres Mathmatiques (CIRM), Luminy-Marseille (France), June 17-21, 1996, Kluwer Academic Publishers, London (1997).

[39] Danford, N. and Schwartz, J. T., *Linear operators* Part 1, General Theory, Intersience Publishers , New York (1958).

[40] Day, M. M., *Normed linear spaces* (third edition) Springer-Verlag, Berlin, Heidelberg (1973).

[41] Demyanov, V. F., Exhausters of a positively homogeneous function, *Optimization* **45** (1999), 13-29.

[42] Demyanov, V. F. and Rubinov, A. M., *Quasidifferential Calculus*, Optimization Software, New York (1986).

[43] Demyanov, V. F. and Rubinov, A. M., *Constructive Nonsmooth Analysis*, Peter Lang, Frankfurt am Main (1995).

[44] Dolecki, S. and Kurcyusz, S., On Φ-convexity in extremal problems. *SIAM J. Contr. Optimiz.* **16** (1978), 277-300.

[45] Dragomir, S. S., Pečarić, J. E. and Persson, L. E., Some inequalities of Hadamard type, *Soochow J. Math* **21** (1995), 335-341.

[46] Dragomir, S. S. and Pearce, C. E. M., Quasi-convex functions and Hadamard's inequality. *Bull. Austral. Mathem. Soc.* **57** (1998), 377-385.

[47] Du, D.-Z. and Pardalos, P. M. (eds), *Minimax and Applications*, Kluwer Academic Publishers, Dordrecht (1995).

[48] Dzalilov, Z., Rubinov, A. M. and Kloeden, P. E., Lyapunov sequences and a turnpike theorem without convexity, *Set-Valued Analysis* **6** (1998), 277-302.

[49] Eberhard, A., Nyblom, M. and Ralph, D., Applying generalized convexity notions to jets. In *Generalized Convexity, Generalized Monotonicity: Recent Results* (Crouzeix J.-P. et al (eds)), Kluwer Academic Publishers, Dordrecht (1998), 111 -175.

[50] Eberhard, A. and Ralph, D., Support functions without subadditivity and the symmetric rank one representer, Research Report No 26, RMIT, Melbourne (1997).

[51] Ehnbat, R., An algorithm for maximizing a convex function over a simple set, *Journal of Global Optimization* **8** (1996), 379-391.

[52] Eaves, B. C. and Zangwill, W. I., Generalized cutting plane algorithm, *SIAM Journal on Control* **9** (1971), 529-542.

[53] Evtushenko, Yu. G., *Numerical Optimization Techniques*, Optimization Software, Inc. Publication Division, New York (1985).

[54] Falconer, K. J., The dimension of the convex kernel of a compact starshaped set, *Bull. London Math. Soc.*, **9** (1977), 313-316.

[55] Fan, K., On the Krein- Milman theorem. In *Convexity, Proc. of Sympos. in Pure Math.* **7**, Amer. Math. Soc., Providence (1963), 211-220.

[56] Fenchel, W., On conjugate convex functions, *Canadian Journal of Mathematics*, **1** (1949), 73-77.

[57] Fiacco, A. V. and McCormick, G. P., *Nonlinear Programming: Sequential Unconstrained Minimization Techniques*, John Wiley, New York (1968).

[58] Floudas, C. A. and Pardalos, P., *A Collection of Test Problems for Constrained Global Optimization Algorithms*, Lecture Notes in Computer Science **455**, Springer-Verlag, Berlin (1990).

[59] Giannessi, F., Theorems of the alternative and optimality conditions, *J. of Optimization Theory and Applications*, **42** (1984), 331-365.

[60] Glover, B. M., A generalized Farkas lemma with applications to quasidifferentiable programming, *Zeitschrift für Operations Research* **26** (1982), 125-141.

[61] Glover, B. M., On quasidifferentiable functions and nondifferentiable programming, *Optimization*, **24** (1990), 253-268.

[62] Glover, B. M., Jeyakumar, V., Ishizuka, Y. and Tuan, H. D., Complete characterizations of global optimality of DC minimization problems, *SIAM J. Optimization*, **6** (1996), 362-372.

[63] Glover, B. M., Jeyakumar, V. and Oettli, W., A Farkas lemma for difference sublinear systems and quasidifferentiable programming, *Mathematical Programming*, **63**(1994(, 109-125.

[64] Glover, B. M., Jeyakumar, V. and Oettli, W., Solvability theorems for classes of difference convex functions, *Nonlinear Analysis, Theory, Methods and Applications*, **22**(1994), 1191-1200.

[65] Glover, B. M., Rubinov, A. M. and Craven, B. D., Solvability theorems involving inf-convex functions. *J. Mat. Anal. Appl.* **191** (1995), 305-330.

[66] Glover, F. and Laguna M. *Tabu Search*, Kluwer Academic Publishers, Boston, 1997.

[67] Goh, C. J. and Yang, X. Q., A sufficient and necessary condition for nonconvex constrained optimization, *Applied Mathematics Letters* **10** (1997), 9-12.

[68] Grossman, Ch. and Kaplan, A., *Straffunktionen und Modifizierte Lagrangefunktionen in der Nichtlinearen Optimierund*, Teubner Texte zur Mathematik, Leipzig (1979).

[69] Gwinner, J., Results of Farkas type, *Numerical Functional Analysis and Optimization*, **9** (1987), 471-520.

[70] Ha Chung-wei, On systems of convex inequalities, *Journal of Mathematical Analysis and Applications*, **68** (1979), 25-34.

[71] Hadwiger, H., *Vorlesungen über Inhalt, Oberfläche und Isoperimetrie*, Springer, Berlin (1957).

[72] Hansen, P., Jaumard, B. and Lu, S. H., An analytical approach to global optimization, *Mathematical Programming* **52** (1991), 227-254.

[73] Hansen, P., Jaumard, B. and Lu, S. H., Global optimization of univariate Lipschitz functions: II. New algorithms and computational comparison, *Mathematical Programming* **55** (1992), 273-292.

[74] Hiriart-Urruty, J.-B., From convex to nonconvex minimization: Necessary and sufficient conditions for global optimality, in *Nonsmooth Optimization and Related Topics*, Plenum, New York (1989), 219-240.

[75] Hiriart-Urruty, J.-B. and Lemarechal, C., Testing necessary and sufficient conditions for global optimality in the problem of maximizing a convex quadratic function over a convex polyhedron, Preliminary Report, University of Paul Sabatier, Toulouse (1990).

[76] Hiriart-Urruty, J.-B. and Lemarechal, C., *Convex Analysis and Minimization Algorithms*, Volumes I and II, Springer-Verlag, Berlin, Heidelberg (1993).

[77] Hock, W. and Schittkowski, K.,*Test Examples for Nonlinear Programming Codes*, Lecture Notes in Economic and Mathematical Systems **187**, Springer-Verlag, Berlin, (1981).

[78] Hoffman, A., The distance to the intersection of two convex sets expressed by the distance to each of them, *Math. Nachr.* **157** (1992), 81-98.

[79] Horst, R. and Pardalos, P. (eds), *Handbook on Global Optimization*, Kluwer Academic Publishers (1996).

[80] Hörmander, L., Sur la fonction d'appui des ensembles convexes dans un espace localement convexe, *Arkiv för Mat.***3** (1955), 181-186.

[81] Horst, R. and Tuy, H., *Global Optimization (Deterministic Approaches)*, third edition, Springer-Verlag, Berlin, New York (1996).

[82] Horst, R. and Tuy, H., On an outer approximation concept in global optimization, *Optimization* **20** (1989), 255-264.

[83] Intriligator, M. D., *Mathematical Optimization and Economic Theory*, Prentice-Hall, Englewood Cliffs, N.J. (1971).

[84] Ioffe, A. and Tikhomirov, V., *Theory of Extremal Problems*, Nauka, Moscow (1974) (In Russian); English translation: North-Holland, Amsterdam, New York (1979).

[85] Ishizuka, Y., Farkas' theorem of nonconvex type and its application to a min-max problem, *Journal of Optimization Theory and Applications* **57** (1988), 341-354.

[86] Ishizuka, Y., Optimality conditions for quasidifferentiable programs with application to two-level optimization, *SIAM Journal of Control and Optimization*, **26(6)** (1988), 1388-1398.

[87] Jeyakumar, V. and Glover, B. M., A new Farkas lemma and global convex maximization, *Applied Mathematics Letters*, **6(5)** (1993), 39-43.

[88] Jeyakumar, V. and Glover, B. M., Nonlinear extensions of Farkas' lemma with applications to global optimization and least squares, *Mathematics of Operations Research* **20** (1995), 818-837.

[89] Jeyakumar, V., Rubinov, A. M., Glover, B. M. and Ishizuka, Y., Inequality systems and global optimization, *Journal of Mathematical Analysis and Applications* **202** (1996), 900-919.

[90] Jeyakumar, V. and Wolkowicz, H., Generalizations of Slater's constraint qualifications for infinite convex programming, *Mathematical Programming, Series A*, **57** (1992), 85-101.

[91] Karmarkar, N., A new polynomial-time algorithm for linear programming, *Combinatorica*, **4** (1984), 375-395.

[92] Kelley, J., The cutting plane method for solving convex programs, *SIAM Journal* **8(4)** (1960), 703-712.

[93] Khamisov O., On optimization properties of functions, with a concave minorant, *Journal of Global Optimization* **14** (1999), 79-101.

[94] Konno, H. J. and Kuno, T., Generalized linear multiplicative and fractional programming, *Annals of Operations Research* **25** (1990), 147-162.

[95] Konno, H., Thach, P. T. and Tuy, H., *Optimization on Low Rank Nonconvex Structures*, Kluwer Academic Publishers, Dordrecht/Boston/London (1997).

[96] Korovkin P. P., *Lincar Opcrators and Approximation Theory*, Moscow, Fizmatgiz (1959) (In Russian).

[97] Kowalik, J., Osborn, M. R. and Ryan, D. M., A new method for constrained optimization problems, *Oper. Research*, **17** (1969), 973-989.

[98] Krasnosel'skii, M. A., Sur un critère pour qu'un domain soit étoilé, Matematicheskii Sbornik **19** (1946), 309-310.

[99] Kutateladze, S. S. and Rubinov, A. M., Minkowski duality and its applications, *Russian Mathem. Surveys* **27** (1972), 137-191.

[100] Kutateladze, S. S. and Rubinov, A. M., *Minkowski Duality and its Applications*, Nauka, Novosibirsk (1976) (In Russian).

[101] Lasdon, L. S., *Optimization Theory for Large Systems*, Macmillan, London (1970).

[102] Levin, V. L., Semiconical sets, semi-homogeneous functions, and a new duality scheme in convex analysis, *Dokl. Akad. Nauk* **354** (1997), 597 - 599. (In Russian)

[103] Levin, V. L., The existence and the uniqueness of a measure preserving optimal map in general Monge-Kantorovich problem. *Functional Analysis and its Applications* **32** (1998), 79-82 (In Russian).

[104] Li, D., Zero duality gap for a class of nonconvex optimization problems, *Journal of Optimization Theory and Applications* **85(2)** (1995), 309-324.

[105] Makarov, V. L. and Rubinov, A. M., *Mathematical Theory of Economic Dynamic and Equilibria*, Moscow, Nauka (1973) (In Russian); English translation: Springer-Verlag, Berlin, New York (1977).

[106] Makarov, V. L., Levin M. J. and Rubinov, A. M. *Mathematical Economic Theory: Pure and Mixed Types of Economic Mechanisms*, Elsevier, Amsterdam, (1995).

[107] Martinez-Legaz, J E., Quasiconvex duality theory by generalized conjugation methods. *Optimization* **19** (1988), 603-652.

[108] Martinez-Legaz, J E., General conjugation and related topics, in: *Generalized Convexity and Fractional Programming with Economic Applications*, Lecture Notes in Economics and Mathematical Systems, Springer-Verlag, Berlin, **345** (1990), 198-218.

[109] Martinez-Legaz, J E. and Singer, I., On conjugation for functions with values in extensions of ordered groups. *Positivity* **1** (1997), 193-218.

[110] Matveenko, V. D., On a dual representation of CRS functions by use of Leontiev functions. In *Proceedings of the First International Conference on Problems of Mathematical Economics, Nonsmooth Analysis and Informatics*, Elm, Baki, 1977, 160-166.

[111] Matveenko, V. D., Homogeneous functions with operations of taking maximum and minimum, Unpublished manuscript (in Russian).

[112] Mladineo, R. H., An algorithm for finding the global maximum of a multimodal, multivariate function. *Mathematical Programming* **34** (1986), 188-200.

[113] Minoux, M., *Programmation Mathematique, Theorie et Algorithmes Dunod*, Bordas et G.N.E.T.-E.N.S.T., Paris (1989).

[114] Moore, E. H., *Introduction to a Form of General Analysis*, American Mathematical Society Colloqiuim Publications, **2**, AMS, 1910.

[115] Moreau, J. J., Inf-convolution des fonctions numerique sur un espace vectorial, *C. R. Acad. Sci. Paris* **256** (1963), 5047-5049.

[116] Moreau, J. J., Inf-convolution, sous-additivite, convexite des fonctions numeriques. *J. Mathem. Pures Appl.* **49** (1970), 109-154.

[117] Morrison, D. D., Optimization by least squares, *SIAM J. Numer. Analysis* **5** (1968), 83-88.

[118] Pallaschke, D. and Rolewicz, S., *Foundations of Mathematical Optimization (Convex analysis without linearity)* Kluwer Academic Publishers, Dordrecht (1997).

[119] Pardalos, P. M. and Rosen, J. B., *Constrained Global Optimization: Algorithms and Applications*, Lecture Notes in Computer Science, **268**, Springer-Verlag, Berlin, (1987).

[120] Pardalos, P. M., Ye, Y. and Han, Ch-G., Algorithms for the solution of quadratic knapsack problems, *Linear Algebra and its Applications* **62** (1991), 69-91.

[121] Pearce, C. E. M. and Rubinov, A. M., *P*-functions, quasiconvex functions and Hadamard-type inequalities, *Journal of Mathematical Analysis and Applications* **240** (1999), 92-104.

[122] Penot, J-P., Duality for radiant and shady programs, *Acta Mathematica Vietnamica* **22** (1997), 541-566.

[123] Penot, J-P., Duality for anticonvex programs, to appear.

[124] Penot, J-P. and Volle, M., Another duality scheme for quasiconvex problems . In: *Trends in Mathematical Optimization* (K.-H. Hoffmann et al (eds)), Birkhauser, Basel (1988), 259-275.

[125] Penot, J-P. and Volle, M., On quasiconvex duality. *Math. Oper. Research* **15** (1990), 597-625.

[126] Penot, J-P. and Zalinesku, C., Harmonic sum and duality, *Convex Analysis*, to appear.

[127] Penot, J-P. and Zalinesku, C., Elements of quasiconvex subdifferential calculus, *Convex Analysis*, to appear.

[128] Phelps, R., *Lectures on Choquet's Theorem*, D.Van Nostrand Company, Inc., Princeton, 1966

[129] Pijavski, S. A., An algorithm for finding the absolute extremum of a function, *USSR Comput. Math. and Math. Phys.*, **2** (1972), 57-67.

[130] Pintér, J., *Global Optimization in Action. Continuous and Lipschitz Optimization: Algorithms, Implementations and Applications*, Kluwer Academic Publishers (1996).

[131] Plastria, F., Lower subdifferentiable functions and their minimization by cutting planes, *Journal of Optimization Theory and Applications* **46** (1985), 37-53.

[132] Ralph, D., *Rank-1 support functionals and the rank-1 generalized Jacobians, Piecewise Linear Homeomorphisms.* Ph. D. Thesis, Computer Science Technical Report No. 938, University of Wisconsin, Madison (1990).

[133] Rastrigin, L. A., *Systems of Extremal Control*, Nauka, Moscow, 1974. (In Russian).

[134] Rockafellar, R. T., *Convex Analysis.* Princeton University Press, Princeton, New Jercey (1970).

[135] Rockafellar, R. T., *Conjugate Duality and Optimization*, No 16 in Conference Board of Math. Sciences Series, SIAM Publications (1974).

[136] Rockafellar, R. T., Lagrange multipliers and optimality, *SIAM Review* **35** (1993), 183-238.

[137] Rolewicz, S., Convex analysis without linearity, *Control and Cybernetics* **23** (1994), 247-256.

[138] Rolewicz, S., On paraconvex multifunctions, in *Proc. III Symp. Oper. Research*, Mannheim (1978), 539-546.

[139] Rubinov, A. M., Minimizing a norm over a compact set, *Vestnik Leningrad Univ.* **20(1)** (1965), 140-142.

[140] Rubinov, A. M., *Superlinear Multivalued Mappings and their Applications to Problems of Mathematical Economics*, Leningrad, Nauka (1980) (In Russian).

[141] Rubinov, A. M., Anti-homogeneous conjugacy operators in convex analysis, *J. Convex Anal.* **2** (1995), 291-307.

[142] Rubinov, A. M., Supremal generators of spaces of homogeneous functions, in *Progress in Optimization: Contribution from Australasia*, Eberhard, A., Glover, B. and Ralph, D. (eds), Kluwer Academic Publishers (1999), 91-100.

[143] Rubinov, A. M., Some properties of increasing convex-along-rays functions, *Proceedings of the Centre for Mathematics and its applications* **36** (1999), 153-167.

[144] Rubinov, A. M., Towards monotonic analysis, *Nonlinear Analysis and Related Topics*, Proceedings of the Institute of Mathematics, National Academy of Sciences of Belarus, **2** (1999)), 147-154

[145] Rubinov, A. M., Radiant sets and their gauges, in: *Quasidiferentiability and Related Topics*, Demyanov, V.F. and Rubinov, A.M. (eds), Kluwer Academic Publishers, to appear.

[146] Rubinov, A. M., Abstract convexity: examples and applications, *Optimization*, to appear.

[147] Rubinov, A. M. and Andramonov, M. Yu., Minimizing increasing star-shaped functions based on abstract convexity. *J. of Global Optimization* **15** (1999), 19-39.

[148] Rubinov A. M. and Andramonov M. Yu., Lipschitz Programming via increasing convex-along-rays functions. *Optimization Methods and Software* **10** (1999), 763-781.

[149] Rubinov, A. M. and Glover, B. M., Characterizations of optimality for homogeneous programming problems with applications, *Recent Advances in Nonsmooth Optimization*, Du, D.-Z., Qi, L. and Womersley, R.S.(eds), (1995), 351-380.

[150] Rubinov, A. M. and Glover, B. M., Duality for increasing positively homogeneous functions and normal sets, *RAIRO-Operations Research* **32** (1998), 105-123.

[151] Rubinov, A. M. and Glover B. M., On generalized quasiconvex conjugation, *Contemporary Mathematics* **204** (1997), 199-217.

[152] Rubinov, A. M. and Glover, B. M., Increasing convex-along-rays functions with applications to global optimization. *J. of Optimization Theory and Appl.*, **102** (1999), 615-642.

[153] Rubinov, A. M. and Glover, B. M., Quasiconvexity via two step functions. In; *Generalized Convexity, Generalized Monotonicity: Recent Results* (Crouzeix, J.-P. et al (eds.), Kluwer Academic Publishers (1998), 159-183.

[154] Rubinov, A. M. and Glover, B. M., Toland-Singer formula cannot distinguish a global minimizer from a choice of stationary points. *Numer. Funct. Analysis and Optimization* **20** (1999), 99-119.

[155] Rubinov, A. M., Glover, B. M. and Jeyakumar, V., A general approach to dual characterization of solvability of inequality systems with applications. *J. Convex Analysis* **2** (1995), 309-344.

[156] Rubinov, A. M., Glover, B. M. and Yang, X. Q., Extended Lagrange and penalty functions in continuous optimization, Research Report 19/97, SITMS, University of Ballarat (1997).

[157] Rubinov, A. M., Glover, B. M. and Yang, X. Q., Decreasing functions with application to penalization, *SIAM Journal Optimization*, to appear.

[158] Rubinov, A. M. and Shimshek, B., Dual problems of quasiconvex Maximization, *Bul. Austral. Math. Soc.* **51** (1995), 139 - 144.

[159] Rubinov, A. M. and Shimshek, B., Conjugate quasiconvex nonnegative functions, *Optimization* **35** (1995), 1 - 22.

[160] Rubinov, A. M. and Shveidel, A., Separability of star-shaped sets with respect to infinity, in: *Progress in Optimization: Contribution from Australasia*, Kluwer Academic Publishers, to appear.

[161] Rubinov, A. M. and Singer, I., Best approximation by normal and conormal sets, Research Report 32/99, SITMS, University of Ballarat (1999), submitted for publication.

[162] Rubinov, A. M. and Singer, I., Distance to the intersection of normal sets and applications, Research Report 39/99, SITMS, University of Ballarat (1999), submitted for publication.

[163] Rubinov, A. M., Tuy, H. and Mays, H., Algorithm for a nonconvex global optimization problem, *Optimization*, to appear.

[164] Rubinov, A. M. and Vladimirov, A. A., Convex-along-rays functions and star-shaped sets. *Numerical Functional Analysis and Optimization* **19** (1998), 593-614.

[165] Rubinov, A. M. and Yagubov, A. A., The space of star-shaped sets and its application in nonsmooth optimization, *Mathematical Programming Study* **29** (1986), 176-202.

[166] Rubinov, A. M., Yang, X. Q. and Glover, B. M., Extended Lagrange and penalty functions in continuous optimization, *Optimization* **46** (1999), 327-351.

[167] Schaible, S. and Ziemba, W. T. (eds), *Generalized Concavity in Optimization and Economics*, Proceedings of the First Conference on Generalized Convexity, held at University of British Columbia, Vancouver (British Columbia, Canada), August 4-15, 1980, Academic Press, New York (1981).

[168] Semadeni, Z., *Banach spaces of continuous functions*, vol. 1, PWN, Warszawa, (1971).

[169] Sikorski, R., *Boolean Algebras* Springer, Berlin, New York (1969).

[170] Singer, I., *The theory of best approximation and functional analysis*. CBMS Reg. Confer. Ser. Appl. Math. **13**, SIAM, Philadelphia, 1974.

[171] Singer, I., A Fenchel-Rockafellar type theorem for maximization. *Bul. Austral. Math. Soc.* **20** (1979), 193-198.

[172] Singer, I., Generalized convexity, functional hulls and applications to conjugate duality in optimization. In: *Selected topics in Operation Research and Mathematical Economics*, (Hammer, G. and Pallaschke, D. (eds)). Lect. Notes in Econ. and Mathem. Syst., **226** (1984), 49-79.

[173] Singer, I., Conjugation operators. In: *Selected topics in Operation Research and Mathematical Economics*, (Hammer, G. and Pallaschke, D. (eds)), Lect. Notes in Econ. and Mathem. Syst., **226** (1984), 80-97.

[174] Singer, I., Generalization of convex supremization duality. In: *Nonlinear and Convex Analysis*, (Lin, B.-L. and Simons, S. (eds)) Lecture Notes in Pure and Applied Mathematics, Marcel Dekker, New York **107** (1987), 253-270.

[175] Singer, I., *Abstract Convex Analysis*. Wiley-Interscience Publication, New York (1997).

[176] Singer, I., Duality in quasi-convex supremization and reverse convex infimization via abstract convex analysis, and application to approximation. *Optimization* **45** (1999), 255-308.

[177] Strekalovsky, A., *On Global Optimality conditions for D.C. Programming Problems*, Irkutsk State University (1997).

[178] Shor, N. Z., *Methods of Minimizing Nondifferentiable Functions*, Springer-Verlag, Berlin (1985).

[179] Shor, N. Z., Dual estimates in multi-extremal problems, *Journal of Global Optimization* **7** (1995), 75-91.

[180] Shubert, B. O., A sequential method seeking the global maximum of a function, *SIAM Journal of Numerical Analysis* **9** (1972), 379-388.

[181] Shveidel, A., On supremal generators of a space of continuous functions *Application of Functional Analysis to Approximation Theory* **2** (1974), Kalinin State Unversity, Kalinin. (In Russian).

[182] Shveidel, A., Separability of star-shaped sets and its application to an optimization problem, *Optimization* **40** (1997), 207-227.

[183] Stavrakis, N. M., The dimension of the convex kernel and points of local nonconvexity, *Proceedings of the Amer. Math. Soc.*, **34** (1972), 222-224.

[184] Törn, A. and Žilinskas, A., *Global Optimization*, Lecture Notes in Computer Science **350**, Springer Verlag (1987).

[185] Toland, J. F., Duality in nonconvex optimization. *J. Math. Anal. Appl.* **66** (1978), 399-415.

[186] Tuy, H., A general deterministic approach to global optimization via d.c. programming, in *Fermat Days 1985: Mathematics for Optimization*, Hiriart-Urruty, J.-B. (ed.), North-Holland, Amsterdam, 137-162.

[187] Tuy, H., D.C. Optimization: Theory, Methods and Algorithms, in: Horst, R. and Pardalos, P.M. (eds.), *Handbook on Global Optimization*, Kluwer Academic Publishers, Dordrecht (1995), 149-216.

[188] Tuy H., Canonical D.C. Programming: Outer Approximation Methods Revisited, *Operations Research Letters*, **18** (1995), 99-106.

[189] Tuy, H., *Convex Analysis and Global Optimization*, Kluwer Academic Publishers, Dordrecht (1998).

[190] Tuy, H., Monotonic optimization: problems and solution approach, submitted paper.

[191] Tuy, H. and Oettli, W., On necessary and sufficient conditions for global optimality, *Rev. Math. Appl.* **15** (1994), 39-41.

[192] Valentine, F. A., *Convex sets,* McGraw-Hill, New York (1964).

[193] Volle, M., Conjugaison par tranches. *Ann. Mat. Pura Appl.* **139** (1985), 279 -312.

[194] Wood, G. R., Multidimensional bisection applied to global optimization, *Computers and Mathematics with Applications* **21** (1991), 161-172.

[195] Wood, G. R., The bisection method in higher dimensions. *Mathematical Programming* **55** (1992), 319-337.

[196] Yang, X. Q. and Li, D., Successive global optimization method via parametric monotone composition formulation, *Journal of Global Optimization*, to appear.

[197] Yevtushenko, Yu. G. and Zhadan, V. G., Exact auxiliary functions in optimization problems, *U.S.S.R Comput. Maths. Math. Phys.* **30** (1990), 31-42.

[198] Zabotin, Ya. I., A minimax method for solving problems of mathematical programming, *Soviet Mathematics (Iz VUZ)* **19** (1975), 28-33.

[199] Zaslavski, A. J., Least closed supremal generating sets, in *Applications of Functional Analysis to Approximation Theory* **5** (1977), Kalinin State University, Kalinin, 30 -39. (In Russian).

[200] Zhadan, V. G., Methods of centres in nonlinear programming, *Lect. Notes in Control and Information Sciences* **143** (System Modelling and Optimization), Springer-Verlag (1989).

[201] Zhang, B., Wood, G. R. and Baritompa, W. P., Multidimensional bisection: the performance and the context, *Journal of Global Optimization* **3** (1993), 337-358.

Index

Nonconvex Optimization and Its Applications

1. D.-Z. Du and J. Sun (eds.): *Advances in Optimization and Approximation.* 1994
ISBN 0-7923-2785-3
2. R. Horst and P.M. Pardalos (eds.): *Handbook of Global Optimization.* 1995
ISBN 0-7923-3120-6
3. R. Horst, P.M. Pardalos and N.V. Thoai: *Introduction to Global Optimization* 1995
ISBN 0-7923-3556-2; Pb 0-7923-3557-0
4. D.-Z. Du and P.M. Pardalos (eds.): *Minimax and Applications.* 1995
ISBN 0-7923-3615-1
5. P.M. Pardalos, Y. Siskos and C. Zopounidis (eds.): *Advances in Multicriteria Analysis.* 1995 ISBN 0-7923-3671-2
6. J.D. Pintér: *Global Optimization in Action.* Continuous and Lipschitz Optimization: Algorithms, Implementations and Applications. 1996 ISBN 0-7923-3757-3
7. C.A. Floudas and P.M. Pardalos (eds.): *State of the Art in Global Optimization.* Computational Methods and Applications. 1996 ISBN 0-7923-3838-3
8. J.L. Higle and S. Sen: *Stochastic Decomposition.* A Statistical Method for Large Scale Stochastic Linear Programming. 1996 ISBN 0-7923-3840-5
9. I.E. Grossmann (ed.): *Global Optimization in Engineering Design.* 1996
ISBN 0-7923-3881-2
10. V.F. Dem'yanov, G.E. Stavroulakis, L.N. Polyakova and P.D. Panagiotopoulos: *Quasi-differentiability and Nonsmooth Modelling in Mechanics, Engineering and Economics.* 1996 ISBN 0-7923-4093-0
11. B. Mirkin: *Mathematical Classification and Clustering.* 1996 ISBN 0-7923-4159-7
12. B. Roy: *Multicriteria Methodology for Decision Aiding.* 1996 ISBN 0-7923-4166-X
13. R.B. Kearfott: *Rigorous Global Search: Continuous Problems.* 1996
ISBN 0-7923-4238-0
14. P. Kouvelis and G. Yu: *Robust Discrete Optimization and Its Applications.* 1997
ISBN 0-7923-4291-7
15. H. Konno, P.T. Thach and H. Tuy: *Optimization on Low Rank Nonconvex Structures.* 1997 ISBN 0-7923-4308-5
16. M. Hajdu: *Network Scheduling Techniques for Construction Project Management.* 1997 ISBN 0-7923-4309-3
17. J. Mockus, W. Eddy, A. Mockus, L. Mockus and G. Reklaitis: *Bayesian Heuristic Approach to Discrete and Global Optimization.* Algorithms, Visualization, Software, and Applications. 1997 ISBN 0-7923-4327-1
18. I.M. Bomze, T. Csendes, R. Horst and P.M. Pardalos (eds.): *Developments in Global Optimization.* 1997 ISBN 0-7923-4351-4
19. T. Rapcsák: Smooth Nonlinear Optimization in R^n. 1997 ISBN 0-7923-4680-7
20. A. Migdalas, P.M. Pardalos and P. Värbrand (eds.): *Multilevel Optimization: Algorithms and Applications.* 1998 ISBN 0-7923-4693-9
21. E.S. Mistakidis and G.E. Stavroulakis: *Nonconvex Optimization in Mechanics.* Algorithms, Heuristics and Engineering Applications by the F.E.M. 1998
ISBN 0-7923-4812-5

22. H. Tuy: *Convex Analysis and Global Optimization*. 1998 ISBN 0-7923-4818-4
23. D. Cieslik: *Steiner Minimal Trees*. 1998 ISBN 0-7923-4983-0
24. N.Z. Shor: *Nondifferentiable Optimization and Polynomial Problems*. 1998 ISBN 0-7923-4997-0
25. R. Reemtsen and J.-J. Rückmann (eds.): *Semi-Infinite Programming*. 1998 ISBN 0-7923-5054-5
26. B. Ricceri and S. Simons (eds.): *Minimax Theory and Applications*. 1998 ISBN 0-7923-5064-2
27. J.-P. Crouzeix, J.-E. Martinez-Legaz and M. Volle (eds.): *Generalized Convexitiy, Generalized Monotonicity: Recent Results*. 1998 ISBN 0-7923-5088-X
28. J. Outrata, M. Kočvara and J. Zowe: *Nonsmooth Approach to Optimization Problems with Equilibrium Constraints*. 1998 ISBN 0-7923-5170-3
29. D. Motreanu and P.D. Panagiotopoulos: *Minimax Theorems and Qualitative Properties of the Solutions of Hemivariational Inequalities*. 1999 ISBN 0-7923-5456-7
30. J.F. Bard: *Practical Bilevel Optimization*. Algorithms and Applications. 1999 ISBN 0-7923-5458-3
31. H.D. Sherali and W.P. Adams: *A Reformulation-Linearization Technique for Solving Discrete and Continuous Nonconvex Problems*. 1999 ISBN 0-7923-5487-7
32. F. Forgó, J. Szép and F. Szidarovszky: *Introduction to the Theory of Games*. Concepts, Methods, Applications. 1999 ISBN 0-7923-5775-2
33. C.A. Floudas and P.M. Pardalos (eds.): *Handbook of Test Problems in Local and Global Optimization*. 1999 ISBN 0-7923-5801-5
34. T. Stoilov and K. Stoilova: *Noniterative Coordination in Multilevel Systems*. 1999 ISBN 0-7923-5879-1
35. J. Haslinger, M. Miettinen and P.D. Panagiotopoulos: *Finite Element Method for Hemivariational Inequalities*. Theory, Methods and Applications. 1999 ISBN 0-7923-5951-8
36. V. Korotkich: *A Mathematical Structure of Emergent Computation*. 1999 ISBN 0-7923-6010-9
37. C.A. Floudas: *Deterministic Global Optimization: Theory, Methods and Applications*. 2000 ISBN 0-7923-6014-1
38. F. Giannessi (ed.): *Vector Variational Inequalities and Vector Equilibria*. Mathematical Theories. 1999 ISBN 0-7923-6026-5
39. D.Y. Gao: *Duality Principles in Nonconvex Systems*. Theory, Methods and Applications. 2000 ISBN 0-7923-6145-3
40. C.A. Floudas and P.M. Pardalos (eds.): *Optimization in Computational Chemistry and Molecular Biology*. Local and Global Approaches. 2000 ISBN 0-7923-6155-5
41. G. Isac: *Topological Methods in Complementarity Theory*. 2000 ISBN 0-7923-6274-8
42. P.M. Pardalos (ed.): *Approximation and Complexity in Numerical Optimization: Concrete and Discrete Problems*. 2000 ISBN 0-7923-6275-6
43. V. Demyanov and A. Rubinov (eds.): *Quasidifferentiability and Related Topics*. 2000 ISBN 0-7923-6284-5

Nonconvex Optimization and Its Applications

44. A. Rubinov: *Abstract Convexity and Global Optimization*. 2000
ISBN 0-7923-6323-X

KLUWER ACADEMIC PUBLISHERS – DORDRECHT / BOSTON / LONDON

MIX
Papier aus verantwortungsvollen Quellen
Paper from responsible sources
FSC® C105338

FSC
www.fsc.org

If you have any concerns about our products,
you can contact us on
ProductSafety@springernature.com

In case Publisher is established outside the EU,
the EU authorized representative is:
Springer Nature Customer Service Center GmbH
Europaplatz 3, 69115 Heidelberg, Germany

Printed by Libri Plureos GmbH
in Hamburg, Germany